Reliable Seismic Hazard Assessment and Earthquake Engineering

Editors:

Junbo Jia
Senior Principal Engineer-Odfjell Technology
Kokstad, Norway

Zhongliang Wu
Formerly President
International Association of Seismology and Physics of the Earth's Interior (IASPEI)
Formerly Director
Institute of Geophysics and Institute of Earthquake Forecasting
China Earthquake Administration

Mihaela Petrova Kouteva-Guentcheva
Full Professor - University of Architecture, Civil Engineering and Geodesy
Sofia, Bulgaria
Scientific Committee Chair of the National Earthquake Engineering Centre

CRC Press
Taylor & Francis Group
Boca Raton London New York

CRC Press is an imprint of the
Taylor & Francis Group, an **informa** business
A SCIENCE PUBLISHERS BOOK

First edition published 2027
by CRC Press
2385 NW Executive Center Drive, Suite 320, Boca Raton FL 33431

and by CRC Press
4 Park Square, Milton Park, Abingdon, Oxon, OX14 4RN

Library of Congress Cataloging-in-Publication Data (applied for)

ISBN: 978-1-041-11298-3 (hbk)
ISBN: 978-1-041-16350-3 (pbk)
ISBN: 978-1-003-68412-1 (ebk)

DOI: 10.1201/9781003684121

Typeset in Times New Roman
by Prime Publishing Services

To Rita and Giuliano, companions in both science and soul,
whose exemplary partnership stands as a testament to the
power of intellect, integrity, and mutual devotion

Preface

This book has been prepared to celebrate the eightieth birthday of Professor Giuliano Francesco Panza, whose distinguished career has contributed prominently to the fields of seismology, lithospheric studies, and physics-based seismic hazard assessment. Over the course of more than five decades, his work has advanced the scientific understanding of seismic-wave propagation, Earth structure, and the physical processes that govern earthquake ground motion.

A significant aspect of Professor Panza's scientific legacy is the development of Neo-Deterministic Seismic Hazard Assessment (NDSHA). Grounded in the explicit modelling of earthquake sources and seismic-wave propagation, NDSHA provides a physically transparent framework for estimating ground shaking, particularly in regions characterized by complex tectonics or limited instrumental and historical data. Its emphasis on methodological clarity and reproducibility has made it a valuable complement to established probabilistic approaches, and it has been applied across a wide range of geological environments worldwide.

Professor Panza's longstanding association with the Abdus Salam International Centre for Theoretical Physics (ICTP) has further amplified the reach and impact of his scientific contributions. Through sustained engagement in training programmes, collaborative research initiatives, and capacity-building activities, he has supported the development of scientific expertise in numerous countries, contributing meaningfully to the growth of an internationally connected research community.

Prof. Panza is a world-renowned seismologist who stands among the true pioneers whose innovative seismic hazard assessment methodology has transformed our understanding of earthquake risk at both regional and global scales. Through his visionary leadership and unwavering dedication to scientific excellence, he built an extraordinary international team uniting researchers from all continents—an intellectual community where professional collaboration effortlessly evolved into enduring friendships. This unique human and scientific synergy became a defining strength of the group. With his groundbreaking ideas and his ability to inspire independent, innovative thinking, Prof. Panza established a distinguished global school that introduced a comprehensive, interdisciplinary approach to seismic-risk reduction. His work continues to guide and elevate the work of generations of scientists worldwide, as evidenced, for example, by the Joint Working Group on NDSHA (JWG-NDSHA) established under the African Seismological Commission (AfSC) and the Asian Seismological Commission (ASC).

Divided into two major parts—Part I: Theoretical Frameworks (comprising six chapters), and Part II: Practical Implementations and Case Studies (comprising thirteen chapters), this volume reflects the breadth of research topics and methodological perspectives shaped by Professor Panza's scientific vision. Within each part, the chapters are organized according to the alphabetical order of the surname of the corresponding author. Together, they address a wide range of themes, including lithospheric structure, seismic-wave modelling, scenario-based seismic hazard assessment, engineering applications, and multi-hazard interactions. While diverse in scope, these contributions share a common foundation in the principles that have guided his career: rigorous physical reasoning, a deep commitment to transparency, and the pursuit of scientific approaches that ultimately enhance societal safety and resilience.

The editors extend their sincere appreciation to all contributors for their dedication in preparing the chapters included in this volume. Their work reflects a broad range of expertise and perspectives that together form the scientific foundation of this book. We are grateful to the reviewers from around the world, many working in regions with significant seismic risk, who generously devoted their time to evaluating the submissions. Their careful and constructive feedback has been essential in ensuring the clarity, rigor, and overall quality of the volume.

We also express our deep gratitude to Professor Giuliano F. Panza for his sustained scientific leadership, his generosity toward colleagues and students, and his unwavering dedication to advancing the geosciences. It is our hope that this volume will serve both as a tribute to his accomplishments and as a resource that supports continued progress in the fields to which he has devoted his life's work.

Contents

List of Contributors

Aiqun Li
School of Civil and Transportation Engineering, Beijing University of Civil Engineering and Architecture, Beijing, 100044, China. and
School of Civil Engineering, Southeast University, Nanjing, 21009, China.

Antonella Peresan
National Research Institute of Astronomy and Geophysics, Helwan, Cairo 11421, Egypt. and African Disaster Mitigation Research Center ADMiR, Cairo 11421, Egypt.

B. Moreno
Centro Nacional de Investigaciones Sismológicas (CENAIS). Ministerio de Ciencia, Tecnología y Medio Ambiente, Cuba. and
Géosciences Environnement Toulouse (GET), Observatoire Midi-Pyrénées, Université Toulouse III Paul Sabatier, France

Bekoa Ateba
Institut de Recherches Geologiques et Minieres (IRGM), Yaounde, Cameroon.

Bo Liu
School of Mechanics and Civil Engineering, China University of Mining and Technology-Beijing, Beijing 100083, China.

Chiara Faraone
University "G. d'Annunzio" of Chieti-Pescara, Department of Engineering and Geology.

E. Arango
Centro Nacional de Investigaciones Sismológicas (CENAIS). Ministerio de Ciencia, Tecnología y Medio Ambiente, Cuba. and
2Géosciences Environnement Toulouse (GET), Observatoire Midi-Pyrénées, Université Toulouse III Paul Sabatier, France.

E. Calais
École normale supérieure, Département de Géosciences, Paris, France.

Elena Dumova Jovanoska
Cyril and Methodius University, Faculty of Civil Engineering, Skopje, Republic of North Macedonia.

Fabio Romanelli
Department of Mathematics, Informatics and Geosciences, University of Trieste, Trieste, Italy.

Fabrizio Mollaioli
Department of Structural and Geotechnical Engineering, Sapienza University of Rome, Rome, Italy.

Franco Vaccari
Department of Mathematics, Informatics and Geosciences, University of Trieste, Trieste, Italy.

Georgi Boyadzhiev
Institute of Mathematics and Informatics, Bulgarian Academy of Sciences, 8 Acad. G. Bonchev str, 1113 Sofia, Bulgaria.University of Architecture, Civil Engineering and Geodesy, 1 Hristo Smirnenski Blvd, 1046 Sofia, Bulgaria.

Giovanna Vessia
University "G. d'Annunzio" of Chieti-Pescara, Department of Engineering and Geology.

Giuliano F. Panza
Accademia Nazionale dei Lincei, Rome, Italy and
Earthquake Administration, Beijing, China and
International Seismic Safety Organization (ISSO), Arsita, Italy and
Beijing University of Civil Engineering and Architecture (BUCEA), Beijing, China.

Gorshkov A.
Institute of Earthquake Prediction Theory and Mathematical Geophysics, Russian Academy of Sciences, Moscow, Russian Federation.

H.M. Hassan
National Institute of Oceanography and Applied Geophysics, OGS, Via Treviso 55, 33100 Udine, Italy. and
African Disaster Mitigation Research Center ADMiR, Cairo 11421, Egypt.

Jiajun Shu
School of Mechanics and Civil Engineering, China University of Mining and Technology-Beijing, Beijing 100083, China.

Jiangdi Fu
School of Civil and Transportation Engineering, Beijing University of Civil Engineering and Architecture, Beijing 100044

Katalin Gribovszki
Institute of Earth Physics and Space Science, Hungarian Research Network and
Institute of Geomatics and Civil Engineering, Faculty of Forestry, University of Sopron

Kossobokov V.
Institute of Earthquake Prediction Theory and Mathematical Geophysics, Russian Academy of Sciences, Moscow, Russian Federation. and
Accademia Nazionale delle Scienze detta dei XL, Rome, Italy. and
International Seismic Safety Organisation, Arsita, Italy.

Kristina Milkova
Cyril and Methodius University, Faculty of Civil Engineering, Skopje, Republic of North Macedonia.

Lili Czirok
Hungarian Meteorological Service

Lyuba Dimova
Department of Meteorology and Geophysics, Sofia University

Milen Tsekov
Department of Meteorology and Geophysics, Sofia University

Mohamed ElGabry
National Institute of Oceanography and Applied Geophysics, OGS, Via Treviso 55, 33100 Udine, Italy. and
National Research Institute of Astronomy and Geophysics, Helwan, Cairo 11421, Egypt.

Mustapha Meghraoui
Institut de Physique du Globe, EOST - Université de Strasbourg 5, rue René Descartes 67084 Strasbourg cedex, France.

Nekrasova A.
Institute of Earthquake Prediction Theory and Mathematical Geophysics, Russian Academy of Sciences, Moscow, Russian Federation

Nicola Venisti
University "G. d'Annunzio" of Chieti-Pescara, Department of Engineering and Geology. and
Osservatorio Sismologico - Università degli Studi di Bari "Aldo Moro", Campus Universitario.

Novikova O.
Beijing University of Civil Engineering and Architecture (BUCEA), Beijing, China.

O. González
Centro Nacional de Investigaciones Sismológicas (CENAIS). Ministerio de Ciencia, Tecnología y Medio Ambiente, Cuba. and
Géosciences Environnement Toulouse (GET), Observatoire Midi-Pyrénées, Université Toulouse III Paul Sabatier, France. and
Observatoire Volcanologique et Sismologique de Martinique. Institut de Physique du Globe de Paris, France.

Panza G.F.
Institute of Earthquake Prediction Theory and Mathematical Geophysics, Russian Academy of Sciences, Moscow, Russian Federation. and
Accademia Nazionale dei Lincei, Rome, Italy. OR Earthquake Administration, Beijing, China. OR International Seismic Safety Organization (ISSO), Arsita, Italy.

Paolo Rugarli
Castalia srl, Via XX Settembre, 39 – Lodi – Italy.

Paulina Amponsah
School of Nuclear and Allied Sciences, Ghana Atomic Energy Commission, Accra, Ghana.

Péter Varga
Institute of Earth Physics and Space Science, Hungarian Research Network

Petia Dineva
Institute of Mechanics, Bulgarian Academy of Sciences (BAS), Sofia, Bulgaria.

Prince Amoah
Nuclear Regulatory Authority, Ghana.

Edmund Amartey
National Nuclear Research Institute, Ghana Atomic Energy Commission.

Reneta Raykova
Department of Meteorology and Geophysics, Sofia University

Ruilin Long
School of Civil and Transportation Engineering, Beijing University of Civil Engineering and Architecture, Beijing, 100044, China.

Sándor Szalai
Institute of Earth Physics and Space Science, Hungarian Research Network

Sonia Parvanova
Department of Structural Engineering, University of Architecture, Civil Engineering and Geodesy (UACEG), Sofia, Bulgaria

Tao Li
School of Mechanics and Civil Engineering, China University of Mining and Technology-Beijing, Beijing 100083, China.

Tony S. Nemer
Department of Geology, American University of Beirut, Riad El-Solh, Beirut 1107 2020, Lebanon.

V. Clouard
Géosciences Environnement Toulouse (GET), Observatoire Midi-Pyrénées, Université Toulouse III Paul Sabatier, France.

Wenbo Wang
School of Mechanics and Civil Engineering, China University of Mining and Technology-Beijing, Beijing 100083, China.

Y. Rojas-Agramonte
Institut für Geowissenschaften, Ruprecht-Karls-Universität Heidelberg, Germany.

Yan Zhang
Institute of Geophysics, China Earthquake Administration, Beijing, China.

Yifeng Wu
School of Civil and Transportation Engineering, Beijing University of Civil Engineering and Architecture, Beijing, 100044, China.

Zijian Wang
School of Civil and Transportation Engineering, Beijing University of Civil Engineering and Architecture, Beijing, 100044, China.

The 80th Anniversary of Professor Giuliano F. Panza
A Scholarly Odyssey in Seismology and the Evolution of NDSHA

Junbo Jia

This celebrative volume marks the eightieth anniversary of Professor Giuliano F. Panza, whose pioneering contributions have profoundly contributed to modern seismology and seismic hazard assessment. Tracing his intellectual journey from formative studies at the University of Bologna to his leadership in Trieste and at the Abdus Salam International Centre for Theoretical Physics (ICTP), the celebrative volume highlights the scientific principles that have guided his career: physical realism, methodological transparency, and a deep commitment to global scientific development.

Central to his legacy is the creation of Neo-Deterministic Seismic Hazard Assessment (NDSHA), a physics-based methodology that models realistic earthquake scenarios and seismic-wave propagation to produce reproducible, physically interpretable ground-shaking estimates. Applied across continents, NDSHA offers a robust hazard assessment. Its integration of structural seismology, source characterization, and computational modelling has significantly advanced both scientific understanding and practical risk-mitigation strategies.

The celebrative volume also outlines the structure of the volume, which brings together nineteen contributions that extend, apply, or engage with themes related to Professor Panza's work. Together, they illustrate the breadth of ongoing research connected to his scientific vision and international collaborations.

1. Introduction

The eightieth anniversary of Professor Giuliano F. Panza marks not only a personal milestone but also an occasion to reflect on the evolution of several foundational ideas in contemporary seismology and earthquake engineering. Across more than five decades, his work has significantly influenced how scientists conceptualize the lithosphere, how seismic hazard is evaluated, and how international collaboration can strengthen scientific capacity worldwide. In the broad landscape of the Earth sciences—where conceptual clarity, methodological innovation, and interdisciplinary cooperation remain indispensable—Professor Panza's influence has been profound and enduring.

This celebrative volume celebrates his distinguished career while offering a scholarly introduction to the scientific themes that underpin this volume. We trace the development of his intellectual journey, from his formative experiences and academic training to the conceptual breakthroughs that culminated in the creation of NDSHA. These scientific contributions, grounded in physical modelling and transparent methodology, have provided an essential counterpoint to traditional probabilistic approaches and have inspired applications across continents.

Equally significant is his role in cultivating a global community of seismologists and earthquake engineers. Through sustained mentorship, institutional leadership, and long-standing engagement with the Abdus Salam International Centre for Theoretical Physics (ICTP), he has helped nurture a generation of scientists from regions where seismic risk is high, yet research infrastructure is limited. This legacy reinforces his conviction that science must serve societies equitably, particularly where natural hazards pose persistent threats to life and development.

The following sections present a comprehensive portrait of Professor Panza's life, scientific achievements, and continuing impact. They also introduce the structure of this volume, which gathers diverse contributions from researchers whose work intersects with—builds upon—or has been inspired by his scientific vision. Together, they illustrate how the principles he championed continue to shape the evolution of seismology and earthquake engineering today.

2. A Journey Rooted in Curiosity and Driven by Purpose

Giuliano F. Panza was born in 1945 in Faenza, a town with deep cultural traditions in the Emilia–Romagna region of Italy. His childhood unfolded in Bologna, where an atmosphere of learning and curiosity gently shaped his early years. Surrounded by books, ideas, and thoughtful conversation, Giuliano developed a natural fascination with the workings of the physical world. Even simple phenomena captured his attention, stirring questions that revealed an instinctive desire to understand nature more deeply. These quiet beginnings—marked by curiosity, reflection, and an early sense of wonder—formed the foundation of the scientific path he would later follow with dedication.

However, his early life was marked by a profound personal challenge. At just six months old, Giuliano suffered a severe intestinal infection of unknown origin. The illness was so critical that survival hung on the most delicate thread. His parents, Giuseppina Liverani, a paediatrician, and Giuseppe Panza, an obstetrician-gynaecologist, devoted themselves to his care with tireless professional commitment and parental devotion. For nearly three years, their continuous, expert medical attention sustained his fragile life. It is not an exaggeration to say that they gave him life twice—first by birth, and then by preserving it through extraordinary medical dedication.

Fig. 1 Childhood photograph of Giuliano with his mother and father

This early chapter—marked by vulnerability, resilience, and the life-saving application of scientific knowledge—left a lasting imprint on Giuliano's worldview. It nurtured in him a profound appreciation of the value of life and a conviction that scientific understanding carries a moral obligation: to protect communities, reduce suffering, and safeguard societies exposed to natural hazards. In retrospect, it is not difficult to see how this formative experience contributed to his lifelong commitment to seismic hazard assessment, risk mitigation, and the advancement of tools designed explicitly to preserve human life.

His early training in physics established the foundations for the research trajectory that would later define his career: an insistence on first principles, a commitment to physical realism, and an enduring belief that the behaviour of the Earth must be understood through rigorous models rather than empirical shortcuts. These values would become central to his scientific philosophy and ultimately underpin the development of NDSHA.

3. Academic Formation and the Seeds of a Scientific Vision

Giuliano pursued his university studies at the Alma Mater Studiorum—Università di Bologna, the world's oldest university still in operation, where he received his laurea in Physics in 1967. His university years coincided with an inflection point in the physical sciences: advancements in numerical modelling, the maturation of plate tectonics theory, and the emergence of computational seismology. While many pursued incremental progress within their disciplines, Giuliano gravitated toward fundamental questions: How does elastic energy propagate within the Earth? What controls the structure of the lithosphere? Can earthquake hazard be predicted using first-principles physics rather than statistical extrapolation? These questions, rooted in a life forged by resilience, curiosity, and purpose, would define his life's work.

His academic journey continued across Europe and North America. Postdoctoral appointments at:

- University of Uppsala (Sweden)
- Federal Institute of Technology, Zürich (Switzerland)
- University of California, Los Angeles (USA)

Exposed him to diverse scientific traditions and cutting-edge computational techniques. These experiences cemented a global scientific perspective that later became a defining feature of his career. They also enriched his ability to synthesize theory, numerical modelling, and empirical data—skills central to the later development of NDSHA.

From early on, Giuliano's scientific philosophy was unmistakably clear: seismology must be grounded in physical realism and methodological transparency. This conviction, combined with his insistence on global inclusivity in research, became the cornerstone of a scientific legacy that remains distinctive.

The intellectual seeds planted during his formative years would eventually culminate in several major scientific achievements—including the development of structural seismology, the refinement of lithospheric models for numerous regions around the world, and ultimately the creation of NDSHA. These conceptual foundations, established early in his training, shaped a career devoted to advancing both scientific understanding and societal resilience.

4. Leadership Across Institutions and Across the World

Professor Panza's academic appointments demonstrate the breadth of his influence. His early role at the University of Bari introduced him to academic leadership, mentoring, and curriculum development. However, it was his long-standing professorship at the University of Trieste that positioned him at the heart of an internationally connected research ecosystem.

At Trieste, he developed a profound and enduring collaboration with the Abdus Salam International Centre for Theoretical Physics (ICTP), an institution founded to foster scientific advancement in developing countries. As leader of the Structure And Nonlinear Dynamics of the Earth (SAND) group, and under the intellectual influence of mathematical geophysics pioneer Vladimir Keilis-Borok, Panza built a vibrant, interdisciplinary community. ICTP became a magnet for scholars from around the world, many of whom are now distinguished scientists contributing to this volume.

The "Trieste school" of seismology, shaped significantly by Panza, was characterized by:

- rigorous application of physical principles,
- strong integration of geodynamics, geology, and seismology,
- emphasis on scenario-based hazard modelling,
- commitment to training scientists from developing nations, and
- openness to cross-disciplinary scientific exchange.

This ecosystem, unique in its global reach and scientific culture, produced countless collaborations, doctoral theses, and international research projects. Today, the NDSHA methodology is applied globally, in large part due to the networks Panza established at ICTP.

His international leadership is also reflected in his involvement with organizations such as:

- the International Lithosphere Program (ILP),
- the UNESCO programs on disaster mitigation,
- the European Seismological Commission,
- and global initiatives in seismic microzonation and hazard evaluation.

He has also served as Professor at the University of Trieste, Visiting Professor at international institutions, scientific advisor for government agencies, and mentor to multiple generations of seismologists and engineers. Many of the authors contributing to this volume trace their academic lineage to the environment Panza cultivated.

His scientific and institutional leadership has been recognized globally through a remarkable series of honours and awards. Among the most notable ones are the Beno Gutenberg Medal (2000) from the European Geophysical Society, his appointment as Knight of the Order of Merit of the Italian Republic (2015), and the AGU International Award (2018)—each reflecting his profound international impact on seismology and hazard assessment. His election to several prestigious academies further underscores his standing in the global scientific community, including the Accademia Nazionale dei Lincei (1987), the Academia Europaea (1990), The Academy of Sciences for the Developing World (1997), the Russian Academy of Sciences (2003), and the Accademia Nazionale delle Scienze detta dei XL (2004).

Collectively, these distinctions attest not only to his scientific excellence but also to his sustained dedication to global collaboration, scientific development, and the application of fundamental research to societal resilience. Through his institutional leadership, intellectual generosity, and lifelong commitment to advancing seismology worldwide, Professor Panza has shaped a truly international legacy—one that continues to influence research communities and hazard-mitigation strategies across continents.

5. Scientific Contributions of Lasting Influence

Throughout his distinguished career, Professor Panza has made enduring contributions to the fields of seismology, geophysics, and earthquake engineering. His scientific achievements span theoretical advances, methodological innovations, computational modelling, and practical applications to seismic hazard assessment. Collectively, these contributions have helped redefine how scientists understand the structure of the lithosphere, how they model seismic-wave propagation, and how they evaluate the intensity of ground shaking associated with major earthquakes.

5.1 Elastic Wave Propagation

One of Panza's earliest and most influential scientific achievements lies in his work on the propagation of elastic and surface waves. At a time when computational resources were limited and many studies relied on simplified models, he demonstrated how surface-wave dispersion and waveform characteristics could be exploited to infer detailed lithospheric structure. His contributions included the refinement of dispersion-curve analysis, the interpretation of group and phase velocities, and the contribution to the development of non linear inversion algorithms for retrieving Earth structure from seismic observations.

This work laid the foundation for what would later become a major branch of seismological research. By emphasizing the diagnostic power of surface waves and their sensitivity to subsurface structure, Panza helped establish surface-wave tomography as a core methodology for regional and global Earth studies. His pioneering analyses remain widely cited and have informed decades of subsequent research.

5.2 Structural Seismology

Building on these early advances, Panza played a central role in shaping the field of structural seismology, an approach that integrates seismological data, geophysical modelling, and geological constraints to develop comprehensive models of the Earth's crust and upper mantle. His insistence on physical realism—grounding hazard assessment in accurate models of the Earth's structure—helped distinguish his work from purely statistical approaches.

Through collaborations across Europe, Asia, Africa, and Latin America, Panza oversaw the development of lithospheric models for dozens of regions worldwide. These models captured key variations in crustal thickness, seismic-velocity gradients, anisotropy, and lithosphere—asthenosphere interactions. Importantly, they demonstrated that lithospheric properties vary significantly across relatively small spatial scales, often in ways that strongly influence seismic-wave amplification and attenuation.

The resulting body of work—spanning large parts of the world—has become a benchmark for modern seismic hazard studies. Many of the structural models produced under his leadership continue to be used as reference models for hazard assessment and geodynamic interpretation.

5.3 Birth and Evolution of NDSHA: A Paradigm Shift in Seismic Hazard Assessment

The culmination of decades of work in wave propagation, seismotectonics, and structural modelling resulted in his signature contribution: NDSHA.

NDSHA emerged not as an incremental improvement, but as a conceptual reorientation of seismic hazard assessment, which redefines the role of physical modelling in understanding earthquake ground motion.

Before examining its core innovations, it is useful to distinguish NDSHA from the earlier Deterministic Seismic Hazard Analysis (DSHA). The primary difference between DSHA and NDSHA lies in their approach to ground-motion prediction and the balance between empirical relationships and physical modelling. Traditional DSHA typically estimates hazard through empirical ground-motion prediction equations (GMPEs) and simplified source-to-site distances. In contrast, NDSHA replaces these empirical steps with numerical simulations of seismic-wave propagation, enabling physically consistent modelling of source, path, and site effects. While DSHA provides a deterministic estimate of shaking, NDSHA offers a physically coherent, scenario-based representation of ground motion rooted directly in geophysics, geology, and mathematical physics.

This shift resulted in several transformative advances that now define the NDSHA framework:

- Use of scenario earthquakes representing the maximum credible events consistent with tectonics.
- Generation of synthetic seismograms through numerical wave-propagation modelling.

- Direct computation of ground-motion time series, rather than reliance on empirical ground-motion prediction equations (GMPEs).
- Explicit modelling of site and path effects using detailed structural models.
- Emphasis on physical transparency, reproducibility, and engineering relevance.

Whereas probabilistic seismic hazard analysis (PSHA) relies on historical catalog statistics and assumptions of ergodicity, NDSHA is solidly rooted in first principles. The method draws directly on geophysics, geology, and mathematical physics to construct hazard scenarios that reflect the actual geological structure and wave-propagation characteristics of a region.

Today, NDSHA is widely recognized across continents (Jia Junbo, 2018). It has been applied in contexts ranging from national hazard maps to nuclear-facility design and seismic microzonation. The methodology has inspired refinement, debate, and progressive developments—signs of a healthy scientific paradigm that challenges and elevates the standards within the field.

5.4 Global Applications and Interdisciplinary Impact

Increasingly, governments, engineers, and researchers have recognized the value of NDSHA for:

- seismic microzonation studies in major urban centres
- critical infrastructure design
- earthquake scenario planning
- risk communication
- disaster-preparedness simulations

The methodology is highlighted in many publications (Guiliano et al., 2021; Jia Junbo, 2018), such as the book Earthquakes and Sustainable Infrastructure (Jia Junbo, 2018), which emphasizes how deterministic and physics-based ground motion modelling improves infrastructure resilience and reduces epistemic uncertainty.

Panza's work demonstrates that scientific rigor and societal impact are not mutually exclusive, but inseparable responsibilities of the modern seismologist. His scientific vision continues to inspire researchers worldwide, many of whom contribute to this volume.

Many of the chapters in this volume reflect these diverse impacts that are testament to his role as both a scientist and a catalyst for global cooperation.

Fig. 2 Selected scientific milestones and collaborations during the career of Professor Giuliano F. Panza.

6. A Life Enriched by Partnership and Collaboration

No celebration of Professor Panza's legacy would be complete without acknowledging the vital role of his wife, Rita Zoccoli. Her unwavering support, intellectual companionship, and shared love for culture and learning have been a constant source of strength throughout his career.

Beyond her personal influence, Rita has also been a partner in scholarly creativity. Their joint work commemorating the 700th anniversary of Dante Alighieri illustrates the remarkable fusion of scientific insight and literary sensibility that characterizes their partnership. This collaboration reflects a deeper truth about Professor Panza: his intellect has always been enriched by a profound appreciation for the humanities, history, and philosophy. He cultivated communities of practice rather than hierarchical networks, emphasizing that the best scientific ideas flourish through open dialogue.

Fig. 3 Giuliano and his wife Rita Zoccoli

In addition to family life, Giuliano's career has been profoundly shaped by the many colleagues, students, and collaborators with whom he has worked over several decades. From Trieste to Bergen, from Mexico City to Beijing, and from Ljubljana to New Delhi, he has formed lasting relationships rooted in scientific integrity and a genuine commitment to shared discovery. His openness to collaboration—irrespective of geography, ideology, or academic tradition—has created an international network of researchers who continue to advance the scientific principles he championed.

Through these partnerships, Panza has cultivated an academic lineage characterized by intellectual rigour, curiosity, and a deep sense of global responsibility. Many of his former students now hold leadership positions in universities, observatories, research institutes, and government agencies. Their continued contributions are a testament not only to his role as a mentor but also to the collaborative ethos that has shaped his scientific outlook from the beginning.

In celebrating his scientific achievements, it is equally fitting to acknowledge the relationships that have supported, accompanied, and inspired him. His legacy is therefore not only one of scientific innovation but also one of human connection— reflecting a life enriched by partnership, mentorship, and the countless collaborations that have helped advance seismic hazard science around the world.

7. Purpose and Structure of this Volume

This edited book is both a tribute and a scholarly contribution. The nineteen chapters presented here were written by researchers whose careers intersect with Professor Panza's scientific domains or who have been influenced directly by his mentorship and methodology. The diversity of topics mirrors the wide-ranging nature of his work, and the global representation of authors underscores the international networks he fostered.

The chapters are organized around thematic areas central to Panza's contributions, though the boundaries between these areas are fluid—reflecting his own interdisciplinary ethos. Together, they form a narrative of scientific progress shaped by decades of innovation, collaboration, and intellectual courage.

This edited volume aims to:

- Honour the scientific and human legacy of Giuliano F. Panza;
- Present recent advances in fields he helped to shape;
- Showcase global applications of NDSHA and structural seismology;
- Highlight emerging ideas that build on his foundational work;
- Foster dialogue among researchers internationally.

8. Overview of the Chapters

This volume brings together nineteen chapters that collectively illustrate the breadth and depth of Professor Panza's scientific legacy. They span structural seismology, waveform modelling, seismic hazard assessment, geodynamics, multi-hazard analysis, earthquake engineering, and emerging methodologies inspired or influenced by the neo-deterministic framework. Each chapter contributes a distinct perspective while remaining anchored in the guiding principles that have shaped Professor Panza's work: physical realism, methodological transparency, and interdisciplinary collaboration.

To provide clarity and coherence, the chapters are grouped into two main parts—Part I: Theoretical Frameworks, and Part II: Practical Implementations and Case Studies. Within each part, the chapters are arranged according to the alphabetical order of the surname of the corresponding author, an editorial choice that reflects both the diversity and equality of scholarly contributions.

Below is an overview of each chapter, presented in the sequence in which they appear within the book.

Chapter 1: Earthquake Forecast/Prediction: What's Next for Science?

This reflective chapter examines earthquake forecasting and prediction from philosophical, methodological, and operational perspectives. Reviewing algorithms such as M8 and real-time testing practices, the author discusses limits of predictability, proper experimental protocols, and the role of intermediate-term forecasting in disaster preparedness. The chapter advocates rigorous testing, transparent uncertainty communication, and pragmatic use of forecast information in preparedness planning.

Chapter 2: Unravelling Fractal Characteristics of Seismogenic Nodes

A pilot study combining PREPA and the Unified Scaling Law for Earthquakes (USLE), this chapter explores fractal signatures of seismogenic nodes across multiple tectonic provinces. Statistical tests on USLE coefficients at identified nodes reveal region-specific patterns that may refine node classification and improve source identification. The work proposes a novel synergy between morphostructural recognition and scaling laws to enhance deterministic hazard inputs.

Chapter 3: Seismic Wave Propagation and Soil-Structure Interaction via Boundary Integral Equations

Presenting advanced boundary-element formulations, this chapter analyses wave propagation in anisotropic, layered media with topographic relief and embedded underground structures. The method quantifies diffraction, scattering, and local stress concentration effects and explicitly models soil–structure interaction for buried and surface facilities. The results stress the necessity of comprehensive mechanical models in design codes for complex geological settings.

Chapter 4: Selection of Proper Signals for Nonlinear Time Histories of Structures, in the Frame of NDSHA

Recognizing the large variability of structural responses across many deterministic realizations, this chapter proposes robust criteria for selecting representative input signals for nonlinear time-history analysis. By ranking simulated NDSHA ensembles using elastic energy metrics and DEEDEM procedures, the authors provide a physics-based procedure to identify worst-case subsets for detailed nonlinear checks. The chapter offers practical guidance for reliable structural verification.

Chapter 5: How Many Sources Must a Man Write Down Before You Call Him a Seismologist?

A playful yet thought-provoking reflection, this chapter interrogates the epistemic relationship between modellers and their seismic sources. Through a mix of wry narrative and conceptual exploration, it highlights modelling assumptions, complexity limits, and the human dimension of seismological practice. The piece encourages sober reflection about the role of simulation in scientific inference.

Chapter 6: Can the Profile of a High-Profile Seismologist Perturb a Ground-Shaking Scenario?

With tongue-in-cheek humour, this chapter uses wave-propagation modelling to playfully examine hypothetical perturbations of a ground-motion field by an object near the free surface. Beyond the levity, the analysis underlines the strong sensitivity of surface wavefields to near-surface heterogeneities and topography, thereby reinforcing the necessity of accurate site characterization in deterministic modelling.

Chapter 7: Recent Seismicity of the West African Region

This chapter revisits the seismic record of West Africa, challenging the perception of the region as uniformly tectonically quiescent. By combining historical reports, recent instrumental data, and tectonic mapping, active lineaments are identified, including links to the Romanche Fracture Zone and the Cameroon Volcanic Line. And events demonstrating non-negligible seismic potential are documented. The study highlights catalog incompleteness and station sparsity, stresses the need for expanded monitoring, and argues for regionally tailored hazard-mitigation measures in line with international disaster-risk frameworks.

Chapter 8: Application of Bicharacteristic Curves of Body and Surface Waves in Geophysics

This chapter presents a new 3-D algorithm for modelling elastic wave propagation in piecewise homogeneous media founded on bicharacteristic curves. The method derives characteristic sets and formulates reflection/diffraction rules at internal boundaries, offering tools for inverse problems in crustal modelling. Developed in collaboration with the University of Trieste, the algorithm has direct applications to tomography and Earth-structure verification from multi-station observations.

Chapter 9: NDSHA (Fault- and Physics-Based) Ground Motion Modelling for the City of Plovdiv, Bulgaria

Adopting a neo-deterministic, fault-based approach, this chapter models kinematic ruptures on several faults near Plovdiv to produce response spectra and synthetic accelerograms for urban sites. The scenarios include historical 1928 events and hypothetical Mw≈7 ruptures, revealing that some modeled ruptures could produce spectral accelerations comparable to or exceeding the worst documented events. The results underscore the utility of synthetic time histories where instrumental records are absent and call for targeted microzonation and risk mitigation in the city.

Chapter 10: Neo-Deterministic Seismic Hazard Assessment (NDSHA) Applied to Seismic Response Analyses: The Case Study of Chieti's Hill (Italy)

This case study applies NDSHA to characterize local seismic response along a 2-D cross section through Chieti, accounting for seven tectonic sources and realistic stratigraphy. Wavefield simulations reveal strong spatial variability in amplification, with pronounced foothill amplification factors and frequency-dependent effects. The chapter demonstrates NDSHA's capacity to resolve source-path-site interactions for urban microzonation and to inform resilient urban planning and engineering design.

Chapter 11: Lithospheric Structure Below Seismic Stations in the East of Cuba, from Cross-Correlation Dispersion Analysis and Its Joint Inversion with Receiver Functions

Using ambient-noise cross-correlation and receiver-function joint inversion, the authors derive new dispersion relations and Rayleigh-wave tomographies that illuminate crustal and upper-mantle variations in eastern Cuba. The models reveal accretionary crustal segments, ophiolitic complexes, and lithospheric low-velocity channels consistent with past subduction and obduction processes. The chapter demonstrates how station-based structural constraints improve regional hazard interpretation in complex Caribbean tectonics.

Chapter 12: Seismogenic Nodes in South Europe

Applying the PREPA (Pattern Recognition of Earthquake-Prone Areas) methodology, this chapter systematically identifies morphostructural nodes likely to host strong earthquakes in southern Europe. The authors validate PREPA predictions against historical seismicity and show PREPA's usefulness where instrumental records are sparse. The resulting catalog of potential nodes is proposed as a concrete input to NDSHA scenarios, filling gaps in seismic history and improving source characterization for deterministic hazard assessments.

Chapter 13: Deterministic Seismic Hazard Computations (Hybrid Method) Applied for Two Cities in the Territory of Hungary

This retrospective summarizes hybrid deterministic computations performed for Budapest and Debrecen using synthetic seismograms along representative profiles. The work produced response-spectra ratios, PGA maps and "Effective Peak Acceleration" curves, and special seismic-risk maps coupling building eigenfrequencies with synthetic motion. The chapter illustrates validation against historical macroseismic intensities and demonstrates practical workflows for urban seismic risk mapping.

Chapter 14: Seismic Hazard Computation for Egypt Based on the Design Magnitude (M_{design})

Introducing the M_{design} concept, this chapter shows how design magnitudes can be derived from truncated historical catalogs to predict post-catalog ground motions. Applying M_{design} to Egypt highlights regions where historical records suffice and areas where nodes or adjusted magnitudes are required. The method offers a falsifiable, physics-informed framework to bridge catalog gaps and improve national hazard mapping.

Chapter 15: Research on Three-Dimensional Elastoplastic Response Spectra for Seismic Isolated Structures

Addressing the challenges of 3-D seismic isolation, this chapter develops elastoplastic response spectra for near-fault pulse-like motions using parametric SDOF studies and large ground-motion suites. The analysis clarifies how isolation parameters (period, ductility, stiffness ratio) influence horizontal and vertical inelastic responses and offers design guidance for controlling excessive displacements and vertical accelerations. The results have direct implications for advancing 3-D isolation technologies.

Chapter 16: Dynamic Response and Shear Characteristics of Rubber Fiber Reinforced Calcareous Sand

This experimental study investigates the efficacy of rubber-fibre reinforcement in improving the dynamic behaviour of calcareous sands common in marine/island settings. Triaxial cyclic tests show that optimal fibre lengths and contents substantially increase liquefaction resistance, delay pore-pressure buildup, and enhance shear stiffness under cyclic loading. The chapter provides practical recommendations for ground-improvement strategies in coastal engineering.

Chapter 17: Energy-Based Engineering Analysis Applied to Physics-Based Seismic Time Histories: A Case Study

This chapter advocates an energy-centric perspective for performance-based earthquake engineering, demonstrating correlations between seismic energy metrics and structural damage indicators. Using NDSHA-generated time histories, the authors show how energy-demand parameters can complement conventional spectral measures in nonlinear analyses. The chapter argues for integrating energy-based criteria into codes and for further research to standardize energy-based design metrics.

Chapter 18: The Effect of the Eastern Mediterranean Offshore Seismogenic Zones on the Seismic Hazard Assessment of the Lebanese Restraining Bend

This parametric investigation evaluates how offshore seismogenic zones influence ground shaking across the Lebanese restraining bend. Using NDSHA scenario tests, the authors demonstrate the sensitivity of coastal hazard to source parameterization and offshore source geometry, and they recommend focused local modelling to inform civil protection and planning measures. The chapter highlights the value of deterministic parametric studies for coastal risk assessment.

Chapter 19: Time-Dependent Seismic Hazard Assessment and Its Application in Earthquake Early Warning System: A Case Study in the China Seismic Experimental Site (CSES)

This chapter outlines a time-dependent NDSHA (T-NDSHA) framework integrating intermediate-term forecasting with physics-based scenario simulations to enhance EEW performance. The case study for the China Seismic Experimental Site evaluates strategies to reduce EEW blind zones by coupling annual forecast consultations with scenario-based ground-motion prescriptions and adaptive triggering schemes. The approach illustrates how deterministic scenario ensembles can support operational preparedness and warning strategies.

9. A Legacy for the Future

Among Professor Panza's many scientific accomplishments, the development and global dissemination of NDSHA stand out as a transformative contribution that continues to shape contemporary approaches to seismic risk evaluation. Conceived as an alternative— and complement—to traditional probabilistic methods, NDSHA offers a physically grounded framework capable of generating realistic scenarios of earthquake ground motion, even in regions with limited historical data or complex tectonics.

The legacy of NDSHA is anchored in its fundamental commitment to physical realism, which ensures that hazard estimates remain tied to geophysical constraints rather than statistical extrapolation.

Over the past three decades, NDSHA has been successfully applied in diverse geological and tectonic settings, and has produced ground-shaking scenarios that later proved consistent with observed motions during real earthquakes—demonstrating the reliability and robustness of the method (Guiliano et al., 2021). Moreover, its capacity to generate time-series ground motions has made it particularly valuable for engineering applications, including the design of critical infrastructure and urban microzonation.

The methodological clarity of NDSHA has also led to its use in sensitivity studies, historical earthquake reconstructions, tsunami modelling, and multi-hazard assessments involving volcanic and tectonic interactions. Because the framework is modular, researchers can incorporate new source models, updated structural profiles, or refined numerical techniques as scientific understanding evolves. This adaptability has enabled NDSHA to remain at the forefront of physics-based seismic hazard assessment, even as computational capabilities and data availability continue to advance.

Perhaps most importantly, NDSHA has played a foundational role in scientific capacity building. By insisting on transparency, reproducibility, and accessibility, Panza and his collaborators have ensured that researchers in countries with limited resources can apply the method effectively. Training courses at ICTP and other institutions have helped expand its global use, contributing to a generation of scientists who prioritize physically meaningful and evidence-based hazard modelling. Through this educational dimension, the influence of NDSHA extends far beyond its scientific achievements, shaping the intellectual frameworks through which young researchers approach seismic risk.

As seismic hazard assessment continues to evolve—incorporating new datasets, computational tools, and interdisciplinary insights—the foundational principles embodied in NDSHA remain essential. The method's emphasis on physical modelling, transparency, and reproducibility continues to inspire researchers, reinforce scientific rigour, and improve the capacity of societies to prepare for major earthquakes. In this sense, NDSHA stands not only as a scientific method but also as a testament to Professor Panza's broader vision: that rigorous science, grounded in first principles, can and must contribute to the protection of human life. This volume is dedicated not only to celebrating his lifetime of remarkable achievement but also to inspiring the next generation of seismologists and earthquake engineers to carry forward his spirit of inquiry, intellectual courage, and collaborative engagement.

With profound respect and admiration, we dedicate this book to:

Professor Giuliano F. Panza

A pioneer, mentor, visionary, and enduring guiding light of NDSHA.

References

Jia Junbo. (2018). *Soil Dynamics and Foundation Modelling: Offshore and Earthquake Engineering*. Springer.

Panza Giuliano F., Kossobokov Vladimir G., Laor Efraim and De Vivo Benedetto. (2021). *Earthquakes and Sustainable Infrastructure: Neodeterministic (NDSHA) Approach Guarantees Prevention Rather Than Cure*. Elsevier.

Part I
Theoretical Frameworks

CHAPTER

1

Earthquake Forecast/Prediction
What's Next for Science?

Kossobokov V.[1,2,3*] and Nekrasova A.[1]

After the two decades of digital revolution, Science can disclose Natural Hazards, assess Risks, and deliver state-of-the-art Knowledge of looming disaster in advance catastrophes along with useful Recommendations on the level of risks for decision-making in regard to engineering design, insurance, and emergency management. Science cannot as yet remove people's favour for illusion regarding Reality, as well as political denial, ignorance, and negligence among decision-makers. Nowadays, we are living in a world of numbers and calculations with an enormous amount of user-friendly software producing colourful outputs that may lead to discovery or mislead to deceptive conclusions, erroneous claims, and predictions.

According to popular perception, earthquakes are random unpredictable phenomena. Many people, including scientists, do not distinguish between "unpredictable", "random", and "haphazard", which distinction is crucial for scientific reasoning and conclusions. Many are misled by concluding a model "verified" or "proved" when their test statistic falls in a reasonable part of the distribution. Best test scoring does not guarantee any adequacy, even if it is "the best" among the available sample of unrealistic models.

Earthquakes are rather haphazard, lacking any obvious principle of organization, but abbey Unified Scaling Law that generalizes the fundamental Gutenberg-Richter relationship accounting for fractal properties of seismicity. Earthquakes do cluster in space and time. They reoccur as swarms, direct cascades of aftershocks, and indirect cascades of foreshocks. These properties of an apparently chaotic hierarchically organized non linear seismic system allow for predictability, although up to a certain limit and after substantial averaging. Therefore, operational seismic hazard assessment and earthquake forecast/prediction claims can be useful for disaster preparedness aimed at reducing impacts from earthquakes, if reliable, but not necessarily perfect.

Should the scientific community continue to widespread unreliable claims, myths, forecasts, and predictions about earthquakes or use reliable knowledge to warn people of looming disaster?

1 Institute of Earthquake Prediction Theory and Mathematical Geophysics, Russian Academy of Sciences, Moscow, Russian Federation.
2 Accademia Nazionale delle Scienze detta dei XL, Rome, Italy.
3 International Seismic Safety Organisation, Arsita, Italy.
* Corresponding author: volodya@mitp.ru

DOI: 10.1201/9781003684121-2

1. Introduction

Earthquakes are sudden fractures of the Earth's crust that radiate seismic waves and cause ground shaking. Earthquakes are not random but haphazard (i.e., lacking any obvious principle of organization), apparently, self-organized phenomena of the Earth's hierarchy of blocks (from tectonic plates to grains of rock) that move relative to each other along their boundaries (Keilis-Borok, 1990). The abruptness along with apparent irregularity and infrequency of significant damaging earthquakes facilitate formation of a common perception that earthquakes are random unpredictable phenomena.

Many people do not distinguish between "unpredictable", "random", and "haphazard", which is crucial for reliable realistic conclusions and recommendations aimed at reducing impacts from disastrous earthquakes (Stark, 2022; Mulargia et al., 2017). In fact, making quantitative probabilistic claims in the frames of the most popular objectivists' viewpoint on probability (Mosteller et al., 1961) requires a long series of recurrences, which cannot be obtained at local scale from the existing catalogues of earthquakes. Therefore, too often the "probabilities" of future earthquakes reported to public officials and citizens in widely-advertised forecasts are subjective, being derived from essentially unverified model hypotheses. The results of comparison with actual seismic data available allow concluding, beyond any doubt, that these "probabilistic" forecasts appear to be grossly misleading and are evidently unacceptable for any kind of responsible risk evaluation and knowledgeable disaster prevention (Kossobokov & Nekrasova, 2012; Wyss et al., 2012).

Digital revolution started two decades ago provides unprecedented opportunities for enhancing studies of the Earth System, including seismic phenomenon. Significant steps have already been made towards assessing earthquake space-time-magnitude relationships and recognition of multifactorial patterns, showing the potential for reproducible, testable, and reliable operational earthquake forecasting. Nevertheless, the following statement, originally attributed to seismic hazard assessment, four decades ago applies to the present-day situation in analyses of potential losses due to earthquakes and associated risks: "However, ignorance still exists on the seismic severity (usually expressed in intensity values) a site may expect in the future as well on the damage a structure may sustain for a given seismic intensity" (Boissonnade & Shah, 1984).

2. Seismic Evidence

Many comprehensive statistical analyses reject hypotheses of random recurrence of earthquakes. Earthquake sequences appear to be lacking any analytically tractable principle of organization. Mathematically, the characteristics of such haphazard, apparently chaotic systems are nevertheless predictable up to a certain limit and after substantial averaging. Therefore, a "success" in forecasting disastrous earthquakes necessarily implies a successive step-by-step determination that narrows down the time interval, location area, and magnitude range of an incipient earthquake.

Earthquake location area is rather stable: A high level of the spatial predictability of earthquakes is evident from their hypocentre determinations. The boundaries of the tectonic plates are clearly defined (Figure 1). Seismicity is particularly intense in subduction zones, a broad band of seismicity extends from Southern Europe to Southeast Asia, which is associated with collision zones between the Eurasian plate and the African, Arabian, and Indian plates. The range of the annual average number of M 4.0 or larger earthquakes in nonempty $1^\circ \times 1^\circ$ cells is more than three decimal orders. Comparing maps on left and right in Figure 1, it is clearly seen that there is a strong correlation between the value of the average annual number derived from 32 years of data and the largest earthquake to have occurred in each cell in 98 years of data (Kossobokov et al., 2000).

Moreover, the results of pattern recognition of earthquake prone areas (PREPA) in numerous regions worldwide (Gorshkov et al., 2003; Gorshkov & Soloviev, 2021) have proved reliable regional classifications of morphostructural nodes into D (*dangerous*) and N (*not dangerous*) classes

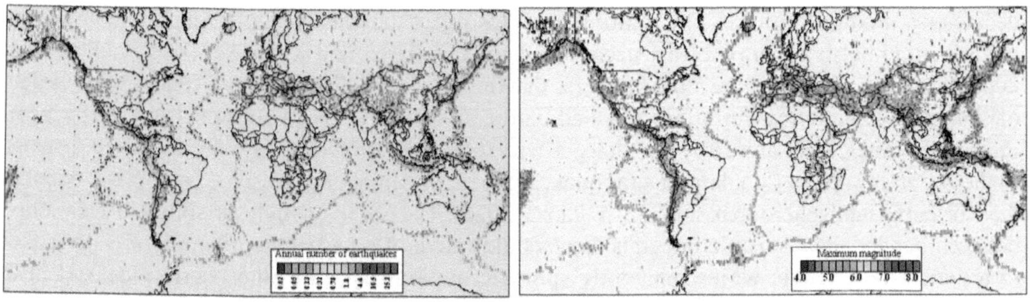

Fig. 1 Annual number (left) of magnitude $M \geq 4.0$ during the period 1964–1995 and maximum observed magnitude M_{max} (right) during the period 1900–1997 (Kossobokov et al., 2000).

with respect to specified magnitude M_0. Since the first publication of a series (Gelfand et al., 1972) the empirical error in classification, as the percentage of post publication target earthquakes in nodes N, accounts for 4/106 = 3.8%, which estimate favours high level of reliability and generally accepted confidence (above 95%) of the PREPA results.

Seismic activity is self-similar: Earthquakes abbey Unified Scaling Law (USLE) that generalizes the fundamental Gutenberg-Richter relationship accounting for the fractal properties of earthquake loci (Kossobokov & Mazhkenov, 1988; Nekrasova & Kossobokov, 2002; Nekrasova et al., 2020; Kossobokov, 2021):

$$\log_{10}N = A + B \cdot (5 - M) + C \cdot \log_{10}L,\ M_{-} \leq M \leq M^{-}$$

where, $N = N(M, L)$ is the expected annual number of earthquakes with magnitude M in an earthquake-prone area of linear dimension L. In addition to A and B, analogous to the *a*- and *b-values* of the Gutenberg-Richter relationship, the coefficient C shows how the number of earthquakes, N, is changing with linear dimension, L, of an enveloping area. Analysing seismic activity at a regional scale, Nekrasova et al. (2011) discovered a well-organized attractor in the 3D domain of the A, B, C possible values, which does not suggest any uniform correlation between pairs of the USLE coefficients.

One of the very first practical conclusions drawn from USLE (Kosobokov & Mazhkenov, 1992) demonstrated that in case of the Lake Baikal region with the area of 1,500,000 km^2 and $C = 1.25$, the inclusion of aseismic areas leads to underestimation of widespread seismic activity measure A_{10} (Riznichenko, 1982) in 1000 km^2 by a factor of 15, and to its overestimation by a factor greater than 2 when a characteristic of seismic activity over an area of 1000 km^2 is computed for a grid cell of 10 km × 10 km. A comprehensive analysis of seismicity at 1,140 cities and urban agglomerations in seismic regions worldwide due to scaling in proportion to the area (Nekrasova & Kossobokov, 2005) reveals significant underestimation of seismic hazard, too large for being ignored in earthquake loss evaluations necessary for knowledgeable disaster mitigation. Such a discrepancy propagates nonlinearly into erroneous estimations of risks. An error in rate of a factor of 6 corresponds to an error about 0.8 in magnitude, which explodes exponentially in the underestimation of seismic energy release by a factor 15, corresponding to a factor above 4.5 for expected fatalities (Wyss et al., 2012).

Note that the distribution of inter-event times depends only on the USLE control parameter $\eta = 10^{-BM} \cdot L^C$ (Bak et al., 2002), which variability at global and local scales evidences the existence of different, intermittent levels of rather steady long-term seismic activity characterized with a near constant value of η (Nekrasova et al., 2011; Kossobokov & Nekrasova, 2017, 2019; Bukchin et al., 2020; Kossobokov et al., 2023; Kossobokov & Nekrasova, 2024). Usually, but not necessarily, the region switches from one level to another in a shorter-term transition associated with strong, catastrophic events. During such a switch, seismic activity may follow very different scenarios with inter-event scaling of different kinds, such as constant, logarithmic, power law, exponential rise/

decay, or a mixture of those. The uniform analyses reject the existence of a single one universal scenario of impulsive energy release in earthquakes, solar flares, and starquakes (Kossobokov et al., 2008; Lepreti et al., 2009); the main finding is that the associated distributions display a wide variety of shapes and cannot be uniformly rescaled onto a single curve by two-parametric transforms of inter-event times, even at nearby locations. There are neither preferred days nor preferred distance to the Moon for the world's largest earthquakes ($M \geq 7.5$), although a seasonal pattern is evident in the Northern Hemisphere (Kossobokov & Panza, 2020). The observed high variability of correlation between geodetic and seismic integrals appears indicative of a partial contribution of earthquakes to a generally aseismic while apparently sporadic motion of small lithospheric blocks (Liu & Kossobokov, 2021).

Geller et al. (1997) have claimed that "earthquakes cannot be predicted". This is certainly true in a sense of the exact time, location, and magnitude of an earthquake reported well in advance of the catastrophe. The temporal predictability of large earthquakes requires special comment. No modern theory of seismic dynamics can answer this question. Inevitably, a negative statement asserting a non-trivial limitation of predictability is only an assumption. On the other hand, direct testing of a reproducible forecast/prediction method in real-time, and nothing else, can unambiguously establish a certain degree of earthquake predictability.

Regretfully, most, if not all, of earthquake prediction claims are "invented" due to very small, if any, sample of clearly defined evidence (Kossobokov & Panza, 2022). Most of the precursors aimed at short-term prediction are, in fact, not tested possibilities recognized or documented only after main shock occurrences (Nagao et al., 2021). The recent volume of special topic studies (Freund et al., 2021) confirms that most, if not all, claims of non-seismic precursors of large earthquakes are lacking the results of any statistical testing required for reliable operational forecasting. The necessity and possibility of applying simple tools of Earthquake Prediction Strategies – Error Diagram (Molchan, 1987, 2003) and Seismic Roulette null-hypothesis (Kossobokov, 1998; Kossobokov et al., 1999; Kossobokov & Shebalin, 2003) – are evident. The results of the on-going global testing of the earthquake prediction algorithm M8 started in 1992 (Healy et al., 1992) have proved that Seismic Roulette is not perfect! It was reported for the first time in 1999, and then confirmed in the semi-annual updates till January 2025 (Table 1, Figure 2).

Table 1 The performance of the M8 and M8-MSc predictions.

Period	*Target Earthquakes*			*Space-Time Volume, μ*		*Confidence*	
	Total	*M8*	*M8-MSc*	*M8*	*M8-MSc*	*M8*	*M8-MSc*
*Magnitude range M*8.0+							
1985–2024	31	18	11	26.94%	12.54%	99.97%	99.91%
1992–2024	29	16	9	23.85%	10.56%	99.97%	99.77%
*Magnitude range M*7.5+							
1985–2024	101	43	17	26.93%	8.30%	99.95%	99.61%
1992–2024	89	33	11	23.33%	7.49%	99.75%	93.12%

Notes: "Target Earthquakes" are earthquakes of the magnitude range *M*8.0+ or *M*7.5+, "Total" refers to their total number in the study "Period", and "M8" and "M8-MSc" refer to the number of those earthquakes in the space-time volumes covered by the M8 and M8-MSc alerts, respectively. "Space-Time Volume" of the M8 and M8-MSc predictions is given in respect to the Seismic Roulette null-hypothesis. "Confidence" level tells how sure one can be that the achieved performance does not arise in the binomial trials.

Unlike the rigid settings of the global testing of M8 and M8-MSc algorithm predictions (Healy et al., 1992; Ismail-Zadeh & Kossobokov, 2021), many investigators overlook an option of spatial and temporal modes and concentrate their efforts on finding in short-term the "exact" fault segment to rupture, which is by far a more difficult task and might be an unsolvable problem. Being related to the rupture size $L = L(M)$ of the incipient earthquake of magnitude M, such modes could be summarized in a classification of location of a source zone from a wider prediction range (Table 2). Moreover, the USLE urges on limiting magnitude range of prediction, $Mm+ = [m, m + \Delta m]$, to about one unit or less (i.e., $\Delta m < 1$), otherwise, the statistics would essentially be related to dominating smallest earthquakes and not to the larger ones, which are of particular interest in mitigation of seismic risk.

Table 2 Classification of earthquake predictions.

Temporal, in Years		*Spatial, in Source Zone Size L*	
Long-term	10	Long-range	Up to 100 L
Intermediate-term	1	Middle-range	5–10 L
Short-term	0.01–0.1	Narrow	2–3 L
Immediate	0.001	Exact	1 L

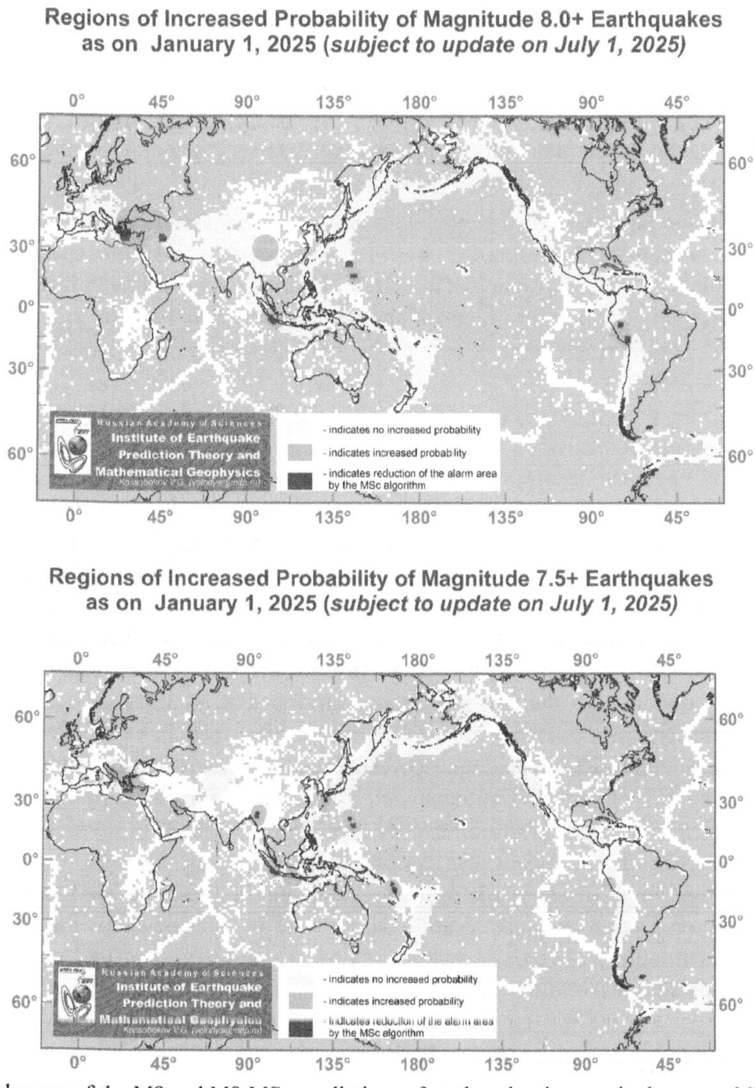

Fig. 2 Global maps of the M8 and M8-MSc predictions of earthquakes in magnitude ranges *M8.0+* (top) and *M7.5+* (bottom).

The results of the above-mentioned Global test of the M8 and M8-MSc algorithms did confirm already at the turn of the 20th century (Kossobokov et al., 1999) a positive statement on predictability of the world largest earthquakes at the intermediate-term scale with the middle-range accuracy of location. The potential of hierarchical approach to operational earthquake prediction can be illustrated with the case history of the 1999 Hector Mine earthquake in California. Specifically, the seismicity patterns that preceded the October 16, Mw 7.1 main shock allowed us to predict it with a middle-range, intermediate-term accuracy, i.e., 100 km and 1.5 years (Romashkova et al., 1999). The location of the earthquake in space and time satisfies the following three steps of gradually improving the accuracy of prediction:

- **Step 1 (zero approximation):** The epicentre is located 37.5 km from one of 73 nodes D prone to magnitude 6.5 or larger earthquakes in California and Nevada (Gelfand et al., 1976; since this publication, 17 earthquakes of $M \geq 6.5$ occurred in the region, all but one in a narrow vicinity of nodes D; the July 6, 2019, M7.1 Ridgecrest epicentre is the exceptional failure-to-predict at 60 km off the nearest node D).
- **Step 2 (first approximation):** The M8 algorithm (Keilis-Borok & Kossobokov, 1990) aimed at M7.0+ earthquakes in a semi-annual update on July 1, 1999 confirmed the alarms diagnosed for the period from July 1998 to July 2003 at 280 km distance from the two out of 73 nodes D.
- **Step 3 (second approximation):** The MSc algorithm (Kossobokov et al., 1990) narrowed down the area of alarm to 33.82–34.68°N × 116.17–117.23°W, as on July 1, 1999. The epicentre of the 1999 Hector Mine main shock and most of its aftershocks occurred within this 96 km × 96 km area.

Moreover, on the background of these intermediate-term narrow-range forecasts, a series of small-magnitude earthquakes began on October 15, the strongest of which occurred 7 hours before the Hector Mine main shock at a distance of less than 1 km from its epicentre, which could have been certainly considered an argument in favour of short-term or even immediate warning (Kosobokov & Shebalin, 2003), similar to the famous case of the 1975 Haicheng earthquake prediction (Adams, 1976; Anonymous, 1977).

Decades later, a golden opportunity was lost to advise the citizens of L'Aquila, Italy on low-key safety measures, when Commisione Grandi Rischi (CGR) issued the politicized statement on 31 March 2009 could have actually provided a much more reliable forecast than the 1975 Haicheng success story, if seismological knowledge had not been misrepresented by CGR. According to the CGR members (Boschi et al., 1995; Dolce & Martinelli, 2005), L'Aquila was at the highest seismic risk in Italy in the near future and buildings in medieval L'Aquila were extremely vulnerable. The precursory earthquake swarm in November 2008–March 2026 was freighting civilians and indeed was a foreshock sequence of the devastating M 6.3 L'Aquila earthquake on 6 April 2009 (Papadopoulos et al., 2010). Many of the 309 victims would not have returned to their homes to sleep after being awakened by the two premonitory tremors of M 3.9 and M 3.5 just 3 hours in advance of the disaster.

Thus, summing up, seismic evidence per se rejects hypotheses of random recurrence of earthquakes. Earthquake hazard is time dependent; its realistic forecast assessment may consider different time scales from decades to months (or even weeks, or days) at regional to local scales, while risks are specific for any object of exposure from a national economy to personal losses of an individual. Mitigation of disastrous losses from earthquakes and associated landslides, tsunami, fires, liquefaction, etc., requires realistic and reliable analysis of potential impact on the natural environment, infrastructures, and social systems. Is this requirement fulfilled nowadays? How do societies prepare for and respond to real, potential, or imminent earthquake disasters?

3. Global Seismic Hazard Maps

"One is well advised, when traveling to a new territory, to take a good map and then to check the map with the actual territory during the journey." (Wasserburg, 2010)

The Global Seismic Hazard Assessment Program (GSHAP) was launched in 1992 by the International Lithosphere Program (ILP) with the support of the International Council of Scientific Unions (ICSU), and endorsed as a demonstration programme in the framework of the United Nations International Decade for Natural Disaster Reduction (UN/IDNDR). The GSHAP project terminated in 1999 (Giardini, 1999) with publication of the final GSHAP Global Seismic Hazard Map. A number of publications (e.g., Krinitzsky, 1993a, 1993b, 1995; Castanos & Lomnitz, 2002; Klügel, 2007) criticizing the paradigm of Probabilistic Seismic Hazard Analysis (PSHA) accepted in GSHAP was challenging its scientific merit. After the deadly 2010 Haiti earthquake and then the 2011 Tōhoku mega-thrust and tsunami a systematic comparison of the GSHAP peak ground acceleration (PGA) estimates with those related to actual earthquakes disclosed gross inadequacy of the GSHAP Final Map (Kossobokov & Nekrasova, 2012; Wyss et al., 2012). In particular, (i) all the 12 earthquakes that caused population losses above 1,000 fatalities, as well as (ii) all the 56 significant earthquakes ($M \geq 7.5$) in 2000–2009 have "surprised" the Final Map by the excessive macroseismic impact differences with an average and median of 2 units on MMI scale. Moreover, for 50% of more than a thousand strong ($M \geq 6.0$) earthquakes, the PGA values on the GSHAP PGA map were surpassed by 0.17 g or more within ten post publication years, which evidence contradicts the predicted "10% probability of exceedance in 50 years" (10% *poe* in 50 years). The conclusion was drawn that the GSHAP results, as well as the methodologies that underlie them, are deeply flawed and therefore unacceptable for responsible risk assessments that help in preventing earthquake disasters. It should be noted that such inadequacy of the Final Map could have been discovered by GSHAP participants if they had checked the earthquake statistics for the period 1990–1999.

A decade ago, Wyss et al. (2012) urged "that the international project Global Earthquake Model (http://www.globalquakemodel.org/) is on the wrong track, if it continues to base seismic risk estimates on the standard method to assess seismic hazard". Apparently, the contributors to GEM keep misleading interested parties by providing erroneous seismic hazard maps (Pagani et al., 2018). It should be noted, however, that, as a matter of fact, the GEM Disclaimer does characterize their Global Seismic Hazard Map as absolutely useless (https://maps.openquake.org/map/global-seismic-hazard-map/):

> *"The information included in this map must not be used for the design of earthquake-resistant structures or to support any important decision involving human life, capital and movable and immovable properties."*

Table 3 lists all the World's deadliest earthquakes since the year 2000. The last three columns show up the differences between the macroseismic intensity at epicentre of the observed earthquake ($I_{0\ EVENT}$) and that predicted by the GSHAP and GEM Maps ($I_{0\ GSHAP}$ and $I_{0\ GEM}$) followed by the ratio $\boldsymbol{R_{PGA} = mPGA_{GEM}/mPGA_{GSHAP}}$ between the predicted maximum PGA values with 10% *poe* in 50 years on the GEM versus GSHAP maps.

The macroseismic intensity differences, sampled by deadliest earthquakes, are almost all positive with their average and median of 2 units on MMI scale. Note that an underestimation of 2 or more units on the MMI scale corresponds to a transition from *STRONG* (VI MMI – *Felt by all. Many were frightened. Some heavy furniture moved; a few instances of fallen plaster. Slight damage to poorly constructed buildings.*) to *SEVERE* (VIII MMI – Damage slight in specially designed structures; considerable *damage in ordinary substantial buildings with partial collapse. Damage great in poorly built structures. Fall of chimneys, factory stacks, columns, monuments, walls. Heavy furniture overturned.*) damage.

The values of $\boldsymbol{R_{PGA}}$ (in the last column of Table 3) indicate an apparent upgrade of the GEM hazard map in respect to the one of GSHAP; the 20 sampled values with the average of 1.8 and median of 1.6 disclose a tendency to increase maximum PGA values with 10% *poe* in 50 years at the sites of the occurred (in our case, deadliest) earthquakes. Interestingly, in comparison to GSHAP, the seismic hazard at sites of the 2003 Bam and 2002 Hindu Kush earthquakes is downgraded by

14% and 26%, respectively, while it is about the same at locations of the 2015 Nepal and the recent 2023 earthquakes in Turkey.

Table 3 Top deadliest earthquakes since 2000.

Region	*Date*	*M*	*Fatalities*	*ΔI0*	*ΔI0'*	*RPGA*
Sumatra-Andaman Islands	26.12.2004	9.0	227,898	4	4*	1.20*
Port-au-Prince (Haiti)	12.01.2010	7.3	222,570	2	1	3.10
Wenchuan (Sichuan, China)	12.05.2008	8.1	87,587	3	3	1.60
Kashmir (India-Pakistan northern border)	08.10.2005	7.7	87,351	2	2	2.06
Nurdağı (Turkey)	06.02.2023	7.8	59,359+	1	2	1.08
Ekinözü (Turkey)		7.5		1	2	1.11
Bam (Iran)	26.12.2003	6.6	26,271	0	1	0.86
Bhuj (Gujarat, India)	26.01.2001	8.0	20,085	3	2	2.30
Off the Pacific coast of Tōhoku (Japan)	11.03.2011	9.0	19,759+	3	3*	1.22*
Bharatpur (Nepal)	25.04.2015	7.8	8,964	2	2	1.01
Yogyakarta (Java, Indonesia)	26.05.2006	6.3	5,782	0	0	1.57
Sulawesi Island (Indonesia)	28.09.2018	7.5	4,340	2	2	1.62
Southern Qinghai (China)	13.04.2010	7.0	2,968	2	1	1.68
Oukaïmedene (Morocco)	08.09.2023	6.8	2,946	3	2	2.09
Boumerdes (Algeria)	21.05.2003	6.8	2,266	2	0	3.75
Nippes (Haiti)	14.08.2021	7.2	2,248	2	1	3.39
Zindah Jān, Herāt (Afghanistan)	07.10.2023	6.3, 6.3	1,482	0	0	1.22
Nias Island (Indonesia)	28.03.2005	8.6	1,313	3	3	2.22
Padang (Southern Sumatra, Indonesia)	30.09.2009	7.5	1,117	1	1	1.38
Hindu Kush (Afghanistan)	25.03.2002	6.1	1,000+	−1	−1	0.74

Notes: At least 1,000+ fatalities including victims of tsunami and other associated effects. The increments $\Delta I_0 = I_{0\ EVENT} - I_{0\ GSHAP}$ and $\Delta I_0' = I_{0\ EVENT} - I_{0\ GEM}$ are computed as in (Kossobokov & Nekrasova, 2012) rounded to the closest integer. The values obtained from the GEM interactive Global Seismic Hazard Map marked *.

It deserves mentioning that GEM appears to ignore ground shaking resulted by many earthquakes with epicentres off shore, including the 26 December 2004 Indian Ocean Disaster and the 11 March 2011 Great Tōhoku earthquake and tsunami; as a matter of fact, only six out of 24 magnitude 8 or larger earthquakes since the year 2000 have a determination of PGA at their epicentres. This restriction, compared to the GSHAP Final Map, reduces the test sample of strong magnitude $M \geq 6.0$ earthquakes of the GEM Global Seismic Hazard Map by a factor larger than 3 and, formally, the number of $M \geq 6.0$ "surprises" from 30–40% to 20–25%. Thus, the results of comprehensive cross-comparisons of Probabilistic Seismic Hazard Analysis of GSHAP and GEM "10% *poe* in 50 years" versus earthquake reality sampled by 10322 strong ($M \geq 6.0$, depth of 70 km or less) earthquakes in 1900–2024 confirm that, same as the one of GSHAP, the GEM "probabilistic" product appears to be harmful, surely, not useful for a design of earthquake-resistant structures and cannot support any important decision involving human life, capital, and movable and immovable properties.

Finally, it is worth reiterating how unreliable, misleading, and harmful PSHA can be, as evidenced by the more than 785,000 lives lost between 2000–2025 (Table 3). So why do scientists continue to use and contribute to PSHA (e.g., Hamdache et al., 2025) when they clearly understand and acknowledge the fatal limitations, assumptions, and unacceptable practices of this probabilistic model?

Unlike PSHA (Giardini, 1999; Pagani et al., 2018) the Neo-deterministic seismic hazard assessment (NDSHA) studies (Panza et al., 2021) do not cover yet the entire seismicity of the Earth. Nevertheless, in a dozen of regional applications worldwide NDSHA outscore the widespread PSHA by taking advantage of a synergy between the to-date available PREPA, USLE, Intermediate-Term Earthquake Prediction (ITEP) of different spatial accuracy, Geodetic Data Analysis (GDA), Maximum Credible Seismic Input (MCSI), and Scenario-based Seismic Hazard Analysis (SSHA). The synergy of PREPA × USLE × ITEP × GDA × MCSI × SSHA applies to a pretty wide spectrum of geophysical observables and allows delivering user-specific NDSHA products, which are tested to be reliable, realistic, and useful evaluation and mapping of apparently time-dependent seismic hazard and associated risks. Of course, it is MCSI and SSHA that provide a quantitative output of seismic input for the structural engineer community at the final stage of NDSHA.

4. Discussion and Conclusions

"Many people, especially ignorant people, want to punish you for speaking the truth, for being correct, for being you. Never apologize for being correct, or for being years ahead of your time. If you're right and you know it, speak your mind. Even if you are a minority of one, the truth is still the truth." (Mahatma Gandhi)

The history of instrumental seismology is rather short, so that many well-known quantitative relations of earthquake size and recurrence are being suggested at the times of lacking data and require a serious revision. In particular, seismic catalogues do not support the presence of "universal" time scaling or space-time-magnitude pattern in sequences of earthquakes either in advance or after a main shock (Kossobokov & Nekrasova, 2017; Kossobokov & Nekrasova, 2019; Bukchin et al., 2020; Kossobokov et al., 2022; Kossobokov & Nekrasova, 2024). Neither have they supported regular periodic recurrence (Sykes & Menke, 2006; Griffin, 2020). However, there is enough evidence to claim (with high confidence) the following paradigms (Kossobokov & Shebalin, 2003; Ismail-Zadeh & Kossobokov, 2021):

- Precursory seismicity patterns exist.
- The dimension of an area where precursory seismicity patterns appear is by far (about ten times) larger than that of the source zone of the incipient large earthquake.
- Many precursory seismicity patterns are similar in regions of fundamentally different seismic and tectonic environments.
- Some precursory seismicity patterns are universal. Their analogues take place in advance of extreme catastrophic events in other complex nonlinear systems.

Days after the 26 December 2004 Great Asian Mega Earthquake and Tsunami during the UN World Conference on Disaster Reduction (18–22 January 2005, Kobe, Hyogo, Japan), a Statement (Kossobokov, 2005) at the "Special Session on the Indian Ocean Disaster: risk reduction for a safer future" was urging on a possibility of a few mega earthquakes of about magnitude 9.0 occurring within the next 5–10 years. This prediction was based on the already encouraging results of the first decade of real-time global testing of earthquake prediction algorithm M8 and good evidence suggesting that seismic events including mega-earthquakes cluster: "All four mega-earthquakes of the 20th century (Kamchatka, 1952/11/04, Mw9.0; Andreanoff Islands, 1957/03/09, Mw9.1; Chile, 1960/05/22, Mw9.5; Alaska, 1964/03/28, Mw9.2) happened within a narrow interval of time. Such a cluster is unlikely with a 99% confidence for uniformly distributed independent events." The prediction was eventually confirmed by the 27 February 2010, Mw 8.8 mega-thrust offshore Maule, Chile, and the 11 March 2011, Mw 9.1 mega-thrust and tsunami off the Pacific coast of Tōhoku, Japan (Kossobokov, 2011; Ismail-Zadeh & Kossobokov, 2021).

An opportunity to reduce the impacts from these mega-earthquakes was missed. Davis et al. (2012) showed how the information on expected world's largest earthquakes provided by the M8 and MSc algorithms (Keilis-Borok and Kossobokov, 1990; Kossobokov et al., 1990), although limited to the intermediate-term span of years and middle-range location of a thousand km can be used to reduce future disasters. The primary reasons for not using the prediction for improving preparations in advance of the Tōhoku earthquake "include: (1) inadequate links between emergency managers and the earthquake prediction information; and (2) no practicing application of existing methodologies to guide emergency preparations and policy development on how to make decisions based on information provided for an intermediate-term middle-range earthquake prediction having limited but known accuracy".

A decade ago an opinion on operational earthquake forecast/prediction problems presented a viewpoint of experienced practitioners (Kossobokov et al., 2015). "The proof of the pudding is in the eating": practitioners subject their tools to rigid testing, first in retrospect, and then in an experimental real-time mode to judge on the ultimate success or failure of the prediction tool. A desirable high confidence assigned to forecast/prediction may arise either from a subjective judgment of an expert (probability concept) or from an objective score achieved in a real-time testing (experiment concept). The uncertainties intrinsic to the natural earthquake generating system of block-and-faults are naturally transferred to the epistemic (systematic due to model neglecting certain effects) and aleatory (statistical due to random noise effects) uncertainties in theoretical and/or computer modelling. Therefore, an observable of a certain physical dimension cannot be expressed by probability, a dimensionless number from 0 to 1. On the other hand, ground-motion models of NDSHA can be combined to generate site-specific hazard curves. In fact, large ensembles of deterministic simulations are feasible to achieve (Panza et al., 2001, 2012), which statistics allows for probabilistic estimates of expected consequences needed for implementation of risk-mitigation actions. Practitioners are positive that any reliable forecasting information can be: (i) effective, (ii) complementary to design and construction of seismically resistant infrastructure, and (iii) well appreciated by the population as a timely precautious reminder and warning (Kossobokov et al., 2015).

We do not live in a black-and-white world. We know quite well that "all models are wrong but some are useful" (Box, 1979), but often forget that some models are useless and some are really harmful, especially when viewed as substitutes for natural phenomena. Not a surprise that our beliefs in initial basic principles may lead to erroneous models and disasters. "And the better the model, the worse it becomes for its applications. Indeed, the pressure of snatching 'initial principles' will lead us to use the model well beyond the possibilities of its application" (Gelfand, 1991). The underlying problem of a widespread scientific crisis in many disciplines lies in misuse of statistics well beyond its possibilities due to superficial education, blind application of available software, and peer review policy of some scientific journals (Stark, 2018; Stark & Saltelli, 2018).

We are living in a risky world and are doomed to making predictions and actions. Years after the 2005 Hyogo and 2015 Sendai Frameworks for Disaster Risk Reduction, countries continue to follow a range of different approaches to earthquake hazard reduction and mitigation strategies; however, so far, no country has achieved an effective policy for dealing with the rising costs of natural disasters. Regretfully, the deceptive widespread paradigm of PSHA remains the core of national building codes in many countries worldwide. However, NDSHA is gaining momentum in spreading worldwide a Paradigm of Reliable Seismic Hazard Assessment (Panza & Bela, 2020; Panza et al., 2021). NDSHA results are based on reliable seismic evidence, exhaustive statistical testing, and computer scenario-based modelling of ground shaking. These provide a realistic and reliable quantitative input for the structural engineer community. The NDSHA methodology has garnered significant attention within both the seismological and engineering communities. The book (Panza et al., 2021) represents a significant milestone, compiling developments of this approach over the past 25 years (Wen & Wang, 2024). The recent progress in modern seismology has significantly enhanced the deterministic seismic hazard assessment approach, shifting it from an empirical paradigm to a physics-based numerical one (SCEC, 2018; SSA, 2023).

Recently, Gupta et al. (2020) emphasized the necessity to develop an earthquake resilient society in the Himalayan region and described in brief the two scenario-based projects conducted by the National Disaster Management Authority. The deterministic physics-based scenarios for the repeat of the 1905, Mw 8.6, Kangra and the 1897, Mw 8.7, Shillong earthquakes were practised in Multi-State Preparedness Campaign drills in 2013 and 2014, respectively. These were very important landmarks in enhancing the earthquake disaster preparedness of the Republic of India. As a matter of fact, the government and population of the Himalayan foothills have achieved a clear understanding that "if the mitigation and preparedness measures are adequate, the impact of any great earthquake can be contained". The practice of earthquake scenario-based "Mega Mock Drills", as a proactive measure to enhance emergency response capabilities, focused on earthquakes and associated hazards is systematically carried out now.

So, should the scientific community continue to exploit the serious fatal flaws of PSHA, or use reliable knowledge to prepare people for mitigating the unavoidable catastrophes?

References

Adams, R. (1976). The Haicheng, China, earthquake of 4 February, 1975: The first successful predicted major earthquake. *Earthquake Engineering and Structural Dynamics, 4*: 423–437. https://doi.org/10.1002/eqe.4290040502.

Anonymous. (1977). Prediction of the Haicheng earthquake. *Eos, Transactions American Geophysical Union, 58* (5): 236–272. https://doi.org/10.1029/EO058i005p00236.

Bak, P., Christensen, K., Danon, L. and Scanlon, T. (2002). Unified Scaling Law for Earthquakes. *Physical Review Letters, 88*: 178501–178504. https://doi.org/10.1103/PhysRevLett.88.178501.

Boissonnade, A.C. and Shah, H.C. (1984). Seismic Vulnerability and Insurance Studies. *The Geneva Papers on Risk and Insurance: Issues and Practice, 9*(32): 223–254. https://doi.org/10.1057/gpp.1984.13.

Boschi, E., Gasperini, P. and Mulargia, F. (1995). Forecasting where larger crustal earthquakes are likely to occur in Italy in the near future. *Bulletin of the Seismological Society of America, 85*(5): 1475–1482. https://doi.org/10.1785/BSSA0850051475.

Bukchin, B.G., Fomochkina, A.S., Kossobokov, V.G. and Nekrasova, A.K. (2020). Characterizing the Foreshock, Main Shock, and Aftershock Sequences of the Recent Major Earthquakes in Southern Alaska, 2016–2018. *Front. Earth Sci., 8*: 584659 https://doi.org/10.3389/feart.2020.584659.

Castanos, H. and Lomnitz, C. (2002). PSHA: Is it science? *Engineering Geology, 66*(3–4): 315–317. https://doi.org/10.1016/S0013-7952(02)00039-X.

Dolce, M. and Martinelli, A. (2005). *Inventario E Vulnerabilità Degli Edifici Pubblici E Strategici dell'Italia Centro-Meridionale* (Vol. II) *Analisi Di Vulnerabilità E Rischio Sismico* (187 pp). INGV/GNDT - Istituto Nazionale di Geofisica e Vulcanologia / Gruppo Nazionale per la Difesa dai Terremoti, L'Aquila, https://emidius.mi.ingv.it/GNDT2/Att_scient/Prodotti_consegnati/Dolce_Zuccaro/Volume%202.pdf.

Freund, F., Mignan, A., Ouillon, G. and Sornette, D. (Eds). (2021). The global earthquake forecasting system: Towards using non-seismic precursors for the prediction of large earthquakes. *European Physical Journal Special Topics, 230*(1): 490. https://link.springer.com/journal/11734/volumes-and-issues/230-1.

Gelfand, I., Guberman, Sh., Keilis-Borok, V., Knopoff, L., Press, F., Ransman, E., Rotwain, I. and Sadovsky A. (1976). Pattern recognition applied to earthquakes epicenters in California. *Phys. Earth and Planet. Inter., 11*: 227–283. https://doi.org/10.1016/0031-9201(76)90067-4.

Gelfand, I.M., Guberman, S.A., Izvekova, M., Keilis-Borok, V.I. and Rantsman, E.Y. (1972). Criteria of high seismicity determined by pattern recognition. *Tectonophysics, 13* (1-4): 415–422 https://doi.org/10.1016/0040-1951(72)90031-5.

Gelfand, I.M. (1991). Two archetypes in the psychology of man. *Nonlinear Science Today, 1*(4): 11–16. https://www.kyotoprize.org/en/speech/the-1989-kyoto-prize/,https://www.kyotoprize.org/wp-content/uploads/2019/07/1989_B.pdf.

Geller, R.J., Jackson, D., Kagan, Y., and Mulargia, F. (1997). Earthquakes Cannot Be Predicted. Science 275, 1616–1616. DOI:10.1126/science.275.5306.1616

Giardini, D. (1999). The global seismic hazard assessment program (GSHAP) – 1992/1999. *Ann. Geophys., 42*(6): 957–974. https://doi.org/10.4401/ag-3780.

Gorshkov, A., Kossobokov, V. and Soloviev, A. (2003). Recognition of earthquake prone areas. In: Keilis-Borok, V.I., and Soloviev, A.A. (Eds.), *Nonlinear Dynamics of the Lithosphere and Earthquake Prediction*. Springer, Heidelberg, 239–310. https://link.springer.com/chapter/10.1007/978-3-662-05298-3_6.

Gorshkov, A. and Soloviev A. (2022). Chapter 7 - Morphostructural zoning for identifying earthquake-prone areas. In: Panza, G.F., Kossobokov, V.G., Laor, E., and De Vivo, B. (Eds.), *Earthquakes and Sustainable Infrastructure: Neodeterministic*

(NDSHA) Approach Guarantees Prevention Rather than Cure1 (135–149). Elsevier. https://doi.org/10.1016/B978-0-12-823503-4.00009-9

Griffin, J.D., Stirling, M.W. and Wang, T. (2020). Periodicity and clustering in the long-term earthquake record. *Geophysical Research Letters, 47*, e2020GL089272. https://doi.org/10.1029/2020GL089272.

Gupta, H.K., Sabnis, K.A., Duarah, R., Saxena, R.S. and Saurabh Baruah. (2020). Himalayan Earthquakes and Developing an Earthquake Resilient Society. *J. Geol. Soc. India, 96*, 433–446. https://doi.org/10.1007/s12594-020-1581-2.

Hamdache, M., Peláez, J.A. and Henares J. (2025). Contribution to Probabilistic Seismic Hazard Assessment and Its Uncertainty for Northeastern Algeria. *Pure and Applied Geophysics, 182*: 439–471. 5. https://doi.org/10.1007/s00024-025-03676-4.

Healy, J.H., Kossobokov, V.G. and Dewey J.W. (1992). *A Test to Evaluate the Earthquake Prediction Algorithm*. M8. USGS Open-File Report has number 92-401 and is of 23 pages + 6 Appendices. https://doi.org/10.3133/ofr92401.

Ismail-Zadeh, A.T. and Kossobokov, V.G. (2021). Earthquake Prediction, M8 Algorithm. In: Gupta, H.K. (Eds.), *Encyclopedia of Solid Earth Geophysics* (pp. 204–208). *Encyclopedia of Earth Sciences Series.* Springer, Cham. https://doi.org/10.1007/978-3-030-58631-7_157.

Keilis-Borok, V.I. and Kosobokov, V.G. (1990). Premonitory activation of seismic flow: Algorithm M8. *Phys. Earth Planet Inter., 61*(1–2), 73-83. https://doi.org/10.1016/0031-9201(90)90096-G.

Keilis-Borok, V.I. (1990). The lithosphere of the Earth as a nonlinear system with implications for earthquake prediction. *Rev. Geophys., 28*: 19–34, 1990. https://doi.org/10.1029/RG028i001p00019.

Klügel, J-U. (2007). Error inflation in probabilistic seismic hazard analysis. *Engineering Geology, 90*(3–4): 186–192. https://doi.org/10.1016/j.enggeo.2007.01.003.

Kosobokov, V.G. (1998). A spatial measure of seismicity. In: *Proc. Uni. Alaska Workshop on Seismicity Patterns Their Statistical Significance and Physical Meaning* (Nikko, Japan, 11/12 May 1998), 13 pp.

Kosobokov, V.G. and Mazhkenov S.A. (1992). On Similarity in the Spatial Distribution of Seismicity. In: D.K. Chowdhury, N.N. Biswas, A.T. Hsui, B. Lasgari, M.E. Wysession, V.I. Keilis-Borok, A.L. Levshin and G.M. Molchan (Eds.), *Selected Papers From Volumes 22 and 23 of Vychislitel'naya Seysmologiya.* AGU. https://doi.org/10.1029/CS001p0006.

Kosobokov, V.G. (2005). The 26 December 2004 Greatest Asian Quake: When to expect the next one? Statement at Special Session on the Indian Ocean Disaster: Risk reduction for a safer future. In: *UN World Conference on Disaster Reduction*, 18–22 January 2005, Kobe, Hyogo, Japan. https://doi.org/10.13140/RG.2.2.11464.14087.

Kossobokov, V (2011). Are Mega Earthquakes Predictable? *Izvestiya, Atmospheric, and Oceanic Physics, 46* (8): 951–961. https://doi.org/10.1134/S0001433811080032.

Kossobokov, V., Nekrasova, A., Filippova, A. and Fomochkina A. (2023). Analyzing the recent major earthquakes in Southern Alaska. In: *AGU Annual Meeting*, 11–15 December 2023, San Francisco, USA, S33I-0491 https://agu.confex.com/agu/fm23/meetingapp.cgi/Paper/1324803.

Kossobokov, V. and Nekrasova A. (2024). Myths about Earthquakes: Quo vadis? *Acta Geologica Sinica* (English Edition), *98*(Sup. 1): 30–32. https://onlinelibrary.wiley.com/doi/epdf/10.1111/1755-6724.15231.

Kossobokov, V. and Panza, G. (2020). A myth of preferred days of strong earthquakes? *Seismological Research Letters, 91*(2A): 948–955. https://doi.org/10.1785/0220190157.

Kossobokov, V. and Panza G. (2022). Seismic Roulette: Hazards and Risks. *Terra Nova, 34*(6): 475–494 https://doi.org/10.1111/ter.12617.

Kossobokov, V., Peresan, A. and Panza, G.F. (2015). On operational earthquake forecast and prediction problems. *Seismological Research Letters, 86*(2): 287–290 https://doi.org/10.1785/0220140202.

Kossobokov, V. and Shebalin P. (2003). Earthquake Prediction. In: Keilis-Borok, V.I. and Soloviev, A.A. (Eds.), *Nonlinear Dynamics of the Lithosphere and Earthquake Prediction.* Springer Series in Synergetics. Springer, Berlin, Heidelberg. https://doi.org/10.1007/978-3-662-05298-3_4.

Kossobokov, V. (2021). Unified Scaling Law for Earthquakes that generalizes the fundamental Gutenberg-Richter Relationship. In: Gupta, H.K. (Eds.), *Encyclopedia of Solid Earth Geophysics* (pp. 1893–1896). *Encyclopedia of Earth Sciences Series.* Springer, Cham. https://doi.org/10.1007/978-3-030-58631-7_257.

Kossobokov, V.G., Keilis Borok, V.I. and Smith, S.W. (1990). Localization of intermediate term earthquake prediction. *J. Geophys. Res., 95* (B12): 19763–19772. https://doi.org/10.1029/jb095ib12p19763.

Kossobokov, V.G., Keilis Borok, V.I. Turcotte, D.L. and Malamud, B.D. (2000). Implications of a statistical physics approach for earthquake hazard assessment and forecasting. *Pure and Applied Geophysics, 157*: 2323–2349. https://doi.org/10.1007/PL00001086.

Kossobokov, V.G., Lepreti, F. and Carbone, V. (2008). Complexity in sequences of solar flares and earthquakes. *Pure and Applied Geophysics, 165*: 761–775. https://doi.org/10.1007/s00024-008-0330-z.

Kossobokov, V.G. and Mazhkenov, S.A. (1994). On similarity in the spatial distribution of seismicity. In: Chowdhury, D.K., Biswas, N.N., et al. (Eds.), *Computational Seismology and Geodynamics* (Vol. 1, pp. 6–15). AGU. https://agupubs.onlinelibrary.wiley.com/doi/10.1029/CS001p0006.

Kossobokov, V.G. and Mazhkenov, S.A. (1988). Spatial characteristics of similarity for earthquake sequences: Fractality of seismicity. In: *Lecture notes of the ICTP Workshop on Global Geophysical Informatics with Applications to Research in Earthquake Prediction and Reduction of Seismic Risk*, Trieste, Italy, 15 Nov.–16 Dec. 1988. ICTP. http://indico.ictp.it/event/a02153/contribution/3/material/0/0.pdf.

Kossobokov, V.G. and Nekrasova, A. (2019). Aftershock sequences of the recent major earthquakes in New Zealand. *Pure and Applied Geophysics, 176*: 1–23 https://doi.org/10.1007/s00024-018-2071-y.

Kossobokov, V.G. and Nekrasova A. (2017). Characterizing Aftershock Sequences of the Recent Strong Earthquakes in Central Italy. *Pure Appl. Geophys., 174*: 3713–3723. https://doi.org/10.1007/s00024-017-1624-9.

Kossobokov, V.G. and Nekrasova, A.K. (2012). Global Seismic Hazard Assessment Program Maps are Erroneous. *Seismic Instruments, 48* (2): 162–170. https://doi.org/10.3103/S0747923912020065.

Kossobokov, V.G. and Nekrasova, A.K. (2024). Myths about earthquakes. In: *AGU Annual Meeting*, Washington DC, USA, NH31E. https://agu.confex.com/agu/agu24/meetingapp.cgi/Paper/1615674.

Kossobokov, V.G., Nekrasova, A.K. and Schepalina, P.D. (2022). Seismic dynamics in advance of and after the largest earthquakes, 1985–2020. *Surveys in Geophysics, 43*: 423–436. https://doi.org/10.1007/s10712-021-09674-0.

Kossobokov, V.G., Romashkova, L.L., Keilis-Borok, V.I. and Healy, J.H. (1999). Testing earthquake prediction algorithms: Statistically significant real-time prediction of the largest earthquakes in the Circum-Pacific, 1992–1997. *Phys. Earth Planet. Inter., 111* (3-4): 187–196. https://doi.org/10.1016/S0031-9201(98)00159-9.

Krinitzsky, E.L. (1993a). Earthquake probability in engineering. Part 1. The use and misuse of expert opinion. The third Richard H. Jahns distinguished lecture in engineering geology. *Engineering Geology, 33*(4), 257–288. https://doi.org/10.1016/0013-7952(93)90030-G.

Krinitzsky, E.L. (1993b). Earthquake probability in engineering: Part 2. Earthquake recurrence and limitations of Gutenberg-Richter b- values for the engineering of critical structures: The third Richard H. Jahns distinguished lecture in engineering geology. *Engineering Geology, 36*(1–2): 1–52. https://doi.org/10.1016/0013-7952(93)90017-7.

Krinitzsky, E.L. (1995). Problems with logic trees in earthquake hazard evaluation. *Engineering Geology, 39*(1–2), 1–3. https://doi.org/10.1016/0013-7952(94)00060-F.

Lepreti, F., Kossobokov, V.G. and Carbone, V. (2009). Statistical properties of solar flares and comparison to other impulsive energy release events. *International Journal of Modern Physics B, 23*(28-29): 5609–5618. https://doi.org/10.1142/S0217979209063894.

Liu, T. and Kossobokov, V.G. (2021). Displacements before and after great earthquakes: Geodetic and seismic viewpoints. *Pure and Applied Geophysics, 178*: 1135–1155. https://doi.org/10.1007/s00024-021-02694-2.

Molchan, G.M. (1977). Earthquake prediction as a decision-making problem. *Pure and Applied Geophysics, 149*, 233–247. https://doi.org/10.1007/BF00945169.

Molchan, G.M. (2003). Earthquake Prediction Strategies: A Theoretical Analysis. In: Keilis-Borok, V.I., and Soloviev, A.A. (Eds.), *Nonlinear Dynamics of the Lithosphere and Earthquake Prediction* (209-237)., Springer Series in Synergetics, Springer, Berlin, Heidelberg, Germany. https://doi.org/10.1007/978-3-662-05298-3_5.

Mosteller, F., Rourke, R.E.K. and Thomas, G.B. Jr. (1961). Probability: A First Course (xii + 319 pp.). Addison-Wesley Publishing Co. Inc., London, England.

Mulargia, F., Stark, P.B. and Geller R.J. (2017). Why is Probabilistic Seismic Hazard Analysis (PSHA) still used? *Physics of the Earth and Planetary Interiors, 264*: 63–75. https://doi.org/10.1016/j.pepi.2016.12.002.

Nagao, T., Kamogawa, M. and Uyeda S. (2021). Earthquake Precursors and Prediction. In: Gupta, H.K. (Eds.), *Encyclopedia of Solid Earth Geophysics. Encyclopedia of Earth Sciences Series* (pp. 193–204). Springer, Cham. https://doi.org/10.1007/978-3-030-58631-7_4.

Nekrasova, A., Kossobokov, V., Aoudia, A., Perezan, A. and Panza, G.F. (2011). A Multiscale Application of the Unified Scaling Law for Earthquakes in the Central Mediterranean area and Alpine region. *Pure and Applied Geophysics, 168*: 297-327. https://doi.org/10.1007/s00024-010-0163-4.

Nekrasova, A. and Kossobokov, V. (2002). Generalizing the Gutenberg-Richter scaling law. *EOS Trans. AGU, 83* (47): Fall Meet. Suppl., Abstract NG62B-0958.

Nekrasova, A. and Kossobokov V. (2005). Unified Scaling Law for Earthquakes: Mega-cities and urban agglomerations. *EOS Trans. AGU, 86*(52): Fall Meet. Suppl., Abstract S23A-0229.

Nekrasova, A. and Kossobokov. (2023). Seismic risk assessment for the infrastructure in the regions adjacent to the Russian Federation Baikal–Amur Mainline based on the Unified Scaling Law for Earthquakes. *Natural Hazards, 116*: 1995–2010. https://doi.org/10.1007/s11069-022-05750-9.

Pagani, M., Garcia-Pelaez, J., Gee, R., Johnson, K., Poggi, V., Styron, R., Weatherill, G., Simionato, M., Viganò, D., Danciu, L. and Monelli D. (2018). Global Earthquake Model (GEM) Seismic Hazard Map (version 2018.1 - December 2018). https://doi.org/10.13117/GEM-GLOBAL-SEISMIC-HAZARD-MAP-2018.1.

Panza, G., Kossobokov, V., De Vivo, B. and Laor, E. (Eds). (2021). *Earthquakes and Sustainable Infrastructure: Neodeterministic (NDSHA) Approach Guarantees Prevention Rather than Cure* (xxv, 672 pp,). Elsevier. Paperback ISBN: 9780128235034, eBook ISBN: 9780128235416. https://doi.org/10.1016/C2020-0-00052-6.

Panza, G.F. and Bela J. (2020). NDSHA: A new paradigm for reliable seismic hazard assessment. *Engineering Geology, 275*: 105403. https://doi.org/10.1016/j.enggeo.2019.105403.

Panza, G.F., La Mura, C., Peresan, A., Romanelli, F. and Vaccari F. (2012). Seismic Hazard Scenarios as Preventive Tools for a Disaster Resilient Society. *Advances in Geophysics, 53*: 93–165. https://doi.org/10.1016/B978-0-12-380938-4.00003-3.

Panza, G.F., Romanelli, F. and Vaccari F. (2001). Seismic wave propagation in laterally heterogeneous anelastic media: Theory and applications to seismic zonation. *Advances in Geophysics, 43*: 1–95. https://doi.org/10.1016/S0065-2687(01)80002-9.

Papadopoulos, G.A., Charalampakis, M., Fokaefs, A., Minadakis, G. (2010). Strong foreshock signal preceding the L'Aquila (Italy) earthquake (Mw 6.3) of 6 April 2009. *Nat. Hazards Earth Syst. Sci., 10*: 19–24. https://doi.org/10.5194/nhess-10-19-2010.

Riznichenko, YuV. (1982). On the study of seismicity. *Izvestiya AN SSSR, Fizika Zemli*, *12*: 3–18 (in Russian).

Romashkova, L.L., Kossobokov, V.G. and Healy, J.H. (1999). The 1999 Hector Mine earthquake was expected. *EOS transactions, American Geophysical Union, 80* (46, November Supplement, 19).

SCEC. (2018). The 2018 SCEC annual meeting. In: *Proceedings Volume XXVIII*, 217pp. https://files.scec.org/s3fs-public/SCEC2018Proceedings.pdf.

SSA. (2023). *Future Directions: Physics-based Ground-motion Modelling*. Vancouver (10–13 October 2023), Seismological Society of America. https://www.seismosoc.org/wp-content/uploads/2023/09/SSA-Ground-Motion-Modeling-2023-FINAL-Program.pdf.

Stark, P.B. (2018). Before reproducibility must come preproducibility. *Nature, 557*: 613. https://doi.org/10.1038/d41586-018-05256-0.

Stark P.B. (2022). Pay no attention to the model behind the curtain. *Pure and Applied Geophysics*, 179: 4121–4145. https://doi.org/10.1007/s00024-022-03137-2.

Stark, P.B. and Saltelli A. (2018). Cargo-cult statistics and scientific crisis. *Significance, 15*(4): 40–43. https://www.significancemagazine.com/593.

Sykes, L.R. and Menke W. (2006). Repeat times of large earthquakes: Implications for earthquake mechanics and long-term prediction. *Bulletin of the Seismological Society of America, 96*(5): 1569–1596. http://doi.org/10.1785/0120050083.

Wasserburg, G.J. (2010). Comment on "AGU statement: Investigation of scientists and officials in L'Aquila, Italy, is unfounded". *Eos, Trans Am Geophys Union, 91*(42): 384. https://doi.org/10.1029/2010E O420006.

Wen, Z.P. and Wang, G.X. (2024). Earthquakes and Sustainable Infrastructure–Neodeterministic (NDSHA) Approach Guarantees Prevention Rather Than Cure. *Earthquake Science, 37*(5): 494–497. https://doi.org/10.1016/j.eqs.2024.06.010.

Wyss, M., Nekrasova, A. and Kossobokov V. (2012). Errors in expected human losses due to incorrect seismic hazard estimates. *Natural Hazards, 62*(3): 927–935. https://doi.org/10.1007/s11069-012-0125-5._

CHAPTER

2

Unraveling Fractal Characteristics of Seismogenic Nodes

Nekrasova A.[1*], Gorshkov A.[1] and Kossobokov V.[1,2,3]

The study exploits the synergy of two approaches within the Neo-Deterministic Seismic Hazard Assessment (NDSHA): the Unified Scaling Law for Earthquakes (USLE) and the Pattern Recognition of Earthquake-Prone Areas (PREPA). The seismic regime in the Greater Caucasus, Altai-Sayan, Lake Baikal, the Commander Islands and adjacent territories is considered based on long-term distributions of USLE coefficients in the vicinity of the two classes D and N of seismogenic nodes identified by PREPA as "dangerous" and "not dangerous" in respect to occurrence of $M \geq M_0$ earthquakes.

Systematic analysis of the distributions of the USLE coefficients A, B, and C at D and N nodes by the non-parametric two-sample Kolmogorov-Smirnov test unravels characteristic patterns of seismic regime in the four regions considered. The obtained results indicate once again that earthquakes are complex natural phenomena and provide important regional relationships of seismic regime at earthquake-prone sites.

The study highlights the need for further investigation in other regions to better justify interpretation of the original findings of the pilot research.

1. Introduction

The lithosphere of the Earth is a hierarchical nonlinear dissipative system. It consists of blocks of different size separated by fractured boundary zones, where earthquakes occur due to the relative motion of the blocks (Keilis-Borok, 1994). Earthquake sequences are not random, but rather lacking any obvious principle of organization (Kossobokov & Soloviev, 2021). Despite apparent randomness, the quantitative characteristics of such systems can be predicted to a certain extent, especially after significant averaging. Contemporary Neo-deterministic seismic hazard assessment (NDSHA) studies (Panza et al., 2021) make use of earthquake haphazard nature by a stepwise

[1] Institute of Earthquake Prediction Theory and Mathematical Geophysics, Russian Academy of Sciences, Moscow, Russian Federation.

[2] Accademia Nazionale delle Scienze detta dei XL, Rome, Italy.

[3] International Seismic Safety Organisation, Arsita, Italy.

Email: gorshkov@mitp.ru[2]; volodya@mitp.ru[3]

https://orcid.org/0000-0003-1639-1088[1]; https://orcid.org/0000-0001-5843-8468[2]; https://orcid.org/0000-0002-3505-7803[3]

* Corresponding author: nastia@mitp.ru

DOI: 10.1201/9781003684121-3

approach, including: (i) identifying areas where catastrophic earthquakes may occur, (ii) analysing the dynamics of the seismic regime to detect periods of increased probability of strong events, and (iii) reliably assessing the potential impact of such events by physics-based scenario modelling (Panza et al., 2001, 2011; Magrin et al., 2016).

Morphostructural zoning (*MSZ*) is a well-developed methodology for identifying locations susceptible to seismic events, even in the absence of previous records of large earthquakes. It is an essential part of the NDSHA synergy (Panza et al., 2021). The MZS defines an unambiguous hierarchical block structure of the region, defining blocks of different ranks, the lineaments that limit them, and nodes formed at the intersections of these lineaments (Gorshkov et al., 2003, 2009). These nodes, formed around the intersections of two or more lineaments, exhibit a mosaic of different topographic forms and an increased number of linear landforms, indicating regional fracturing and instability (Peresan et al., 2015; Gorshkov et al., 2003). On the basis of morphological and geophysical parameters PREPA allows to divide seismogenic nodes into D and N classes with respect to specified earthquake target magnitude M_0 (Gorshkov & Soloviev, 2021). The high level of the PREPA reliability has been confirmed by statistics of the target earthquakes that occurred in the study regions of the world after the original publications (Peresan et al., 2015; Gorshkov & Novikova, 2018); the large number of correct identifications of morphostructural nodes, including those where target earthquakes had not yet been recorded at the time of the original classification, supports the evident robustness of this method.

Unified Scaling Law for Earthquakes (*USLE*) that accounts for fractal distribution of seismic occurrence is another essential part of the NDSHA synergy (Panza et al., 2021). It generalizes the classical Gutenberg-Richter relationship as follows (Kossobokov & Mazhkenov, 1988, 1994; Nekrasova & Kossobokov, 2020; Kossobokov, 2021):

$$\log_{10} N(M, L) = A + B \cdot (5 - M) + C \cdot \log_{10} L,\ M_{-} \leq M \leq M^{-}$$

where, $N(M, L)$ is the expected annual number of earthquakes of magnitude M within an earthquake prone area of diameter L; A, B, and C are constants. The coefficients A and B are analogous to the *a*- and *b-values* of the Gutenberg-Richter relationship, while the coefficient C, quantifies the local fractal dimension of seismic loci. Note: (i) the distribution of inter-event times depends only on the control parameter $10^{-BM} \cdot L^{C}$ (Bak et al., 2002; Christensen et al., 2002) and (ii) complex relationship between the estimates of A, B, and C disclosing a not obvious well-organized attractor in 3-D (Nekrasova et al., 2011).

The integration of spatial data on earthquake-prone areas with space-time information derived from seismic flow analysis facilitates the identification of priority areas, where the probability of strong earthquakes is relatively high (Panza et al., 2011; Peresan et al., 2011).

In the following, we analyse the specific properties of the USLE coefficients at the D and N seismogenic nodes in the four tectonically different regions: the Greater Caucasus, Altai-Sayan, Lake Baikal, and the Commander Islands.

2. Data

Our analysis utilises the data and results obtained from the independent regional studies summarized in Table 1.

In the region of Greater Caucasus between 40°–46°N and 36°–51.5°E, MSZ outlined 107 seismogenic nodes, of which 49 were recognized as D nodes having the potential for earthquakes with $M \geq 5.0$ (Gorshkov et al., 2003). For the same territory, the reliable USLE coefficient estimations were derived from the regional earthquake catalogue spanning 1998–2014 at 102 out of 107 cells of 1/4° size centred at seismogenic nodes (Figure 1). The regional earthquake catalogue was compiled from annual periodicals "Earthquakes in Northeastern Asia" (1998–2008) and "Earthquakes in Russia" (2004–2014), with a minimum magnitude threshold (M_{-}) of 3.6. The calculations were performed using hierarchical scales with cell sizes of 2°, 1°, 1/2°, and 1/4° (Kossobokov & Nekrasova, 2018a).

Table 1 Summary of the PREPA and USLE data and parameters used in this study.

Region	*Greater Caucasus*	*Altai–Sayan*	*Lake Baikal*	*Commander Islands*
Location	40°–46°N; 36°–51.5°E.	46°–56°N; 82°–99°E		48°–58°N; 99–122°E
PREPA publication	Gorshkov et al., 2003	Gorshkov et al., 2018		In preparation
M_0	5.0	6.0		7.75
Number of SG-nodes	107	82	73	19
Number of D-nodes	49	27	22	7
USLE publication	Kossobokov & Nekrasova, 2018a	Kossobokov & Nekrasova, 2018b	Nekrasova & Kossobokov, 2023	In review
Period	1998–2014	1997–2015	1994–2019	1996–2022
$M\_$	3.6	2.2	2.6	3.5
USLE hierarchy	2°, 1°, 1/2°, 1/4°	2°, 1°, 1/2°, 1/4°	2°, 1°, 1/2°, 1/4°, 1/8°	2°, 1°, 1/2°, 1/4°, 1/8°
Number of regional USLE estimations (ALL)	102	207	1715	585
Data source	"Earthquakes in the Northern Eurasia" (1998– 2008) and "Earthquakes in Russia" (2004–2014)	"Earthquakes in the Northern Eurasia" (1997– 2009) and "Earthquakes in Russia" (2003–2015)	Baikal Division GS RAS http://www.seis-bykl.ru/modules.php?name=Data&da=	Kamchatka Branch GS RAS http://sdis.emsd.ru/info/earthquakes/catalogue.php

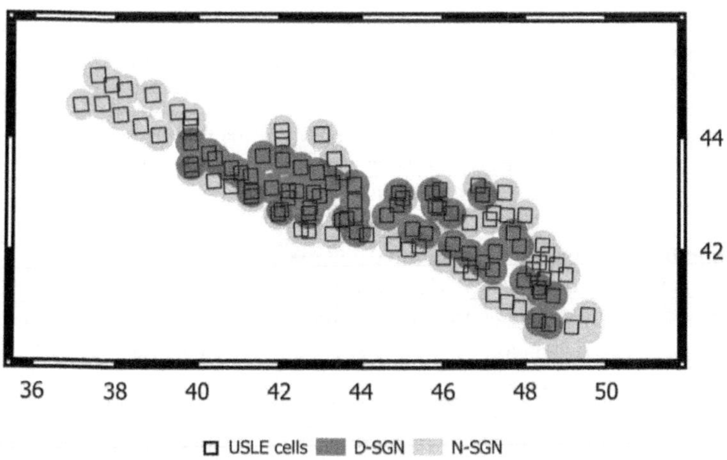

Fig. 1 Greater Caucasus: The cells of reliable determination of USLE coefficients (Kosobokov & Nekrasova, 2018a) on top of the 30 km radius circles centred at D and N ($M \geq 5.0$) seismogenic nodes (Gorshkov et al., 2003).

The Altai–Sayan region between 46°–56°N and 82°–99°E includes 82 seismogenic nodes, of which 27 were classified as D ones in respect to strong magnitude $M \geq 6.0$ earthquakes (Gorshkov et al., 2018). The reliable estimates of the USLE coefficients for the Altai–Sayan region were obtained in 207 cells of a regular 1/4° grid (Figure 2, on the left) using the same, as in the Greater Caucasus, hierarchical scales of 2°, 1°, 1/2°, and 1/4° (Kossobokov & Nekrasova, 2018b). For USLE parameter estimation, we used the regional earthquake catalogue compiled from the annual periodicals "Earthquakes of Northern Eurasia" (1997–2009) and "Earthquakes in Russia" (2003–2015).

In the Lake Baikal region within 48°–58°N and 99°–122°E, MSZ identifies 22 D nodes prone to occurrence of strong magnitude $M \geq 6.0$ earthquakes (Gorshkov et al., 2018). In this region the catalogue of magnitude 2.5 or larger earthquakes at shallow depths of less than 70 km covering the period 1994–2019 compiled by the Baikal Division of the Geophysical Survey, Federal Research Centre of the Russian Academy of Sciences allowed to consider a finer hierarchical grids with cells of 2°, 1°, 1/2°, 1/4°, and 1/8° size. Reliable USLE coefficient estimations were determined for 1715 cells of a regular 1/8° grid (Nekrasova & Kossobokov, 2023). The right map in Figure 2 shows the 30-km zones surrounding D and N nodes, as well as the grid cells where reliable USLE estimations of *A*, *B*, and *C* were obtained.

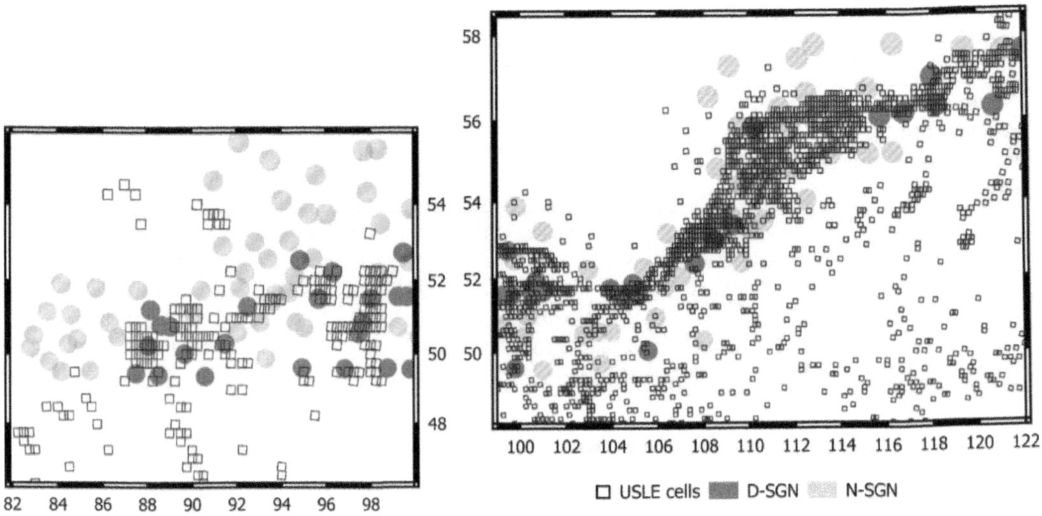

Fig. 2 Altai–Sayan (left) and Lake Baikal (right): the cells of reliable determination of USLE coefficients (Kosobokov & Nekrasova, 2018b, Nekrasova & Kossobokov, 2023) on top of the 30 km radius circles centred at D and N ($M \geq 6.0$) seismogenic nodes (Gorshkov et al., 2018).

The Commander Islands region is located between 51.95°–56.8°N and 162.15°–172.0°E. Totally, 19 nodes have been mapped by MSZ in this area, of which 7 are identified as D prone to $M \geq 7.75$. Reliable USLE coefficient estimations for the Commander Islands region were obtained in 585 grid cells with a linear dimension of 1/8°. USLE parameter calculations were performed using hierarchy of 2°, 1°, 1/2°, 1/4°, and 1/8° cells. This was possible thanks to the regional catalogue of magnitude 3.5 or larger earthquakes at shallow depths of less than 70 km covering the period 1996–2022 compiled by the Kamchatka Branch of the Geophysical Survey, Russian Academy of Sciences. Figure 3 illustrates the 30-km zones surrounding D and N nodes, as well as the grid cells where reliable USLE parameter estimations were obtained. The results of PREPA and USLE analyses for the Commander Islands region have not been published yet.

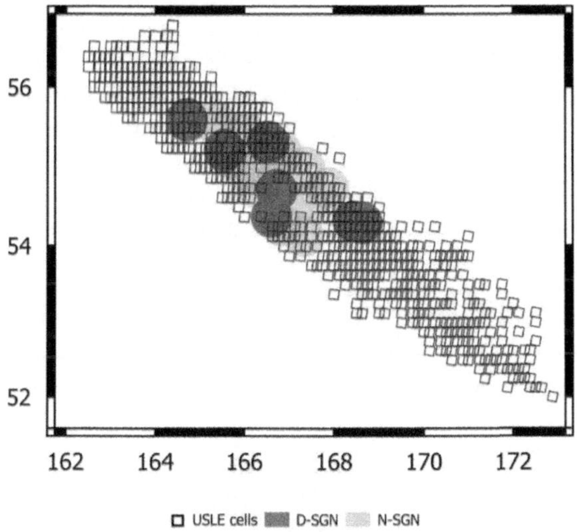

Fig. 3 Commander Islands: The cells of reliable determination of USLE coefficients on top of the 30 km radius circles centred at D and N ($M \geq 7.75$) seismogenic nodes.

3. Methods

The non-parametric two sample Kolmogorov-Smirnov criterion (Kolmogorov, 1933; Smirnov, 1948) was applied to compare in each of the four regions the empirical distributions of the USLE coefficients at seismogenic nodes classified as D and N by PREPA.

The two sample Kolmogorov-Smirnov statistic $\lambda_{K\text{-}S}$ is defined as:

$$\lambda_{K\text{-}S}(D, n, m) = [nm/(n + m)]^{1/2} \times D$$

where, $D = \max|P_{1,n}(x) - P_{2,m}(x)|$ is the maximum value of the absolute difference between the cumulative distributions $P_{1,n}(x)$ and $P_{2,m}(x)$ of the two samples, whose sizes are n and m, respectively. Asymptotically, when n and $m \to \infty$, the cumulative probability distribution of $\lambda_{K\text{-}S}$ converge to $\alpha_{K\text{-}S}$ defined as:

$$\text{Prob}(\lambda_{K\text{-}S} \leq x) = 1 - 2\sum_{k=1}^{\infty}(-1)^{k-1}\exp(-2k^2 x^2)$$

If the encircled $\lambda_{K\text{-}S}$ statistic equals 1.36 or larger the null-hypothesis of the same distribution of the two samples is rejected at significance level $\alpha_{K\text{-}S} < 5\%$, which confirms the difference of the two distributions with confidence $(1 - \alpha_{K\text{-}S}) > 95\%$.

For each of the selected regions, to study the spatial characteristics of the seismic regime we analysed the empirical distributions of USLE coefficients A, B, and C within 30 km around D nodes (*D-SGNs*) and N nodes (*N-SGNs*) as identified by the PREPA, as well as the overall distributions for the entire study region (*ALL*). Note that areas of 30 km radius around nodes may partially overlap.

4. Results

Figure 4 presents the empirical probability density distribution functions (p.d.d.f.'s) of the USLE coefficients A, B, and C (from left to right) estimated in the Greater Caucasus, Altai–Sayan, Lake Baikal, and Commander Islands regions (from top to bottom). The evident differences of regional distributions are confirmed by the results of the Kolmogorov-Smirnov testing summarized in Table 2. The p.d.d.f. for each of the USLE coefficients was formed by aggregating the values at grid cells located within the vicinity of D or N nodes. Figure 5 presents the empirical probability distribution functions (p.d.f.'s) of the regional USLE coefficients formed using this aggregation.

A comparison of the USLE *A* and *C* parameter distributions at 49 D and 53 N seismogenic nodes in the Greater Caucasus region (Figure 5 upper row, Table 2) does not indicate any significant difference in the seismic regime between these locations, as follows from $\lambda_{K\text{-}S}$ values close to 1 in both cases. In contrast, the value of $\lambda_{K\text{-}S} = 1.69$ for distributions of the *B* coefficient (analogous to the Gutenberg-Richter *b-value*) reveals a significant difference at high confidence level exceeding 99%. In Figure 5 the p.d.f. for the *B* coefficient at N nodes (*N-SGN* curve) detected above the one

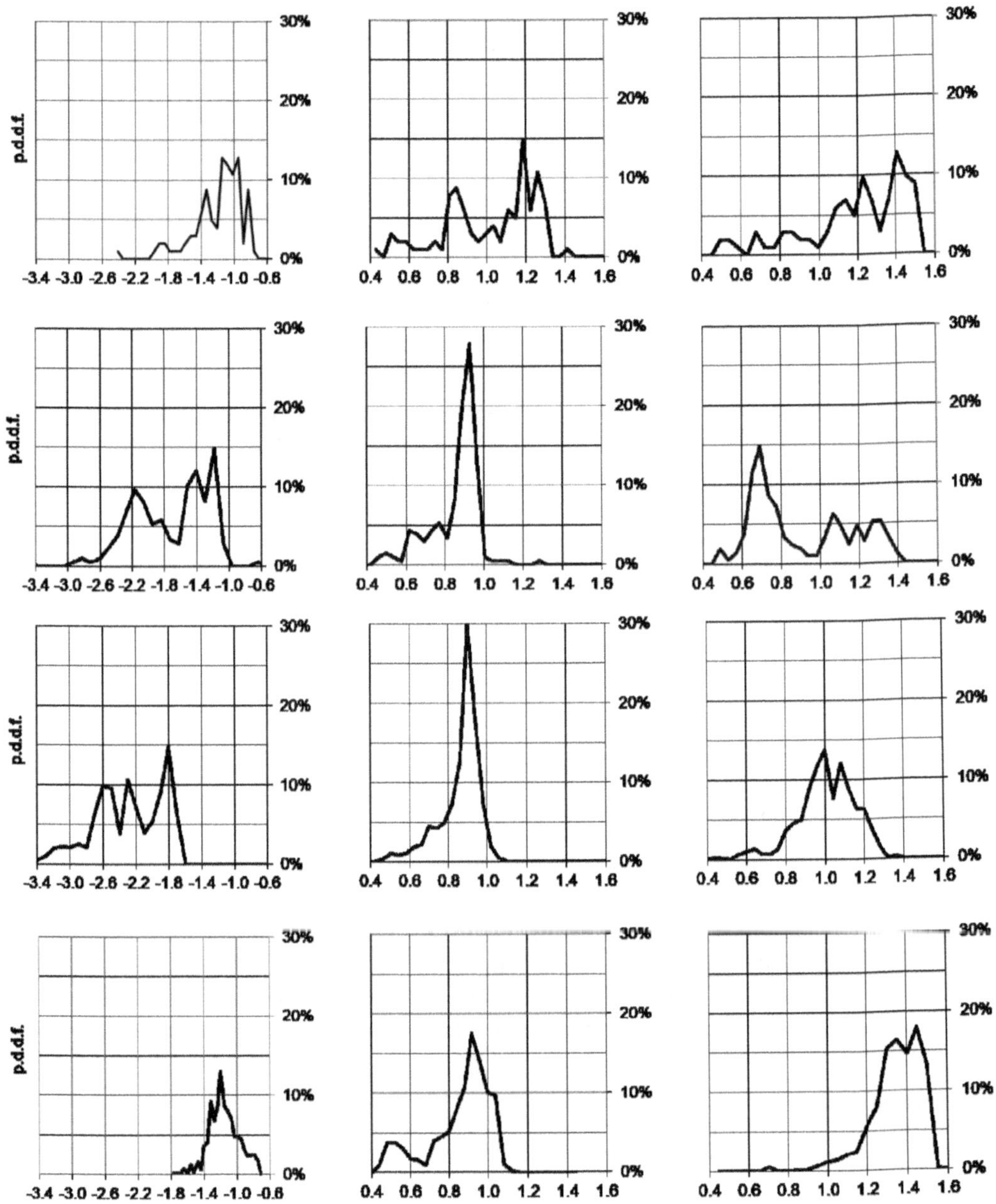

Fig. 4 Empirical probability density distribution functions (p.d.d.f.'s) of the USLE coefficients *A*, *B*, and *C* (panels from left to right) estimated in the Greater Caucasus, Altai–Sayan, Lake Baikal, and Commander Islands regions (from top to bottom).

at D nodes (*D-SGN* curve) indicating a more balanced difference between the number of small and large earthquakes. Although the two *C* coefficient distributions exhibit a clear visual 20% difference for about 40% of seismogenic nodes characterized by fractal dimension below 1.2, this is not enough to reject the null-hypothesis of the same statistical distribution (Table 2).

For the Altai–Sayan region with 38 and 19 USLE grid cells at D and N nodes, respectively, the significant differences are clearly observed in all the three p.d.f.'s of coefficients *A*, *B*, and *C* (Figure 5). Specifically, the determined $\lambda_{K\text{-}S} = 1.78$, 1.41, and 1.69 in Table 2 implies confidence $(1 - \alpha_{K\text{-}S}) > 99\%$, 96%, and 99%, respectively. Moreover, D nodes determined by PREPA are significantly different from the entire set of USLE cells in the region in respect to each of the USLE coefficients *A*, *B*, and *C*: $\lambda_{K\text{-}S} = 2.04$, 1.47, and 2.04 corresponds to confidence $(1 - \alpha_{K\text{-}S}) > 99\%$, 97%, and 99%, respectively.

As evident from Figure 5 (second row from top) in Altai–Sayan region the D nodes compared to N nodes are characterized by a higher seismic activity (coefficient *A*), slightly more imbalanced ratio between small and large earthquakes (coefficient *B*), and highly clustered seismicity (coefficient *C*).

For the Lake Baikal region with 238 and 265 USLE grid cells within 30 km distance from 22 and 51 D and N nodes, respectively, the significant differences are clearly observed in p.d.f.'s of coefficient *A* only (Figure 5, third row from top). However, for all the three coefficients *A*, *B*, and *C*, the Kolmogorov-Smirnov test rejects the null-hypothesis of the same statistical distribution (Table 2) with $\lambda_{K\text{-}S} = 5.34$, 2.30, and 2.15 that correspond to confidence level $(1 - \alpha_{K\text{-}S})$ much higher than 99.9% in each case. It is notable that p.d.f.'s associated with either D or N nodes are very different from those of all 1715 grid cells where USLE coefficients were determined (Nekrasova & Kossobokov, 2023); $\lambda_{K\text{-}S} \geq 1.50$ corresponds to confidence $(1 - \alpha_{K\text{-}S}) > 99.9\%$ each. Unlike the Altai-Sayan region the D nodes compared to N nodes are characterized by a lower seismic activity (coefficient *A*), which can be attributed to the significant difference of tectonics in the two neighbouring regions.

Table 2 The Kolmogorov-Smirnov two-sample statistics $\lambda_{K\text{-}S}$ (above the diagonal), the sample size (at the diagonal), and $D = \max|P_{1,n}(x) - P_{2,m}(x)|$ (below the diagonal) for the regional USLE coefficients (*A, B, C*) sampled at the vicinity of D and N seismogenic nodes in the four regions. The $\lambda_{K\text{-}S} > 1.36$ marked bold corresponds to the confidence level $(1 - \alpha_{K\text{-}S}) > 95\%$.

SGN	*USLE Coefficient A*			*USLE Coefficient B*			*USLE Coefficient C*		
	D	*N*	*All*	*D*	*N*	*All*	*D*	*N*	*All*
Greater Caucasus Region									
D	49	1.12	0.78	49	**1.698**	1.01	49	0.99	0.59
N	0.22	54	0.51	0.33	54	0.94	0.20	54	0.55
All	0.14	0.09	103	0.18	0.16	103	0.10	0.09	103
Altai–Sayan Region									
D	38	**1.78**	**2.04**	38	**1.41**	**1.47**	38	**1.69**	**2.04**
N	0.50	19	0.72	0.40	19	1.10	0.47	19	0.64
All	0.36	0.17	209	0.26	0.25	209	0.36	0.15	209
Lake Baikal Region									
D	239	**5.34**	**2.97**	239	**2.30**	**1.75**	239	**2.15**	**1.50**
N	0.48	265	**4.17**	0.21	265	**3.47**	0.19	265	**1.95**
All	0.20	0.27	1715	0.12	0.23	1715	0.10	0.13	1715
Commander Islands Region									
D	112	1.28	**2.23**	112	1.22	**1.83**	112	**2.80**	0.77
N	0.15	189	**2.76**	0.15	189	**2.17**	0.33	189	**4.02**
All	0.23	0.23	585	0.19	0.18	585	0.08	0.34	585

The analysis of PREPA versus USLE results for the Commander Islands region was conducted for 112 and 189 grid cells associated with 7 D and 12 N seismogenic nodes. By inspection of Figure 5 (bottom row) one may find rather similar the p.d.f.'s of coefficients *A* and *B*, while those of coefficient *C* are clearly different. This observation is clarified or even confirmed in part by the Kolmogorov-Smirnov test applied to distributions of the USLE coefficients *A*, *B*, and *C*: $\lambda_{K\text{-}S} = 1.28$, 1.22, and 2.80 in Table 2 corresponds to confidence level $(1 - \alpha_{K\text{-}S}) > 92\%$, 90%, and much larger than 99.9%, respectively. Similar to the Lake Baikal region, the seismogenic nodes identified by

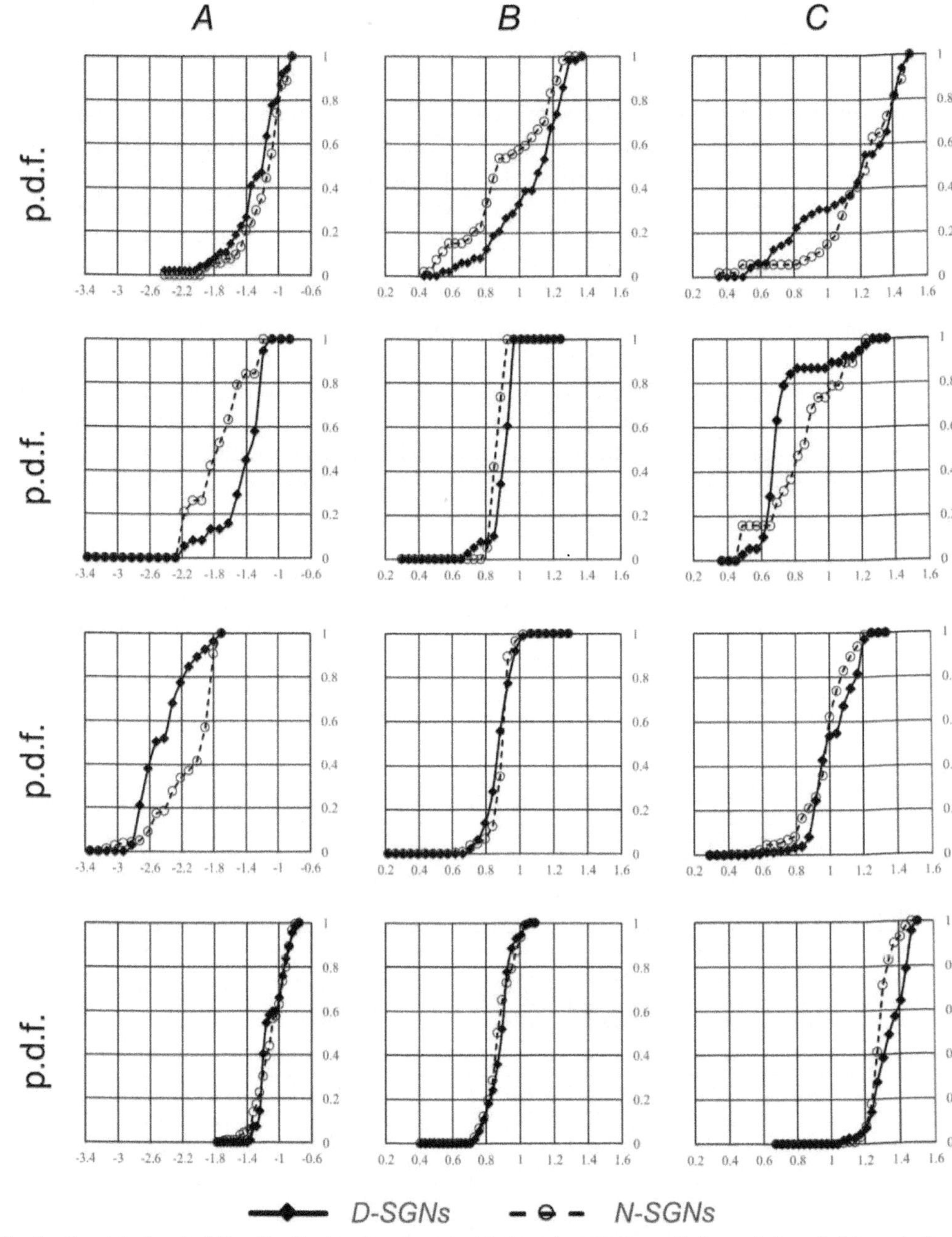

Fig. 5 Empirical probability distribution functions (p.d.f.'s) of the USLE coefficients *A*, *B*, and *C* (panels from left to right) estimated at D and N seismogenic nodes (D-SGNs and N-SGNs, respectively) in the Greater Caucasus, Altai–Sayan, Lake Baikal, and Commander Islands regions (from top to bottom).

PREPA appear to have very different seismic characteristics than at all 585 grid cells of reliable USLE determinations: except for $\lambda_{K\text{-}S} = 0.77$ (about 41% confidence) at N nodes, $\lambda_{K\text{-}S}$ values imply the confidence larger than 99.7%.

5. Discussion and Conclusions

Neo-Deterministic Seismic Hazard Assessment is an innovative multi disciplinary scenario and a physics-based approach, which is gaining momentum in spreading worldwide a Paradigm of Reliable Seismic Hazard Assessment (Panza & Bela, 2020; Panza et al., 2021). NDSHA results are based on reliable seismic evidence, PREPA, implications of the USLE, and exhaustive scenario-based modelling of ground shaking.

In our pilot study we have investigated for the first time the relationship between the PREPA classification of morphostructural nodes based on topographic, geologic, geomorphologic, and other geophysical characteristics excluding seismicity and the USLE coefficients based entirely on the regional earthquake catalogue. Our choice of the four regions was predetermined by their difference in tectonics: (i) Greater Caucasus characterized by underthrusting and stacking of tectonic units (Mosar et al., 2022), (ii) Altai–Sayan represented by a complex aggregate of units of different geodynamic affinity (Buslov et al., 2013), (iii) a series of continental rifts centred at Lake Baikal (Radziminovich et al., 2013), and (iv) the Commander Islands at the western edge of Aleutian subduction zone (Newberry et al., 1986; Bird, 2003). Moreover, we have used two different kinds of the USLE coefficients *A*, *B*, and *C* estimation: (i) centred at seismogenic nodes (Figure 1) in Greater Caucasus and (ii) at all regular grid points within 30 km distance from PREPA nodes (Figures 2 and 3) in the other three regions.

As expected, the results of our pilot study have revealed the difference of potential statistically reliable characteristics of regional seismic patterns (Table 2) that may contribute efficiently to the PREPA identification of D and N nodes. The observed distinctions appear region-specific in line with tectonic differences of the study regions.

From inspection of Figure 5, one may note that unlike Altai-Sayan region, seismic activity (coefficient *A*) at D nodes in Lake Baikal and the Greater Caucasus was found to be lower than at N nodes, while hardly different in the Commander Islands region. The distributions of *B* coefficient are indicative of the apparent imbalance between small and large earthquakes in Greater Caucasus and Altai-Sayan regions, which distinction is not evident in Lake Baikal and the Commander Islands, where a comparable frequency of strong and weak earthquakes is observed in both classes of seismogenic nodes. The distributions of the USLE *C* coefficient characterizing fractal dimension of epicentre loci at D and N nodes are apparently different in all the four regions, although the maximal deviation $D = 20\%$ in Table 2 does not guarantee statistically reliable difference for Greater Caucasus. Thus each region demonstrates unique USLE characteristics at PREPA classes that distinguish D and N nodes.

However, visual inspection could be misleading in the absence of statistical confirmation. Quantifying statistically significant differences may be affected by the level of the earthquake catalogue completeness and sample sizes (given at the diagonal of Table 2). The earthquake datasets for the Lake Baikal and Commander Islands regions appear more comprehensive than for the Greater Caucasus and Altai–Sayan regions. Compare also PREPA class sample sizes of 239 + 265 and 112 + 189 nodes versus 49 + 54 and 38 + 19, respectively. Nevertheless, the pilot analysis provides a broad foundation for further investigations into the seismic regime characteristics at the vicinities of morphostructural nodes and/or intersections of fault zones (Gelfand et al., 1976; King, 1986). In particular, the observed differences in the USLE coefficients at the two classes of nodes represent a broad avenue for future research, including expansion to analysing temporal variability of the USLE control parameter (Nekrasova & Kossibokov, 2020; Kossobokov et al., 2022).

We believe that the proposed combination of the two approaches, which so far have been used separately in the NDSHA synergy, can provide an improved reliable interpretation of the actual state of seismic regime and serve another step towards better understanding earthquake occurrence in a complex nonlinear system of blocks and faults.

References

Bak P., Christensen K., Danon L. and Scanlon T. (2002). Unified Scaling Law for Earthquakes. *Phys. Rev. Lett.*, *88*, pp. 178501–178504. https://doi.org/10.1103/PhysR evLett.88.178501.

Bird P. (2003). An updated digital model of plate boundaries., *Geochem. Geophys. Geosyst.*, *4*, 1027. https://doi.org/10.1029/2001GC000252.

Buslov M.M., Geng H., Travin A.V., Otgonbaatar D., Kulikova A.V., Ming C., Stijn G., Semakov N.N., Rubanova E.S., Abildaeva M.A., Voitishek E.E. and Trofimova D.A., (2013). Tectonics and geodynamics of Gorny Altai and adjacent structures of the Altai–Sayan folded area. *Russ. Geol. Geophys.*, *54*(10), 1250–1271. https://doi.org/10.1016/j.rgg.2013.09.009.

Christensen K., Danon L., Scanlon T. and Bak P. (2002).Unified scaling law for earthquakes. *Proc. Nat. Acad. Sci.*, *99* (Suppl 1), 2509–2513.

Gelfand I.M., Guberman S.A., Keilis-Borok V.I., Knopoff L., Press F., Ranzman E.Y., Rotwain I.M. and Sadovsky A.M. (1976). Pattern recognition applied to earthquake epicenters in California. *Physics of the Earth and Planetary Interiors*, *11*(3), 227–283. https://doi.org/10.1016/0031-9201(76)90067-4.

Gorshkov A., Kossobokov V. and Soloviev A. (2003). Recognition of earthquake prone areas. In: Keilis-Borok V.I., Soloviev A.A. (Eds.), *Nonlinear Dynamics of the Lithosphere and Earthquake Prediction* (pp. 239–310)., Springer, Heidelberg. https://link.springer.com/chapter/10.1007/978-3-662-05298-3_6.

Gorshkov A.I., Panza G.F., Soloviev A.A., Aoudia A. and Peresan A. (2009). Delineation of the geometry of the nodes in the Alps-Dinarides hinge zone and recognition of seismogenic nodes ($M \geq 6.0$). *Terra Nova*, *21* (4), 257–264. https://doi.org/10.1111/j.1365-3121.2009.00879.x.

Gorshkov A. and Novikova O. (2018). Estimating the validity of the recognition results of earthquake-prone areas using the ArcMap. *Acta Geophysica*, *66* (5), 843–853. https://doi.org/10.1007/s11600-018-0177-3.

Gorshkov A.I., Soloviev A.A., Zharkikh and Ju I. (2018). Recognition of Strong Earthquake Prone Areas in the Altai–Sayan–Baikal Region. *Doklady Earth Sciences*, *479* (1), 412–414. https://doi.org/10.1134/S1028334X1803025X.

Gorshkov A. and Soloviev A. (2021). Morphostructural zoning for identifying earthquake-prone areas. In: Panza G.F., Kossobokov V.G., Laor E., and De Vivo B. (Eds.), *Earthquakes and Sustainable Infrastructure: Neodeterministic (NDSHA) Approach Guarantees Prevention Rather than Cure* (pp. 135–149.) Elsevier.

Keilis-Borok V.I. (1994). Symptoms of instability in a system of earthquake-prone faults. *Physica D: Nonlinear Phenomena*, *77*, I, 1–3, 193–199. https://doi.org/10.1016/0167-2789(94)90133-3.

King G.C.P. (1986). Speculations on the geometry of the initiation and termination processes of earthquake rupture and its relation to morphology and geological structure. *Pure Appl. Geophys*, *124*, 567–585. https://doi.org/10.1007/BF00877216.

Kolmogorov A. Sulla. (1933). *Determinazione Empirica di Una Legge di Distribuzione*. Giornale dell'Istituto Italiano degli Attuari, *4*, 83–91.

Kossobokov V. (2021). Unified Scaling Law for Earthquakes that Generalizes the Fundamental Gutenberg-Richter Relationship. In: Gupta H.K. (Eds.), *Encyclopedia of Solid Earth Geophysics* (pp. 1893–1896). *Encyclopedia of Earth Sciences Series*. Springer, Cham. https://doi.org/10.1007/978-3-030-58631-7_257.

Kossobokov V.G. and Mazhkenov S.A. (1988). Spatial characteristics of similarity for earthquake sequences: Fractality of seismicity. In: *Lecture Notes of the ICTP Workshop on Global Geophysical Informatics with Applications to Research in Earthquake Prediction and Reduction of Seismic Risk*, Trieste, Italy, 15 Nov.–16 Dec., ICTP—International Centre for Theoretical Physics. http://indico.ictp.it/event/ a0215 3/contr ibuti on/3/mater ial/0/0.pdf.

Kossobokov V.G. and Mazhkenov S.A. (1994). On similarity in the spatial distribution of seismicity. In: Chowdhury D.K., and Biswas N.N., et al. (Eds.), *Computational Seismology and Geodynamics*, *1*, 6–15. American Geophysical Union. https://agupubs.onlinelibrary.wiley.com/doi/10.1029/CS001p0006.

Kossobokov V.G. and Nekrasova A. 2018a. Earthquake Hazard and Risk Assessment based on Unified Scaling Law for Earthquakes: Greater Caucasus and Crimea. *Journal of Seismology*, *22*, 1157–1169. https://doi.org/10.1007/s10950-018-9759-4.

Kossobokov V.G. and Nekrasova A. 2018b. Earthquake Hazard and Risk Assessment based on Unified Scaling Law for Earthquakes: Altai-Sayan Region. *Natural Hazards*, *93*(3), 1435–1449. https://doi.org/10.1007/s11069-018-3359-z.

Kossobokov V.G., Nekrasova A.K. and Schepalina P.D. 2022. Seismic dynamics in advance of and after the largest earthquakes, 1985–2020. *Surveys in Geophysics*, *4*, 423–436. https://doi.org/10.1007/s10712-021-09674-0.

Kossobokov V.G. and Soloviev A.A. (2021). Testing Earthquake Prediction Algorithms, *Journal of the Geological Society of India*, *97* (12), 1514–1519. https://doi.org/10.1007/s12594-021-1907-8.

Magrin A., Gusev A.A., Romanelli F., Vaccari F. and Panza G.F. 2016). Broadband NDSHA computations and earthquake ground motion observations for the Italian territory. *Int. J. Earthq. Impact Eng.*, *1*(4), 395–420. https://doi.org/10.1504/IJEIE.2016.083253.

Mosar J., Mauvilly J., Koiava K., Gamkrelidze I., Enna N., Lavrishev V. and Kalberguenova V. (2022). Tectonics in the Greater Caucasus (Georgia –Russia): From an intracontinental rifted basin to a doubly verging fold-and-thrust belt. *Marine and Petroleum Geology*, *140*, 105630. https://doi.org/10.1016/j.marpetgeo.2022.105630.

Nekrasova A., Kossobokov V., Aoudia A., Perezan A., Panza G.F. (2011). A Multiscale Application of the Unified Scaling Law for Earthquakes in the Central Mediterranean area and Alpine region. *Pure Appl. Geophys.*, *168*, 297–327 Springer Basel AG. https://doi.org/10.1007/s00024-010-0163-4.

Nekrasova A.K. and Kossobokov V.G.. (2020)., The Unified Scaling Law for Earthquakes. *Journal of Volcanology and Seismology*, *14*(6), 353–372. https://doi.org/10.1134/S0742046320060056.

Nekrasova A. and Kossobokov V. (2023). Seismic risk assessment for the infrastructure in the regions adjacent to the Russian Federation Baikal–Amur Mainline based on the Unified Scaling Law for Earthquakes. *Nat. Hazards*, *116*, 2. https://doi.org/10.1007/s11069-022-05750-9.

Newberry J.T., Laclair D.L. and Fujita K. (1986). Seismicity and tectonics of the far Western Aleutian Islands. *Journal of Geodynamics*, *6*(1–4), 13–32. https://doi.org/10.1016/0264-3707(86)90030-X.

Panza G.F., Romanelli F. and Vaccari F. (2001). Seismic wave propagation in laterally heterogeneous anelastic media: Theory and applications to seismic zonation. *Advances in Geophysics*, *43*, 1–95.

Panza G.F., Peresan A., Magrin A., Vaccari F., Sabadini R., Crippa B., Marotta A.M., Splendore R., Barzaghi R., Borghi A., Cannizzaro L., Amodio A, and Zoffoli S. (2011). The SISMA prototype system: Integrating geophysical modeling and Earth observation for time-dependent seismic hazard assessment. *Nat. Hazards*, *69*, 1179–1198. https://doi.org/10.1007/s11069-011-9981-7

Panza G.F. and Bela J. (2020). NDSHA: A new paradigm for reliable seismic hazard assessment. *Engineering Geology*, *275*, 105403. https://doi.org/10.1016/j.enggeo.2019.105403.

Panza G.F., Kossobokov V., De Vivo B. and Laor E. (Eds.) (2021). *Earthquakes and Sustainable Infrastructure: Neo-deterministic (NDSHA) Approach Guarantees Prevention Rather than Cure* (pp. xxv, 672 pp). Elsevier, Paperback. ISBN: 9780128235034, eBook ISBN: 9780128235416. https://doi.org/10.1016/C2020-0-00052-6.

Peresan A., Zuccolo E., Vaccari F., Gorshkov A. and Panza G.F. (2011). Neo-deterministic seismic hazard and pattern recognition techniques: Time dependent scenarios for North-Eastern Italy. *Pure Appl. Geophys.*, *168*, 583–607. https://doi.org/10.1007/s00024-010-0166-1.

Peresan A., Gorshkov A., Soloviev A. and Panza G.F. (2015). The contribution of pattern recognition of seismic and morphostructural data to seismic hazard assessment. *Boll. Geofis. Teorica ed Applicata*, *56*, 295–328. https://doi.org/10.4430/bgta0141.

Radziminovich N.A., Gileva N.A., Melnikova V.I., Ochkovskaya M.G. (2013). Seismicity of the Baikal rift system from regional network observations. *Journal of Asian Earth Sciences*, *62*, 141–161. https://doi.org/10.1016/j.jseaes.2012.10.029.

Smirnov N. (1948). Table for Estimating the Goodness of Fit of Empirical Distributions. *Annals of Mathematical Statistics*, *19*, 279–281. https://doi.org/10.1214/aoms/1177730256.

CHAPTER

3

Seismic Wave Propagation and Soil-structure Interaction via Boundary Integral Equations

Sonia Parvanova[1] and Petia Dineva[2*]

Dedicated to Professor Giuliano F. Panza in the occasion of his 80th birthday

This study examines the seismic response of an elastic, isotropic or anisotropic, homogeneous or arbitrarily layered half-space with surface relief, additionally containing buried tunnels (either lined or unlined) and embedded foundations. The incident waves include elastic SH, P, or SV waves, as well as waves radiated by an embedded line source, a scenario commonly encountered in geotechnical engineering for infrastructure projects, such as tunnels and foundations in complex geological environments. The aim is to understand how key factors, such as soil topography, anisotropy, layering, and the interaction between soil, tunnels, foundations and relief affect the seismic wave field and the resulting structural responses. The direct boundary element method (BEM) is employed as a computational tool, using a frequency-dependent fundamental solution for a 2D anisotropic solid derived via the Radon transform. The numerical scheme was verified, and several parametric studies were conducted to examine the effects of different model parameters. The results clearly demonstrate the sensitivity of the seismic wave field to soil anisotropy, relief peculiarities, soil layering, and dynamic soil–underground structures interactions. This highlights the necessity of more comprehensive models that account for the entire wave path, from the seismic source through geological deposits to the local site with geotechnical infrastructure under consideration.

1. Introduction

The interaction between underground structures and adjacent surface area during earthquakes is a critical field of research. As urban populations grow and land resources become increasingly scarce, many cities have developed extensive underground infrastructure. As a result, the construction of subway tunnels beneath buildings has increased, particularly in densely populated areas, which poses a risk in regions prone to strong earthquakes.

Understanding the mutual influence between subsurface structures, soil and overlying buildings during seismic events is vital for enhancing urban resilience and ensuring the safety of both surface

[1] Department of Structural Engineering, University of Architecture, Civil Engineering and Geodesy (UACEG), Sofia, Bulgaria. ORCID: 0000-0002-4372-8734.
[2] Institute of Mechanics, Bulgarian Academy of Sciences (BAS), Sofia, Bulgaria.
Email: slp_fce@uacg.bg[1]
* Corresponding author: petia@imbm.bas.bg

DOI: 10.1201/9781003684121-4

and underground developments. Despite the crucial role tunnels play in European infrastructure, standardized design guidelines or regulations at the European level are currently lacking, see Terzi and Manolis (2023). There is a limited body of research focused on soil-tunnel systems. An excellent review of the research on the seismic behavior of tunnels and the dynamic interaction effects between soil, tunnels, adjacent underground and above ground structures is presented in Wang (1993), Tsinidis et al. (2020), Terzi and Manolis (2023), Yu et al. (2017), Chen et al. (2023).

The available mechanical models differ depending on various factors, including but not limited to: (i) type of the loading, i.e., static or dynamic (Bobet, 2003; Yu & Chen, 2021); (ii) nature of the dynamic load: earthquake–induced ground motion (Liu et al., 2021; Panji & Mojtabazadeh-H., 2021) or explosion (Wu et al., 2011; Wang et al., 2013), (iii) type of the geological material (Guo et al., 2010); (iv) soil-tunnel interaction effects (Pitilakas et al., 2014; Wang et al., 2017; Hashash et al., 2018; Alielahi & Fizzi, 2021); (v) analyses in either the time or the frequency domains, see Manolis et al. (1995), Karakostas and Manolis (2000); (vi) geometrically and physically linear or nonlinear analyses (Alielahi & Fizzi, 2021; Liang & Zhu, 2019); (vii) type of the interface conditions between the liner and the surrounding geological bulk material (Fang et al., 2023); (viii) whether the tunnel is flexible and follows the soil deformations (Kuesel, 1969; Monsees, 1991; Wang & Munfakh, 2001) or is modelled as a rigid one in a soft soil, i.e., as a beam on elastic foundation (Sanchez-M. et al., 2009); (ix) analyses in antiplane/inplane state, see Panji and Mojtabazadeh (2021), Manolis et al. (2015), Vasilev et al. (2015) or in three dimensions (Manolis et al., 1995; Kouretzis et al., 2006). Numerical methods applied by the researchers for modelling the soil-structure interaction are so numerous, that it is impossible to cover them objectively even briefly in this introduction. For that reason, we mention only the recent developments devoted to the most commonly applied numerical techniques. It deserves to begin with: (a) Finite difference method (FDM) (Mayoral & Mousqueda, 2020; Oliaei & Basirat, 2022); (b) Finite element method (FEM) (Alielahi & Fizzi, 2021; Wang, 2018; Abate & Massimino, 2017; Fabozzi et al., 2018; Patil et al., 2018; Wang et al., 2022); (c) BEM, (Liu et al., 2021; Panji & Mojtabazadeh-H., 2021; Karakostas & Manolis, 2000; Alielahi et al., 2015; Li et al., 2023; Parvanova et al., 2014a, b; Dineva et al., 2014 a, b); (d) Hybrid methods: FDM and FEM in El Omari et al. (2021), multi scale method in Yu et al. (2013), FDM-BEM in Manolis et al. (2015, 2013), FEM-BEM in Parvanova and Vasilev (2016), Boumaiza and Aour (2022). The literature review indicates that studies considering soil anisotropy in soil-structure interaction problems are rather limited (Ba et al., 2018a,b). A universal property of natural soils is transverse isotropy which takes into account stiffness differences between the horizontal and vertical direction due to some preferred orientation of the soil particles in deposition. Depending on the computational method employed, the results obtained for the dynamic behavior of rigid massless foundations resting on elastic half-space can be classified into the following groups: (a) analytical solutions (Luco & Westmann,1971; Pais & Kausel, 1988; Gazetas, 1991); (b) among the extensive work done by numerical techniques, we highlight papers by BEM (Wong & Luco, 1978; Spyrakos & Beskos, 1986; Dominguez, 1993; Karabalis & Rizos, 1993; Qian & Beskos, 1995; Celebi, 2006; Shahi & Noozad, 2011); FEM (Kausel et al., 1975, 1978; Kausel & Rosset, 1981; Lin & Tassoulas, 1987; Freisinger et al., 2022) and hybrid techniques (Spyrakos & Xu, 2003; Mita & Luco, 1989).

The BEM and various hybrid techniques based on it are among the most effective methods in computational elastodynamics. The following section provides a brief overview of current research on seismic wave propagation and soil-structure interaction models, emphasizing the role of BEM as a key approach in this study.

The authors' initial research on applying the BEM to develop hybrid models for seismic wave propagation in complex geological media was inspired by two significant visits made by the second author to the Department of Earth Sciences at the Istituto di Geodesia e Geofisica, Università di Trieste, Italy, where she collaborated with *Professor Panza's research group*. These visits were made possible through the following six-month scholarships: (a) NATO-CNR Science Fellowship Program 2002, research work on the subject "Hybrid approach of BIE-Modal summation methods for site effects estimation of a seismic region in a laterally varying medium"; (b) CEI-

ICTP Fellowship Program 2006, investigation on the subject, "Site effects analysis of laterally inhomogeneous geological region by hybrid Modal summation-BIE methodology. The case study with Sofia structural model". Under the academic guidance of *Prof. Panza* and in collaboration with his research group, the hybrid *Modal Summation–Boundary Element Method* was developed, verified, and applied to real seismic scenarios in a scientific, creative, and collegial environment. For further details, see Dineva et al. (2003, 2008) and Panza et al. (2009). In line with a core principle in seismic mechanics that the model must accurately represent the entire system, including the seismic source, the wave path from the source to the local region, and the local site with its unique geometrical and mechanical properties, we have further developed several types of hybrid models based on BEM. These include: (a) Wave number integration-BEM in Schanz et al. (2009), Wuttke et al. (2011), Dineva et al. (2012); (b) Finite difference-BEM in Manolis et al. (Manolis et al., 2013, 2015); (c) Finite element method in absolute coordinates-BEM, see Zachariew and Dineva (2019); BE-FEM based on the macro-element concept in Parvanova and Vasilev (2016), Vasilev et al. (2015).

Most of the BEM work in elastodynamics, however, concerns elastic isotropic solids. There is a lack of studies taking into account the material anisotropy, with a few exceptions, see: (a) Gazetas (1981) first presented a semi-analytical solution for dynamic responses of a rigid strip footing on an incompressible transversely isotropic soil stratum; (b) Wang and Rajapakse (1991) modelled the dynamic response of an arbitrary shaped rigid strip foundation embedded in an orthotropic elastic half-plane by indirect BEM; (c) Barros (2006) dealt with the dynamics of cylindrical rigid structures embedded in transversely isotropic elastic half-space by indirect BEM; (d) dynamic impedance of rigid foundation embedded in anisotropic multilayered soil was studied by indirect BEM in Han et al. (2015) and in Ba et al. (2018a,b, 2022); (e) Eskandari-Ghadi et al. (2010, 2014) investigated the vertical vibration of a rigid circular disc attached to a transversely isotropic homogeneous and layered half-space, respectively; (f) Parvanova and Dineva (2022) studied 2D dynamic problem in frequency and time domains for a flexible foundation resting in an elastic anisotropic half-plane with arbitrary geometry.

The following conclusions could be made based on the above state of the art:

(a) There is a paucity of high-performance methodologies for treating the dynamic interaction between different underground structures, taking into account all important components along the wave path from the embedded dynamic source till the local geological region with its specific geometry, complex mechanical properties and in addition containing engineering structures;

(b) The BEM is an attractive candidate for modelling wave propagation in semi-infinite media due to several certain advantages, namely: (i) formulation via boundary integral equations is equivalent to the original boundary-value problem (BVP) described by partial differential equations. This fact, coupled with the use of the analytically derived fundamental solution for an anisotropic continuum guarantees a high level of accuracy; (ii) the fundamental solution obeys the Sommerfeld radiation condition and thus infinitely extended boundaries are automatically accounted for without necessity of special types of viscous boundaries; (iii) since only surfaces need to be modelled, there is reduction of the problem dimensionality, with a corresponding reduction in the size of the algebraic system as compared with domain-type numerical methods; (iv) selective solution at internal points in the domain of interest is possible once the BVP has been solved by the usage of the well-known integral representation formulae, which obviates the large scale volume discretization; (v) flexibility in modelling of arbitrary geometry; (vi) high accuracy in the evaluation of stress concentration fields.

Despite the advantages just discussed, there is a lack of the BEM results taking into account the dynamic soil-underground structures interaction accounting for material anisotropy. All above motivate the authors to focus on the following aim: dynamic BEM analysis of multi-layered anisotropic soil-underground structures interaction taking into account: (a) the soil topography and non-parallel soil layering; (b) the material anisotropy; (c) the characteristics of the dynamic source; (d) the characteristics of the whole wave path from the embedded dynamic source till the free surface with relief peculiarities.

The structure of the paper is as follows: Section 2 provides the general problem statement. In Section 3, the problem is reformulated using boundary integral equations along existing boundaries. Section 4 verifies the numerical procedure proposed by the authors. Section 5 presents and discusses a series of simulation results concerning different illustrative examples. Concluding remarks are summarized in Section 6.

2. Problem Statement

An elastic isotropic or anisotropic heterogeneous half-plane Ω_{hp} is considered in a Cartesian coordinate system $Ox_1x_2x_3$. The heterogeneity is due to the presence of surface or subsurface relief, underground tunnel constructions Ω_t, and soil layering. The half-plane is presented by N homogeneous elastic isotropic (or anisotropic) layers, defined with Lame constants and density λ_i, μ_i, ρ_i, $i = 1, 2,...N$ and with non-horizontal interfaces Λ_i rested on the seismic bed with material properties λ_0, μ_0, ρ_0, see Figure 1. In the case of general anisotropy and in-plane wave motion the stiffness tensor of the soil medium is presented by the contracted notation of Voigt (1906) and the respective six stiffness coefficients: c_{11}, c_{12}, c_{16}, c_{22}, c_{26}, c_{66}. In the case of orthotropic material c_{16}, $c_{26} = 0$, the number of material constants reduces to c_{11}, c_{12}, c_{22}, c_{66}. The orthotropic case is transformed into isotropic one if $c_{11} = c_{22} = \lambda + 2\mu$; $c_{12} = \lambda$; $c_{66} = \mu$. The underground structures are presented by two circular lined or unlined tunnels in Ox_1x_2 plane with $x_3 = 0$ and they are infinite along Ox_3 axis. Both tunnels have the same embedment depth relative to the flat free surface and are separated by a specified distance. The external radius of the tunnel wall or the radius of the soil-tunnel interface is a, the internal radius of the tunnel wall respectively of the traction free tunnel contour is a_{int}. The wall thickness is $a - a_{\text{int}}$. Both the in-plane wave motion in the plane $x_3 = 0$, and anti-plane wave motion along Ox_3 axes are studied. Two types of seismic loads are under consideration: (a) vertically incident transient or time-harmonic longitudinal P or shear SH and SV waves; (b) waves radiated by an embedded in the half-plane line source. The embedded dynamic source is presented by body force $f_i(\mathbf{x}_0, t) = f_{0i}f(t)\delta(\mathbf{x}, \mathbf{x}_0)$ with magnitude f_{0i}, time history function $f(t)$ and source position vector $\mathbf{x}_0 = (x_{01}, x_{02})$, where δ is the Dirac's delta function, see Figure 1.

The half-plane traction free surface is denoted by Γ_{hp}. The contours of the tunnel walls are $\Gamma_L = \Gamma_{L,\text{ ext}} \cup \Gamma_{L,\text{ int}}$ and $\Gamma_R = \Gamma_{R,\text{ ext}} \cup \Gamma_{R,\text{ int}}$ for the left and right tunnel respectively, the subscript 'ext' stands for soil-tunnel interface, while subscript 'int' stands for internal traction free tunnel wall contour. The soil-tunnel interfaces are denoted by $\Gamma_{L,s-t}$ and $\Gamma_{R,s-t}$ for the left and right tunnel, respectively. We note that the geometry of the tunnels, free surface topography of the half-plane, and soil layers interfaces could be arbitrary.

In the considered cases of anti-plane and in-plane strain state, the only nonzero field quantities are: (a) for anti-plane wave motion – displacement $u_3(x_1, x_2)$, stresses σ_{i3}, and tractions $t_3 = \sigma_{3i}n_i$; (b) for in-plane wave motion-displacements $u_i(x_1, x_2)$, stresses σ_{ij} and tractions $t_i = \sigma_{ij}n_j$. All field quantities depend on radius-vector $\mathbf{x} = (x_1, x_2)$ and the time t or the wave frequency ω, where $i = 1, 2$, n_j is the outward pointing unit normal vector to the surface under consideration.

The aim of the posed mechanical problem is to determine the scattered and diffracted wave field in the region under consideration, as well as to evaluate the local dynamic stress concentration zones.

The boundary-value problem (BVP) is defined by the governing equation of motion in time-domain–Eq. (1), for out-of-plane and in-plane wave motion, respectively, initial and compatibility and equilibrium conditions along the defined boundaries.

$$\sigma_{i3,i}(\mathbf{x}, t) + \psi f_3(t)\,\delta(\mathbf{x}, \mathbf{x}_0) = \rho\ddot{u}_3, (\mathbf{x}, t), \quad \text{for anti-plane strain problem}$$

$$\sigma_{ij,j}(\mathbf{x}, t) + \psi f_i(t)\,\delta(\mathbf{x}, \mathbf{x}_0) = \rho\ddot{u}_i, (\mathbf{x}, t), \quad \text{for in-plane strain problem}$$

$$\mathbf{x} \in \Omega_k, \Omega_t \quad k = 0,...N, \qquad ...(1)$$

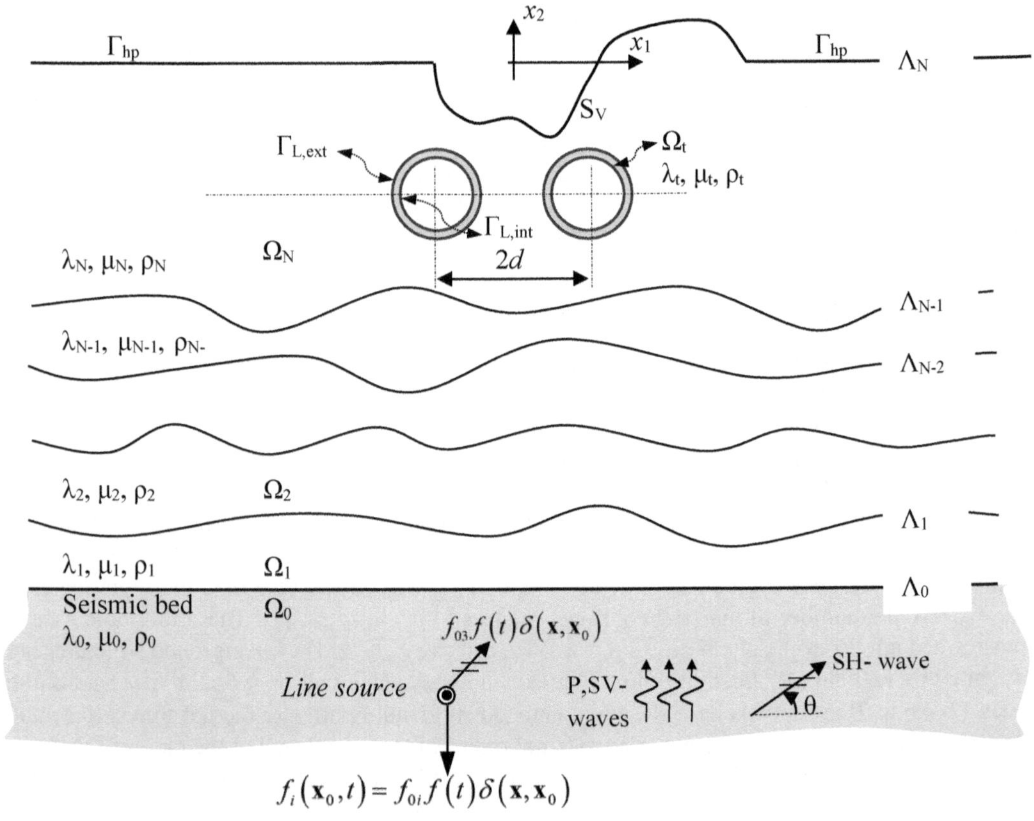

Fig. 1 Problem geometry.

where: $\mathbf{x}(x_1, x_2)$ is the position vector of the observer point; $\mathbf{x}_0(x_{01}, x_{02})$ is the position vector of the embedded dynamic source; $u_3(\mathbf{x}, t)$ $u_i(\mathbf{x}, t)$ are the displacement components of anti-plane and in-plane wave motion; $\ddot{u}_i(\mathbf{x}, t)$ are the acceleration components, displacements are connected with the strain tensor ε_{ij}, $i, j = 1, 2$ by the equation of the geometry $\varepsilon_{ij} = 1/2\,(u_{i,j} + u_{j,i})$; σ_{ij} are the components of the stress tensor. For anti-plane wave motion $\sigma_{i3} = \mu u_{3,i}$ where μ is the shear modulus. For in-plane wave motion $\sigma_{ij} = C_{ijkl}\,\varepsilon_{kl}$, where C_{ijkl} is the fourth-order stiffness tensor for elastic general anisotropic continuum. Comma subscripts denote partial differentiation with respect to the spatial coordinates, while the summation convention over repeated indices is implied. Zeros initial displacements and velocities are assumed. Coefficient ψ is 1 for $\mathbf{x} \in \Omega_0$ when the wave propagates due to an embedded at $\mathbf{x}_0 \in \Omega_0$ seismic source and $\psi = 0$ in the case of an incident wave coming from infinity.

The following well-known numerical procedure is applied for solution of the above defined initial boundary-value problem: (a) the fast Fourier transform (FFT) is applied in respect to time to the field variables in the governing equation (1); (b) the corresponding boundary-value problem is solved in frequency domain; and (c) the inverse fast Fourier transform (IFFT) is applied to the solutions in frequency domain and finally solutions in time domain are obtained. The equations of motion in frequency domain for anti-plane and in-plane wave motion have the following form:

$$\sigma_{i3,i}\,(\mathbf{x}, \omega) + \rho\omega^2 u_3\,(\mathbf{x}, \omega) = \psi f_{03}\, g\,(\omega)\,\delta\,(\mathbf{x}, \mathbf{x}_0),$$

$$\sigma_{ij,j}\,(\mathbf{x}, \omega) + \rho\omega^2 u_i\,(\mathbf{x}, \omega) = \psi f_{0i}\, g\,(\omega)\,\delta\,(\mathbf{x}, \mathbf{x}_0),\ \mathbf{x}\,(x_1, x_2) \in \Omega_{hp}, \Omega_t \qquad ...(2)$$

Here, $u_i\,(\mathbf{x}, \omega)$ are the Fourier transform of the field displacements, $g(\omega)$ is the Fourier transform of the time-history function $f(t)$.

The boundary conditions along the free surface are $t_i(\mathbf{x}, \omega) = 0$ for $\mathbf{x} \in \Gamma_{hp}$, $i = 1, 2$ for in-plane wave motion and $i = 3$ for anti-plane wave motion. Additionally, compatibility conditions of total displacements $u_i^{(k)}$ and equilibrium conditions of the corresponding tractions are satisfied along all the interfaces Λ_k, $k = 1, 2... N - 1$. Sommerfeld's radiation condition is satisfied at infinity. Note that in the case of incident elastic waves coming from the infinity: the total wave field $u_i^{(0)}$, $t_i^{(0)}$ in the seismic bed region Ω_0 comes from superposition of the free-field motion (ff) and the waves scattered (sc) by the relief peculiarities, the tunnels and soil layers, as follows: $u_i^{(0)} = u_i^{\text{ff}(0)} + u_i^{\text{sc}(0)}$; $t_i^{(0)} = t_i^{\text{ff}(0)} + t_i^{\text{sc}(0)}$. The total wave field inside tunnels is denoted by u_i^t, t_i^t, and it is equal to the scattered field transmitted through their interfaces. Free field wave motion solutions present the wave field in a half-plane without any heterogeneities such as relief, tunnels and foundation and they are given in Garcia-Sanchez (2005). For the soil-tunnel interfaces $\Gamma_{L,s-t}$ and $\Gamma_{R,s-t}$ the traction equilibrium and displacement compatibility are: $t_i^{t,\text{ext}}(\mathbf{x}, \omega) = -t_i^{\text{hp}}(\mathbf{x}, \omega)$; $u_i^{t,\text{ext}}(\mathbf{x}, \omega) = u_i^{hp}(\mathbf{x}, \omega)$ and $\mathbf{x}(x_1, x_2) \in \Gamma_{L,s-t} \cup \Gamma_{R,s-t}$. Finally, the traction free boundary conditions are satisfied along the internal tunnel wall interfaces $\Gamma_{L,\text{int}} \cup \Gamma_{R,\text{int}}$. In case of unlined tunnels the respective cavity contours are traction free.

The goal of the defined problem is to determine the seismic stress-strain state caused by out-of-plane and in-plane wave motions in an elastic isotropic or anisotropic, homogeneous or layered half-plane with an arbitrary relief, which in addition contains underground tunnels and arbitrary soil layering. The synthesis of wave fields, considering various heterogeneities such as relief, underground structures, and soil layers in the wave path, is crucial for computational geophysics and earthquake engineering. Understanding the non-uniform stress distribution and identifying zones of stress concentration near underground structures are also crucial for civil engineering practitioners.

3. Reformulation of the Posed Problem via Boundary Integral Equations

The direct boundary integral equation method (BIEM) has become quite popular over the last decades, among the currently used numerical approaches, because of its efficient handling of heterogeneous materials and semi-infinite regions. This is achieved by reformulating the corresponding boundary-value problem via integral equations along the existing boundaries of the object under consideration. This technique is developed based on a transformation of the governing partial differential equations, which involve unknown fields inside and along the boundary of the domain, into an integral equation involving fields on the boundary only.

The mechanical model defined in frequency domain via governing Eq. (2) can be reformulated by a system of displacement boundary integral equations (3) along the domain boundaries throughout an efficient combination of elastodynamic reciprocity theorem and specific properties of the fundamental solution of the governing equation (2), see Manolis and Beskos (1988), Dominguez (1993), Manolis et al. (2017):

$$C_{ij}u_j(\mathbf{x}, \omega) = \int_{\Gamma_\Omega} U_{ij}^*(\mathbf{x}, \xi, \omega)\, t_j(\xi, \omega)\, d\Gamma(\xi) - \int_{\Gamma_\Omega} T_{ij}^*(\mathbf{x}, \xi, \omega) u_j(\xi, \omega) d\Gamma(\xi) + \psi f_{0j} g(\omega) U_{ij}^*(\mathbf{x}, \mathbf{x}_0, \omega) \quad ...(3)$$

Here: C_{ij} is the jump term depending on the local geometry at the source point $\mathbf{x} = (x_1, x_2)$, the couple $\mathbf{x}$, ξ presents the position vectors of the source and receiver point $\xi = (\xi_1, \xi_2)$, respectively; $U_{ij}^*(\mathbf{x}, \xi, \omega)$ is the displacement fundamental solution of the governing elliptic partial differential equation (2) for anti-plane or in-plane wave motion, respectively. Three different fundamental solutions (FS) are adopted in the current programming. FS for elastic isotropic medium presented via Bessel's functions for in-plane and out-of-plane wave motions and frequency-dependent FS derived by Radon transform for general anisotropic continua, see Wang and Achenbach (1994), Garcia-Sanchez (2005), Dineva et al. (2005), Manolis et al. (2017). The corresponding traction fundamental solution is $T^*_{ij}(\mathbf{x}, \xi, \omega) = C_{ijq}U^*_{qk,l}(\mathbf{x}, \xi, \omega)n_k$. The boundary T_Ω is the contour of the domain under consideration, respectively Ω is consecutively, Ω_t and $\Omega_{\text{hp}} = \Omega_0 \cup \Omega_1 \cup \Omega_2 ... \cup \Omega_N$.

In the current modeling, the sub-structuring approach is applied, making it particularly suitable for multiple identical heterogeneities and enabling efficient processing of integrals along interface boundaries. This approach will be explained using the example of two identical tunnels in a half-plane with surface relief.

First the domain Ω_t of the left or right tunnel is considered as a single model, see Figure 1. After application of the discretization and collocation procedure, BIE (3) transforms into a matrix equation $\mathbf{G}^t\mathbf{t}^t - \mathbf{H}^t\mathbf{u}^t = \mathbf{0}$, see Dominguez (1993). The system matrices $\mathbf{G}^t$ and $\mathbf{H}^t$ result from numerical integration (using Gaussian quadrature) of integrals containing the products of fundamental solutions times interpolation (or shape) functions used for the field variables. They are fully populated matrices of size $2N^t \times 2N^t$ where N^t is the total number of nodes of the discretized domain Ω_t, while vectors $\mathbf{u}^t$ and $\mathbf{t}^t$ now contain the boundary displacements and tractions along both the boundaries of the Ω_t domain. After satisfying the boundary conditions along the traction free boundary $\Gamma_{R(L),\text{int}}$ the relation between nodal tractions and displacements of the external tunnel interface $\Gamma_{R(L),\text{ext}}$ is obtained as follows (Dineva et al., 2014a):

$$\mathbf{t}^t_{R(L),\text{ext}} = \mathbf{C}\,\mathbf{u}^t_{R(L),\text{ext}},\ \mathbf{C} = -\mathbf{B}_{\text{ext,int}}(\mathbf{B}_{\text{int,int}})^{-1}\,\mathbf{B}_{\text{int,ext}} + \mathbf{B}_{\text{ext,ext}}$$

$$\mathbf{B} = \begin{bmatrix} \mathbf{B}_{\text{int,int}} & \mathbf{B}_{\text{int,ext}} \\ \mathbf{B}_{\text{ext,int}} & \mathbf{B}_{\text{ext,ext}} \end{bmatrix} = (\mathbf{G}^t)^{-1}\,\mathbf{H}^t \qquad ...(4)$$

In terms of matrix notation, subscripts denote interfaces, or 'int' for $\Gamma_{R(L),\text{int}}$ and 'ext' for $\Gamma_{R(L),\text{ext}}$ where all source and the receiver points congregate. The superscript 't' stands for the tunnel domain Ω_t.

In case of no layering next the half-plane domain Ω_{hp} is considered defined by the free surface Γ_{hp} and both the soil-tunnel interfaces $\Gamma_{L,s-t}$ and $\Gamma_{R,s-t}$. Following surface discretization and imposition of a nodal collocation scheme on Eq. (3), integral equation transforms into a new matrix system of equations $\mathbf{G}^{\text{hp}}\mathbf{t}^{\text{hp}} - \mathbf{H}^{\text{hp}}\mathbf{u}^{\text{hp}} = \mathbf{F}$. The influence matrices are of size $2\text{N}^{\text{hp}} \times 2\text{N}^{\text{hp}}$ where N^{hp} is the total number of nodes along the free surface Γ and both soil-tunnel interfaces $\Gamma_{L,s-t}$ and $\Gamma_{R,s-t}$. The vectors $\mathbf{u}^{\text{hp}}$ and $\mathbf{t}^{\text{hp}}$ now contain the boundary displacements and tractions along the same boundaries. In the so formed system of equations the boundary conditions discussed at the end of *Section 2* are satisfied. More specifically, $\mathbf{t}^{\text{hp}} = -\mathbf{t}^t_{R(L),\text{ext}}$ as defined by eq. (4) for the soil-tunnel interfaces $\Gamma_{L,s-t}$ and $\Gamma_{R,s-t}$, for details see Parvanova et al. (2014b). The traction free boundary conditions are satisfied along the free surface contour Γ_{hp}. What follows is the solution of the so-formed algebraic system of equations with respect to the complex-valued unknowns. The same idea can be applied for consecutive exclusion of the soil layers which leads to the reduced system of linear algebraic equations, see Dineva et al. (2014a).

It is instructive to note the treatment of asymptotic behavior of the displacement fundamental solution and the corresponding traction when the distance between the source and field point tends to zero. All singularities in the kernels related with boundary integral equations (3) are overcome following the numerical procedure described in Garcia-Sanchez (2005). When the static fundamental solutions, as a part of the dynamic ones, are calculated over an element containing the collocation point, the displacement fundamental solution-based kernels exhibit a weak singularity of the type O (ln r), while the traction fundamental solution-based kernels exhibit a strong singularity of the type O (1/r). The integrals with weak singularity are computed using appropriate quadrature rules, while the strongly singular ones are evaluated by the rigid body motion concept. The last cannot be directly applied to an open polygon geometry and in that respect the half-plane is supplemented to enclosed geometry in order to apply the rigid body motion concept. Since the strong singularities are only in the static part of the traction fundamental solution, the model with additional boundary elements is used once. All regular parts of the integrals are calculated by using numerical integration and standard Gaussian quadrature.

What follows is the verification of the proposed computational scheme in Section 4. Parametric study results which illustrate the effects of some key model parameters on the resulting scattered wave field and on the local stress concentration zones are summarized in Section 5.

4. Numerical Scheme Verification

The accuracy of the proposed computational technique must first be established before conducting any parametric studies. The developed and verified numerical procedure is implemented in MATLAB (2023). The well-known accuracy criterion states that $\lambda_{SV}/l_{BE} > 10$, where l_{BE} is the length of the corresponding boundary element (BE) and λ_{SV} is the length of the propagating shear wave, is satisfied. To evaluate the accuracy of the proposed numerical scheme four test examples are solved.

The following normalized characteristics are introduced in this section: (a) Normalized frequency presented by the field variable $\eta = \omega a/\pi V_{SV}$ or $\Omega = \omega a/V_{SV}$, where a is the heterogeneity size, ω and V_{SV} are the frequency and shear wave velocity of the propagating elastic shear wave; (b) Normalized displacements: $|u_i/u_0|$, $i = 1, 2$ here u_0 is the amplitude of the incident wave or $|u_i|c_{66}/f_0$, here f_0 is the magnitude of the applied dynamic body force; (c) DSCF (dynamic stress concentration factor): $|\sigma_{\varphi\varphi}|$, $\sigma_{\varphi\varphi}$ is the hoop stress or $|\sigma_{\varphi\varphi}|/\rho\omega V_{SV}$.

***Test Example 1*: Two lined tunnels of radius *a*, buried in a homogeneous elastic and isotropic half-plane under the incident SH wave.**

The first test example is an elastic isotropic half-plane with a flat free surface containing two parallel single wall tunnels of circular cross section under incident SH-waves. A closed form solution by the method of wave function expansion was proposed in Balendra et al. (1984) and the image technique was used to represent the wave signal along the free surface. The mechanical and geometrical data of the numerical model are as follows: the half-plane is represented by loess at natural moisture with density $\rho_{hs} = 1.64$ t/m^3 and shear modulus $\mu_{hs} = 0.111$ GPa; the tunnels' wall is made of concrete with density $\rho_t = 2.41$ t/m^3 and shear modulus $\mu_t = 8.4$ GPa, see Balendra et al. (1984). For radius, a, of the outer tunnel contour the wall thickness is $0.1a$, the centers of the circles are of depth $h = 2.5a$ beneath the flat free surface, and the distance, s, between both the centers is variable, see Fig. 2a. The displacement amplitudes along the circumference of the left tunnel and shear stress amplitudes, σ_{rz} and $\sigma_{\varphi z}$, at the inner surface, for frequency $\omega = 200$ rad/s, incident angle of the SH-wave $\theta = \pi/3$ with respect to Ox_1 axis, and radius $a = 1$ m, are shown in Fig. 2. Apparently, the current results are in good agreement with those obtained in Balendra et al. (1984).

***Test Example 2:* A single circular cavity of radius *a*, buried in a homogeneous elastic and isotropic half-plane subjected to vertically incident time-harmonic SV waves.**

The second validation example is a single circular cavity of radius a, buried in a homogeneous elastic and isotropic half-plane, see Figure 3a. The problem was solved by Luco and de Barros (1994) who applied indirect BEM based on the two-dimensional Green's function for a viscoelastic half-plane. The mechanical properties and geometrical dimensions of the benchmark example are as follows: $(h/a) = 1.5$ (solid line) and $(h/a) = 5.0$ (dashed line), $\eta = \omega a/\pi V_{SV} = 0.5$, damping ratios $\xi = 0$, and longitudinal P-wave velocity $V_P = 2V_{SV}$. The presented results are obtained once by using FS for elastic isotropic medium presented via Bessel's functions and then by frequency-dependent FS derived by Radon transform for general anisotropic continua. In the latter case the anisotropic material stiffness tensor recovers the isotropic one, by assigning $c_{11} = c_{22} = \lambda + 2\mu$, $c_{12} = \lambda$, $c_{66} = \mu$, $c_{16} = c_{26} = 0$ in the respective program code. Constant hysteretic damping of 0.1% is assigned in the numerical model for anisotropic continuum in order to avoid numerical difficulties. The BEM mesh used herein consists of 36 quadratic boundary elements (BE) along the flat free surface,

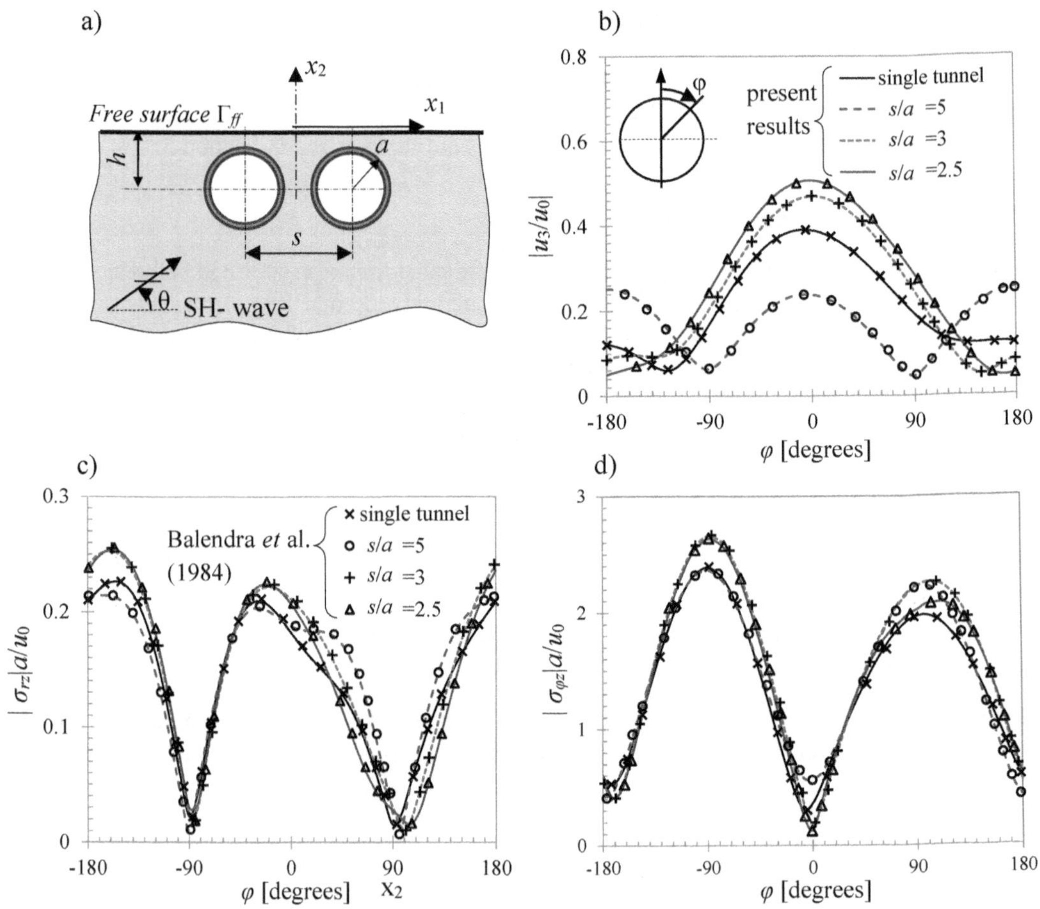

Fig. 2 (a) Problem geometry of test example 1; (b) Variation of the displacement amplitude along the circumvent of the concrete tunnel; (c) Variation of the shear stress amplitude σ_{rz}; (d) Variation of the stress amplitude $\sigma_{\varphi z}$.

and 16 quadratic BE along the cavity contour. The discretized length of the free surface is $+/-5a$ measured left and right with respect to the coordinate system origin. The amplitude variation along the free surface of the total horizontal and vertical displacement components are plotted in Figure 3(b-c). Figure 3(d-e) shows the total horizontal and vertical displacements along the cavity wall. The present set of results is successfully compared with those obtained in Luco and de Barros (1994).

Test Example 3: **Time-harmonic and transient P wave scattering by a semi-circular canyon with radius *a* in an elastic isotropic half-plane.**

The third benchmark example is a semi-circular canyon of radius a, located in a transversely isotropic (TI) half-plane whose material properties are taken from Ba et al. (2017). The excitation is presented by normal incident P- and SV- waves. Following Ba et al. (2017), the material characteristics of the half-plane are presented by Young's modulus in the horizontal and vertical directions E_h and E_v respectively, the shear modulus G_v and the Poisson's ratios ν_h and ν_v characterizing the lateral strain response in the plane of symmetry to stress acting parallel and normal to it. Three different materials are considered, two TI with $E_h/E_v = 0.5$, $E_h/E_v = 2$, and isotropic one with $E_h/E_v = 1$. The respective stiffness tensor components of the isotropic material type with $E_h/E_v = 1$ are: $c_{11} = 1.2\,E_0$, $c_{12} = 0.4\,E_0$, $c_{22} = 1.2\,E_0$, $c_{66} = 0.4\,E_0$, where E_0 and ρ_0 are the reference Young's modulus

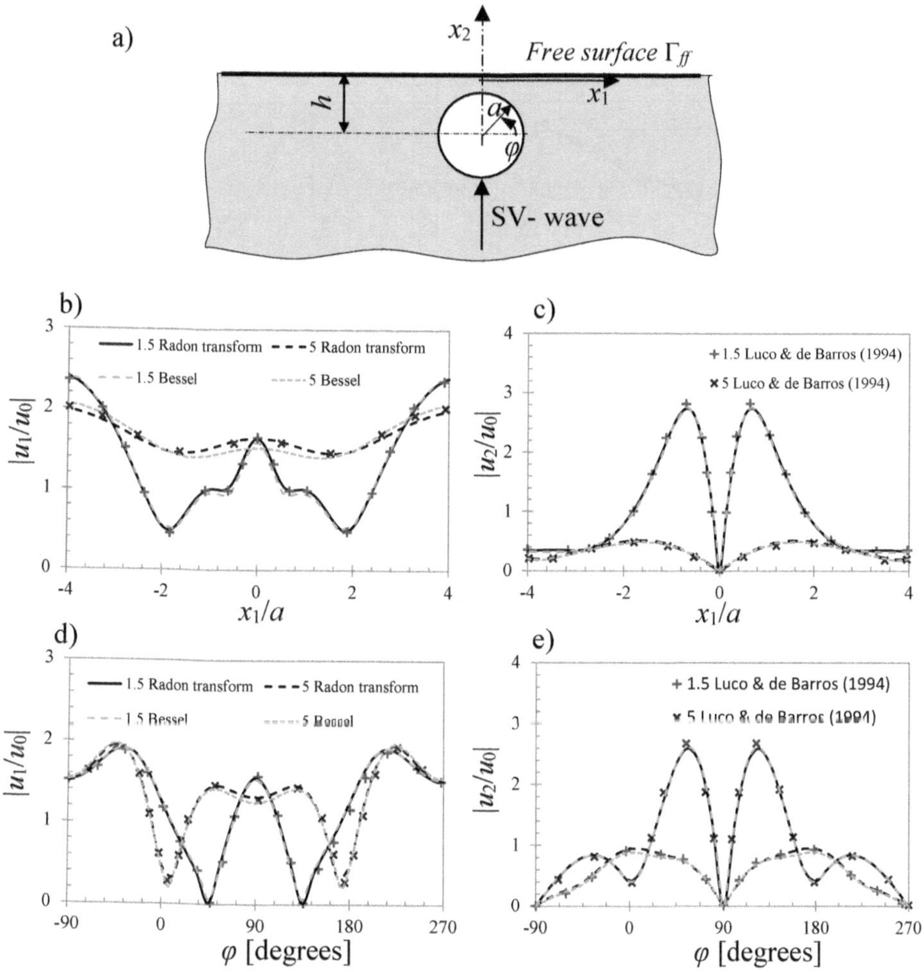

Fig. 3 (a) Problem geometry; Amplitudes of the total horizontal (b) and vertical (c) displacements along the free surface. Amplitude of total horizontal (d) and vertical (e) displacements along the cavity circumvent.

and mass density respectively. The hysteretic material damping is set to 0.1%. The BEM mesh consists of 72 quadratic boundary elements with 24 of them in the canyon arc and the length of the discretized free surface is $5a$ left and right measured to the coordinate system origin. The time-harmonic displacement components are portrayed in Figure 4. More specifically Figure 4a presents normalized surface displacement amplitudes $|u_1/u_0|$ due to normal incident SV-wave with amplitude u_0 for fixed dimensionless frequencies $\eta = \omega a / \left(\pi\sqrt{0.4E_0/\rho_0}\right)$ equal to 0.5 and 1. Since the response is symmetric the left graphics are for $\eta = 0.5$, the right plots are for $\eta = 1$. The normalized vertical surface displacement amplitudes $|u_2/u_0|$ caused by normal incident P-wave are depicted in Figure 4b, the left part for $\eta = 0.5$ and the right part for $\eta = 1$. The results in Ba et al. (2017) for $E_h/E_v = 0.5$ are depicted together with the authors' results for comparison. A good match between both sets of results is observed.

What follows is the time domain response due to vertical incident P-wave of the same problem in the case of isotropic soil type. The time domain results are obtained from frequency domain results by using Inverse Fast Fourier Transform (IFFT) algorithm. The time function of the incident P-wave is the well-known Ricker wavelet defined as:

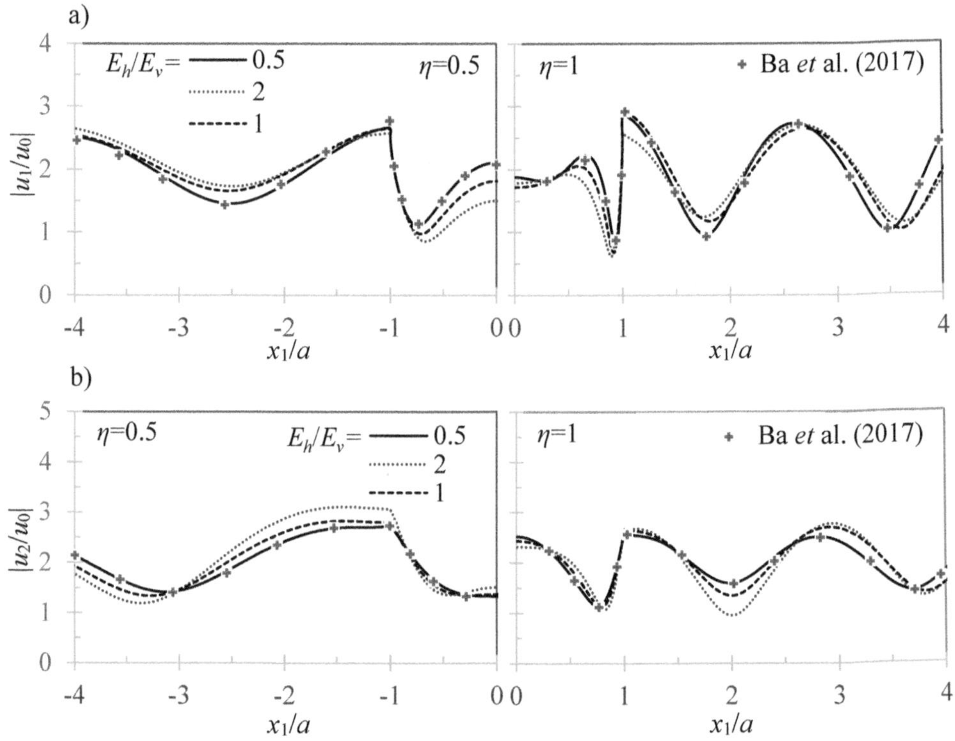

Fig. 4 Normalized amplitudes of surface displacements for a semicircular canyon in TI half-plane and dimensionless frequencies $\eta = 0.5$ (left) and $\eta = 1$ (right) due to: (a) normal incident SV-wave; (b) normal incident P-wave.

$$f(\tau) = f(t) = (2\pi^2\eta_c^2(\tau - \tau_0)^2 - 1)\exp(\pi^2\eta_c^2(\tau - \tau_0)^2) = (2\pi^2 f_c^2(t - t_0)^2 - 1)\exp(\pi^2 f_c^2(t - t_0)^2)$$

In the above equation, $\eta_c = \omega_c a/\left(\pi\sqrt{0.4E_0/\rho_0}\right) = f_c 2a/\left(\sqrt{0.4E_0/\rho_0}\right)$ is the normalized characteristic dimensionless frequency, f_c is the characteristic frequency in Hz, $\tau = t\sqrt{0.4E_0/\rho_0}/2a$ is the normalized time. Note that $\eta_c\tau = f_c t$ and then for the time function is valid that $f(\tau) = f(t)$. In all the calculations η_c is set to 2 and the normalized time $\tau = 0$ corresponds to the peak amplitude of the incident waves located at the depth of $4a$ under the surface of the half-plane, see Ba et al. (2017). To fulfil the last statement t_0 and respectively its dimensionless equivalent τ_0 are obtained as $t_0 = 4a/\sqrt{c_{22}/\rho}$ and $\tau_0 = t_0\sqrt{0.4E_0/\rho_0}/2a$, where $\sqrt{c_{22}/\rho}$ is the velocity of the P-wave. The frequency range in the analysis is $\eta = [0; 7.906]$ divided into 200 equal frequency steps. The normalized vertical displacements (by unit amplitude of the time signal) for representative points of the surface relief are portrayed in Fig. 5. The receiver points are $x_1/a = 0$ or bottom of the canyon, $x_1/a = 3$, and $x_1/a = 2$ from the flat free surface. The results obtained by Ba et al. (2017) are plotted together with the present ones. We note that the benchmark results of Ba et al. (2017) were directly reproduced from their own figures digitally and the dimensionless time is divided by 2 because of different normalization of the absolute time variable in the current paper. An excellent match between both sets of results is observed.

Test Example 4: **A single soil layer resting on an elastic seismic bed subjected to vertically incident P-wave.**

In order to validate layers with different mechanical properties a site consisting of a single soil layer (L) resting on a seismic elastic bed (0) is considered and compared with the analytical solution given by Wolf (1985). The seismic bed and homogeneous layer are characterized by equal Poisson's

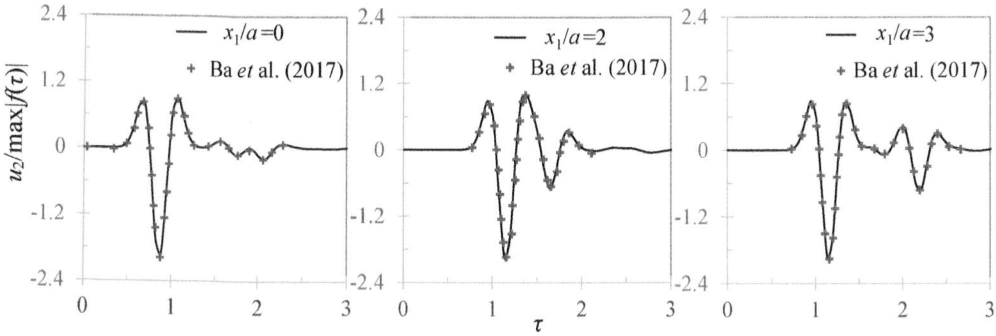

Fig. 5 Normalized vertical displacements at representative points versus normalized time.

ratios $\nu_L = \nu_0 = 1/3$, and equal dimensionless densities $\rho_L = \rho_0 = 1/3$. The hysteretic damping ratio is $\xi = 0.05$. The shear modulus of elastic bed is unity, the shear modulus of the single layer is chosen variable in order to capture limiting cases of the half-plane when $V_{SV,0}/V_{SV,L} = 1$ and single layer on a rigid half plane when $V_{SV,0}/V_{SV,L} = 1000$. The BE mesh in our numerical model consists of 72 quadratic BE over the flat free surface and 40 quadratic BE for the bed-layer interface. The thickness of the layer is assumed $d_L = 2$, and the discretized surface is 20 d_L measured left and right to its center. The amplification of the outcropping motion, for vertical incident P-wave, is plotted and compared against analytical results given by Wolf (1985) in Figure 6.

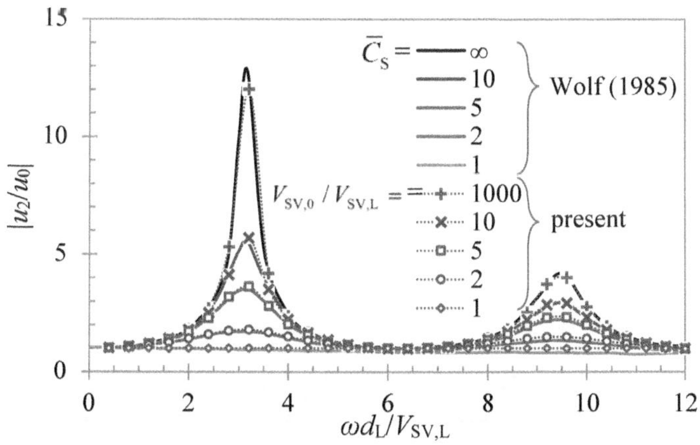

Fig. 6 Amplification outcropping versus normalized frequency due to normal incident P-wave: comparison with solution of Wolf (1985).

5. Simulations and Parametric Study

The aim of this parametric study is to examine how the complex scattered wave field and local stress concentration zones in an elastic, isotropic or anisotropic, semi-infinite homogeneous or multi layered geological region—with surface relief and embedded engineering structures such as a foundation and two unlined or lined tunnels—are influenced by key model parameters. These parameters include: (a) characteristics of the incoming wave, whether incident from infinity or radiated by a dynamic source embedded at a prescribed site; (b) type and location of the source; (c) material anisotropy; (d) dynamic interactions between the soil and the foundation, as well as between the soil and the tunnels (lined or unlined), considering their interaction with the half-plane surface relief features.

Simulation Example 1: **SH wave propagation in a homogeneous half-plane containing a semi-elliptic canyon along its free surface.**

Consider a semi-elliptic canyon with horizontal semi-axis a and vertical semi-axis h, where $0 \le h/a \le 2$, see Figure 7a. The geometrical shape of the canyon varies from vertically elongated ellipse at $h/a > 1$, through semi-circle at $h/a = 1$, to horizontally elongated ellipse at $h/a < 1$, till the case of a flat surface when $h/a = 0$. The seismic load is an out of plane time-harmonic force embedded at a point $\mathbf{x}_0(0, -5a)$ and having magnitude $f_{03} = 10^9$ N. The material properties of the half-plane are $\mu_0 = 1230$ MPa, $\rho_0 = 1.85$ t/m^3. In Figure 7b,c the normalized displacement amplitudes are drawn along the free surface versus the ratio h/a that governs the shape of the canyon for two normalized frequencies $\eta = 0.25$ and $\eta = 1$, where $\eta = 2a/\lambda_{SH}$ (λ_{SH} is the SH-wavelength). The normalization of the displacements is done by the maximal displacement obtained for the considered frequency range or for the relevant time interval. The BEM mesh comprises 72 quadratic boundary elements along the free surface including the boundary of the canyon. The length of the discretized flat surface is $\pm 10a$.

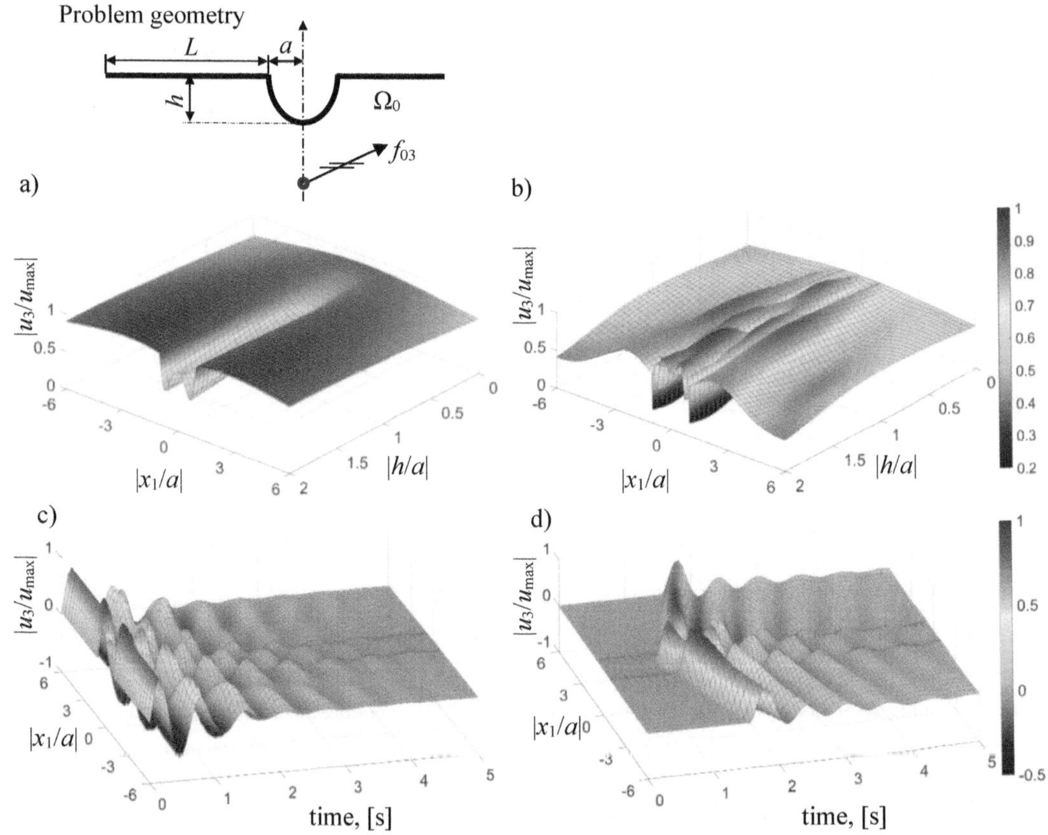

Fig. 7 Normalized displacement amplitude along the free surface of homogeneous half-plane with semi-elliptic canyon with variable vertical semi-axis due to SH wave radiating from a seismic source at a point $\mathbf{x}_0(0, -5a)$ for fixed normalized frequency $\eta = 0.25$ in (a) and $\eta = 1$ in (b). Time-dependent normalized displacement distribution along the free surface due to: (c) wave radiating from transient line seismic source embedded at point $x_0(0, -5a)$; (d) normal transient incident SH-wave.

To examine the transient response of the semi-circular canyon we consider seismic source and vertically incident SH-wave with time-history function $f(t) = e^{-\alpha t}\sin(\omega_0 t)$ and its Fourier transform $g(\omega) = \omega_0/(\omega_0^2 + (\alpha + i\omega)^2)$, where $\alpha = 1$ and $\omega_0 = 10$[rad/s]. The time domain results are obtained from frequency domain results by using the IFFT algorithm. The synthetic seismograms along the free surface of a homogeneous half-plane with a semi-circular canyon of radius a due to transient seismic source embedded at depth $5a$ or by incident plane transient SH-wave are depicted in Figure 7d,e. These figures illustrate how site effects change their specific characteristics over time during excitation and how they depend on its type.

Simulation Example 2: **SH wave propagation in a stratified half-plane containing a surface relief.**

A stratified half-plane with a semi-elliptic canyon or a semi-elliptic hill along the free surface is the current numerical example. The geometry of both the problems is given in Figure 8. The used input data are: $\rho_0 = \rho_1 = \rho_2 = \rho_3$, $\mu_0 = 10^{10}\,\mu_3$, $\mu_1 = 100\,\mu_3$, $\mu_2 = 10\,\mu_3$, μ_3 is the shear modulus of the upper layer, the thickness of layers is $t = 3a$, the vertical to horizontal semi-axis ratio is $0 \leq h/a \leq 2$. The BEM mesh consists of 60 quadratic BE along the free surface, 60 quadratic BE for each of the layer's interfaces and the length of the discretized horizontal boundaries is $L = \pm 30a$ or $L = \pm 10t$. The SH-waves radiate by a line source located at a point $\mathbf{x}_0(0, -10a)$ with magnitude f_{03}. The variation of the normalized displacement amplitudes along the free surface versus ratio h/a for two normalized frequencies $\eta = 0.25$ and $\eta = 1$, is portrayed in Figure 9. The sensitivity of the displacement field and site effects to the shape of the surface relief and the frequency of the dynamic load is evident.

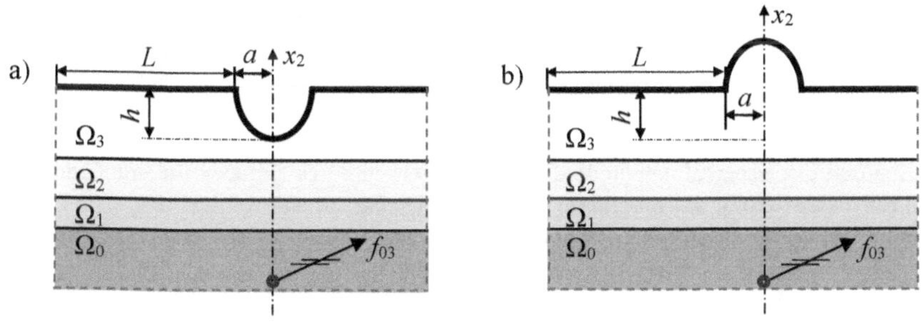

Fig. 8 Geometry of the simulation example 2.

Simulation Example 3: **Two cylindrical tunnels in a homogeneous half-plane with a free surface relief subjected to incident SH-wave.**

Two cylindrical tunnels in a homogeneous half-plane with free surface relief in the form of two elliptic hills with semi-axes a_1 and a_2 is the current example. The shape of the hills varies among three cases (see Figure 10):

Case 1: circular, $a_1 = a_2 = 7.5$ m;
Case 2: horizontally elongated, $a_1 = 11.25$ m, $a_2 = 5$ m;
Case 3: vertically elongated, $a_1 = 5$ m, $a_2 = 11.25$ m.

In all cases, the depth of the tunnel embedment is $h = 9$ m and their center-to-center distance is 15 m. The material properties of the half-plane are $\mu_0 = 18.518$ MPa; $\rho_0 = 1.746$ t/m^3, while we consider an SH-wave of normal incidence, i.e., $\theta = \pi/2$ with respect to axis Ox_1 with a unit amplitude for convenience. In the absence of any heterogeneities, the displacement amplitude doubles along the horizontal free surface of the uniform half-plane, see Achenbach (1973). Obviously, the presence of buried tunnels and free surface relief will cause either amplification or de-amplification effects in the displacement field. The normalized frequency of the wave is defined as $\eta = \omega a / V_{SH}$, where a is the external radius of the tunnels, V_{SH} is the anti-plane shear wave velocity. Finally, regarding BEM modelling, we employ 79 quadratic BE along the free-surface and 48 for each of the two tunnel liners.

Three basic types of tunnel construction are involved in the analysis: (a) Unlined cylindrical circular tunnel (i.e., cavity) of radius $a = 6$ m; (b) Lined tunnel with liner wall consisting of a single ring layer of shotcrete with material properties $\mu_t = 9130$ MPa; $\rho_t = 2.2$ t/m^3 (Figure 10 - left tunnel),

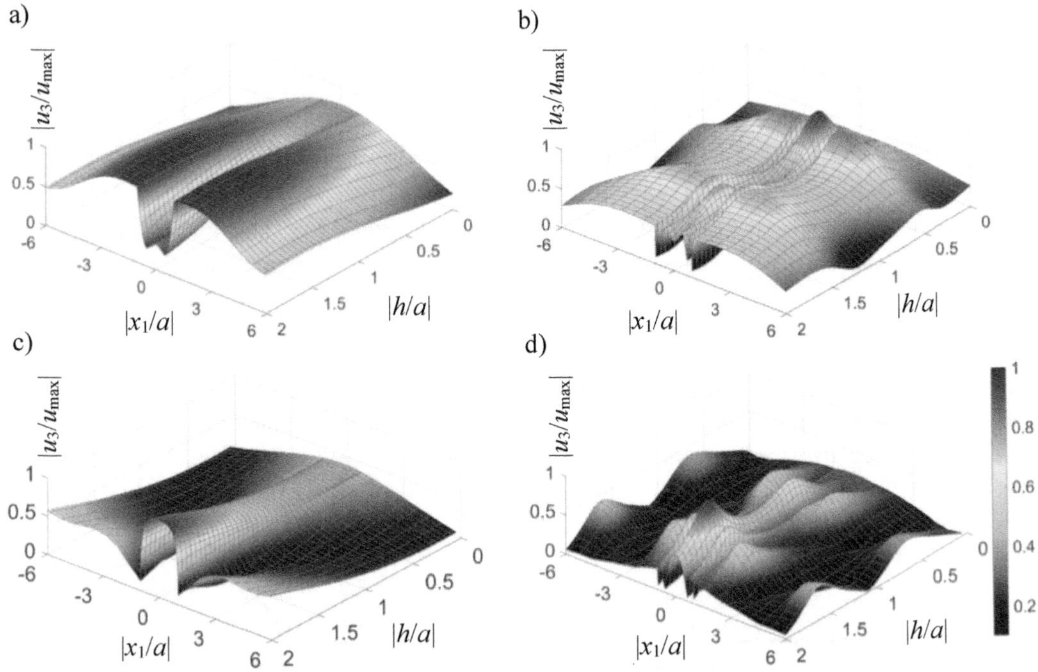

Fig. 9 Normalized displacement amplitude distribution along the free surface of a stratified half-plane with a semi-elliptic canyon in (a,b) or a semi-elliptic hill in (c,d) having variable vertical semi-axis h due to SH wave radiating from a line seismic source embedded at a point $\mathbf{x}_0(0, -10a)$ for a fixed normalized frequency: (a,c) η = 0.25; (b,d) η = 1.

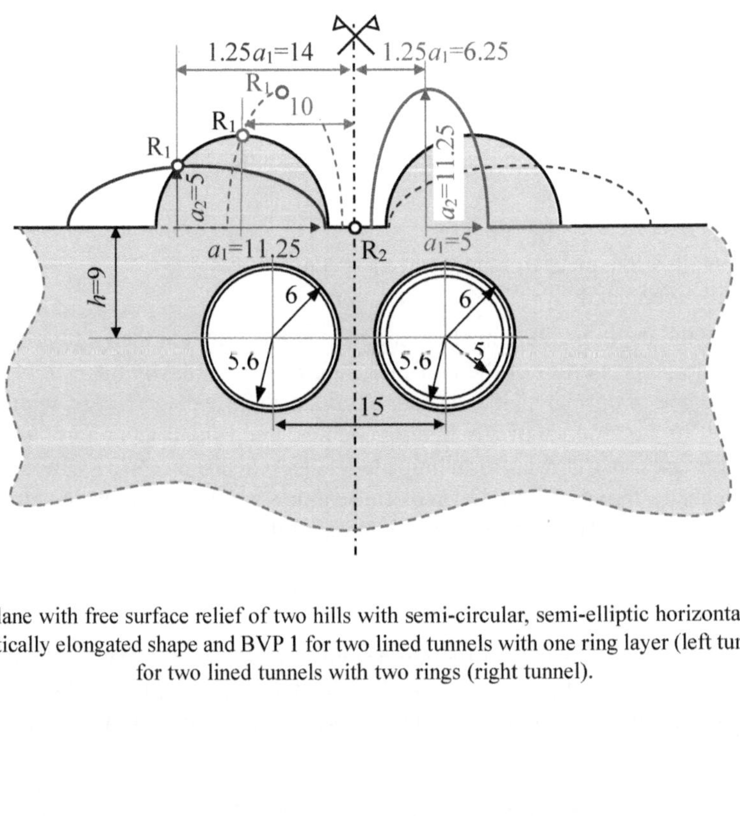

Fig. 10 Half-plane with free surface relief of two hills with semi-circular, semi-elliptic horizontally elongated or semi-elliptic vertically elongated shape and BVP 1 for two lined tunnels with one ring layer (left tunnel) and BVP 2 for two lined tunnels with two rings (right tunnel).

where $a = 6$ m and the wall thickness of 0.4 m; (*c*) Lined tunnel with liner wall comprising two ring layers (Figure 10 - right tunnel): an external layer with air-placed concrete (shotcrete) with the same material properties as before, and an internal prefabricated concrete layer with material properties $\mu_{ti} = 12295$ MPa; $\rho_{ti} = 2.5$ t/m^3. The geometry is as follows: $a = 6$ m, the thickness of the external shotcrete layer is 0.4 m, while the thickness of the internal prefabricated concrete layer is 0.6 m.

The simulation results start with Figures 11–12 that plot the frequency dependence of the displacement amplitude at different receiver points such as *R*1:(–10, 7.5) m for the case of circular hills; *R*1:(–14, 5) m for the horizontally elongated elliptic hills; *R*1:(–6.25, 11.25) m for vertically elongated elliptic hills, all located at the top of the left hill; and *R*2:(0, 0) m located in the middle between both the hills in all cases, see Figure 10. Figures 11–12 compare the variation of displacement amplitude with frequency for all three types of tunnel structures previously mentioned: cavity, single liner, and composite liner. The following conclusions can be summarized: (a) Site effects are most clearly demonstrated in the case of horizontally elongated elliptic hills along the free surface and lined tunnels, see Figures 12b,c; (b) the influence of tunnel construction type is also clearly visible, with marked differences observed between the cavity case and all other lined tunnel cases; (c) the displacement amplitude at any given receiver location strongly depends on the ratio of the tunnel radius *a* and SH-wave length λ_{SH}, keeping in mind that these two quantities filter into the definition of the non-dimensional frequency $\eta = 2\pi a/\lambda_{SH}$.

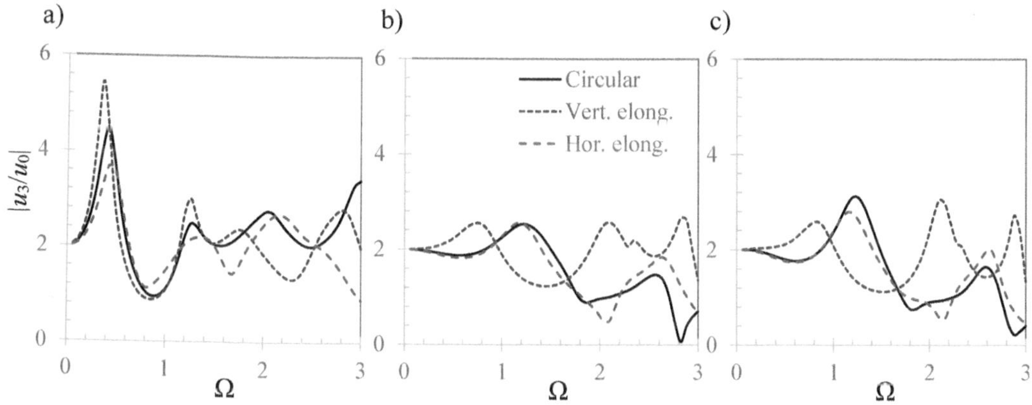

Fig. 11 Displacement amplitude versus non-dimensional frequency Ω at receiver point R1 for different surface topography and the case of: (a) unlined tunnels; (b) tunnel with a single wall; (c) tunnel with two ring layers.

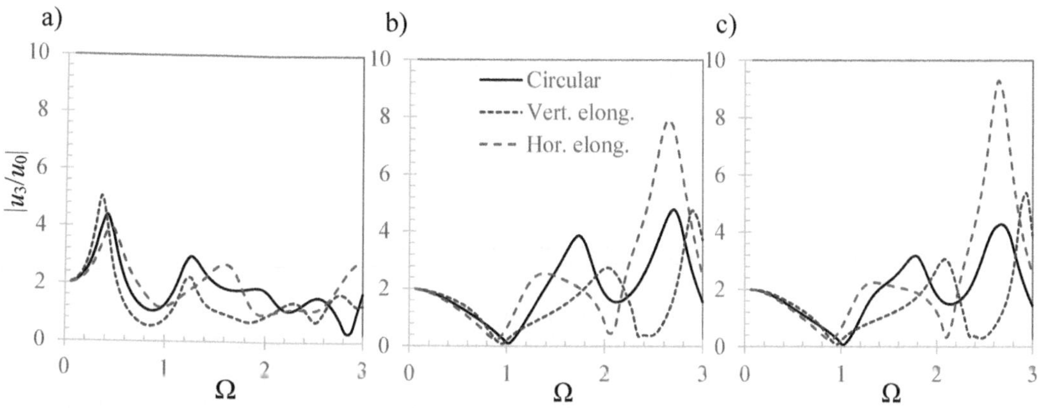

Fig. 12 Displacement amplitude versus non-dimensional frequency Ω at receiver point R2 (0, 0) for different surface topographies and the case of: (a) unlined tunnels; (b) tunnel with a single wall; (c) tunnel with two ring layers.

The next series of results concern Figures 13–14, which depict the displacement amplitude spatial variation (in terms of dimensionless distance x_1/a) at two discrete frequency values of $\Omega_1 = 0.48$; $\Omega_2 = 2.64$. The plots are for the three different geometrical shapes of the free-surface relief hills, i.e., the semi-circular shape (Fig. 13a and 14a), the semi-elliptic, horizontally elongated shape (Figures 13b, 14b) and the semi-elliptic, vertically elongated shape (Figures 13c, 14c). Since the response is symmetric, only half of the free surface is plotted in case of semi-elliptic hills. These plots clearly show how the displacement amplitude distribution along the free surface changes according to the type of the free surface relief and to the presence of underground tunnel construction. Of interest is the fact that the buried cavity and the buried lined tunnels change the surface displacements in quite different ways, with the former producing much larger amplification factors (in excess of 4) compared to the latter.

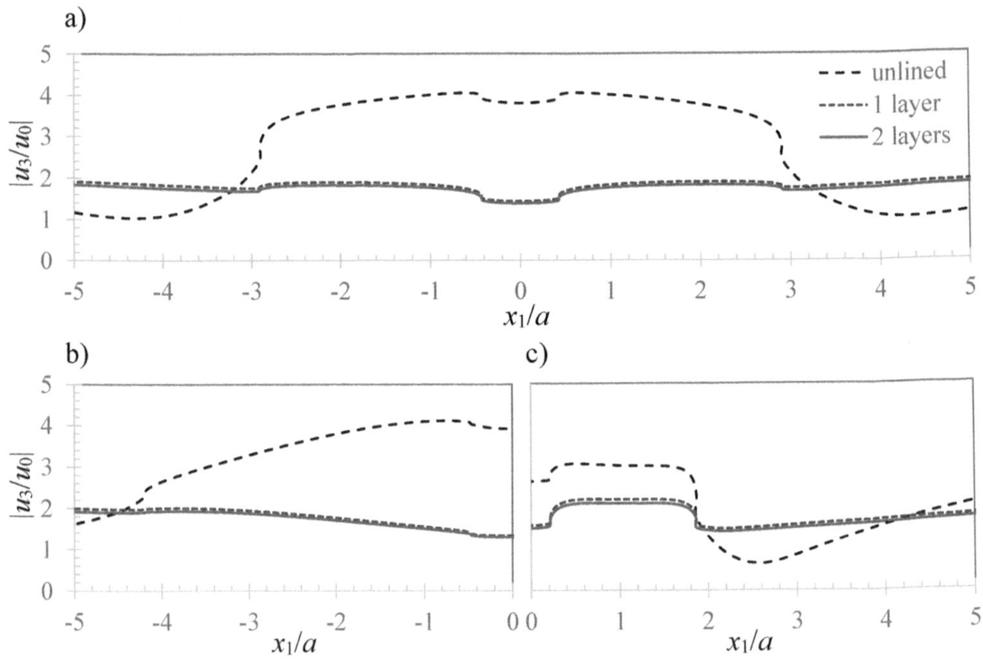

Fig. 13 Displacement amplitude versus distance x_1/a for $\Omega = 0.48$ and: (a) hills with semi-circular shape; (b) hills with semi-elliptic horizontally elongated shape; (c) hills with semi-elliptic vertically elongated shape.

Simulation Example 4: **Two lined tunnels in an elastic layer with a surface relief resting on a supporting elastic homogeneous half-plane under incident time-harmonic or transient P- and SV-wave.**

An elastic layer with free surface relief and two lined tunnels resting on the supporting elastic homogeneous half-plane is considered and subjected to incident time-harmonic or transient in time in-plane wave motion. The geometry of the numerical example is depicted in Figure 15 where R_{int} and R_{ext}, are 5.5 and 6 m respectively, and the segment angle spanning 240° (–30° to 210°, ccw). The BEM mesh used herein consists of 120 quadratic boundary elements along the interface between the layer and the seismic bed, 24 quadratic boundary elements for each of the tunnel interfaces, 96 quadratic elements along the flat part of the free surface and 24 elements along the boundary of the semi-circular canyon. The discretized free surface is 1000 m measured left and right to the axis of symmetry, which is equal to 20d, where d is the thickness of the resting layer. The mechanical properties of the layer are as follows: $E_1 = 5e4$ kN/m^2 ($\mu_1 = 1.852e4$ kN/m^2), $v_1 = 0.35$; $\rho_1 = 1.746$ t/m^3, while those of the tunnel are $E_t = 2.1e7$ kN/m^2 ($\mu_t = 9.13e6$ kN/m^2), $v_t = 0.15$; $\rho_t = 2.2$ t/m^3. The

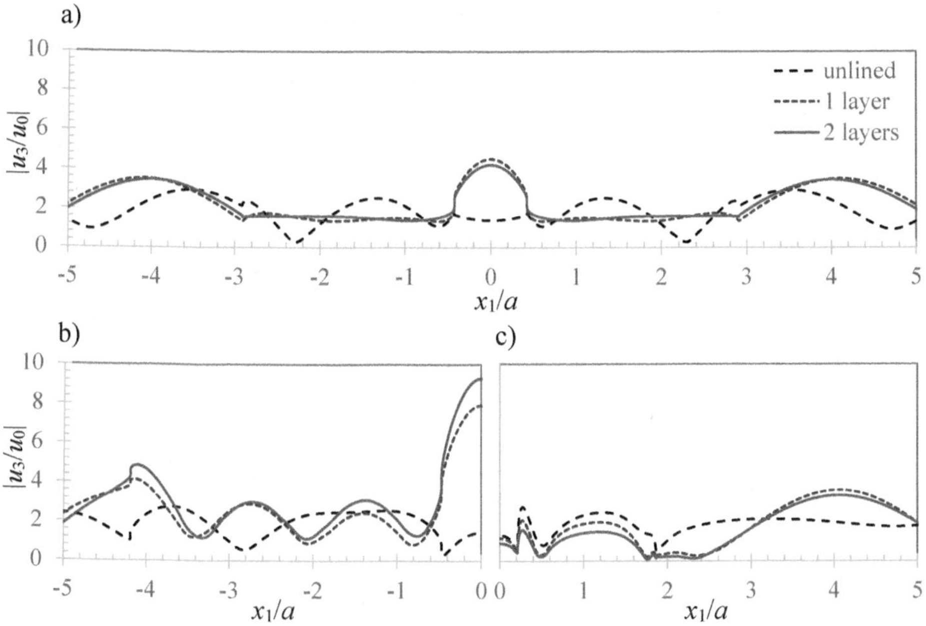

Fig. 14 Displacement amplitude versus distance x_1/a for a fixed frequency $\Omega = 2.64$ and: (a) hills with semi-circular shape; (b) hills with semi-elliptic horizontally elongated shape; (c) hills with semi-elliptic vertically elongated shape.

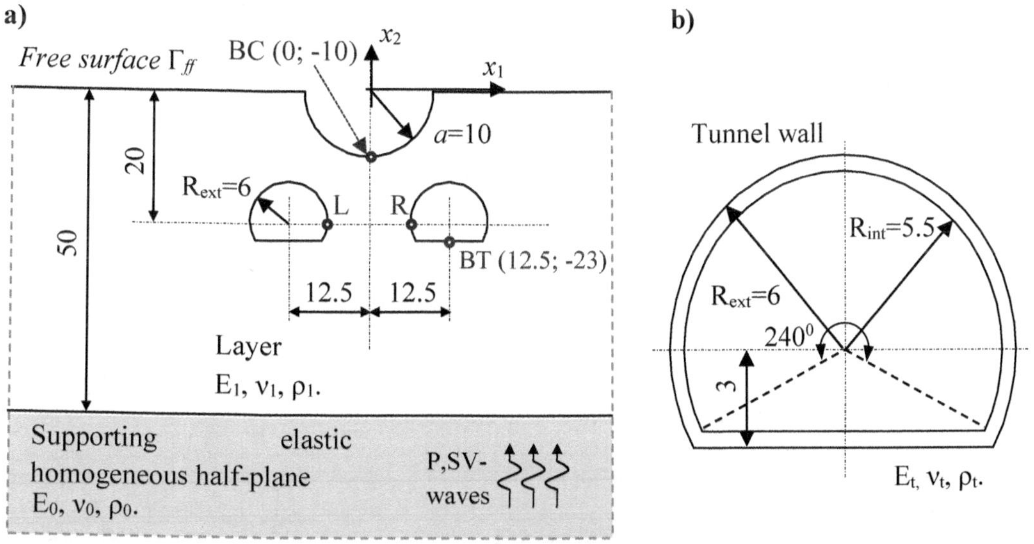

Fig. 15 Half-plane with free surface relief containing circular canyon: (a) geometrical and mechanical data; (b) tunnels' construction.

elastic seismic bed is characterized by $E_0 = 3.0\text{e}7$ kN/m^2 ($\mu_0 = 1.2\text{e}7$ kN/m^2), $\nu_0 = 0.25$; $\rho_0 = 2.0$ t/m^3. The damping ratio for all BE domains is 5%. All geometrical dimensions are given in Figure 15.

The influence of the elastic bed stiffness on the wave field is shown in Figures 16–18. More specifically, Figure 16 draws the horizontal and vertical displacement at the bottom of the canyon caused by vertically incident SV- and P-waves, respectively. The hoop and radial stress at point L and point R (Figure 15) for both incident waves are depicted in Figure 17. As the bed stiffness increases, the seismic response of points along the free surface relief and tunnels also increases significantly. Sensitivity of the displacement component amplitudes across the free surface at a fixed frequency $\eta = 0.63$ to the elastic bed stiffness is demonstrated in Figure 18. Since the response is symmetric, Fig. 18a plots the horizontal displacement components for the case of incident SV-wave, Figure 18b depicts vertical displacement components in case of normal incident P-wave. All the results in Figures 16–18 are for the case of lined tunnels with the geometry shown in Figure 15b.

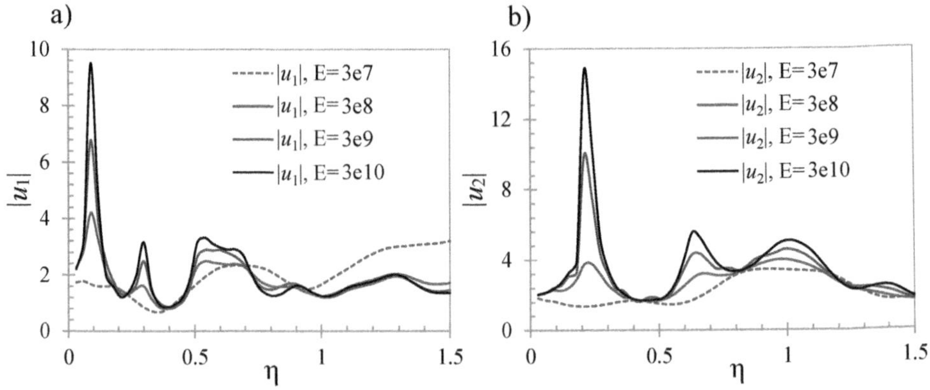

Fig. 16 Horizontal and vertical displacements at bottom of the canyon for variable elastic bed stiffness due to: (a) Incident SV-wave; (b) Incident P wave.

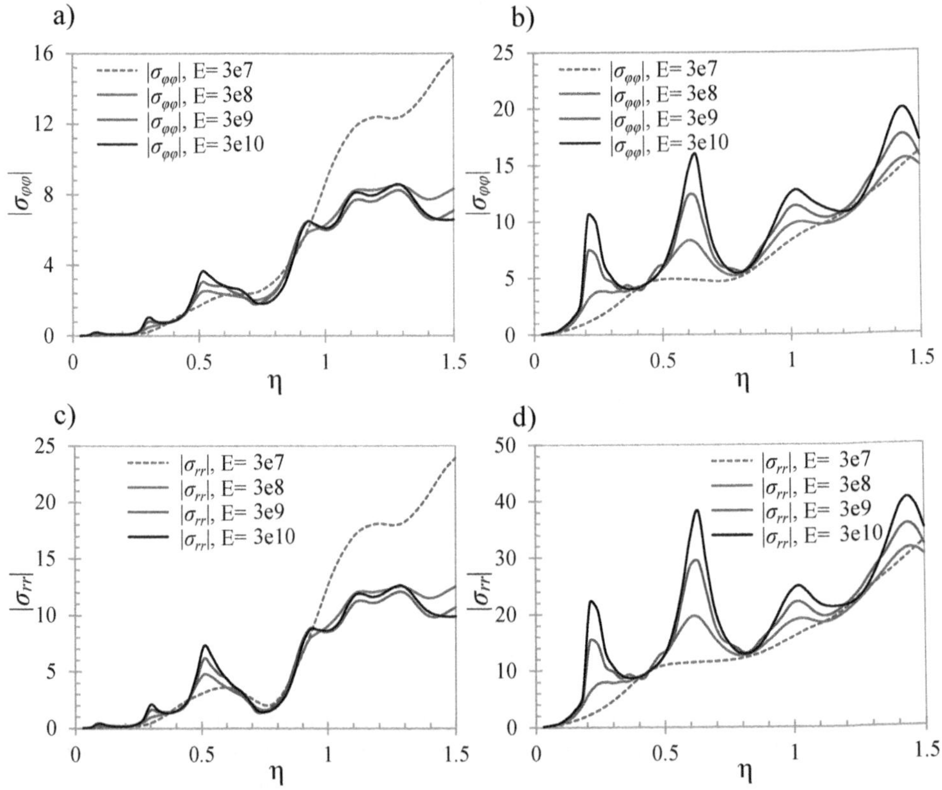

Fig. 17 Hoop and radial stresses at points L (– 6.5, – 20) and R (6.5, – 20) for variable elastic bed stiffness due to: a,c) Incident SV-wave; b,d) Incident P wave.

Figure 19 illustrates how the type of tunnel construction (unlined or lined) influences the displacement along the free surface at the canyon bottom and the stress field at points L and R on

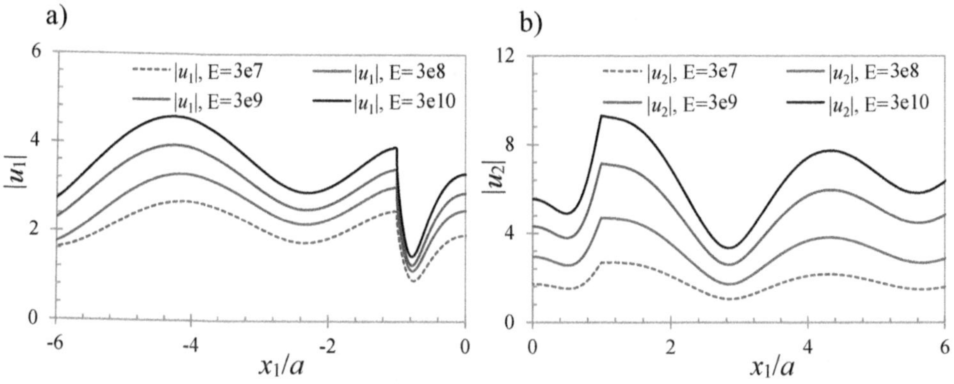

Fig. 18 Influence of the elastic bed stiffness for displacement amplitudes across the free surface at $\eta = 0.63$ in case of: (a) Incident SV-wave; (b) Incident P-wave.

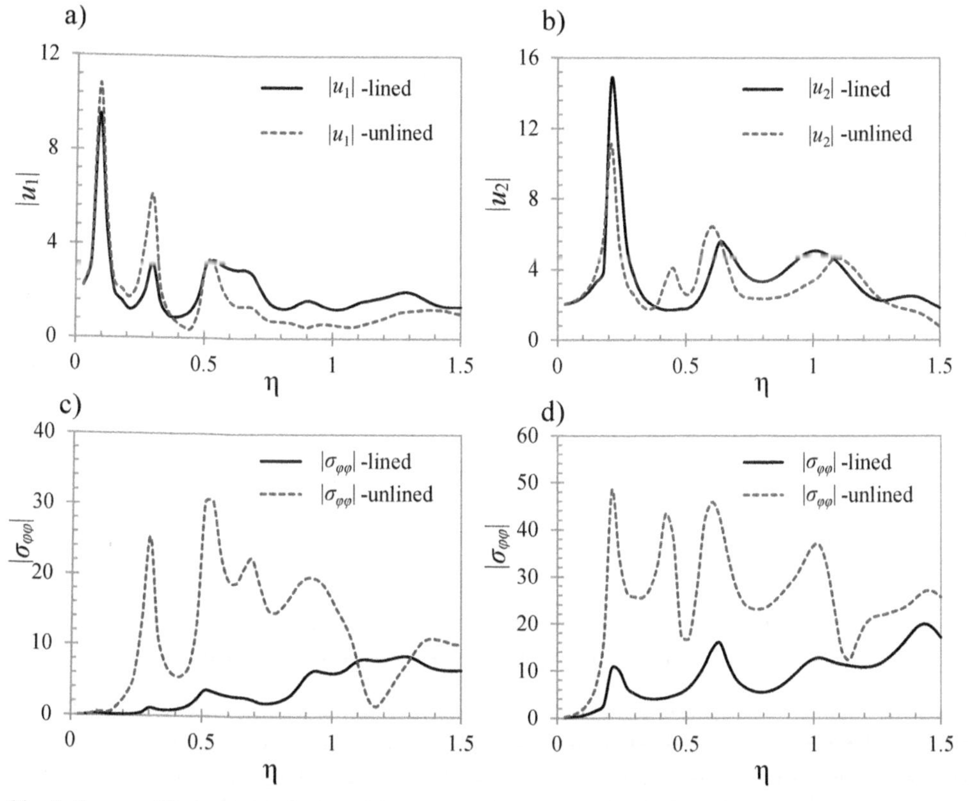

Fig. 19 Influence of the tunnel walls: (a,b) horizontal and vertical displacements at BC (0, – 10) caused by incident SV- and P-wave; (c,d) hoop stresses at points L (– 6.5, – 20) and R (6.5, – 20) due to incident SV- and P-wave.

both the tunnels. More specifically, Figures 19a and 19b plot horizontal and vertical displacement components at the bottom of the canyon caused by normal incident SV- and P-waves, respectively, for lined and unlined tunnels. The hoop stresses at the observer points of the tunnel wall for both incident waves and both the types of tunnel construction are depicted in Figure 19c–d. These plots demonstrate a strong frequency-dependent reduction in the stresses along the tunnels in the case of liners.

The last set of figures is for incident transient P- and SV- waves, where the time-history function is $f(t) = e^{-\alpha t}\sin(\omega_0 t)$, with $\alpha = 1$ and $\omega_0 = 10$[rad/s]. The results are for the lined tunnels and for the seismic bed stiffness equals to 3.0e7 kN/m^2. Figures 20a and 20b plot the horizontal and vertical displacements respectively versus time due to the incident SV- and P-wave at point BC (0, – 10) (bottom of the canyon) and point BT (12.5, –23) (bottom of the tunnel). Hoop and radial stresses at points L(–6.5, –20) and R(6.5, –20) due to normal incident SV- and P-wave are drawn in Figures 20c and 20d, respectively.

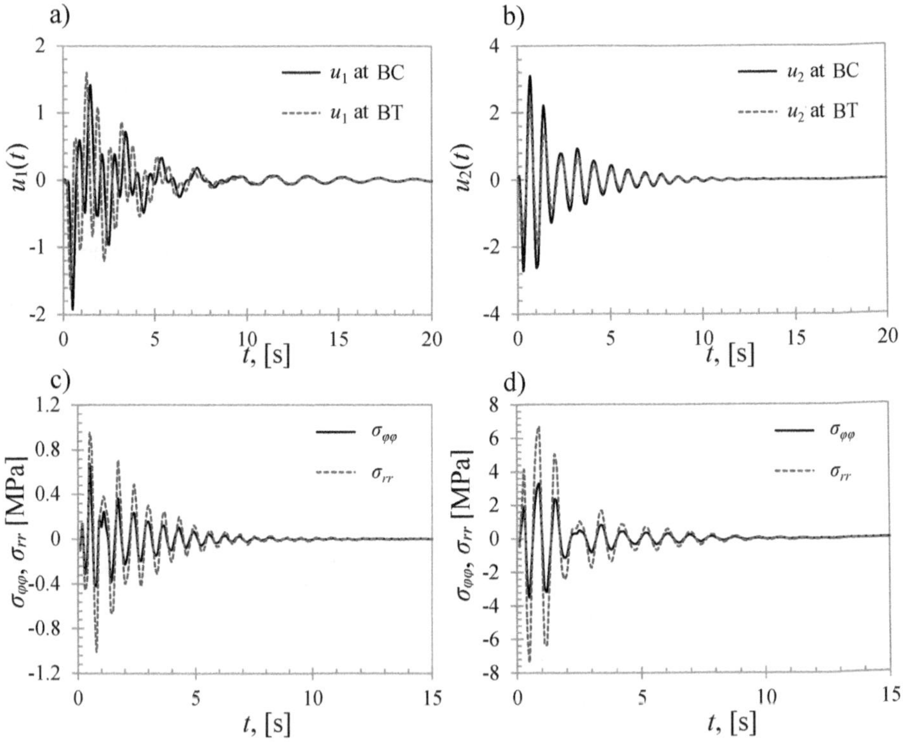

Fig. 20 Response of the soil-tunnel system in time domain: Horizontal (a) and vertical (b) displacement components at the bottom of the canyon (BC) and at the bottom of the tunnel (BT), caused by transient incident SV- and P-wave; Hoop and radial stresses at points L(–6.5, –20) and R(6.5, –20) due to normal transient incident SV-wave (c) and P-wave (d).

Simulation Example 5: **Two unlined or lined tunnels embedded in a TI half-plane with free surface relief subjected to incident time-harmonic P- and SV- wave motion.**

Two circular tunnels entirely embedded in a semi-infinite transversely isotropic (TI) soil medium of depth $2a$ are the subject of the current numerical example. The separation distance between both the tunnels is $3a$. The surface relief is presented by a hill or canyon with the spatial coordinate function $x_2(x_1) = +/\!-a \cdot \cos^2(\pi x_1/4a)$, where the positive sign corresponds to the hill, the negative sign is for the canyon topography (Figure 21). An embedded line source of magnitude $f_{0i} = (0, -1 \text{ kN})$ located at point $\mathbf{x}_0 = (0, -4a)$ is considered, see Figure 21.

The material properties of TI half-plane are chosen similar to those given in Ba et al. (2017). These properties include Young's modulus in the horizontal and vertical directions (E_h and E_v, respectively), the shear modulus G_v and the Poisson's ratios v_h and v_v characterizing the lateral strain response in the plane of symmetry to stress acting parallel and normal to it. Three different material types are considered in the following simulations, all summarized in Table 1. The first material type corresponds to $E_h/E_v = 0.5$ with modulus $G_v = 40500$ kN/m^2 and mass density $\rho_0 = 1.8$ t/m^3. In order to demonstrate the influence of soil anisotropy two additional soil media are involved in the numerical simulations: TI soil medium with $E_h/E_v = 2$ and a shear modulus $G_v = 40500$ kN/m^2

and isotropic soil case with $E_h/E_v = 1$, corresponding to a shear modulus $G_v = 40500$ kN/m^2 and a Poisson's ratio $v = 0.25$. The external radius of the tunnel is $a = 5$ m, the tunnel wall thickness is 0.4 m which results in an internal radius of the tunnel wall $a_{int} = 4.6$ m. The mechanical properties of the lining correspond to a concrete material with shear modulus $\mu = 1.25e7$ kN/m^2, Poisson's ratio $v = 0.2$ and mass density $\rho = 2.5$ t/m^3. The tunnel's geometry is given in Figure 21. The hysteretic material damping is set to 2.5%. In the results presented below the excitation frequency is normalized as $\Omega = \omega a/\sqrt{c_{66}/\rho}$.

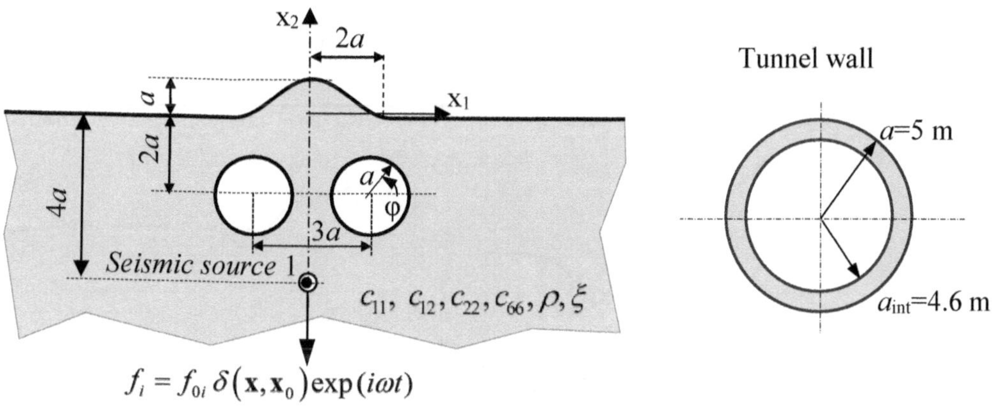

Fig. 21 Problem geometry of simulation example 5.

Table 1 Material constants.

Mat. Type	*Stiffness Tensor Components, [kN/m^2]*	ρ *[t/m^3]*
1 ($E_h/E_v = 0.5$)	$c_{11} = 76090.89$; $c_{12} = 24545.43$; $c_{22} = 147272.7$; $c_{66} = 40500$	1.8
2 ($E_h/E_v = 2$)	$c_{11} = 189000$; $c_{12} = 67500$; $c_{22} = 101250$; $c_{66} = 40500$	1.8
3-isotropic case	$c_{11} = 121500$; $c_{12} = 40500$; $c_{22} = 121500$; $c_{66} = 40500$	1.8

The first set of parametric results concerns two unlined and two lined circular tunnels considering the following types of half-plane free surface: (a) flat boundary; (b) boundary with a hill topography with the spatial coordinate function $x_2(x_1) = a \cdot \cos^2(\pi x_1/4a)$; (c) boundary with a canyon with the spatial coordinate function $x_2(x_1) = -a \cdot \cos^2(\pi x_1/4a)$. The anisotropic material properties correspond to $E_h/E_v = 0.5$ or material type 1 from Table 1. Figure 22 shows the vertical displacement components at representative points: the center of the free surface (top of the hill, bottom of the canyon), the point (0, 0) along the flat surface, and points on the right hole at polar angles $\varphi = \pi/2$, $\varphi = \pi$, and $\varphi = 3\pi/2$, all plotted against normalized frequency. The results are presented in dB based on reference m/kN.

The variation of the normalized displacement amplitudes along the free surface versus normalized frequency Ω for different surface topography is depicted in Figure 23. More precisely, the horizontal displacement amplitudes are plotted in Fig. 23a-c, the vertical ones are drawn in Figure 23d-f. The displacement components for the canyon topography are portrayed in Figures 23a,d. Figure 23b and Figure 23e are for the flat free surface and Figures 23c and Figure 23f concern the surface relief in shape of the hill. All the plots in Figures 22 and 23 are for the unlined tunnels.

The vertical displacement components at the same observer points, center point of the free surface and three points along the right tunnel, in the case of lined tunnels are presented in Figure 24. By comparing Figure 22 with Figure 24, the influence of the lined tunnels on the surface and interface displacements is clearly visible. The effect of the surface relief is highly reduced in the case of lined tunnels compared to unlined ones.

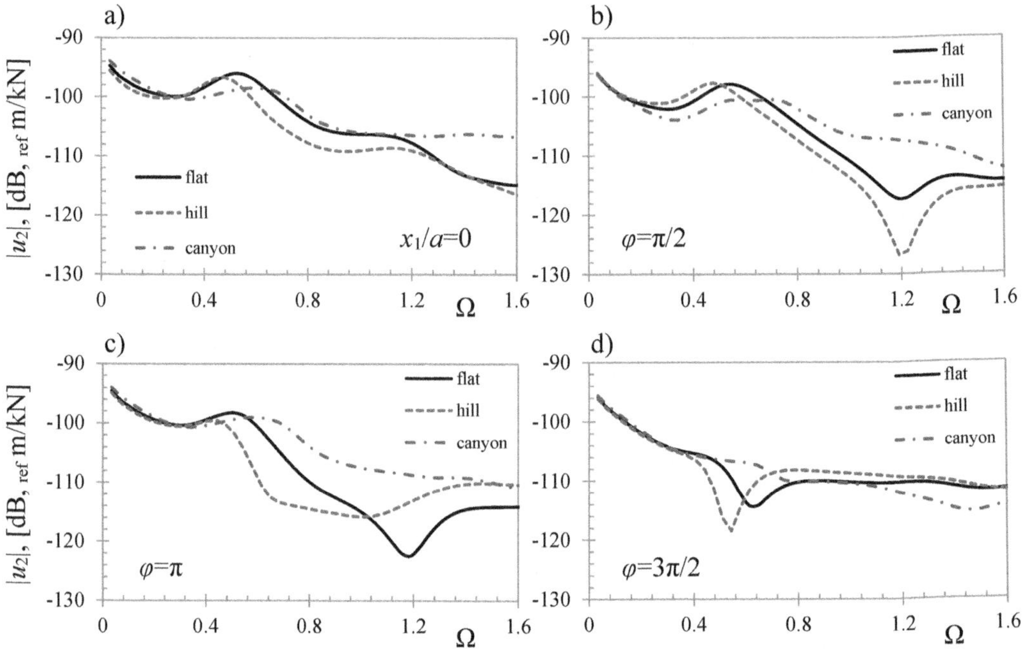

Fig. 22 Amplitudes of vertical displacements for different surface topography versus normalized frequency in case of unlined tunnels: (a) center point of the free surface; points of the right hole with polar angle (b) $\varphi = \pi/2$ (c) $\varphi = \pi$ (d) $\varphi = 3\pi/2$.

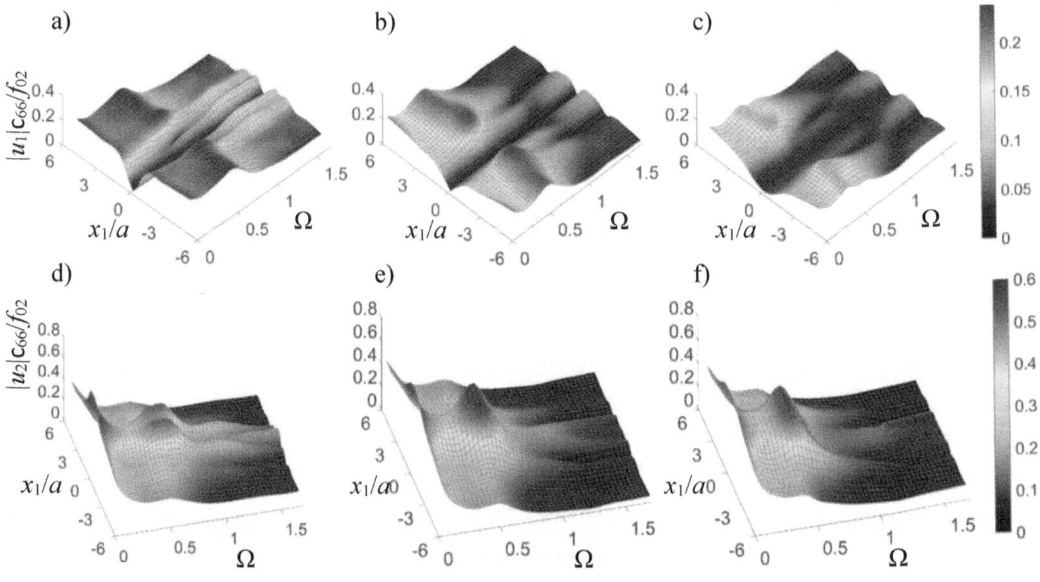

Fig. 23 Influence of the surface topography. Normalized horizontal (a,b,c) and vertical (d,e,f) displacement amplitudes along the free surface of TI half-plane with $E_h/E_v = 0.5$ and: a,d) canyon topography; b,e) flat free surface; c, d) hill topography.

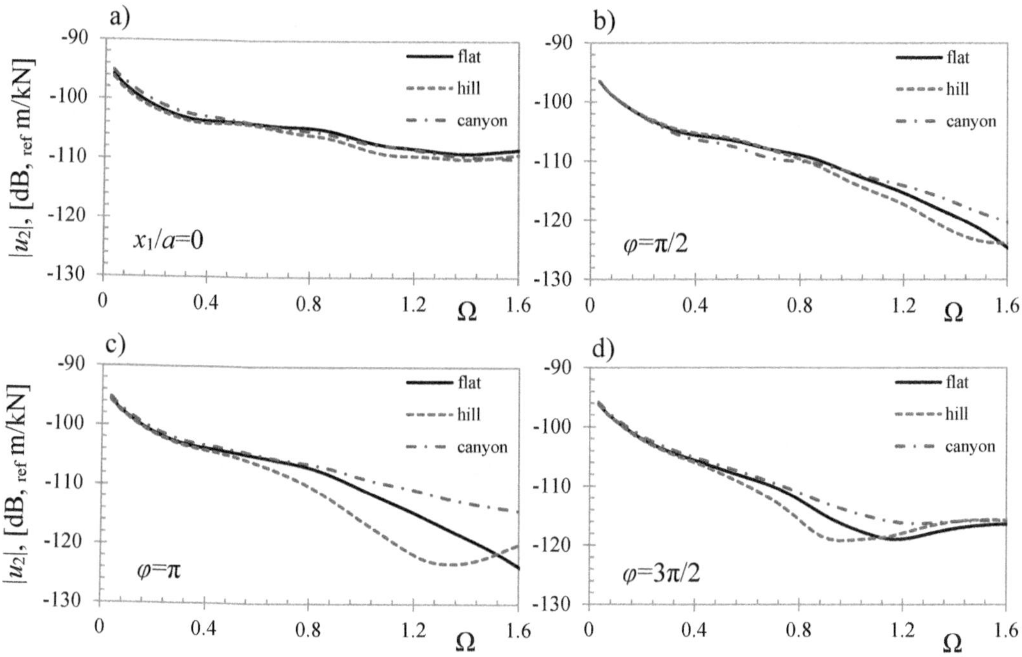

Fig. 24 Amplitudes of vertical displacements for different surface topography versus normalized frequency in case of lined tunnels: (a) center point of the free surface; points of the right hole with polar angle (b) $\varphi = \pi/2$ (c) $\varphi = \pi$ (d) $\varphi = 3\pi/2$.

The second set of results, related to this simulation example, concerns the influence of soil anisotropy in case of unlined tunnels. The surface relief is presented by a hill with the geometry shown in Fig. 21. The soil medium is firstly TI with $E_h/E_v = 0.5$ (mat. type 1), secondly $E_h/E_v = 2$ (mat. type 2), and finally isotropic with $E_h/E_v = 1$ (mat. type 3). The normalized vertical displacement components versus normalized frequency Ω measured at the observer point (0, + a) are shown in Figure 25a for all soil profiles. The hoop stresses along the right cavity contour for fixed normalized excitation frequency Ω = 0.442 are plotted in Figure 25b. The fixed frequency corresponds to the first peak of the vertical displacement plots in Figure 25a.

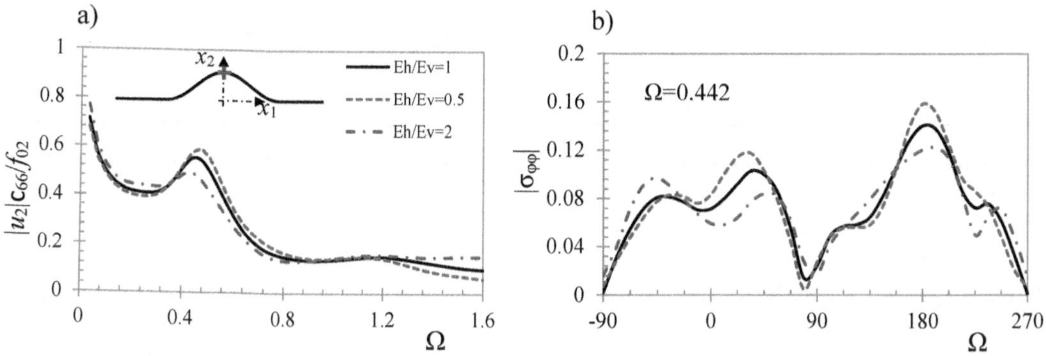

Fig. 25 (a) Amplitudes of vertical displacements versus normalized frequency for the top of the hill (0, +a) and different material types; (b) Distribution of the hoop stress along the right cavity contour for different soil characteristics and fixed frequency Ω = 0.442.

The variation of the normalized displacement amplitudes along the free surface versus normalized frequency Ω for different TI soil mediums is depicted in Figure 26. The horizontal and vertical displacement components for TI soil material with $E_h/E_v = 0.5$ are depicted in Figure 26a and Figure 26d, respectively. Figure 26b and Figure 26e show the displacement components for isotropic soil medium with $E_h/E_v = 1$, and finally Figure 26c and Figure 26f illustrate the variation of displacements along the surface in case of TI soil with $E_h/E_v = 2$.

The obtained results illustrate clearly that the wave field in a semi-infinite anisotropic heterogeneous geological region is a complex result of many key model factors. The evaluation of the specific influence of each parameter individually is a difficult task as far as the final dynamic response is a complex result of a mutual play of all individual factors.

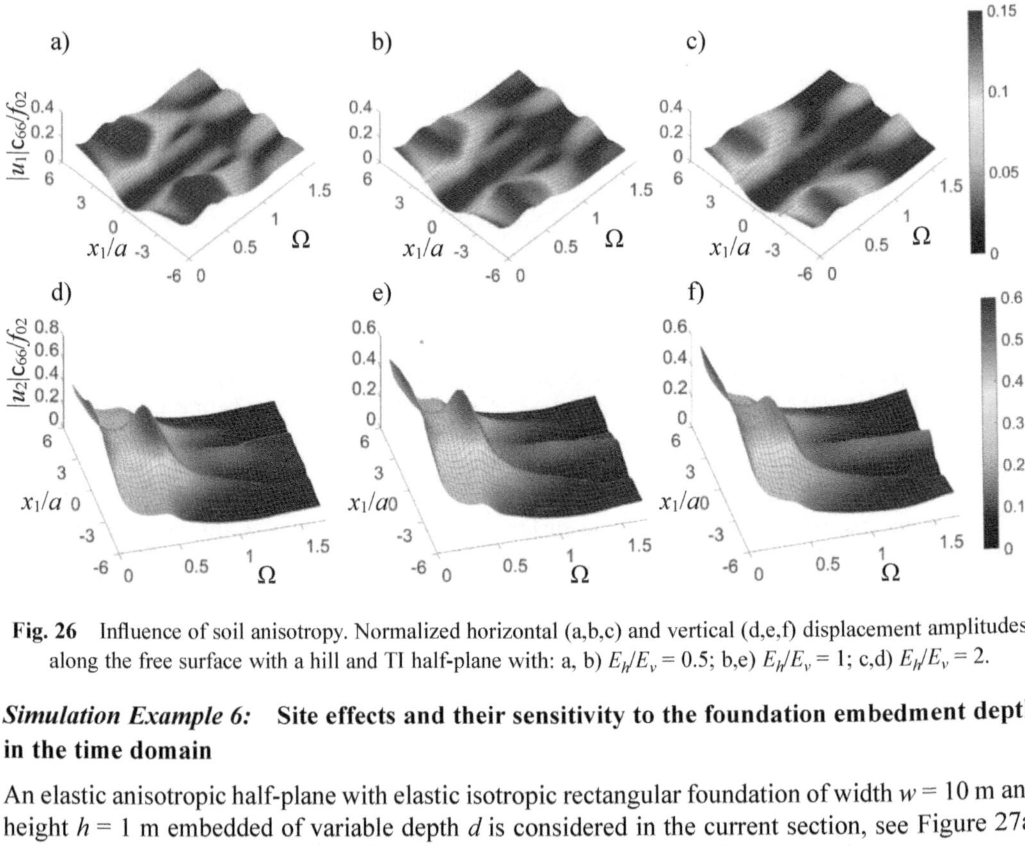

Fig. 26 Influence of soil anisotropy. Normalized horizontal (a,b,c) and vertical (d,e,f) displacement amplitudes along the free surface with a hill and TI half-plane with: a, b) $E_h/E_v = 0.5$; b,e) $E_h/E_v = 1$; c,d) $E_h/E_v = 2$.

Simulation Example 6: **Site effects and their sensitivity to the foundation embedment depth in the time domain**

An elastic anisotropic half-plane with elastic isotropic rectangular foundation of width $w = 10$ m and height $h = 1$ m embedded of variable depth d is considered in the current section, see Figure 27a. The half-plane is made of orthotropic material with the following stiffness tensor components: $c_{11} = 1.57.10^6$ kPa, $c_{12} = 1.26.10^6$ kPa, $c_{22} = 4.17.10^6$ kPa, $c_{66} = 1.10^6$ kPa; the density is 1.57 t/m^3, taken from Chuhan et al. (2004). The foundation has modulus of elasticity $E = 3.10^7$ kPa (shear modulus $\mu = 1.25.10^7$ kPa), Poisson's ratio $v = 0.2$ and density $\rho = 2.5$ t/m^3. Constant material damping of 1% is assumed. The BEM mesh of the half-plane, comprises 72 quadratic boundary elements (BE), 36 of which are for modelling of the excavation contour. The discretized free surface is $5w/2$ measured left and right to the coordinate system origin.

Time domain responses due to the vertical incident P-wave for elastic transversely isotropic half-plane containing an embedded foundation is studied in this example. The time function is Ricker wavelet defined as: $f(t) = (2\pi^2 f_c^2 (t - t_0)^2 - 1) \exp(\pi^2 f_c^2 (t - t_0)^2)$ where, $f_c = 45$ Hz and $t_0 = 0.05$ s. The frequency range in the analysis is $\eta = [0; 1.5]$ divided into 200 equal frequency steps. Two representative observer points are considered with $x_1 = 0$ along the foundation surface, and $x_1/w = 1$ along the flat free surface, where w is the width of the foundation, see Figure 27. The normalized displacement components for representative observer points obtained for two different embedding foundation depths ($d = 4$ and $d = 5$) are portrayed in Figure 27 versus dimensionless time $\tau = t\sqrt{c_{66}/\rho_{hp}}/w$. The time signal is depicted in the graphics in order to visualize its amplification due the foundation-anisotropic soil interaction. The same set of results due to vertically incident SV-waves are presented in Figure 28. Figures 27 and 28 clearly illustrate the sensitivity of the site effects to the foundation's depth. For example, the percentage differences in solutions for the maximum values of normalized horizontal displacement at $d = 4$ m and $d = 5$ m (Figure 27 in the right plot) is 45.6, calculated with respect to their average values. Regarding vertical displacement components due to incident P-waves, the differences are significant at the foundation surface and less pronounced on the flat free surface. Similar conclusions, but in the opposite direction, can be drawn for the vertical and horizontal displacement components in the case of SV waves, as shown in Figure 28.

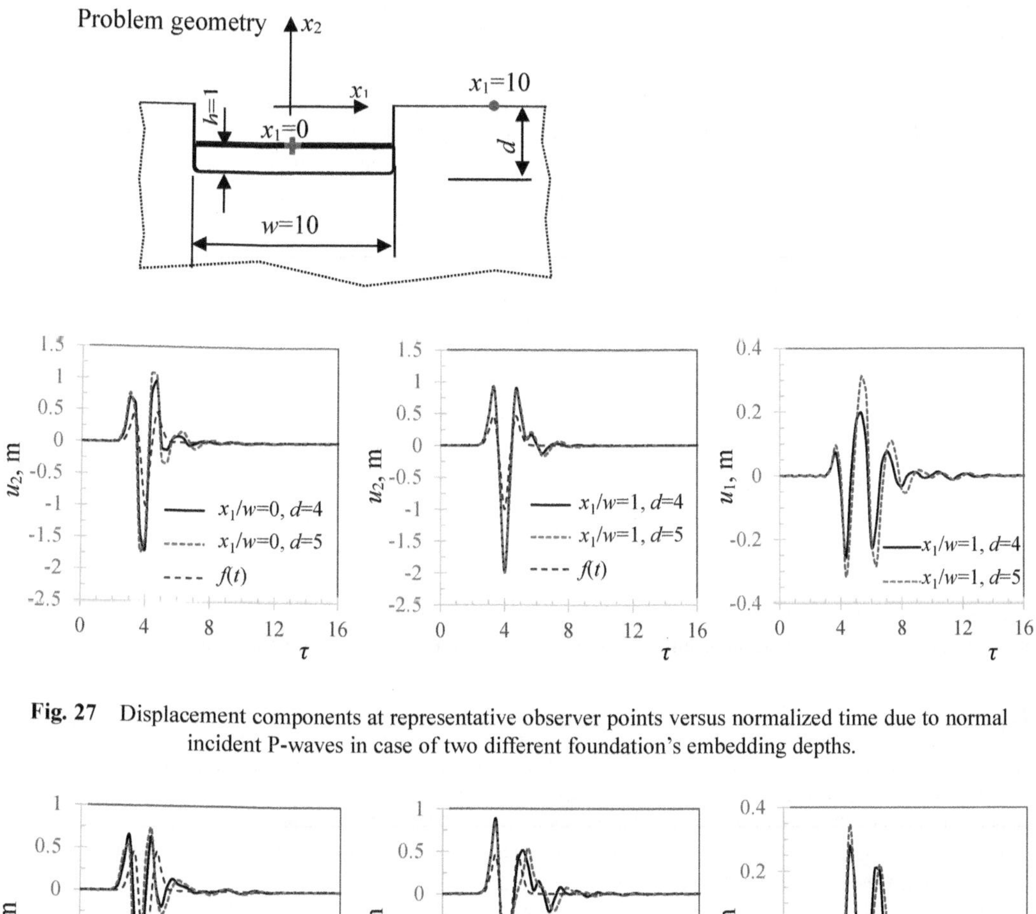

Fig. 27 Displacement components at representative observer points versus normalized time due to normal incident P-waves in case of two different foundation's embedding depths.

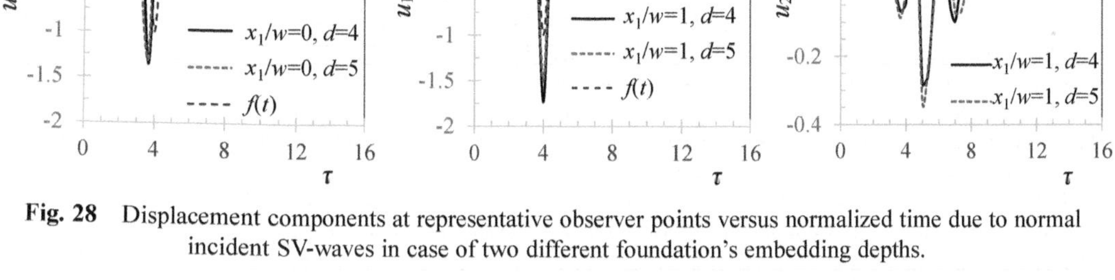

Fig. 28 Displacement components at representative observer points versus normalized time due to normal incident SV-waves in case of two different foundation's embedding depths.

6. Concluding Remarks

This work presents mechanical models for seismic wave propagation in an elastic isotropic or anisotropic, semi-infinite, homogeneous, or layered geological region with topographic relief of arbitrary geometry and containing embedded underground structures such as unlined/lined tunnels and foundations. The computational tool is the direct boundary integral equation method based on the analytically derived by Radon transforms fundamental solution for 2D general anisotropic continua. In addition, the sub-structuring approach is applied, which is particularly suitable for considering multiple heterogeneities (such as foundations, tunnels, layers, relief) as it leads to efficient numerical processing of integrals along the interface boundaries and improves the characteristics of the developed software. The aforementioned methodology, along with the authors' software based on it, has been verified. The software is used to compute complex wave fields comprising waves scattered and diffracted by surface relief, layers and underground structures as well as to evaluate the local dynamic stress concentration zones and local site effects developed in the geological region.

Soil-structure dynamic interaction is influenced by multiple factors, yet existing studies in the literature often consider only one or two of these factors in isolation. This has led to a shortage of mechanical models that comprehensively address complex interactions, such as those involving material anisotropy, unique soil topographies, and the presence of diverse engineering structures within a geological region. The proposed model aims to bridge this gap by integrating these multiple factors. The parametric study presented here demonstrates the sensitivity of the wave field to variations in topography, material anisotropy, and the intricate dynamics of soil–structure interactions. The specific influence of each parameter individually is difficult to isolate, due to the sheer complexity of the underlying boundary-value problem, which precludes the development of simplified mechanical models tailored for such purposes.

The results presented herein, which from a physical point of view are manifestations of elastic wave diffraction, local stress concentration and soil-structure interaction phenomena yield some useful insights on site effects in complex geological deposits, which in turn influence the response of buried and aboveground structures during earthquakes. Finally, these insights are crucial in expanding the range of current seismic design codes, which at present provide limited information on the role of geological complexities in the establishment of site-dependent response spectra. Investigation of the dynamic interaction of underground structures in a semi-infinite anisotropic soil media due to natural, technical or human-made excitations is a subject of fundamental interest in soil dynamics, civil, earthquake and geotechnical engineering, computational and solid mechanics.

References

Abate G. and Massimino, M.R. (2017). Parametric analysis of the seismic response of coupled tunnel-soil-aboveground building systems by numerical modelling. *Bull. Earthq. Eng. 15*: 469–491.

Achenbach, J.D. (1973). *Wave Propagation in Elastic Solids*. North-Holland, Amsterdam.

Alielahi, H. and Feizi, D. (2021). Numerical Study on Dynamic Effects of Soil-Tunnel-Structure Interaction. *International Journal of Civil Engineering, 19*: 1339–1355.

Alielahi H., Kamalian, M. and Adampira, M. (2015). Seismic ground amplification by unlined tunnels subjected to vertically propagating SV and P waves using BEM. *Soil Dynam. Earthq. Eng., 71*: 63–79.

Ba, Z., Lee, V.W., Liang, J. and Yan, Y. (2017). Scattering of plane qP- and qSV-waves by a canyon in a multi-layered transversely isotropic half-space. *Soil Dyn. Earthq. Eng., 98*: 120–140.

Ba, Z., Liang, J., Lee, V.W. and Kang, Z. (2018a). Dynamic impedance functions for a rigid strip footing resting on a multi-layered transversely isotropic saturated half-space. *Engineering Analysis with Boundary Elements, 86*: 31–44.

Ba, Z., Liang, J., Lee, V.W. and Hu, L. (2018b). IBEM for Impedance Functions of an Embedded Strip Foundation in a Multi-Layered Transversely Isotropic Half-Space. *J. Earthq. Eng., 22*: 1415–1446.

Ba, Z., Fu, J., Wang, F. and Wang, Y. (2022). Three-Dimensional Dynamic Response Analysis of Rigid Foundation Embedded in Layered Transversely Isotropic Half-Space. *Journal of Earthquake Engineering, 261*(6): 8611–8628.

Balendra, T., Thambiratnam, D., Kohs, C.G. and Lee, S. (1984). Dynamic response of twin circular tunnels due to incident SH-waves, *Earthquake Engineering and Structural Dynamics, 12*: 181–201.

Barros, P.L.A. (2006). Impedances of rigid cylindrical foundations embedded in transversely isotropic soils. *International Journal for Numerical and Analytical Methods in Geomechanics, 30*(7): 683–702.

Bobet, A. (2003). Effect of pore water pressure on tunnel support during static and seismic loading. *Tunn Undergr Space Technol., 18*: 377–393.

Boumaiza, D. and Aour, B. (2022). On the effectiveness of the iterative coupling FEM–BEM for the analysis of the problems with elastoplastic failure behavior. *Journal of Failure Analysis and Prevention, 22*(3): 1091–1106.

Celebi, E., Firat, S. and Cankaya, I. (2006). The Evaluation of Impedance Functions in the Analysis of Foundations Vibrations Using Boundary Element Method. *Applied Mathematics and Computation, 173*(1): 636–667.

Chen, X., Shen, J., Bao, X., Wu, X., Tang, W. and Cui, H. (2023). A review of seismic resilience of shield tunnels. *Tunn. Undergr. Space Technol., 136*: 105075.

Chuhan, Z., Yuntao, R., Pekau, O.A. and Feng, J. (2004). Time-domain boundary element method for underground structures in orthotropic media, *Journal of Engineering Mechanics, 130*(1): 105–116.

Dineva, P., Parvanova, S., Vasilev, G. and Wuttke, Fr. (2014a). BEM modeling of seismic soil-tunnels interaction. Seismic soil-tunnels interaction via BEM I. Mechanical model (Part I) *Journal of Theoretical and Applied Mechanics, 44*(3): 31–48.

Dineva, P., Parvanova, S., Vasilev, G. and Wuttke Fr. (2014b). Seismic soil-tunnels interaction via BEM II. Numerical results. (Part II). *Journal of Theoretical and Applied Mechanics, 44*(4): 29–50.

Dineva, P., Paskaleva, I., La Mura, C. and Panza, G. (2008). Hybrid MS-BIEM for seismic site response phenomena. The case study of Sofia. In: A. Zaicenco, I. Craifaleanu, I. Paskalev (Eds.), *Harmonization of Seismic Hazard in Vrancea Zone with Special Emphasis on Seismic Risk Reduction* (pp. 163–181). Springer edition ISBN 978-1-4020-9242-8(e book).

Dineva, P., Rangelov, T. and Gross, D. (2005). BIEM for 2D steady-state problems in cracked anisotropic materials. *Eng. Anal. Bound. Elem., 29*: 689–698.

Dineva, P., Vaccari, F. and Panza G. (2003). Hybrid Modal summation-BIE method for site effect estimation of a seismic region in a laterally varying media. *Theoretical and Applied Mechanics, 33*(4): 55–88.

Dineva, P., Wuttke, Fr.. and Manolis G. (2012). Elastic wave scattering and stress concentration effects in non-homogeneous poroelastic geological media with discontinuities. *Soil Dynamic and Earthquake Engineering, 41*: 102–118.

Dominguez, J. (1993). *Boundary Elements in Dynamics*. WIT Press/Computational Mechanics Publications, Southampton.

El Omari, A., Chourak, M., Cherif, S.E., Navarro Ugena, C., Echebba, E.M., Rougui, M. and Chaaraoui, A. (2021). Numerical modeling of twin tunnels under seismic loading using the finite difference method and finite element method. *Mater. Today Proc., 45*(8): 7566–7570.

Eskandari-Ghadi, M., Fallahi, M. and Ardeshir-Behrestaghi, A. (2010). Forced vertical vibration of rigid circular disc on a transversely isotropic half-space. *J. Eng. Mech., 136*: 913–922.

Eskandari-Ghadi, M., Nabizadeh, S.M. and Ardeshir-Behrestaghi, A. (2014). Vertical and horizontal vibrations of a rigid disc on a multilayered transversely isotropic half-space. *Soil Dyn Earthquake Eng., 61–62*(2): 135–139.

Fabozzi, S., Bilotta, E., Yu, H. and Yuan, Y. (2018). Effects of the asynchronism of ground motion on the longitudinal behaviour of a circular tunnel. *Tunnel. Undergr. Space Techn., 82*: 529–541.

Fang, X.Q., Ma, H.M., Zhu, C.S., Ding, Q.L., Zhu, Z.G. and Han, Z.M. (2023). Imperfect interface model and dynamic interaction mechanism around tunnels under seismic waves: A review. *Tunn. Undergr. Space Technol., 137*: 105120.

Freisinger, J., Radisic, M., Taddei, F. and Müller, G. (2022). Dynamic response of three-dimensional rigid and flexible foundations on layered soils with local inhomogeneities. *Soil Dynamics and Earthquake Engineering, 153*: 107007.

Garcia-Sanchez, F.G. (2005). *Numerical Study of Fracture Problems in Elastic Anisotropic and Piezoelectric Solids*. Dissertation, University of Sevilla.

Gazetas, G. (1981). Strip foundations on cross anisotropic soil layer subjected to static and dynamic loading. *Geotechnique, 31*(2): 161–179.

Gazetas G (1991) Formulas and Charts for Impedances of Surface and Embedded Foundations. *J. Geotech. Eng. ASCE, 117*: 1363–1381.

Guo, H., Chen, X., He, M., Xi, S. and Tang, J. (2010). Frictional contact algorithm study on the numerical simulation of large deformations in deep soft rock tunnels. *Min. Sci. Technol. China, 20*(4): 524–529.

Han, Z., Lim, G. and Li, J. (2015). Dynamic impedance functions for arbitrary-shaped rigid foundation embedded in anisotropic multilayered soil. *Journal of Engineering Mechanics, 141*(11): 04015045.

Hashash, Y.M.A., Dashti, S., Musgrove, M., Gillis, K., Walker, M., Ellison, K. and Basarah, Y.I. (2018). Influence of tall buildings on seismic response of shallow underground structures. *J. Geotech. Geoenviron. Eng., 144*(12): 04018097.

Karabalis, D.L. and Rizos, D.C. (1993). Dynamic Analysis of 3D Foundations. In: Manolis, G.D. and Beskos, D.E. (Eds.), *Boundary Element Techniques in Geomechanics* (pp. 177–208). London: Elsevier.

Karakostas, Ch. and Manolis, G. (2000). Dynamic response of unlined tunnels in soil with random properties. *Engineering Structures*, 22: 1013–1027.

Kausel, E., Waas, G. and Roësset, J.M. (1975). Dynamic analysis of footings on layered media. *Journal of the Engineering Mechanics Division, 101*(5): 679–693.

Kausel, E., Whitman, R.V., Morray, J.P. and Elsabee, F. (1978). The spring method for embedded foundations. *Nucl. Eng. Des., 48*: 377–392.

Kausel, E. and Roesset, J.M. (1981). Stiffness matrix for layered soil. *Bull. Seismol. Soc. Am., 72*: 1459–1481.

Kouretzis, G.P., Bouckovalas, G.D. and Gantes, C.J. (2006). 3-D shell analysis of cylindrical underground structures under seismic shear (S) wave action. *Soil Dynam. Earthq. Eng., 26*: 909–921.

Kuesel, T.R. (1969). Earthquake design criteria for subways. *J. Struct. Div. ASCE, 95*: 1213–1231.

Li, Z., Li, J. and Li, W. (2023). Evaluation of seismic response of tunnels near slope surfaces and performance of anti-seismic measures. *Soil Dynam. Earthq. Eng., 165*: 107734.

Liang, J.W. and Zhu, J. (2019). A FE-IBE method for linearized nonlinear soil-tunnel interaction in water saturated, poroelastic half-space: I. Methodology and numerical examples. *Soil Dynam. Earthq. Eng., 120*: 454–467.

Lin, H.T. and Tassoulas, J.L. (1987). A hybrid method for three-dimensional problems of dynamics of foundations. *Earthq. Eng. Struct. Dyn., 14*: 61–74.

Liu, Z., Liu, J., Pei, Q., Yu, H., Li, C. and Wu, C. (2021). Seismic response of tunnel near fault fracture zone under incident SV waves. *Undergr. Space, 6*(6): 695–708.

Luco, J.E. and Barros, C.P. (1994). Dynamic displacements and stresses in the vicinity of a cylindrical cavity embedded in a half-space. *Earthq. Eng. Struct. Dyn., 23*: 321–340.

Luco, J.E. and Westmann, R.A. (1971). Dynamic Response of Circular Footings. *Journal of Engineering Mechanics Division, 97*(4): 1381–1395.

Manolis, G.D. and Beskos, D.E. (1988). *Boundary Element Methods in Elastodynamics*. London: Allen and Unwin.

Manolis, G.D., Dineva, P.S., Rangelov, T.V. and Wuttke, F. (2017). Seismic Wave Propagation in Non-Homogeneous Elastic Media by Boundary Elements. In: *Solid Mechanics and Its Applications* (Vol. 240). Springer International Publishing.

Manolis, G., Makra, K., Dineva, P. *a*nd Rangelov, T. (2013). Seismic motions in a non-homogeneous soil deposit with tunnels by a hybrid computational technique. *Earthquake and Structures, 5*(2): 161–205.

Manolis, G.D., Parvanova, S.L., Makra, K. and Dineva, P.S. (2015) Seismic response of buried metro tunnels by a hybrid FDM-BEM approach. *Bull. Earthq. Eng., 13*: 1953–1977.

Manolis, G.D., Tetepoulidis, P.I., Talaslidis, D.G. and Apostolidis, G. (1995). Seismic analysis of buried pipeline in a 3D soil continuum. *Eng. Anal. Bound. Elem., 15*: 371–394.

Manolis, G., Makra, K., Dineva P. and Rangelov, T. (2013). Seismic motions in a non-homogeneous soil deposit with tunnels by a hybrid computational technique. *Earthquake and Structures, 5*(2): 161–205.

MATLAB. version 23.2.0.2409890 (R2023b) (2023) The MathWorks Inc, Natick, Massachusetts.

Mayoral, J.M. and Mosqueda, G. (2020). Seismic interaction of tunnel-building systems on soft clay. *Soil Dynamics and Earthquake Engineering, 139*: 106419.

Mita, A. and Luco J.E. (1989). Impedance Functions and Input Motions for Embedded Square Foundations. *Journal of Geotechnical Engineering ASCE, 115*: 491–503.

Monsees, J.E. (1991). Underground seismic design. In: *ASCE Structural Group Lecture Series* (pp 1–41). Boston Society of Civil Engineers Section.

Oliaei, M. and Basirat, R. (2022). The Effect of earthquake frequency content on the internal forces in the tunnel permanent lining. *J. Earthq. Eng., 26*(2): 817–836.

Pais, A. and Kausel, E. (1988). Approximate formulas for dynamic stiffnesses of rigid foundations. *Soil Dyn. Earthq. Eng., 7*: 213–227.

Panji, M. and Mojtabazadeh-Hasanlouei, S. (2021). Seismic ground response by twin lined tunnels with different cross sections. *SN Appl. Sci., 3*: 787.

Panza, G., Paskaleva, I., Dineva, P. *a*nd La Mura Cr. (2009). Earthquake site effects modelling by hybrid MS-BIEM: The case study of Sofia Bulgaria. *Rendiconti di Scienze Fisiche by the Accademia dei Lincei, 20*: 91–116. Springer.

Parvanova, S., Dineva, P., Manolis, G. and Wuttke, Fr. (2014a). Seismic response of lined tunnels in the half-plane with surface topography. *Bulletin of Earthquake Engineering, 12*(2): 981–1005.

Parvanova, S., Dineva, P. and Manolis, G. (2014b). Elastic wave fields in a half-plane with free surface relief, tunnels and multiple buried inclusions. *Acta Mechanica, 225*(7): 1843–1865.

Parvanova, S. and Dineva, P. (2022). Dynamic BEM analysis of elastic foundation on anisotropic half-plane. *Archive of Applied Mechanics, 92*: 2617–2636.

Parvanova, S. and Vasilev, G. (2016). An application of BEM-FEM hybrid technique for soil-structure interaction problems in time domain through ANSYS software. *Annual of UACEG*, Sofia, *49*(3): 201–219.

Patil, M., Choudhury, D., Ranjith, P.G. and Zhao, J. (2018). Behavior of shallow tunnel in soft soil under seismic conditions. *Tunn Undergr Space Technol., 82*: 30–38.

Pitilakis, K., Tsinidis, G., Leanza, A. and Maugeri, M. (2014). Seismic behavior of circular tunnels accounting for above ground structures interaction effects. *Soil Dyn Earthq. Eng.*, *67*: 1–15.

Qian, J. and Beskos, D. (1995). Dynamic interaction between 3-D rigid surface foundations and comparison with the ATC-3 provisions. *Earthq. Eng. Struct. Dyn.*, *24*: 419–437.

Sanchez-Merino, A.L., Fernandez-Saez, J. and Navarro, C. (2009) Simplified longitudinal seismic response of tunnel linings subjected to surface waves. *Soil Dynam. Earthq. Eng.*, *29*: 576–582.

Schanz, T., Wuttke, F. and Dineva, P. (2009). *Hybrid* Simulation of Seismic Wave Propagation in Laterally Inhomogeneous Media. In: T. Schanz and R. Iankov (Eds.), *Coupled Site and Soil-structure Interaction Effects with Application to Seismic Risk Mitigation,* (pp. 53–63). Springer edition, ISBN 978-90-481-2(e-book). https://link.springer.com/chapter/10.1007/978-90-481-2697-2_4.

Shahi, R. and Noorzad, A. (2011). Dynamic response of rigid foundations of arbitrary shape using half-space Green's function. *Int. J. Geo.Mech.*, *11*: 391–398.

Spyrakos, C.C. and Beskos, D.E. (1986). Dynamic Response of Rigid Strip-Foundations by a Time-Domain Boundary Element Method. *International Journal for Numerical Methods in Engineering, 23*(8): 1547–1565.

Spyrakos, C.C. and Xu, C. (2003). Dynamic Analysis of Flexible Massive Strip-Foundations Embedded in Layered Soils by Hybrid BEM-FEM. *Computers and Structures, 82*(29–30): 2541–2550.

Terzi, V.G. and Manolis, G.D. (2023). Basic numerical modelling issues in dynamic soil-tunnel interaction. *Soil Dynamics and Earthquake Engineering, 172*: 108060.

Tsinidis, G., De Silva, F., Anastasopoulos, I., Bilotta, E., Bobet, A., Hashash, Y.M.A., He, C., Kampas, G., Knappett, J., Madabhushi, G., Nikitas, N., Pitilakis, K., Silvestri, F., Viggiani, G. and Fuentes, R.. (2020). Seismic behaviour of tunnels: from experiments to analysis. *Tunn Undergr. Space Technol.*, *99*: 103334.

Vasilev, G., Parvanova, S., Dineva, P. and Wuttke, Fr. (2015). Soil-structure interaction using BEM–FEM coupling through ANSYS software package. *Soil Dynamics and Earthquake Engineering, 70*: 104–117.

Voigt, W. (1906). *Lehrbuch der kristallphysik.* Leipzig: Springer.

Wang, C.Y. and Achenbach, J.D. (1994) Elastodynamic fundamental solutions for anisotropic solids. *Geophys. J. Int.*, *118*: 384–392.

Wang, D., Feng, S., Li, J. and Sun, Y. (2013). Dynamic response of underground tunnel to internal explosion loading. *J. PLA University of Science and Technology*, Natural Science Edition, *14*(5): 511–516.

Wang, D., Wang, S., Zhang, H. and Liu, Z. (2022). Seismic response analysis of tunnel with cavity behind lining. *J. Vibroeng.*, *24*(2): 331–345.

Wang, H., Lou, M. and Zhang, R. (2017). Influence of presence of adjacent surface structure on seismic response of underground structure. *Soil Dyn. Earthq. Eng.*, *100*: 131–143.

Wang, H.F. (2018). Structure-Soil-Structure Interaction between Underground Structure and Surface Structure. Open access peer-reviewed chapter. doi: 10.5772/intechopen.76243.

Wang, J.N. (1993). *Seismic Design of Tunnels: A State-of-the-art Approach.* New York: Parsons Brickerhoff Quade and Douglas, Inc.; Monograph 7.

Wang, J.N. and Munfakh, G.A. (2001). Seismic design of tunnels. In: Brebbia, C.A. and Corz, A. (Eds.), *Earthquake Resistant Engineering Structures III* (pp. 589–598). Southampton: WIT Press.

Wang, Y. and Rajapakse, R.K.N.D. (1991). Dynamics of rigid strip foundations embedded in orthotropic elastic soils. *Earthquake Engineering and Structural Dynamics, 20*(10): 927–947.

Wolf, J.P. (1985). *Dynamic Soil-Structure Interaction.* Prentice Hall, Inc., Englewood Cliffs.

Wong, H.L. and Luco, J.E. (1978). Dynamic response of rectangular foundation to oblique incident seismic wave. *Earthquake Eng. Str. Dyn.*, *6*(1): 3–16.

Wu, Y.B., Liu, R.S. and Tian, Z.M. (2011). Analysis on dynamic response of a shield tunnel under internal explosion loading. *J. Wuhan Univ. Technol.* (*Transp Sci. Eng.*) *33*(7): 107–111.

Wuttke, Fr., Dineva, P. and Schanz T. (2011). Seismic wave propagation in laterally inhomogeneous geological region via a new hybrid approach. *Journal of Sound and Vibration, 330*: 664–684.

Yu, H. and Chen, G. (2021). Pseudo-static simplified analytical solution for seismic response of deep tunnels with arbitrary cross-section shapes. *Comput. Geotech.*, *137*: 104306.

Yu, H., Yuan, Y. and Bobet, A. (2017). Seismic analysis of long tunnels: A review of simplified and unified methods. *Undergr. Space*, *2*(2):73–87.

Yu, H., Yuan, Y., Qiao, Z., Gu, Y., Yang, Z. and Li, X. (2013). Seismic analysis of a long tunnel based on multi-scale method. *Eng. Struct.*, *49*: 572–587.

Zahariew, E. and Dineva, P. (2019). Seismic Response of Soil-Structure Systems via BIEM and FEM in Absolute Coordinates. Chapter of the Book, In: Eds: E. Zahariev and J. Cuadrado (Eds.), *IUTAM Symposium on Intelligent Multibody Systems – Dynamics, Control, Simulation* (Vol. 33, pp. 175–203). Springer, ISSN 1875-3507, ISBN 978-3-030-00526-9.

CHAPTER

4

Selection of Proper Signals for Nonlinear Time Histories of Structures

In The Frame of NDSHA

Paolo Rugarli[1*]

This work is related to the appropriate choice of accelerograms to be used for nonlinear time history seismic analyses. Starting from 400 events simulated by means of NDSHA, and all identical in magnitude, epicentral distance, depth, strike, dip and rake, site, soil, and structure, it is shown that the effects of the 400 × 3 signals are completely different and that it is necessary to examine a large number of them to be able to have an idea of the possible structural effects. It is demonstrated, with 400 Nonlinear Time History analyses, that using the elastic energy stored by the dissipative elements, through linear simulations, very fast, is a good way to order the events from the worst to the best. This also guides in the choice of signals to be used for nonlinear checks. Important considerations are then made on the correct way to set the problem of seismic checks, and on the goodness, relevance and reliability of the current regulations, and on the possible future practice of seismic analyses.

1. Introduction and Aim of the Work

In recent decades, many new theoretical, numerical, and technological developments have occurred, which have led to profound changes, still ongoing, in the way earthquake engineering was and is practised.

Both the numerical techniques used for analysis and, substantially, the definition of seismic design loads have changed and are changing.

1.1 Numerical Analysis Techniques

As regards numerical analysis techniques, roughly speaking, we first moved from linear "equivalent static" analyses to response spectrum modal analyses. Then from response spectrum modal analyses to nonlinear time history analyses, with a deviation (not very fruitful nor rigorous) constituted by the nonlinear equivalent static analysis, or the so-called *pushover*.

At the centre of most of the numerical analyses carried out in the last 30–40 years (equivalent static, modal and response spectrum, and pushover) has been the response spectrum analysis.

[1] Castalia srl, Via XX Settembre, 39 – Lodi – Italy. https://steelchecks.com/rugarli.asp, manuscripts received on 23rd of March, 2025

* Corresponding author: paolo.rugarli@castaliaweb.com

DOI: 10.1201/9781003684121-5

This uses the "response spectrum" as a fundamental description of the seismic loads, i.e., the quantification of the maximum expected response of single oscillators to seismic actions.

Through the decomposition of the motion into independent eigenmodes, obtainable in a linear context and with certain (heavy) hypotheses on damping, the response of the structure was then quantified by superimposing, in some way, the asynchronous and non-specific peaks of the modal responses. Here "non specific" means "not referring to the same event".

This was initially done with the SRSS (square root of the sum of the squares) technique, later with the more advanced CQC (complete quadratic combination), which considers the frequency differences between modes in order to choose whether to couple them or not. This is done dosing the mixed products of the effects from 0 (decoupling) to 2 (full coupling) as a function of the distances in the periods: 2 for close periods, 0 for distant periods, intermediate values from 0 to 2 for the others.

All response spectrum analyses (RSA), however, are far from having a complete physical meaning, for at least these reasons:

(i) Considering the response spectrum of a single signal, the peaks of the modes are reached at different instants. The combination of the peaks with gross methodologies that somehow take into account the non-simultaneity (SRSS, CQC), leads to the addition of effects that are not physically contemporaneous, and to obtain responses that violate the equilibrium (due to the square root operation performed on the sum).

(ii) Considering the standardized spectrum, or also the envelope spectrum, the spectral values referring to different periods refer to different events. The combination of the peaks is so not related to any specific event.

(iii) The response spectra of the standards are not true response spectra of true signals, but rather conventional envelopes in which the spectral values associated with the various periods are obtained in general with different signals, or rather by averaging many different signals.

(iv) The shape of the spectra is entirely determined by simplified formulae, and regularized (e.g., the peak is at 2.5 times the PGA), but this is not experimentally verified by the spectra of the accelerograms of real earthquakes.

(v) The spectral ordinates are linearly variable with the PGA (they are "anchored" to it), but, as we will see, this has no scientific basis nor any experimental confirmation.

(vi) For the purposes of the design, the elastic spectral ordinates are reduced by division by a factor called "behaviour", q (R in US literature). This fortuitously and flat-rately takes into account the dissipations that occur in the structure and in the non-structural parts, due to their inelastic nature. The evaluation of this divisor (which can have values from 1 to about 5 substantially modifying the severity of the elastic spectra), occurs in a very schematic way and linked to the structural typologies, not following a rigorous calculation.

As for the pushover analysis (PA), if it is true that it removes the linear limit, it does so with force distributions calculated using a mode of oscillation that is assumed to be dominant. In the absence of a dominant mode, the errors are large, and the techniques attempted to remove the problem appear neither reliable nor convincing.

A significant step forward with respect to RSA is represented by the Linear Time History analysis, which, in the case in which the natural vibration modes of the structure are available following a modal analysis (MA), specializes in the Modal Time History analysis (THM).

Here the problem of the contemporaneity of the peaks is overcome, but the strong limitation linked to the lack of linearity of the real structures subjected to strong shaking remains.

This is due to nonlinearity of material (plasticization and breakages, asymmetric traction/compression behaviours) and to nonlinearity linked to instability phenomena (the so-called second order effects), or also, to contact nonlinearity (one-side constraints).

Furthermore, as the oscillations progress, and with the repetition of plastic loading and unloading cycles, the properties of the materials degrade, so that this further effect implies step-by-step analysis to follow the evolution of the degradation.

Besides, the actions for the TH are described in a much more detailed and complex way, by means of *signals*, that is, accelerograms applied to the foundation soil, to *the structure* in three proper directions, X, Y, and Z.

How many signals, and *which* signals is an open problem, which this work deals with.

In recent decades, the need to perform nonlinear time history analyses, THNL, has become increasingly important. Despite the limitations of the engineering of the constitutive laws, the knowledge of the structural details, and the actual relevance of the accelerograms used, these analyses allow us to take numerous steps forward compared to all the techniques listed above.

Currently open, and particularly important, research fronts are those related to the study of constitutive laws with degradation, to the study of isolation and damping devices, to the study of the relevant signals to be used.

1.2 Seismic Design Loads

In this overview, only a quick reference has been made to the relevance of the seismic design loads. How to design structures for seismic actions? How to take the earthquake into account, that is, in what form and with what criteria to evaluate its effects in the form of actions to be applied to the structure?

In this regard, studies on seismic hazard (SHA) clearly play a central role. And here too, as with regard to numerical analysis techniques, in recent decades there have been, objectively, profound changes.

The probabilistic approach (PSHA), introduced in the late 1960s, has shown considerable theoretical limitations and violations, and has been systematically falsified by earthquakes that actually occurred in various parts of the globe[1].

In the meantime, the deterministic approach (DSHA) has been effectively replaced by the neodeterministic one (NDSHA) that correctly takes into account the tensor nature of the mechanical problem, avoiding the use of attenuation relations (for an extensive treatment of the topic with a historical excursus and a vast list of references, see Panza and Bela (2020)).

In essence, what has happened is that the earthquake has been quantitatively described and simulated as a phenomenon of an exquisitely mechanical nature, appropriately considering the three mechanical systems (MS) that determine its effects:

- MS1: the fault and the kinematics of its rupture.
- MS2: the Earth, which acts as a medium for the propagation of waves from the source to the bedrock of the site. To be precise, the part of the crust below the ground is involved in the wave propagation from the source to the bedrock of the site.
- MS3: the soil, which filters the signal to the bedrock and brings it to the surface.

It was thus possible to generate credible, relevant, and mechanically rigorous signals (accelerograms), simulating what actually happens during an earthquake.

Given the uncertain nature of the kinematics of the "next" rupture, and also given the uncertainty about the mobilized energy (i.e., the magnitude), simulations typically produce *hundreds* of possible simulated events for a given fault. This occurs even for the same magnitude.

This objectively poses a problem: which and how many simulated events should be chosen as input for the THNL seismic analyses? This is a crucial problem as it is not possible to run the THNL analyses for *all* the simulated events.

With regard to magnitude, recent interdisciplinary studies (geophysicists and engineers) have highlighted the perfect adherence of the NDSHA procedures historically followed, with the

1 In reality, PSHA is intrinsically unfalsifiable, because each event has a certain probability of occurrence, not zero. The falsification consists in the fact that the seismic design loads indicated by the regulations that use PSHA as those to be used for the design of buildings, have been greatly exceeded during real seismic events.

procedures provided by the Eurocodes for the treatment of the loads acting on buildings (snow, wind, live loads, etc.).

In particular (Rugarli et al., 2019), it has been shown that amplifying the seismic moment by a certain factor γ_q is equivalent to increasing the reference magnitude by a value ΔM in relation to the previous factor γ_q ($\Delta M = 2/3\ \mathrm{Log}_{10}\ (\gamma_q)$). It has therefore been possible to formulate a general relationship, now known as the "Panza-Rugarli law" (Wen & Wang, 2024), which allows the magnitude of the designed earthquake to be defined in a theoretically solid way. In summary[2]:

$$M_{\mathrm{design}} = M_{\max} + \gamma_{EM}\,\sigma_M = M_{\max} + 0.7 \quad \text{...(1)}$$

Where:

- M_{design} is the magnitude of the design earthquake;
- $M_{\max}$ is the maximum credible magnitude for the fault examined (MCRE, maximum credible earthquake Zhang et al., 2025), estimable from catalogues or by using seismogenic nodes;
- γ_{EM} is a safety factor on the earthquake magnitude;
- σ_M is the typical error value in the instrumental evaluation of the magnitude (0.25).

After the second equal, we have the currently proposed value for $\gamma_{EM}\,\sigma_M$, that is 0.7.

Once a certain fault has been identified, that can give rise, for the site under examination, to significant seismic events, and the design magnitude M_{design} has been set by means of (1), physically relevant signals can be generated, that are in agreement with all three mechanical systems mentioned previously: MS1, MS2, and MS3.

We now need to understand what effects the waves thus simulated have on the fourth mechanical system in series with the other three: the structure, MS4.

2. Engineering Considerations About the Use of Signals *Outside* the NDSHA Frame: Preliminary

THNL is not only used with signals generated by means of NDSHA. The use of accelerograms for THNL is regulated in Europe by Eurocode 8 (CEN 2005), which prescribes the use of at least three signal triplets, three simulated events.

Regarding the signals to be used, Eurocode 8 establishes that:

(i) The signals used must be "adequately qualified with regard to the seismogenic features of the sources and the soil conditions appropriate to the site".
(ii) That their ordinates are scaled so that their PGA (peak ground acceleration), a_g^*, respects the value stated by the legislation for the site in question, taking into account the site effects ($a_g S$).
(iii) That the average PGA of the signals used is not less than the value prescribed for the PGA at the site, amplified by site effects, provided by the standard for the soil in question.
(iv) That the average response spectrum at 5% damping of the signals used in a certain direction, is not less than 90% of the value of the standard spectrum with 5% damping. This must occur in the period range between $0.2T$–$2T$, where T *is* the fundamental period of the structure in the direction where the accelerogram will be applied.

These prescriptions say nothing about compliance with the mechanical systems MS1 (the fault) and MS2 (the deep earth-strata), which are instead typical of the source-site pair and which change from pair to pair. They only refer to the mechanical system MS3 (the soil), which, however, is not explicitly modelled but only identified by qualitative analogy.

The mechanical system MS2 is almost not mentioned ("generated through a physical simulation of source and travel path mechanisms"), almost as if it were irrelevant, while for the MS1 the Code simply asks for "adequate seismogenic features".

[2] This relation, first part, was first proposed in (Rugarli et al., 2018) and later better explained and motivated in (Rugarli et al., 2019).

In practice, this translates into the choice of natural signals taken from the catalogues, which refer to events that have a similar magnitude, epicentral distance, hypo central depth and similar soils compared to what is expected for the site in question, and which have spectral ordinates similar to those forced by the Code or made so by virtue of a scaling.

Therefore, the signals are taken from places completely different from the one being examined (for which there are no recordings). Furthermore they are scaled (up or down) to ensure that their response spectrum does not deviate too much (no more than 10% down) from that rigidly set by the regulations and anchored to the reference PGA.

The reference PGA a_g, imposed by the norm, thus indirectly comes to assume a pivotal role, since it governs the entire spectrum. That is, the accelerograms are forced to respect the probabilistic and simplifying assumptions of the norm.

In fact, the PGA, in the PSHA methodology, is evaluated with "probabilistic" considerations on the basis of a "nominal life" or a "reference life" V_r (in years) and a "probability of exceedance" P set by the legislator.

In current practice, 50 years and 10% are normally used for residential buildings, or 100 years and 5% for strategic buildings.

The "return period" is substantially evaluated (unless substantially irrelevant, or numerically negligible, *make-up) as* V_r/P, or (approximately) 500 years for normal buildings (475). This is equivalent to saying that the six on the dice comes up every six throws (1/(1/6)). Here the dice has 475 faces and the event having "probability" 1/475 occurs on average once every 475 years. According to the probabilistic belief (Rugarli, 2015a, 2015b).

This is what is currently done by engineers at the time of writing.

3. Engineering Considerations about NDSHA Signals

The signals generated with the neodeterministic techniques developed by Prof. Panza's group have the following fundamental merits:

- They represent in a physically correct way the mechanical scenario that gives rise to the shocks.
- They correctly take into account all three mechanical systems that, in the series, precede the structure, MS1, MS2, and MS3. Precisely those associated with the site, without using analogies or unmotivated comparisons. These systems are studied and their mechanical characteristics are pertinently modelled.
- They, therefore, correctly take into account the signal filtering action carried out by the mechanical systems in question.
- They allow us to rigorously consider the attenuation effects that the wave train experiences from the source to the site.
- They are all simulations in which the final effects are mechanically in direct and rigorous connection with the initial event, without introducing "probabilities", "return periods", or other artifices.
- Through (1) the magnitude is evaluated in a way that is perfectly consistent with what is done with other actions in structural engineering (Rugarli et al., 2019).

However, since our knowledge of the mechanical system MS1, i.e., the fault and its failure mode, is limited, each possible scenario (i.e., fault + site) gives rise to hundreds of different possible simulated events, even with the same magnitude.

Therefore, we find ourselves with the need to treat in a mechanically rigorous way the effects on the structure, of hundreds of events and thousands of signals, especially if, as often happens, the faults potentially involved are more than one, giving rise to different possible scenarios, all relevant for the purposes of seismic verifications.

To this end, EDEM (Elastic Deformation Energy Maximization) method comes to fundamental aid. This is a pre-analysis method that was proposed in 2018 and that will be further tested and explored in this work (Rugarli et al., 2018).

4. The Elastic Deformation Energy Maximization (EDEM) for the Quantification of the Severity of a Simulated Event

4.1 Need for Selection of Relevant Simulated Events

Before describing how the method works, it is worth explaining right away that the consequences of the incoming signals on the structure, i.e., on the fourth mechanical system in series, MS4, strictly depend:

- on the spatial orientation of the *X* and *Y* axes of the structure, with respect to the West-East geographic axis;
- on the ways in which the structure oscillates in the elastic range, and therefore from its natural frequencies;
- on the degree of scattering of the participating mass, on the various vibration modes;
- on the nature of the structure and its elastic limits, which determine the stress levels corresponding to the first plasticization and the abandonment of the elastic range;
- on the capacity of the structure to deform without collapsing, and therefore from its ultimate deformations, in a word from its *ductility*.

It is, therefore, not strictly possible to speak in an absolute sense of the severity of a triplet of signals, *because such severity depends on who receives them*. The same triplet can give rise to very different shaking consequences on different structures (MS4).

This means that the four relevant mechanical systems *must all be treated together*, and that there is no point in trying to study the dangerousness of the first three, if the characteristics of the fourth are not known in detail.

If we consider a fairly normal TH, lasting even just 30 seconds, and with a step of 0.05 seconds, a number often used because it is sufficiently small but not too small, we obtain 600 load steps, or 600 load cases to examine, in the evolution of the time history.

Since a structure can contain hundreds or thousands of notable points (nodes) and hundreds or thousands of finite elements that simulate the construction members (beams, columns, braces, walls), each load case, an instant of the TH, corresponds to a huge amount of data. This obviously represents a problem, because this data must not only be stored, but also examined and *understood*.

The severity of a triplet of signals may imply certain maximum actions in one element at one instant, certain other maximum actions in another element at another instant, certain displacements of one node, certain others of another.

The problem we face is, therefore, the following:

- Given hundreds of simulated events, each of which is associated with a triplet of signals (accelerograms in X, Y and Z directions), how can we choose the most dangerous ones for the specific structure under consideration, without having to perform THNLs on all the triplets?

The issue is of capital importance for the following reasons:

- The computational time required to run THNLs on all hundreds of possible events is currently prohibitive, and will remain so for a long time to come.
- *Randomly* choosing a few signal triplets to use, puts you at risk of picking up non-significant signals. This will be clearly demonstrated in the following sections.
- The choice based on the maximum PGA is not precautionary (as will be seen below).
- The choice of signals that have their response spectrum particularly severe for the dominant modes of the structure makes sense, but it does not protect against the very frequent cases in which the structure has more than one significant mode. Furthermore, it cannot take into account the fact that the maximum effects on the structure can be achieved by adding effects that, in each direction, *are not the maximum*. In other words, in general, if EE_x, EE_y, and EE_z (*EE, earthquake effects*) are the effects on the structure of the components *x*, *y*, *z* of a simulated

event i, and i_x is the event that maximizes x direction response, i_y is the event that maximizes y response, i_{max} the event that maximizes the overall sum of the x, y, and z response, in general:

$$i_{max} \neq i_x \neq i_y \neq i_z \qquad ...(1)$$

- Since the structural behaviour is elastic and multimodal, *any technique that does not correctly take into account the superposition of modes during the TH cannot be considered safe from dangerous underestimations or useless overestimations.*
- In the nonlinear field, since the modal decomposition is not available, the TH must be followed step by step.

While it is true that all THNLs imply a machine time and disk space that is currently impractical, at least the machine time issue is removed if THMs are considered instead of THNLs.

Although linear modelling has the fundamental limitation of considering the structure infinitely capable of sustaining stresses, and therefore of not considering plasticity, breakages and equilibrium instability effects, it has the ability to estimate the effects of a triplet of signals, correctly considering all the oscillation modes of the structure, and their different phases during the oscillation.

It, therefore, seems logical to propose to select the accelerograms on the basis of the severity of the effects they cause on the structure under examination, examining it in the linear elastic field using THMs.

This, at least in theory, would imply running hundreds of THMs, with much lower computation times than those needed for THNLs, but still very expensive, and also with the problem of having to manage a monstrously high amount of data on disk.

As we will see shortly, both problems can be solved brilliantly.

4.2 Strain Energy and Its Role as a Demand Predictor

A valuable scalar indicator of the demand required to structure is its deformation energy. In physics, things are often very simple when viewed from an energetic point of view, which is scalar and not vectoral.

In the elastic field, the deformation energy is a state function, which means that if at a certain instant the displacements of a structure are known, its deformation energy at that instant can be uniquely determined.

In the context of the finite element method, which is the ubiquitous technique for structural analysis, if $\mathbf{u}(t)$ is the vector of nodal displacements at some instant t, and $\mathbf{K}$ (not a function of time) is the stiffness matrix of the structure, then the strain energy E, is

$$E(t) = (1/2)\, \mathbf{u}^{\mathbf{T}}\mathbf{K}\mathbf{u} \qquad ...(2)$$

Since E *is* a scalar and, during the TH, variable with time, to each simulated event s, it is possible to associate a curve $E_s = E_s(t)$, to which a maximum $E_{s,max}$ will correspond, at a certain instant for an event, s. Note that during the analysis time, the kinetic energy of the structure is exchanged with that of elastic deformation, while depending on the damping present, a part of the energy is dissipated, or converted into heat.

A THM requires a previous modal analysis, which therefore requires the decoupling of the dynamic behaviour. This limits the description of the damping, which, basically, can be modelled either by means of Rayleigh damping, or by means of approximations such as the Wilson-Penzien one (light damping approximation).

While kinetic energy does not in itself represent a problem for the structure (but the functionality of equipment can be compromised by strong accelerations), deformation energy is stored only at the cost of the onset of stresses and deformations in the material, and in particular in that which has the function of protecting against seismic actions.

Therefore, a greater deformation energy implies overall greater stresses and greater deformations, which, integrated on the sections, imply greater axial forces, elongations, bending moments, rotations, etc.

Obviously, a global measure of the deformation energy of all the structural elements may not be too significant of the real stress state of the structure, because in the presence of many structurally irrelevant elements, their deformation energy could be significant and mask that, added to the previous one, of the more relevant structural elements.

It is therefore quite natural to think that a more pertinent index of the real stress of the structure is obtained by monitoring either the deformation energy of only some particular elements, or, better still, the overall deformation energy of those structurally strategic elements, which are responsible for supporting seismic actions.

The definition of such elements is straightforward, if one considers that the most modern seismic designs make use of the concept of dissipation, and design specific structural elements with the aim of deforming in the plastic regime, dissipating energy: such elements are called *dissipative*.

Therefore, instead of monitoring the total strain energy E, it is more pertinent to monitor the strain energy of the dissipative elements alone, E_d (DEEDE, *Dissipative Elements Elastic Deformation Energy*). The strain energy stored by these elements is in direct mechanical connection with their stress, which, in turn, is in direct mechanical connection with the measured severity of the signal triplet for that specific structure.

It is observed that while for dissipative connecting truss elements the relevant deformation energy is the total one, for dissipative beam elements only the relevant deformation energy associated with the mere bending around (generally) the strong axis is relevant.

Obviously, a *THNL* would give different results, in the sense that the peaks of E or E_d measured with the THM *may not be, and in general will never be, reached.* With the presence of plasticity, the stiffness matrix changes, the instantaneous modes of oscillation change, and the dynamic evolution is different.

What makes sense to investigate, however, is how much the mere elastic deformation energy can be considered a reliable predictor of the severity of the effects that would be measured on the structure, if this were examined in the nonlinear field by means of a rigorous THNL.

4.3 Computationally Efficient Evaluation of E

Imagining to have a certain set of signal triplets, associated with an equal number N of simulated events (in this study: 400, 1,200 signals). It is of interest to determine for each simulated event s, the $E_{s,\max}$ and the $E_{d,s,\max}$. To do this it is not necessary to run N linear modal time histories (THM), but it is sufficient to observe that for the purposes of calculating E and E_d one can proceed as follows.

Given the validity of the modal decomposition in the linear field and with the indicated damping forms, it turns out that the vector of nodal displacements $\mathbf{u}(t)$ for a certain simulated event (the subscript s is omitted here), can be expressed as:

$$\mathbf{u}(t) = \sum \boldsymbol{\Psi}_\mathrm{i} \cdot [u_{ix}(t)\Gamma_{ix} + u_{iy}(t)\Gamma_{iy} + u_{iz}(t)\Gamma_{iz}] \qquad ...(3)$$

Where:

- i is the mode index, between 1 and m
- x, y, z are the reference axes for the structure
- $\boldsymbol{\Psi}_\mathrm{i}$ is the i-th modal form, a vector that does not depend on time t
- Γ_{ix} is the modal participation factor in the x-direction for mode i, a scalar
- Γ_{iy} is the modal participation factor in the y-direction for mode i, a scalar
- Γ_{iz} is the modal participation factor in the z-direction for mode i, a scalar

- $u_{ix}(t)$ is a scalar function of time t. This function of time is the response of a simple oscillator of unit mass having the proper period T_i of the mode i, to the synthetic accelerogram $a_{gx}(t)$ applied in the x direction.
- $u_{iy}(t)$ is a scalar function of time t. This function of time is the response of a simple oscillator of unit mass having period T_i of mode i, to the synthetic accelerogram $a_{gy}(t)$ applied in the y direction.
- $u_{iz}(t)$ is a scalar function of time t. This function of time is the response of a simple oscillator of unit mass having period T_i of mode i, to the synthetic accelerogram $a_{gz}(t)$ applied in the z direction.

If we *define* the function $f_i(t)$ as:

$$f_i(t) \equiv u_{ix}(t)\Gamma_{ix} + u_{iy}(t)\Gamma_{iy} + u_{iz}(t)\Gamma_{iz} \qquad ...(4)$$

At that time,

$$\mathbf{u}(t) = \sum_{i=1}^{m} \mathbf{\Psi_i} \cdot f_i(t) \qquad ...(5)$$

Then, the *total* strain energy of all the elements of the structure, (2), can be expressed as:

$$E(t) = \frac{1}{2}\left(\sum_{i=1}^{m} \mathbf{\Psi_i^T} \cdot f_i(t)\right) \cdot \mathbf{K} \cdot \left(\sum_{j=1}^{m} \mathbf{\Psi_j} \cdot f_j(t)\right) \qquad ...(6)$$

Remembering that the mixed eigenvectors are orthonormal, with respect to the stiffness matrix, the previous relation reduces to:

$$\frac{1}{2}\left(\sum_{i=1}^{m} \mathbf{\Psi_i^T} \cdot \mathbf{K} \cdot \mathbf{\Psi_i} \cdot f_i^2(t)\right) = \frac{1}{2}\left(\sum_{i=1}^{m} E_{ii} \cdot f_i^2(t)\right) \qquad ...(7)$$

where *pseudo modal energy* of mode i has been defined as:

$$E_{ii} = \mathbf{\Psi_i^T} \cdot \mathbf{K} \cdot \mathbf{\Psi_i}$$

From a computational point of view, it is easy and fast to calculate the whole $E(t)$ for a certain triplet of signals, corresponding to a simulated event. The calculation, to be fast and efficient, must be done in the following way:

Step 1 We calculate the 3 m functions of time $u_{ix}(t)$, $u_{iy}(t)$, $u_{iz}(t)$, applying 3 *m times the* Duhamel integral to the synthetic accelerograms corresponding to the simulated event examined. These are stored in memory.

Step 2 m scalars E_{ii} are calculated and saved in memory. They are calculated without forming the global stiffness matrix **K**, but rather by operating at element level: in fact, if n_{dof} is the number of degrees of freedom of the system (for example, 20 thousand) the assembled matrix **K**, even taking into account its sparsity, takes up a lot of memory space. To avoid this, we work at the level of the element's stiffness matrix, and not the global stiffness matrix. In fact, if e is an element index, n_e is the total number of elements, and $\mathbf{K_e}$ is the element's matrix exploded (assembled) to the size of the global stiffness matrix n_{dof}, $\mathbf{k_e}$ is the stiffness matrix of the same element with size equal to (12 × 12) for beam elements and (6 × 6) for truss elements, we have:

$$E_{ii} = \mathbf{\Psi_i^T} \cdot \mathbf{K} \cdot \mathbf{\Psi_i} = \mathbf{\Psi_i^T} \cdot \sum_{e=1}^{ne} \mathbf{K_e} \cdot \mathbf{\Psi_i} = \sum_{e=1}^{ne} \mathbf{\psi_{ie}^T k_e \psi_{ie}}$$

With $\boldsymbol{\psi}_{ie}$ subvector of $\boldsymbol{\Psi}_i$ of order 12 for the beam elements and 6 for the truss elements, containing only the modal displacements relevant to the nodes connected to the element.

Step 3 It is now easily calculated, for each discretized instant t_k,

$$E(t_k) = \frac{1}{2}\left(\sum_{i=1}^{m} E_{ii} \cdot f_i^2(t_k)\right)$$

4.4 Computationally Efficient Evaluation of E_d

For the evaluation of the energy of the dissipative elements only, or of a sub-part of elements, there are some additional computational problems. In fact, if it is true that:

$$\boldsymbol{\Psi}_i^{\mathbf{T}} \cdot \mathbf{K} \cdot \boldsymbol{\Psi}_i = 0$$

It is not generally true that:

$$\boldsymbol{\Psi}_i^{\mathbf{T}} \cdot \mathbf{K_e} \cdot \boldsymbol{\Psi}_j = 0$$

And not even that:

$$\boldsymbol{\psi}_{je}^{\mathbf{T}}\, \mathbf{k_e}\, \boldsymbol{\psi}_{ie} = 0$$

So, the previous steps 2 and 3 must be modified as follows, keeping in mind that n_{de} is the number of elements declared dissipative.

Step 2 For each instant t_k, the displacement vector is created:

$$\mathbf{u}(t_k) = \sum_{i=1}^{m} \boldsymbol{\Psi}_i \cdot f_i(t_k)$$

For this to be done effectively on all t_{ks}, the mode shapes must all be loaded into memory, and not read and reread from disk.

Step 3 The energy at a certain instant is calculated before moving on to the next instant, and stored.

The energy at a certain instant is:

$$E = 0.5\; \mathbf{u}(t_k)^{\mathbf{T}} \cdot \sum_{e=1}^{nde} \mathbf{K_e} \cdot \mathbf{u}(t_k) = \sum_{e=1}^{nde} \mathbf{u}_{\mathbf{e}}^{\mathbf{T}}(t_k)\mathbf{k_e}\, \mathbf{u_e}\,(t_k)$$

where, the summations are extended to dissipative elements only.

If the number of elements is high, the reconstruction of the element's stiffness matrix must be done too many times. Assuming 1,000 elements and 800 time steps, the stiffness matrix of a typical element should be reconstructed 800 thousand times. Much more effective, decisive, from a computational point of view, is to calculate the stiffness matrices of the involved elements (a part of the total) only once, and keep them in memory. This implies a severe reduction in computation times, but requires a certain memory space.

4.5 The EDEM Method

The EDEM method (Elastic Deformation Energy Maximization, Rugarli et al., [2018]), consists in rapidly examining hundreds of triplets of signals in order to extract, for each of them, the maximum total deformation energy, $E_{\max}$, and/or the maximum deformation energy of the dissipative elements only, $E_{d,\max}$.

For brevity, in the case where the method uses the *total energy* as an indicative scalar, it is called TEEDEM (Total Elements Elastic Deformation Energy Maximization), while in the case where the indicative scalar is the deformation energy absorbed by the dissipative elements alone, it takes the name of DEEDEM (Dissipative Elements Elastic Deformation Energy Maximization).

TEEDEM

N triplets of signals, each corresponding to a simulated event *s*, are examined using the methods described in the previous section. The maximum strain energy reached during the THM (i.e. the maximum of *E* as *t* varies) is considered representative of each simulated event.

This maximum $E_{s,\max}$ is stored and associated with the simulated event *s*. Once the function $E_{\max} = E_{max}(s)$ has been obtained, the simulated events (the triplets) are rearranged in decreasing order of *E*, so that the derivative $E'(s)$ is ≤ 0, and therefore the triplets of the simulated events are presented in such a way that they gradually involve decreasing maximum total strain energies.

Therefore, the most dangerous triplets will be first, while the least dangerous ones will be last. That is, if the signals (triplets) are ordered in this way, they are in decreasing order of expected severity for the structure under examination.

DEEDEM

It works in the same way as TEEDEM, but the number stored and compared is not the total energy, but only the maximum energy absorbed by the elements declared as dissipative.

EDEM (in its two possible modes) orders the simulated events rigorously using a mechanically very well-founded and solid energy criterion and with completely affordable computation times. For an example of computation times, see the next section.

However, it remains to be investigated whether and to what extent the sequence predicted by this quick and simplified method is then respected when examining the same structure pre-analysed with linear techniques *in a genuinely nonlinear field.*

To do this, a steel structure from a real project was taken, and after applying DEEDEM and sorting the signals, all the corresponding nonlinear analyses were performed, examining the results in the way that will be described in the following sections.

5. The Structural Case Studied

5.1 Introduction

In order to verify whether DEEDEM in particular is a valid way to establish the severity of a simulated event, and to choose the most severe ones in a large set, a steel structure related to an office building actually built in Italy about 20 years ago was considered. The structure has eccentric bracing (EBF) and is therefore intrinsically dissipative.

The structure was analysed in the nonlinear dynamic field, performing as many nonlinear time history analyses as the available signal triplets. At the end of the analysis cycle (400 nonlinear analyses), the results obtained from the THNL analyses were compared with those predicted by DEEDEM, in the following way.

In particular, the signal triplets were reordered following the evaluation, for each triplet, of the maximum strain energy of the dissipative elements, reached during a linear THM simulated using the techniques already illustrated in the previous section (DEEDEM).

Since each triplet can be associated with a single scalar, the maximum strain energy absorbed during the THM by the elements declared dissipative only (DEEDE), it was possible to re-sequence the triplets in order of decreasing maximum absorbed energy: the most severe triplets first, the least severe ones last, in strict order of decreasing maximum elastic strain energy.

Once this particular order for the triplets had been established, it was verified whether the results of the nonlinear analyses showed a similar trend, that is, whether the severity of the triplets, as assessed with the most rigorous nonlinear dynamic analyses, decreased, as expected, as the triplets advanced.

It should be noted that in both DEEDM and THNL analyses, the pairs of accelerograms provided by NDSHA in radial and transverse directions have been previously transformed, through simple vector composition calculations, to the X and Y direction of the structure.

5.2 The Structure and Its Main Data

The structure (Figure 1) has 5,174 nodes, 328 truss elements, 2,800 beam elements, and 4,050 membrane elements. The membrane elements were applied to the floors and by definition do not give any flexural contribution. Their material is concrete.

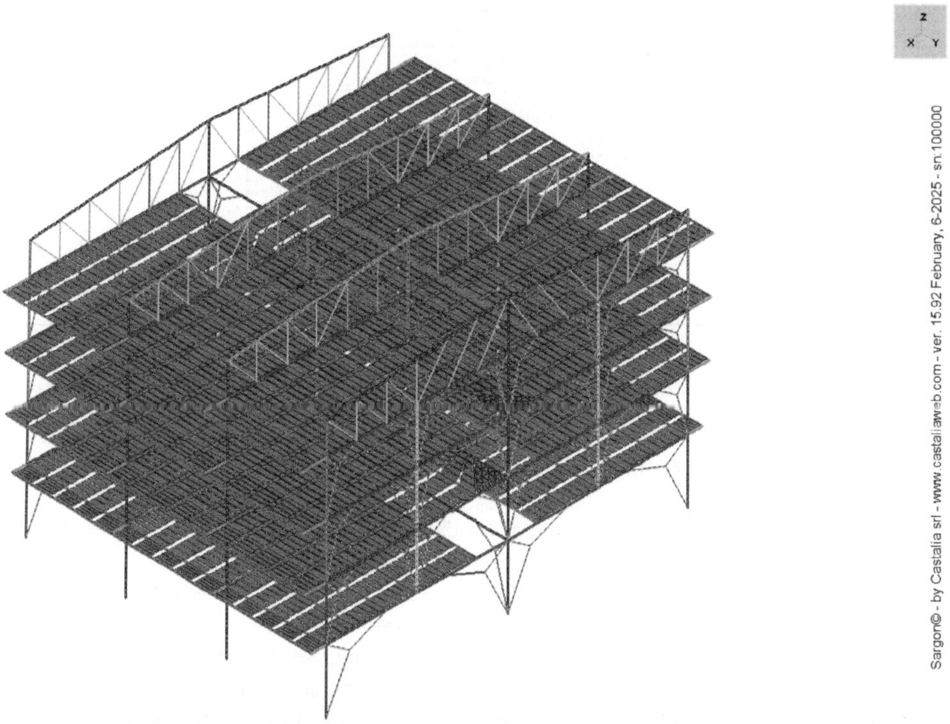

Fig. 1 The finite element model of the structure under consideration.

The membrane elements were kept elastic throughout the nonlinear analyses.

In the nonlinear field, the beam elements were divided into 5 Lobatto integration points, and 250 fibres, on average, on the section.

For columns, the typical length of the *beam element* is 940 mm, for beams 450 mm in the X direction and 500 mm in the Y direction.

The number of degrees of freedom of the finite element model is 20,484.

The total mass present on the model is equal to 1,817 tonnes, including the weight of the structure and the loads carried, as well as a fraction equal to 30% of the variable floor loads.

The plan dimensions are 27 m (X) by 22.5 m (Y). The elevations of the floors are 4.7 m, 8.7 m, 12.7 m, 16.7 m.

The structure supports horizontal loads by means of three bracing frames with eccentric braces and dissipative links, arranged in an H shape: 2 lateral ones in the X direction and one central one in the Y direction (Figure 2).

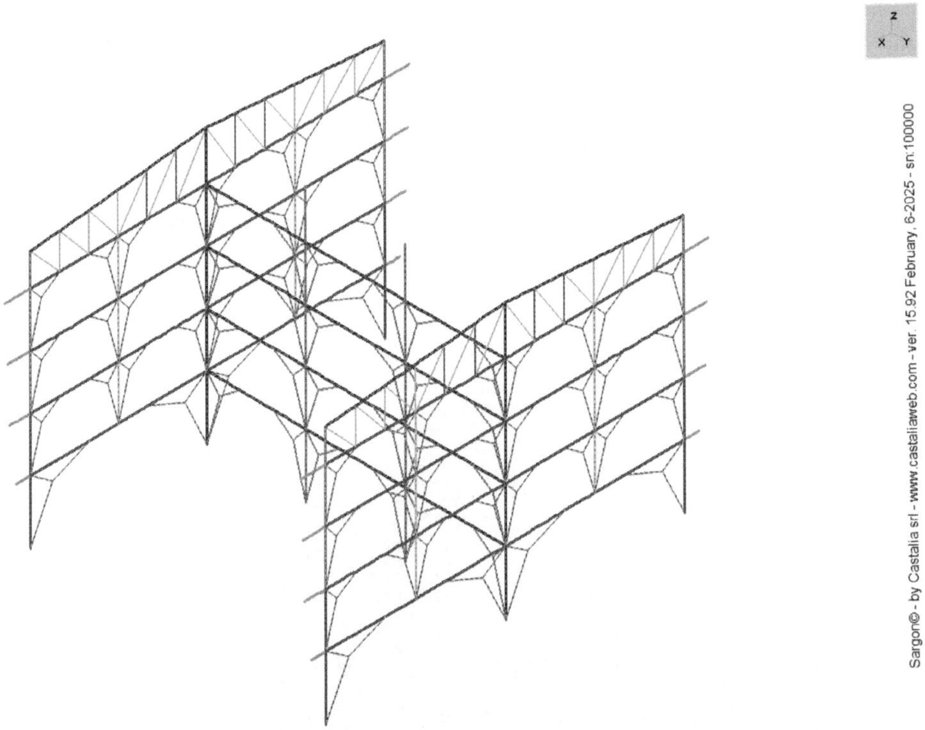

Fig. 2 The bracings of the structure under examination.

To ensure more space, a false column from the upper floor, which does not continue to the ground, is placed on the first level of the bracing beams in the X direction.

In this study, as "dissipative elements" have been considered all the beam finite elements being part of the beam members connected to the trusses in the 3 bracing frames, all the trusses in the bracing frames, and all the first column beam-elements connected to ground with clamped-nodes.

So, not only the "link elements" and the beam-to-column and beam-to-ground elements have been considered dissipative, but also the trusses of the bracing frames. This is because, given the great variability of the energy involved, the plasticization of the trusses could not be excluded *a priori*.

5.3 Modal Analysis

A modal analysis was performed by extracting 50 eigenmodes. The oscillation period ranges from 1.265 sec for the first mode to 0.147 sec for the 50th mode. The overall percentages of participating mass are 98.4% in the X-direction, 99.9% in the Y-direction, and 76.9 % in the Z-direction.

The structure presents the following main modes (participating mass greater than 1%):

- Mode 1, $T_1 = 1.265$ sec, 87.8% in Y direction
- Mode 3: $T_3 = 1.045$ sec, 93.9% in X direction
- Mode 8: $T_8 = 0.512$ sec, 4.8% in Y direction
- Mode 9: $T_9 = 0.496$ sec, 3.1% in Y direction
- Mode 13: $T_{13} = 0.349$ sec, 2.0 % in Y direction
- Mode 20: $T_{20} = 0.259$ sec, 47.3% in Z direction
- Mode 21: $T_{21} = 0.258$ sec, 2.2% in Z direction
- Mode 43: $T_{43} = 0.151$ sec, 23.7% in Z direction

The second mode, torsional, has a period equal to 1.224 sec, but has a zero percentage of participating mass in X and a very small one in Y (0.16%), since the masses have been modelled in the simplest way, without accidental eccentricity.

5.4 Inherent Damping

The "inherent" structural damping was modelled as Rayleigh damping, both in the EDEM method and in the nonlinear dynamic analyses, imposing a damping relative to the critical equal to 5% for the dominant mode in X (the third one T_3 = 1.045 sec), and for the dominant mode in Y (the first one, T_1 = 1.265 sec).

This choice implies modal damping ranging from a minimum of 4.98% for the second mode, to a maximum of 19.8% for the 50th mode.

5.5 Nonlinear Material Properties and Model

The material used for steel is S355 that has a yield strength of 355 MPa. The adopted constitutive model is elastic-perfectly plastic, with maximum strain equal to 0.02536 (2.53%) both positive and negative (in the elastic range such a strain would give rise to a stress of 210,000 × 2.53% = 5,313 MPa).

Beyond this deformation the fibre is declared broken and considered no longer loaded or active.

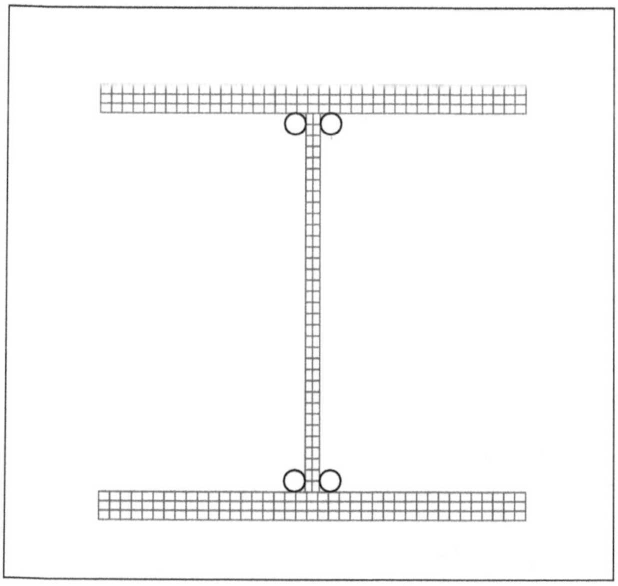

Fig. 3 A fibre-modelled section (HE300B).

The beam elements were subdivided by Lobatto points, 5 along the axis, and by fibres on the section, 250 on average (the exact number depends on the exact shape of the section, Figure 3).

Thus, on average, each beam element of length 450–500 mm for beams and 900 mm for columns, is simulated with 1,250 fibres having a length around 100–150 mm.

In the non linear field, each truss element is associated with a single fibre.

Geometric nonlinearity was not applied in the analyses.

The constitutive model of the material, in these simulations, does not embed degradation except for rupture, which occurs when the maximum ultimate deformation is reached.

Upon reaching a mechanism, the deformations grow indefinitely and cause the fibers to break, and consequently to inactivate the element (both beam and truss). This typically leads to the impossibility of convergence on the residual. Note that convergence on the displacement would allow convergence with unlimited displacements and accelerations, but this would not make any structural or engineering sense. The unlimited plastic excursion is not possible either in the model or in nature. It is therefore not true that the severity of the shaking, in the presence of structures designed to be dissipative, is of little importance[3].

Different results would obviously be obtained with higher (or lower) ultimate strains. Numerical experiments have shown, however, that collapse is reached even using very high values of ultimate strain (20%), taking some severe triplets as examples.

The choice of the elastic perfectly plastic model was made because it maximizes the redistributions, resulting in the substantial hinging of the beams. The redistributions, in turn, move the structure away from its last elastic configuration.

In other words, to evaluate the significance of an elastic energy method, it seemed useful to favour the maximum plastic redistribution and, therefore, the maximum separation of the nonlinear behaviour from the elastic one, which is obtained with the elastic-perfectly plastic model.

5.6 Nonlinear Time Histories Settings

The nonlinear analyses (400) were performed with the DYNCURAN program, written by the author of this work, and part of the Sargon© calculation program, developed since 1991 also by the author of this work.

The following settings were used:

- Newmark method with $\beta = 0.25$ $\gamma = 0.5$
- Full Newton-Raphson
- Time step 0.05 seconds.
- Rayleigh inherent damping as already indicated
- Mass concentrated at the nodes (lumped approach)
- SRSS Error Norm
- Tolerance on displacement error 0.001
- Residual error tolerance 0.001
- Maximum number of iterations 30
- Step width automatically reduced if no convergence
- Lobatto points for beam elements: 5
- Number of fibres on the average section: 250
- Material nonlinearity yes
- Geometrical nonlinearity no
- Tolerance for return to plastic locus 1e-5
- Maximum number of iterations to return to the plastic locus 100
- Duration: variable. Stopped until all accelerogram components remain less than 0.002 g up to the end of triplet data, for all directions (x, y, and z).

In the case of a substantially elastic response, with little plasticity, the typical analysis lasted about an hour and a quarter. In the presence of strong plasticization, and with analyses conducted to the end (absence of collapse) the single analysis can even last three hours.

To solve the 400 THNL analyses on all 400 simulated events took approximately 23 days of continuous calculations (from February, 22, to March, 17, 2025).

3 This statement has been made by some to detract from the invalidations of PSHA made by real earthquakes, arguing that the severity of the shaking is of relative importance where the structure is designed to be dissipative.

A computer with an Intel i7 13700 processor, 2.1Ghz clock, 32Gb RAM, Windows 11 64 bit Professional was used.

5.7 The Sets of Signals Used

Two sets of signals were used.

First set (S1)

The signals were provided by Prof. Fabio Romanelli and Dr. Franco Vaccari (Romanelli & Vaccari, 2025) and refer to the site called P2 in the work (Ricci et al., 2025). These are 400 simulated events (1,200 accelerograms).

All signals are simulations of events related to the exact same fault, the exact same magnitude (M_w 6.9), related to the exact same site, with the exact same soil. Therefore, all signals have the following seismic data:

Magnitude $M_w = 6.9$
Depth 13.1 km
Epicentral distance 29.91 Km
Sampling: 0.01sec
Duration: 49.9 sec
RA direction angle 11° from EAST to SOUTH
Strike: 101°
Dip: 25°
Rake: 90°

The source mechanisms and the kinematics of the failure change.

In particular:

- 100 simulations using a 0° bilateral model (abbreviation B0): from 1 to 100
- 100 simulations using a 90° bilateral model (abbreviation B90): from 101 to 200
- 100 simulations using a 90° unilateral model (abbreviation U90): from 201 to 300
- 100 simulations using a unilateral 180° model (abbreviation U180): from 301 to 400

The numbering from 1 to 400 indicated is the one originally received.

Second set (S2)

These are another 400 simulated events (another 1200 signals, Romanelli & Vaccari, 2025), identical to the previous ones for the seismogenic data and the site, with the only difference that the magnitude M_w is 6.2, that is, 0.7 less than the previous one (see set (1)).

6. Results in Terms of DEEDE, Set S1, No Resequencing of the Triplets

6.1 What Has Been Done

It was assumed that the building had an angle between its X-axis and the radial axis RA (strike), equal to: 0°, 30°, 45°, 60°, and 90°. For brevity, this angle between the strike axis (radial direction) and the X-axis of the building will be indicated with the symbol δ. In other words, a different orientation of the structure with respect to the strike direction was imagined, all other things being equal.

In each of these cases, a DEEDEM analysis was performed, always keeping the signal sequence unchanged.

In each of these cases, the vector transformation of the signals to the reference system of the structure implies different effects.

In particular, since the radial signals are stronger than the transverse ones in this **S1** set, an angle $\delta = 0°$ stresses more the structural direction X (zero angle between the RA direction and the X axis), and therefore the dominant mode in X, having a period of 1.045 sec.

Instead, when the direction between the X-axis of the structure and the RA-axis is equal to 90° the strongest signals go in the Y-direction for the structure, and therefore in the dominant mode in Y which has a period equal to 1.265 sec.

In other cases, there is a different distribution in the two directions.

Thus, as we will see, *the worst simulated events are expected to depend on how the structure is oriented with respect to the strike axis.*

From a computational point of view, a DEEDEM analysis on 400 triplets (1200 signals) took about 40 minutes of computation time, on this model, with the hardware and software described.

6.2 $\delta = 0$

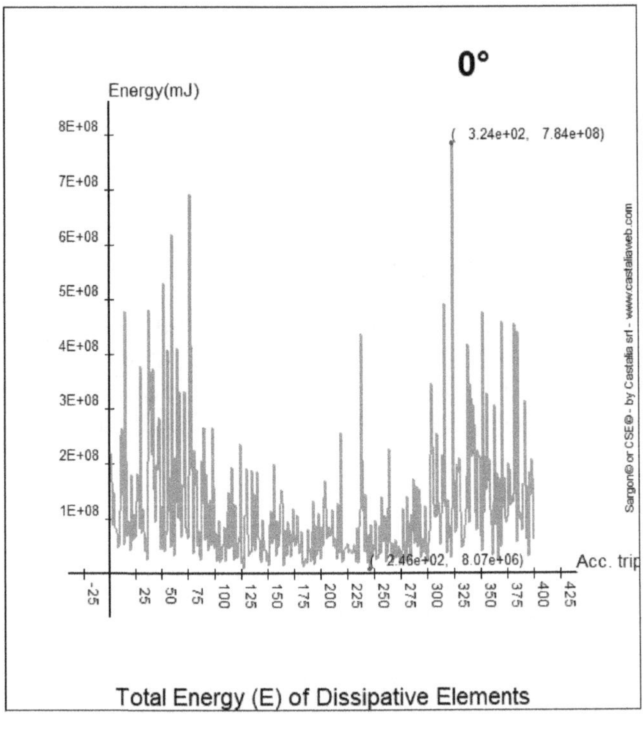

Fig. 4 Angle $\delta = 0°$. Maximum DEEDE results as the simulations vary. On the abscissa the simulation number, from 1 to 400, keeping the triplets in the same sequence in which they were received. On the ordinate the maximum elastic energy accumulated, during the event, by the dissipative elements alone (mJ, Nmm).

Comments

The maximum DEEDE is obtained from simulation 324, and gives an energy of 7.84 e8 mJ.

The minimum DEEDE is obtained from simulation 246, and gives an energy of 8.07 e6 mJ.

The multiplier that goes from the minimum to the maximum is equal to 97.

The B0 (1–100) and U180 (301–400) signals appear on average more severe than the B90 (101–200) and U90 (201–300) signals.

6.3 $\delta = 30°$

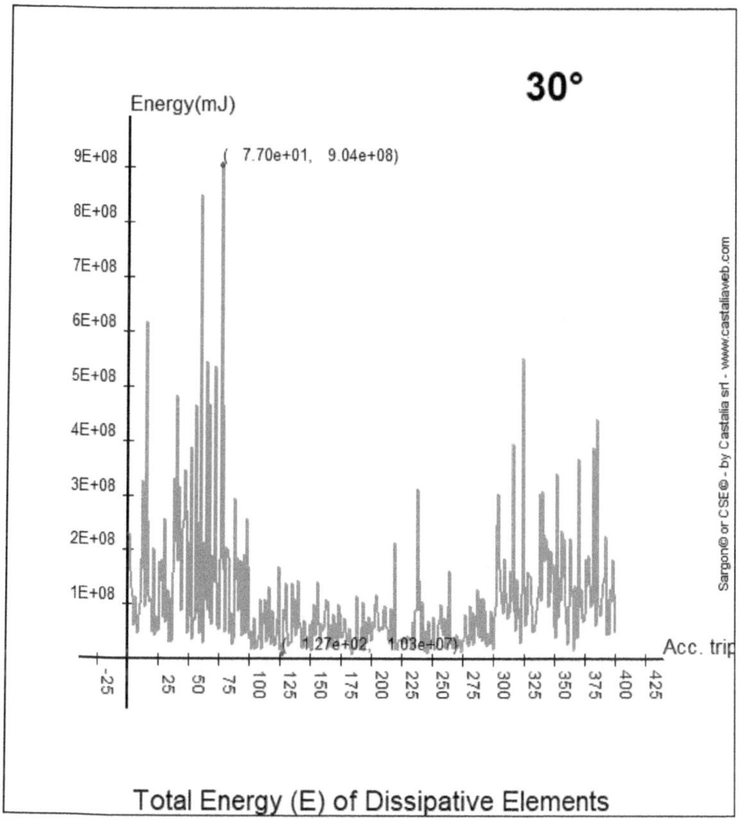

Fig. 5 Angle $\delta = 30°$. Maximum DEEDE results as the simulations vary. On the abscissa the simulation number, from 1 to 400, keeping the triplets in the same sequence in which they were received. On the ordinate the maximum elastic energy accumulated, during the event, by the dissipative elements alone (mJ, Nmm).

Comments

The maximum DEEDE is obtained from simulation 77, and gives an energy equal to 9.04 e8 mJ.

The minimum DEEDE is obtained from simulation 127, and gives an energy equal to 1.03 e7 mJ.

The multiplier that goes from the minimum to the maximum is equal to 88.

The B0 (1–100) and U180 (301–400) signals appear on average more severe than the B90 (101–200) and U90 (201–300) signals.

6.4 $\delta = 45°$

Comments

The maximum DEEDE is obtained from simulation 77, and gives an energy equal to 1.15 e9 mJ.

The minimum DEEDE is obtained from simulation 366, and gives an energy equal to 1.30 e7 mJ.

The multiplier that goes from the minimum to the maximum is equal to 88.

The B0 (1–100) and U180 (301–400) signals appear on average more severe than the B90 (101–200) and U90 (201–300) signals.

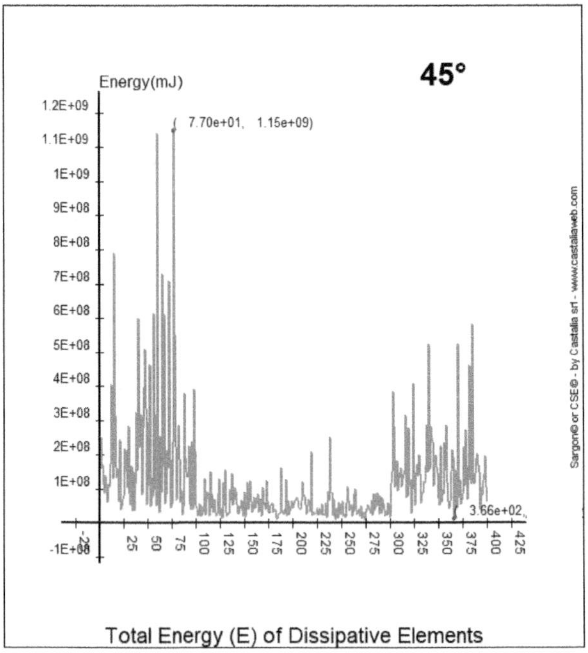

Fig. 6 Angle $\delta = 45°$. Maximum DEEDE results as the simulations vary. On the abscissa the simulation number, from 1 to 400, keeping the triplets in the same sequence in which they were received. On the ordinate the maximum elastic energy accumulated, during the event, by the dissipative elements alone (mJ, Nmm).

6.5 δ = 60°

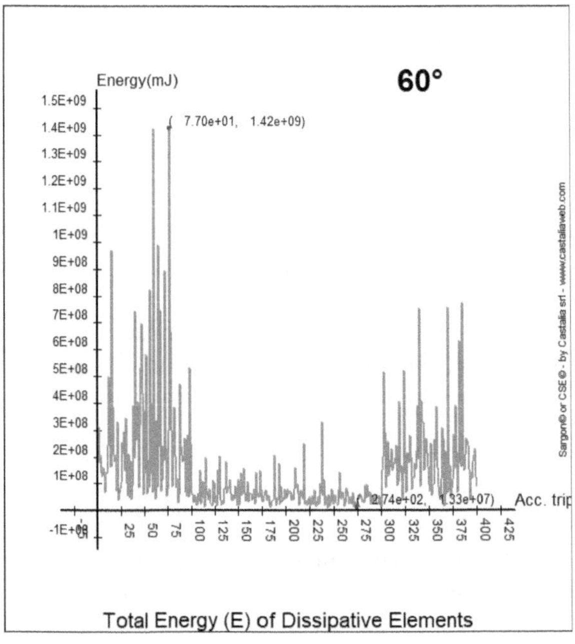

Fig. 7 Angle $\delta = 60°$. Maximum DEEDE results as the simulations vary. On the abscissa the simulation number, from 1 to 400, keeping the triplets in the same sequence in which they were received. On the ordinate the maximum elastic energy accumulated, during the event, by the dissipative elements alone (mJ, Nmm).

Comments

The maximum DEEDE is obtained from simulation 77, and gives an energy equal to 1.42 e9 mJ.

The minimum DEEDE is obtained from simulation 274, and gives an energy equal to 1.33 e7 mJ.

The multiplier that goes from the minimum to the maximum is equal to 107.

The B0 (1–100) and U180 (301–400) signals appear significantly more severe than the B90 (101–200) and U90 (201–300) signals.

6.6 δ = 90°

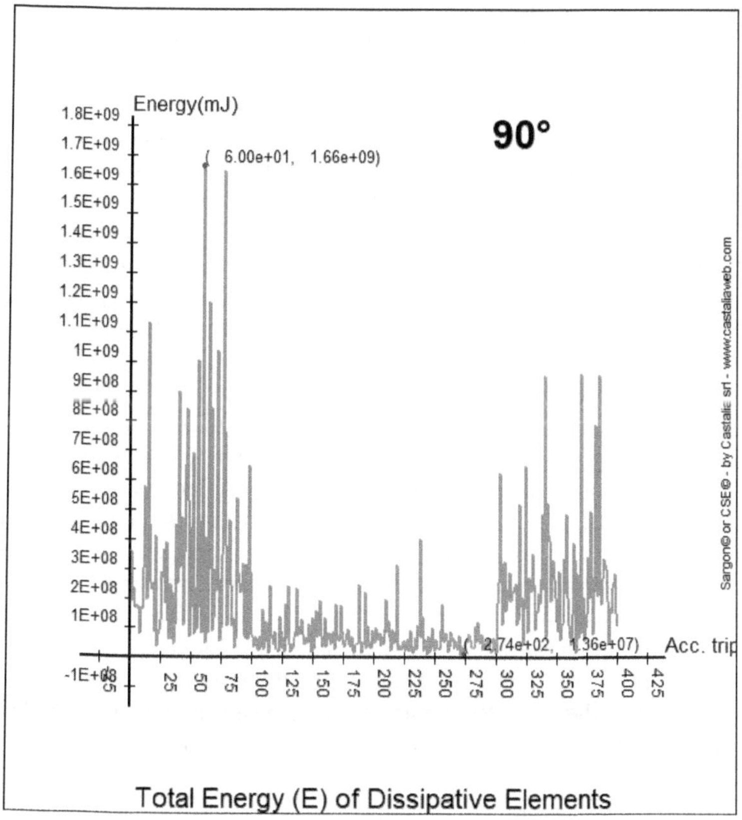

Fig. 8 Angle $\delta = 90°$. Maximum DEEDE results as the simulations vary. On the abscissa the simulation number, from 1 to 400, keeping the triplets in the same sequence in which they were received. On the ordinate the maximum elastic energy accumulated, during the event, by the dissipative elements alone (mJ, Nmm).

Comments

The maximum DEEDE is obtained from simulation 60, and gives an energy of 1.66 e9 mJ.

The minimum DEEDE is obtained from simulation 274, and gives an energy equal to 1.36 e7 mJ.

The multiplier that goes from the minimum to the maximum is equal to 122.

The B0 (1–100) and U180 (301–400) signals appear significantly more severe than the B90 (101–200) and U90 (201–300) signals.

6.7 Engineering Consequences of the EDE Plots

The examination of the diagrams showing how the maximum deformation energy in the dissipative elements varies, as the simulations vary and as the orientation of the building varies with respect to the strike direction, leads to the following important considerations:

(i) The correct orientation of the building with respect to the strike direction is essential to have reasonable estimates. It makes no sense to take the signals without considering their direction with respect to the building. This implies that in the simulations one must necessarily convert the signals to the correct X and Y direction for the structure.

(ii) The variation of maximum strain energy absorbed by the dissipative elements (DEEDE) for the different orientations of the building is very strong. It ranges from 7.84 e8 to 1.66 e9 mJ in this case. In this case, due to the difference in periods in the X and Y directions, and the greater significance of the signals in the radial direction, the maximum energy increases monotonically with the angle δ. The variation of energy is strong (multiplier 2).

(iii) Given the same orientation of the building, the 400 simulations produce substantially different effects, with about two orders of magnitude of difference. And this even if the magnitude, the epicentral distance, the depth of the epicentre and the site and the soil are identical. What is relevant is on the one hand the frequency content of the signal, on the other the phases of the component harmonics. Therefore, it would make no sense to randomly take 3, 7, or 10 triplets, and perform the THNL analyses on them, trusting to have enveloping actions simply because the magnitude or, as we will see, the PGA are high. Only by examining a sufficiently high number of simulated events can one have an idea of the range of possible consequences on the structure. In the absence of a sufficiently extensive analysis, one cannot have any idea about the *relative significance* of the signals examined, for the specific structure at hand.

(iv) The magnitude of an event is not sufficient, in and of itself, to characterize its effects because events of the same magnitude can produce substantially different effects. Only by examining a vast set of simulated events, and only by using the particular structure under examination as a detector, can one have an appropriate idea of the seismic actions.

(v) The same applies to the epicentral distance and the hypocentral depth.

(vi) Considering the shape of the diagrams, and keeping in mind that the original order of events was from 1 to 100 events associated with B0, from 101 to 200 those associated with B90, from 201 to 300 those associated with U90, from 301 to 400 those associated with U180, it is clear that in this case B0 is the most dangerous source mechanism, together with U180, while B90 and U90 have minor effects. Which mechanism is the worst depends on the orientation of the building.

7. Reordering of the Triplets by Decreasing Energy, for Angle $\delta = 45°$

In order to better understand whether EDEM is a good predictor of the real severity of a simulation in the nonlinear field, it is essential to reorder the 400 simulated events in order of decreasing strain energy (in the dissipative elements).

In this way, by ordering the simulated events, we have a criterion to establish what ordering the damage indices read on the structure should have for the 400 simulations in the nonlinear field.

$\delta = 45°$ was taken, that is with the X axis included exactly between the radial and transverse axes of the radiation with a point source.

By sorting the simulated events by decreasing strain energy, a new ordering is found for them, of course. With this ordering, the curve of maximum embedded energies, for the 400 simulations, has the form of Figure 9.

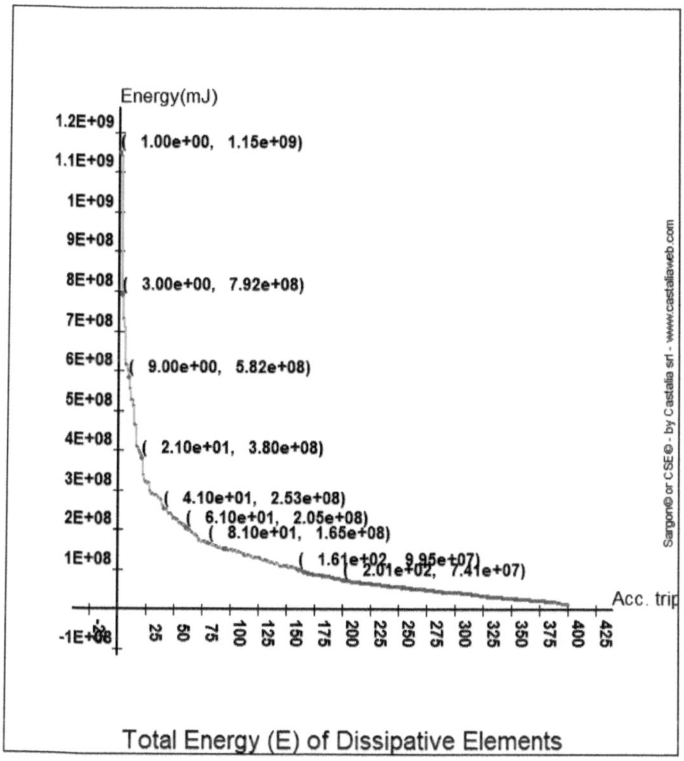

Fig. 9 Maximum elastic strain energy absorbed by the dissipative elements for the 400 simulated events rearranged by decreasing energy, angle $\delta = 45°$.

Of course, the data are the same as in Figure 6, but they are presented in a different order. The figure also shows the values corresponding to different percentiles, according to Table 1.

Table 1 Strain energy values in dissipative elements, for different percentiles.

Event Number	*Percentile*	*Energy Value*
1	100°	1.15e9
21	95°	3.80e8
41	90°	2.53e8
61	85°	2.05e8
81	80°	1.65e8
161	60°	9.95e7
201	50° (median)	7.41e7

If we order the simulations in decreasing energies, if x is the simulation number (between 1 and F where in this case $F = 400$), the percentile p associated with each simulation x is obtained as:

$$p = [1 - (x - 1)/F] \cdot 100 \quad ...(8)$$

The curve has a practically hyperbolic trend (a hyperbolic fitting has $R^2 = 0.966$, almost 1), which makes the huge variability of energy absorbed clearly evident, as the simulations vary.

Now maintaining *this same new sequential order for the simulated events*, the trend of the PGA for the X component as the signals vary is shown in Figure 10.

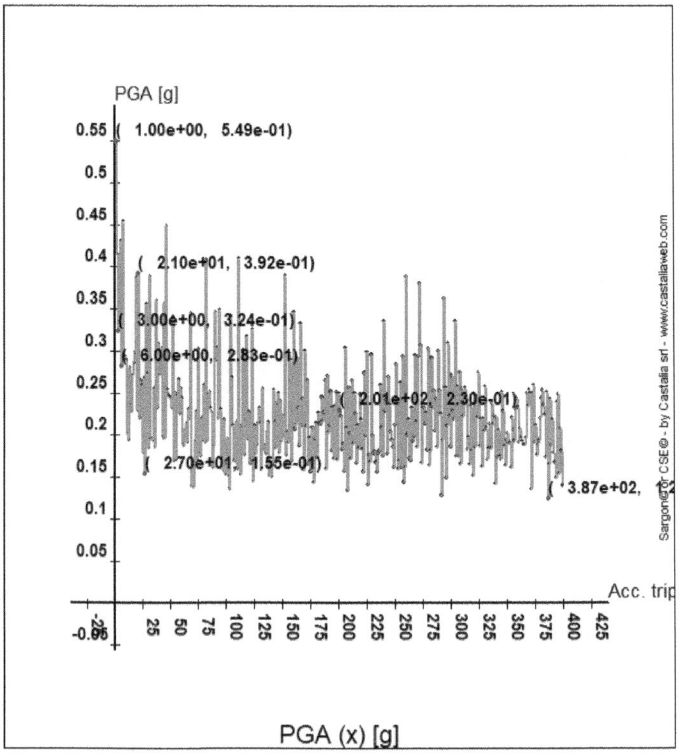

Fig. 10 Trend of the PGA of the signal in the X direction as the simulations vary, with the same sequence as in Figure 9. On the abscissa the simulation number, on the ordinate the PGAx in units g.

As can be seen, the curve does not show any significant regularity, clearly showing that there is no relationship between the PGA of a signal and the severity of the effects that signal will bring on the structure. The PGAy, not represented, has similar characteristics.

For example, we know from Figure 9 that simulation 3 implies an energy of 7.92 e8 mJ, we now see from Figure 10 that this simulation corresponds to a PGAx of 0.324g.

Simulation 21, which has a much lower energy, 3.80 e8 (less than half) results in a PGAx of 0.392g, much larger than 0.324.

In general, the violently irregular trend of the PGAx, compared to the very regular one of the energy, with the same ordering of the events, illustrates the fact that there is no direct relationship between the intensity of the PGA and that of the maximum energy absorbed by the structure. The fact that the maximum energy corresponds to the maximum PGA is to be considered, given the obtained curve, substantially random.

8. Comparison of the Energies with Re-ordered Signals for Sets S1 and S2, $\delta = 45°$

It seemed interesting to compare the decreasing energy curves obtained with different sets of signals.

The comparison is made between the energies associated with the first set of simulations S1, magnitude 6.9, and the energies associated with the second set of simulations S2, magnitude 6.2. In both cases, the simulations are sorted by decreasing strain energies.

The comparison, in Figure 11, leads to the following observations.

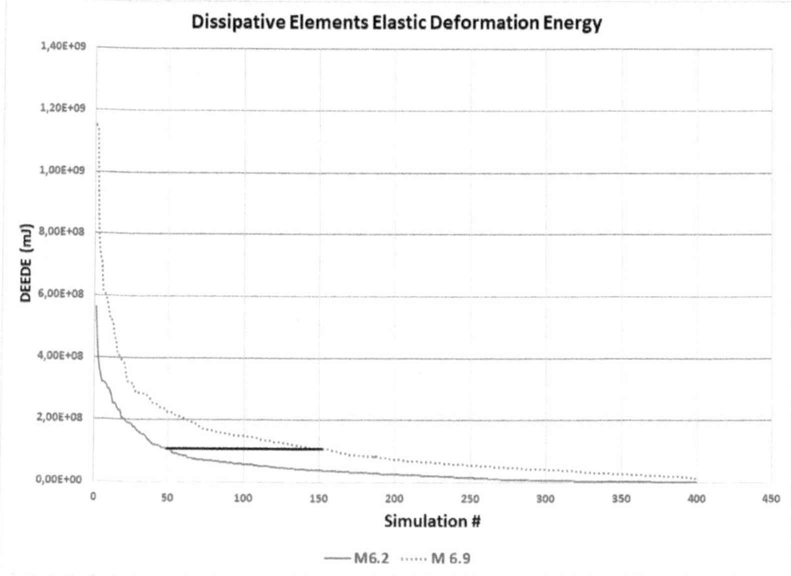

Fig. 11 Energy curves sorted by decreasing energy, Magnitude 6.2 and Magnitude 6.9.

- The curve for magnitude 6.9 is above the curve for magnitude 6.2, which means that it *generally involves* higher energy inputs (Table 2).
- Given the above, it is more stringent to use the curve relating to the largest magnitude for the purposes of verification (which confirms the validity of (1)).
- It is not true, however, that any simulation associated with magnitude 6.9 results in higher embedded energies than any simulation associated with magnitude 6.2. For example, the 50th simulation with magnitude 6.2 results in higher energy than all simulations with magnitude 6.9 starting from approximately the 150th and onwards (Figure 11).
- *In order to properly consider the effects associated with a certain magnitude, it is therefore necessary to consider a sufficiently high number of simulations, and refer to the percentiles.* It makes no sense to take a few simulations and, trusting that the magnitude is high, expect to have subjected the structure to a significant simulation.

Table 2 Comparison of the energies absorbed into the structure by the dissipative elements (in mJ), for different percentiles, relative to the simulations with $M = 6.2$ and $M = 6.9$.

PERCENTILE	*M 6.2*	*M 6.9*
95°	2.02e8	3.80e8
85°	8.45e7	2.05e8
75°	5.77e7	1.49e8
50° (median)	2.61e7	7.41e7

9. Results in Terms of Nonlinear Time Histories for Angle δ = 45°, Set S1

9.1 Physical Expected Consequences of the Simulated Event Reordering

If the reordering done with DEEDEM had been successful in sequencing the signals in order of decreasing severity, what would be obtained, after having done the 400 non-linear analyses corresponding to the 400 simulated events, performed with the same sequence and maintaining the same (re) numbering with respect to the original one, is the following.

By relating the events *s* to one or more damage indexes of the structure (damage index, DI), the curves that give a DI as a function of the simulated event, DI = DI(*s*), with *s* from 1 to 400, should show a trend similar to that of the curve in Figure 9.

Precisely, not much that for a certain DI, the curve DI = DI(*s*) is also exponential, but that it is also substantially decreasing: the first events are the most severe, the last events are the least severe.

In turn this can be expressed by stating that there is a relation between the DEEDE and the DI, that is, that with a good fitting, a relation of the kind DI = DI(DEEDE), can be found.

9.2 Damage Indices Assumed to Monitor Structural Behavior

For each of the 400 THNLs, 4 different DIs have been stored.

Exit code (EC)

It is a conventional code that is valid:

- 3, if the analysis is not completed due to the collapse of the structure (failure to converge caused by the failure of one or more elements);
- 2, if the analysis is terminated having reached convergence at every step, and hence without collapse.

This code is used to distinguish situations in which,

(i) due to the lack of convergence and therefore collapse, the damage indices do not fully describe the severity of the event; from

(ii) the case in which, since the structure was able to resist the entire time history, it is possible to read the effects that the simulated event has produced on the structure.

In case of non-convergence, the values read immediately before the last fatal load step are not in fact significant and show strong dispersions, which however have no physical meaning.

Maximum plastic volume (MPV)

During the THNL associated with a simulated event *s*, the maximum plastic volume of the structure is stored. The plastic volume is the volume of the fibres that at a given instant are in a "loading" state, that is, with their stress state on the plastic locus. During the THNL, this volume can increase if more material has its stress state on the plastic locus, or decrease if a part of the material has returned to the elastic range (unloading).

The maximum volume of metal (in mm^3) that was in the "plastic" state (MPV) at a certain moment during the analysis, is an indicator of how much material simultaneously entered the plastic field (with plastic and therefore irreversible deformations), and therefore also of the severity of the damage suffered by the structure.

Each simulated event corresponds to a different maximum, reached at different times, and with different effects.

Maximum strain in the model (before collapse) (MEPS)

The maximum model strain is a measure of the maximum modulus of the total strain (elastic plus plastic) experienced by all fibers in the model (beam and truss elements).

The maximum strain of the fibres of the entire model vary from instant to instant. For each simulation, at a certain instant, the maximum value over the duration of the analysis (MEPS) is reached.

This number is saved and associated with that simulation. If the strain exceeds the value corresponding to the yield (355 MPa/210,000 MPa = 0.00169), it is evident that the fibre is in the plastic range.

Maximum Inter-storey-drift (MIDRIFT)

Interstorey drift (IDRIFT) is the ratio between the relative horizontal displacement (obtained by composing the X and Y displacements) between a floor and the lower one, divided by the distance between the two floors on the vertical. For example, if at a point a floor moves in x by 3 mm, and in y by 4 mm, while at the upper floor, at that same instant the corresponding point moves in x by 6 mm and in y by 7 mm, the distance between the two floors being 4000 mm, the interstorey drift will be:

$$\text{IDRIFT} = \frac{\sqrt{(6-3)^2 + (7-4)^2}}{4{,}000} = \frac{\sqrt{18}}{4{,}000} = 0.001060660$$

At any instant of the THNL, the IDRIFT can change from point to point and from floor to floor. It is a measure of shaking severity because it measures the relative slippage of a floor on the underlying (or foundation), which, in turn, is an important index of damage.

If during the THNL for an event s, the maximum IDRIFT that the entire structure had (MIDRIFT) is stored, this number is an important indicator of damage.

Damage curves

Based on what has been said, at the end of the 400 THNL of the set **S1** the following 4 curves will be available, depending on the event index s, from 1 to 400:

1. EC = EC(s)
2. MPV = MPV(s)
3. MEPS = MEPS(s)
4. MIDRIFT = MIDRIFT(s)

These curves are particularly significant when taking into account events that did *not* lead to a failure to converge and therefore to collapse.

9.3 Results of Nonlinear Analyses

9.3.1 Exit Codes

Examining the graph in Figure 12, we see that as expected, the analyses that have not reached convergence are the first ones, that is, those that correspond to the highest DEEDE. Then there is a certain area in which analyses that have reached the end appear sporadically (code 2), and from a certain number of triplets onwards, we see that all analyses have reached convergence (with rare exceptions, that is, the opposite of the previous situation).

It is established as a demarcation criterion between the interval of triplets (and of DEEDE) that gave rise to the collapse, and the interval that did not, that the transition triplet between the two groups is the first that, together with the two subsequent ones, involves a code equal to 2. This triplet is number 119. We can therefore say that the analyses before the 119th involve the collapse (relatively rare exceptions, in total 26), while those from the 119th onwards do not imply collapse but only, in the majority of cases, damage, with rare (in total, 14).

Depending on the phases and frequency content of the signals, it is possible, but rare, that a signal with high DEEDE can nevertheless converge. It is also possible, but rare, that despite a lower energy, some triplets can lead to collapse. However, it is clearly legible that the expected fact that the first triplets, those with higher DEEDE, imply more severe effects, while the last ones, less severe effects.

Furthermore, there are large ranges, in which the energies *always imply* collapse, or *always imply* no-collapse.

If we look at Figure 9 again, we can clearly see that the triplets that caused the collapse are those with the highest DEEDE. In correspondence with triplet 119, we have a DEEDE equal to 1.32×10^8 mJ.

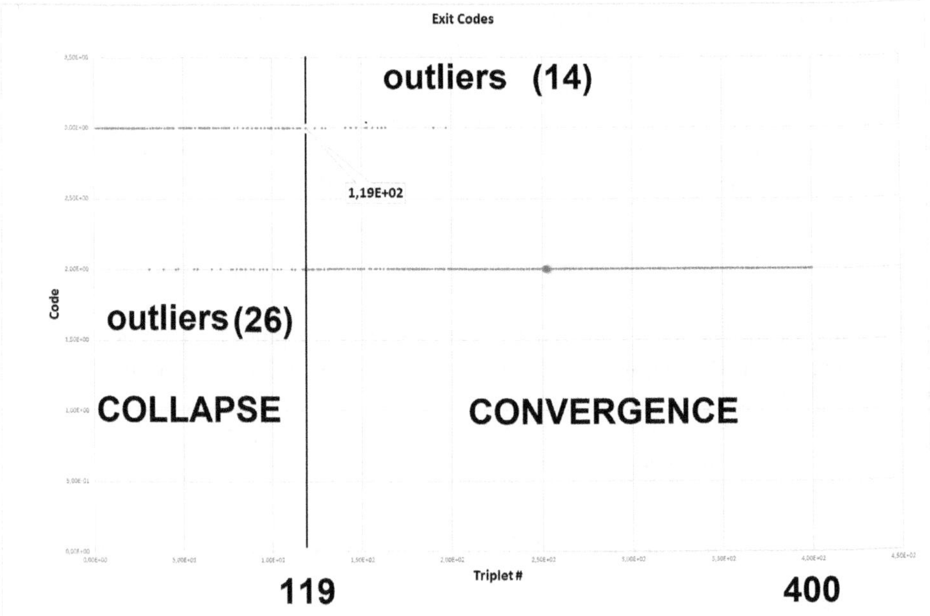

Fig. 12 Exit codes for the 400 THNL runs. Number 3 indicates failure to converge (collapse). Number 2 indicates runs that have completed with convergence at each step.

9.3.2 Damage Indices as a Function of Event Number

In Figures 13, 14, 15, the plots of the DIs versus the simulated event (i.e., the triplet) number is shown.

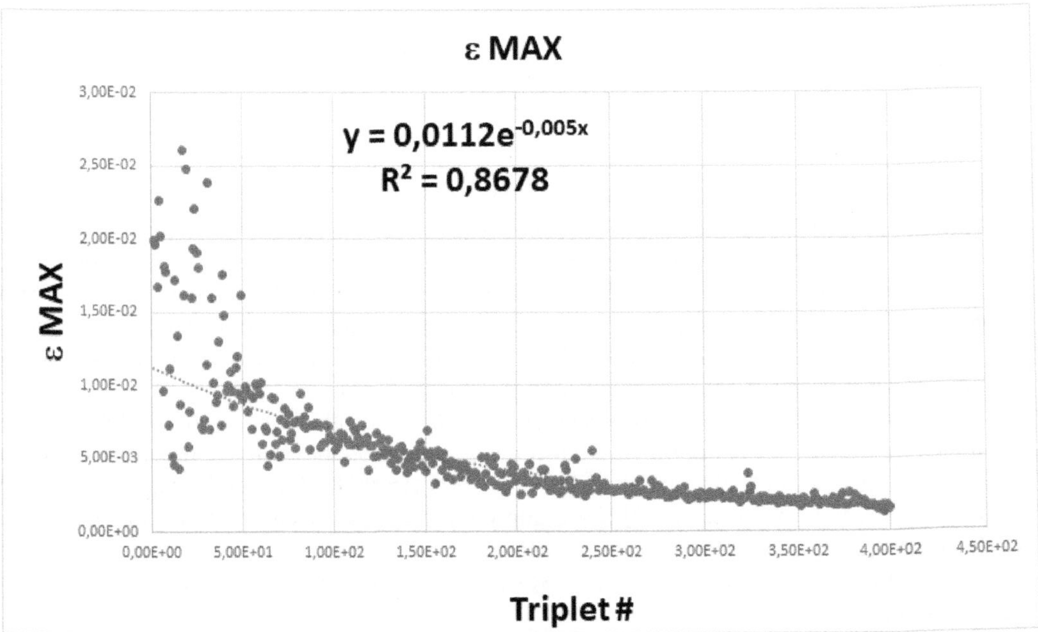

Fig. 13 Maximum strain function of triplet number from 1 to 400. An exponential fitting curve is also shown. $R^2 = 0.868$.

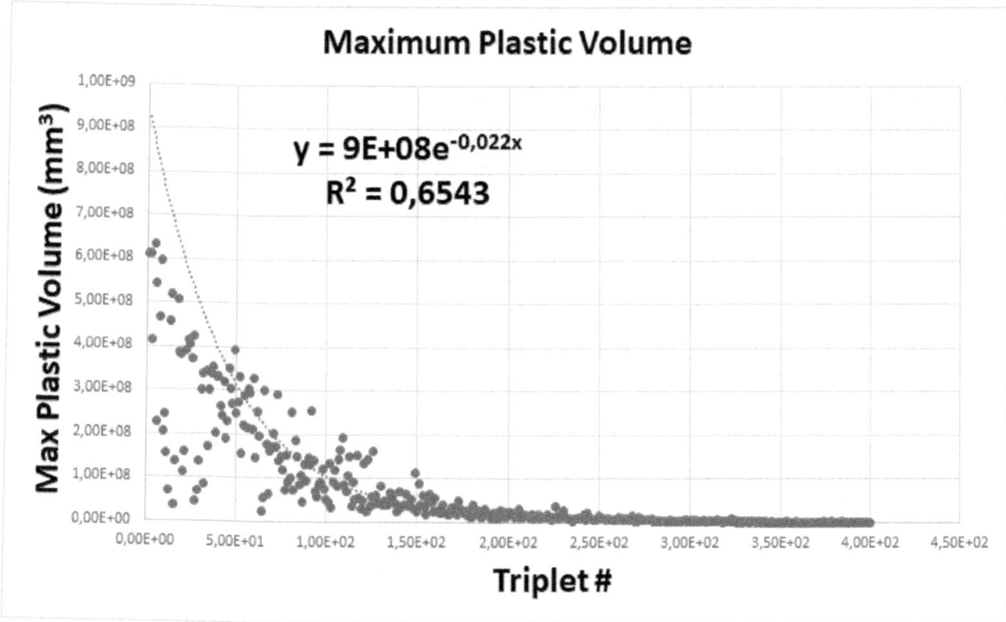

Fig. 14 Maximum plastic volume as function of triplet number from 1 to 400. An exponential fitting curve is also shown. $R^2 = 0.654$.

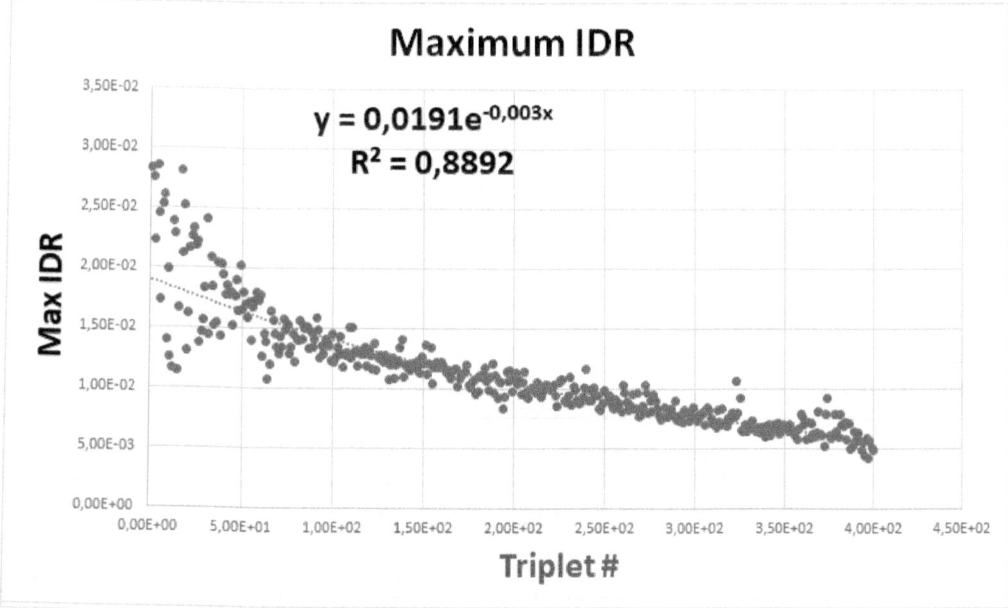

Fig. 15 Maximum interstorey drift as function of triplet number from 1 to 400. An exponential fitting curve is also shown. $R^2 = 0.889$.

The trend of the curves is similar to that of Figure 9, and the exponential fitting for maximum deformation and for maximum interstorey-drift is quite good. The dispersion related to the first 120 triplets, or so, is due to the fact that these events THNL generally ended with collapse, and so the maximum value has not a meaning related to the whole duration of the events.

9.3.3 Damage Indices as a Function of DEEDE

Having established that analyses 1–118 imply collapse (with rare exceptions) and that this is consistent with the fact that these are the analyses with the highest DEEDE, we now want to see whether there is a correlation, and how strong it is, between the DEEDE and the damage indices, for the analyses (119–400) that have converged (with rare exceptions).

To do this, having the two curves DEDEDE = DEEDE(s) and DI = DI(s), where DEEDE is the Dissipative Elements Elastic Deformation Energy calculated with DEEDEM and therefore in the elastic field, s is the simulation index (from 119–400), DI is the generic damage index, it is observed that the curves DI = DI(DEEDE) can be easily obtained, which directly relate the DEEDE to the damage index.

These curves tell us whether DDEDE is a reliable predictor, and how much, of the outcomes of the much more computationally expensive THNLs.

Maximum deformation in the fibres

The first curve, in Figure 16, shows that DEEDE is a good predictor of the severity of the effects of the nonlinear simulations, evaluated with the damage index "maximum strain in the fibers". With a polynomial interpolation, we have an R^2 equal to 0.844. With a linear interpolation, we have $R^2 = 0.843$, while with an exponential interpolation, we have $R^2 = 0.859$.

As the DEEDE increases, the value of the damage index increases, even if there is a certain dispersion, also considering that in simulations from 119 to 400 some simulations actually implied collapse, and therefore the associated deformation values have no physical meaning but only numerical.

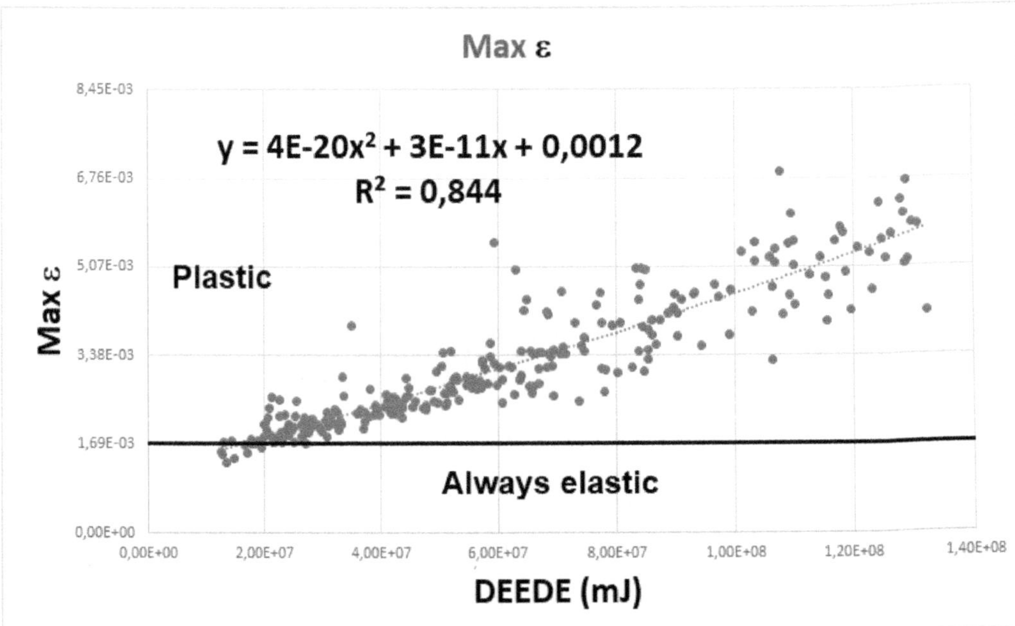

Fig. 16 Maximum fibre deformation during THNL, for analyses 119 and later. Simulations with maximum deformation greater than 0.00169 imply plasticization. Those with maximum deformation lower than this threshold are such as to leave the structure entirely in the elastic range.

If we consider the threshold value corresponding to $\varepsilon_y = 355/210000$, i.e., the yield strain (dark horizontal line in Figure 16), we see that only very few analyses ended without any plasticization. Almost all analyses involved plasticization, up to a maximum strain equal to 0.00689. It should be underlined that in analyses from 1 to 119 this value was even much higher, in some cases close to the ultimate strain equal, in these simulations, to 0.0253, not only in the analyses that involved collapse, but also in the sporadic ones that did not.

Maximum plastic Volume

In Figure 17, the results relating to the damage index "maximum volume in plasticization conditions" are shown, as the THNL varies.

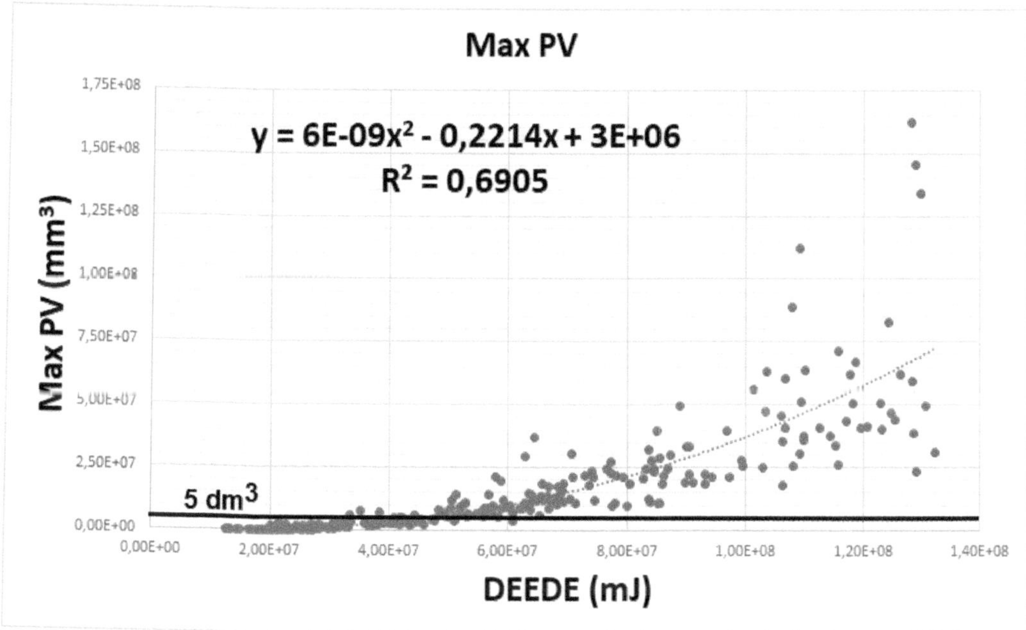

Fig. 17 Maximum volume under loading conditions during THNL, for analyses 119 and later, in mm^3.

Also in this case, DEEDE is correlated with the damage index, but the correlation is less good than in the previous case (see also Figure 14). We have an $R^2 = 0.69$ with a polynomial interpolation, and significantly lower values with other types of interpolation ($R^2 = 0.626$ with linear interpolation, and $R^2 = 0.456$ with exponential interpolation).

Also in this case it should be noted that some simulations (from 119 onwards) did not reach convergence. Only very few analyses have a zero plastic volume, as is consistent with the maximum deformation values.

While it is true that a non-zero plastic volume implies plasticization and therefore irreversible damage, it is also true that in a structure the plasticization of a small volume is not to be considered serious damage. For example, if we consider a negligible plastic volume to be a volume below 5×10^6 $mm^3 = 5$ dm^3, 5 litres, spread throughout the structure, we see that a certain number of analyses, although implying plasticization, imply minimal damage. This can be seen in Figure 17. Examining the tables of the curves, we see that this occurs from the 380th triplet onwards.

Maximum Interstorey Drift

The maximum Interstorey drift value during the THNL is well predicted by the value of DEEDE, as seen in Figure 18. With a polynomial interpolation we have $R^2 = 0.878$, which is a very good value. With linear interpolation, we have $R^2 = 0.853$, and with an exponential interpolation, we have $R^2 = 0.809$.

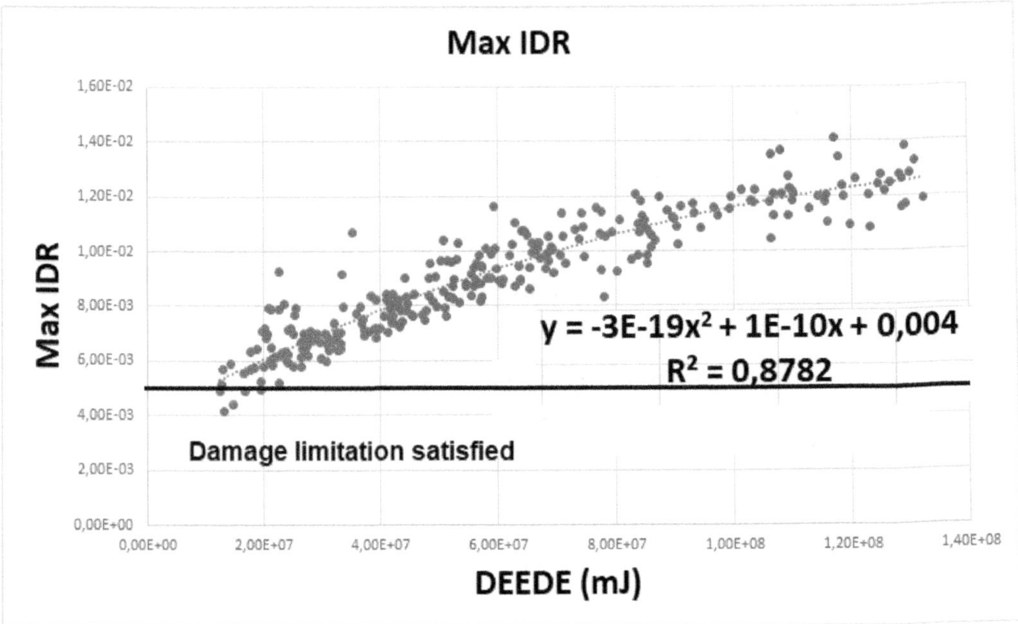

Fig. 18 Maximum Interstorey drift, for analysis from 119 onwards.

Eurocode 8 provides a threshold value for the maximum drift, compatible with "acceptable damage", equal to 0.005 for non-ductile infills (dark horizontal line in Figure 18). In Figure 18, it is seen that very few analyses satisfy this criterion, always with low DEEDE.

10. Proposed New Seismic Design Methodology

Based on the results obtained it can be said (Figure 19):

1. DEEDEM is a reliable method for predicting the severity of triplet effects associated with an event, since the DEEDE value is directly related to both the maximum interstorey drift, and the maximum deformation in the fibers, calculated with THNLs. However, THNLs are much, much more expensive in terms of computation time.
2. There are generally three DEEDE energy ranges in which the structure:
   a. Collapses (for major DEEDEs);
   b. It gets damaged without collapsing (for intermediate DEEDEs).
   c. It does not suffer significant damage (for minor DEEDs).
3. By requiring the homogeneity of the results (collapse/damage/no damage) of 3-5 triplets having contiguous DEEDE (that is, that are close to each other in the ordering of the triplets by decreasing DEEDE), it is possible to determine, after having performed all the THNL associated with the simulated events, the two triplets that separate the three intervals. These triplets are called "sentinel triplets". These two triplets are associated with two very different DEEDE values. In the case studied, these values are $E_d = 1.32 \times 10^8$ mJ, and $E_d = 2.11 \times 10^7$ mJ.
4. The number of sentinel triplets in the triplet ordering allows us to evaluate the energy percentile associated with them. In this case, 119, the threshold between the collapse zone and the

damage zone, corresponds to a percentile of approximately 70% (applying (8) we obtain 70.5): 7 triplets out of 10 have lower energies, and therefore 7 triplets out of 10 do not cause collapse. The value 380, which separates the zone with damage from the one with negligible damage, corresponds to a 5% percentile ((8) gives 5.25), only 5 events out of 100 do not cause significant damage to the structure.

5. The energy percentile associated with the transition between one interval and another can be seen as an index of performance of the structure. In this case the structure has a performance index of 70/100 for the risk of collapse, and 5/100 for the risk of significant damage.

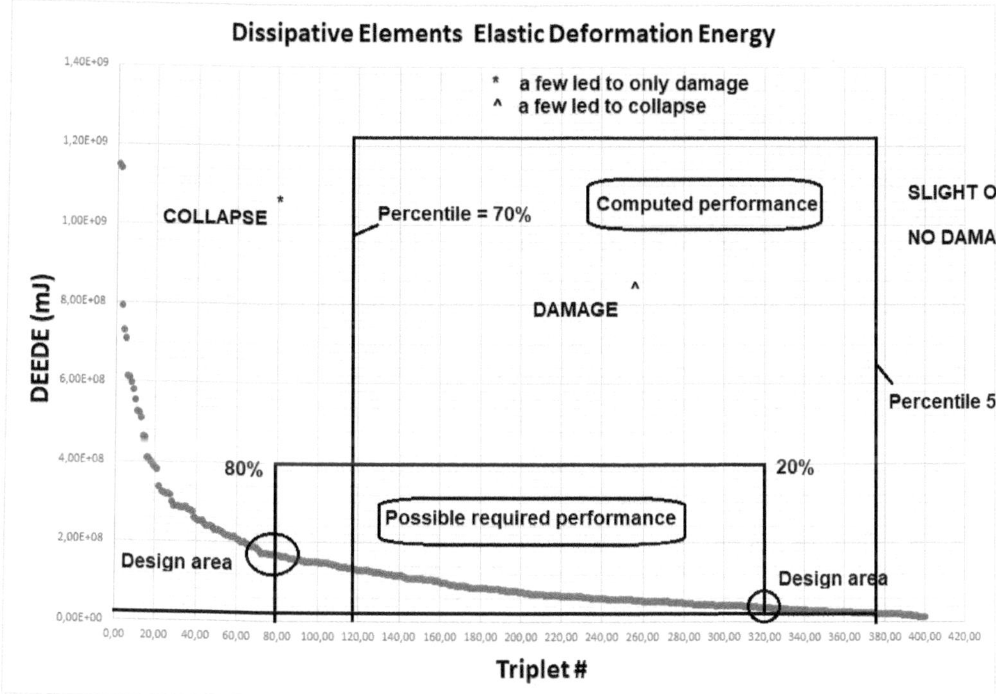

Fig. 19 The relationship that DEEDE has with the results of the 400 THNL analyses.

These arguments, made after 400 THNLs, can be rearranged to provide a method for seismic safety assessment that does not involve performing all THNLs, but instead uses DEEDEM to select a few signal triplets to be used with THNLs. This can be done in the following way.

(i) *Target* percentiles are established for the risk of collapse p_c, and for the risk of permanent damage p_d. Always with reference to Figure 19, these percentiles could be (for example) 80% and 20%, respectively. In a desirable redefinition of the seismic safety verification methods, such target percentiles would take the place of the “spectrum-compatibility” (see the next section). These percentiles should not be understood as *probabilities*, but as mere (dimensionless) *performance indices*. They should be understood as target data, set by law and preceding the analyses. They refer to the distribution of the DEEDE for the structure actually under examination, and with the appropriate simulations for the site under examination, and therefore correctly taking into account all 4 mechanical systems in series.

(ii) N simulated events are generated by NDSHA, considering all the faults pertinent to the site under examination, and the design magnitudes M_{design} associated with them (by (1)). The number N must be high, how high is currently still a subject of study. In this work, $N = 400$ with a single fault.

(iii) N DEEDEs associated with these simulated events are calculated and the events, i.e. the accelerogram triplets, are reordered by decreasing DEEDE. This operation requires very affordable calculation times (40 minutes for this structure).

(iv) Corresponding to the percentile triplets p_c and p_d (80/100 and 20/100 in the example in Figure 19), the two simulated events are immediately found, having numbers s_c and s_d in the new ordering of events that respects the decreasing DEEDE. In the example studied here, $N = 400$ and therefore through (8) we find $s_c = 81$ and $s_d = 312$.

(v) For the collapse check, M THNLs are performed considering the triplet s_c and the immediately following ones: s_c, $(s_c + 1)$, $(s_c + 2)$ ... $(s_c + M - 1)$. Reasonable values for M can be 3, 5 or 7. These are the triplets in the "design area" on the left in Figure 19. If all M THNLs give rise to an analysis that does not lead to collapse, the check is satisfied. Otherwise, if even just one THNL leads to collapse, the check is not satisfied. This would in fact indicate that the structure is either in the collapse zone, or in the collapse-damage transition zone. This requires a few hours of calculation.

(vi) For damage verification purposes, M THNLs are performed considering the triplet s_d and the immediately following ones: s_d, $(s_d + 1)$, $(s_d + 2)$,... $(s_d + M - 1)$. These are the triplets in the "design area" on the right in Figure 19. If all the M THNLs give rise to an analysis that does not involve damage (the structure remains in the elastic range or the maximum interstorey drift does not exceed a certain threshold value), then the verification is satisfied. Otherwise, if even just one THNL causes unacceptable damage, the verification is not satisfied. This would in fact indicate that one is either in the damage zone, or in the damage-no damage transition zone. Even in this case, a few hours of calculation are required. It should be noted that it would be entirely logical, for the damage verifications, to use the same system described here, but with signals generated using a magnitude *lower* than that determined with (1). How much lower is a problem of a political-economic nature.

Such a verification would require a few hours of computation, instead of 23 days. And it would take into account 400 NDSHA simulations. Most of the triplets would be evaluated with DEEDEM, only for some, chosen on the basis of an energy criterion, THNL should be used.

11. Discussion on the Proposed Methodology

The methods currently provided for by the regulations for seismic checks, all take a target response spectrum as a reference.

This methodology, as we have seen, presents numerous problems, especially but not only, due to the PSHA approach which makes use of concepts, methods and hypotheses lacking experimental evidence and solid theoretical foundations.

An important step forward is to use (Fasan et al., 2016, 2017) the envelope spectrum of the simulated signals, specific for the site under consideration ($MCSI_{SS}$, Maximum Credible Seismic Input Site Specific). In this case the limitations related to the use of PSHA are removed, and a set of simulated events, generated by means of NDSHA, and consequently in accordance with the first three mechanical systems in series (SM1, SM2, SM3), is usefully and pertinently used.

Although this methodology is a considerable step forward, it retains the typical problems of the response spectrum approach. In particular, the fact that in the RSA analysis, on a certain specific structure, modal peaks that are reached *in different events* are combined using SRSS or CQC techniques, and therefore using an SRSS of the effects on the individual modes of oscillation.

This *causes any solid physical meaning to what is being done to be lost, because effects that cannot be contemporary are combined.* The phases are lost, and the results obtained from the analysis, in terms of member forces, displacements and deformations, are not univocally attributable to any particular simulated event, but rather, derive from the mathematical mixing of effects obtained with different events. Furthermore, due to the square roots, they are no longer balanced.

By the way, it should be noted that, following such a procedure, it makes no sense to require excessive precision (the three or four digits of the PGA of the Italian standards (Rugarli, 2008, 2014) are a striking example, but not the only one: everything that is calculated following such a brutal approximation should, for this reason alone, be conveniently truncated (Rugarli, 2015c)).

The method proposed here does not refer to different events, mixing their effects. In this case, in the end what must be done is to perform a certain number (2 M) of THNLs, each of which uses a triplet of accelerograms, *uniquely associated with a well specified simulated event.*

Therefore, unlike what happens with RSA, it is possible to associate the effects read on the structure with a very specific simulated event, a very specific source mechanism. The method is used to choose which events are significant for the particular structure under examination, but it does not use response spectra, but rather the DEEDE, which is a mechanical quantity obtained with linear methods, firmly anchored to the physics of the problem, without any combination of the effects of different events.

One might be tempted to transform percentiles into probabilities, but this would be neither correct nor justified, nor necessary.

Percentiles could become "probabilities" only if there were good reasons to consider the simulated events from which they originate as exchangeable (De Finetti, 1970). This means "equally probable". But, in reality, although all simulated events are *possible*, it is not possible to say that they are *equally probable*, and therefore *exchangeable*. Our current knowledge is insufficient for such an assumption.

On the other hand, having obtained, through curves such as the one in Figure 9, a sufficiently exhaustive representation of the possible effects that earthquakes originating from faults significant for the site in question can cause, there is no reason to forcefully introduce the concept of *probability*. The performance is not related to the probability but is related to the degree of coverage of the range of DEEDEs that could originate in the structure.

The percentiles indicate this degree of coverage.

Instead of prescribing spectrum-compatibility, with all the constraints and limitations described, what needs to be prescribed is that the signals to be used for the THNLs of a certain particular structure, refer to certain predetermined degrees of coverage of the range of possible strain energies.

One could therefore speak not of spectrum-compatibility of the signals, but of DEEP-compatibility of the signals, where the acronym DEEP stands for *Dissipative Elements Energy Percentile*.

12. Engineering Considerations about the Use of Signals *Outside* the NDSHA Frame: Conclusions

The examination of the consequences of the four hundred simulated events on a generic structure, chosen without any particular specific characteristics with respect to them, has clearly shown that:

(i) PGA is not related to the severity of the expected effects of a triplet of signals. There may be signals with PGAs half or less than others, and which instead cause worse effects due to their specific frequency content and the phases of the constituent harmonics.

(ii) Magnitude is related to the severity of the effects but only in a medium sense. There may be events with lower magnitudes that give rise to significantly greater structural effects.

(iii) The regularized response spectrum provided by the regulations does not take into account the very strong amplification effects that the response spectra of natural or synthetic signals show for certain particular frequencies. Using such spectra does not therefore appear to be in favor of safety.

(iv) It follows that the spectrum compatibility of signals used for THNL, imposed by the regulations, has no basis other than conventional. It has no physical and mechanical basis.

13. Conclusions

After an extensive campaign of numerical tests, using a large set of simulated signals, on a real structure, modeled in a realistic and detailed way, it was seen that the effects on the structure of the simulated events are very different from each other, both in the elastic field and in the elasto -plastic field.

Because of this fact, it is not in favour of safety to use a few events for THNL analyses, chosen at random, trusting that the seismogenic characteristics *per se* imply a sufficient, or consistent, severity.

Nor is the use of the response spectrum method a rigorous method, because it involves combining effects obtained from different events: the physical meaning of what is being done is lost.

It is necessary to examine a large set of signals and choose those that imply a certain level of strain energy, evaluated by certain percentiles. Otherwise, one risks taking insignificant events.

The significance of events must be viewed in a relative sense, reliably considering the set of events that can be generated, given the design magnitude, *and the structure at hand.*

The numerical results, after having performed 400 Non-Linear Time History analyses, on a model with more than 20 thousand degrees of freedom, have confirmed that the elastic deformation energy absorbed by the dissipative elements alone (DEEDE), and evaluated with very rapid Modal Time History techniques, is a reliable predictor of the relative severity of the simulated events, for the specific structure under examination, in the specific site under examination.

The use of the DEEDEM technique allows to choose in a physically appropriate way the few signals to be used for the detailed structural checks, in the non linear field, by means of Non-Linear Time Histories.

14. Further Developments

This study indicates the need to further investigate some aspects that have been considered but that require further investigation and confirmation. The aspects to be studied further are:

- To verify whether the spread over two orders of magnitude of the deformation energies as the simulated events vary, at equal magnitude, a circumstance fraught with consequences, is also confirmed using other simulations, relating to other sites and other faults, as well as other structures.
- Understand whether the strictly exponential trend of the energies collected for the equi-scenario events has a specific physical reason or is somehow related to how the events were generated.
- Understand which source mechanisms are more frequent or plausible and which are less so, based on geophysical evidence.
- Verify the good behaviour of DEEDEM also with structures having a greater spread of the participating mass on different modes, with hardening constitutive laws, with even higher ultimate deformations (10–15%), with second order effects.
- Establish how much the energy percentiles vary with the number of simulations, and find a reference number for simulations that is meaningful. Based on previous studies done on the stability of the envelope spectrum of signals (Fasan et al., 2017), it seems that the number of 100 for each source mechanism is correct (i.e., 4×100 for each scenario).

Acknowledgements

The author would like to thank the editors for the opportunity to participate in the volume for the 80th birthday of Prof. Panza, Giuliano. The author had the privilege of being able to work a lot with Giuliano, with whom an open and fruitful exchange of ideas was always possible. This led to useful results and was extremely interesting. Thank you Giuliano! Best wishes!

The author wishes to thank Prof. Romanelli and Dr. Vaccari, both for providing the 800 signal triplets (2400 accelerograms) used in this study, and for the fruitful exchange of ideas regarding the results obtained, and future research directions.

The methods described in this work have all been implemented and are available in the structural calculation software Sargon©, of which the author of this work is the author, already in use since 1991 by many structural engineers in Italy.

References

CEN EN (2005, March). 1998-1:2005. Eurocode: Design of Structures for Earthquake Resistance.

De Finetti, B. 1970. *Theory of Probability*. Einaudi.

Fasan, M., Magrin, A., Amadio, C., Romanelli, F., Vaccari, F. and Panza, G.F. (2016). A seismological and engineering perspective on the 2016 Central Italy earthquakes. *International Journal of Earthquake and Impact Engineering, 1*: 395–420.

Fasan, M., Magrin, A.. Amadio, C.. Panza, G.F.. Romanelli, F. and Vaccari, F. (2017). A possible revision of the current seismic design process. In: *6th World Conference on Earthquake Engineering* (pp. 1–121). Santiago, Chile.

Panza, G.F. and Bela, J. (2020). NDSHA: A new paradigm for reliable seismic hazard assessment. *Engineering Geology* , *275*: 105403. (Engineering Geology 2023 Best Paper Award.)

Ricci, A., Romanelli, F., Vaccari, F., Boncio, P., Venisti N., Faraone, C., Vessia, G. and Panza, G.F. (2025). Comparison between the Neo-deterministic Seismic Hazard and FEM approach to assessing 2D local seismic response at Chieti's city site (Abruzzo, Italy), *Engineering Geology, 347*.

Romanelli, F. and Vaccari, F. (2025). Personal Communication.

Rugarli, P. (2008). In Italian: Zone Griglie o... Stanze? *Ingegneria Sismica*

Rugarli, P. (2014). Validazione Strutturale. EPC Libri, Rome.

Rugarli, P. (2015a). The role of standards in the *invention of truth*. Acta, Conference *The Resilience of Cities of Art to Earthquakes*, 3–4 November 2015. National Academy of Lincei, Rome.

Rugarli, P. (2015b). Primum: Do no harm. Acta, Conference *The Resilience of Cities of Art to Earthquakes*, 3–4 November 2015. Accademia Nazionale dei Lincei, Rome.

Rugarli, P. (2015c). *Analisi Modale Ragionata*. EPC Libri, Rome.

Rugarli, P., Amadio, C., Peresan, A., Fasan, M., Vaccari, F., Magrin, A., Romanelli, F. and Panza, G.F. (2018). Neo-deterministic scenario-earthquake accelerograms and spectra: An NDSHA approach to seismic analysis. In: Jia, J., and Paik, J.K. (Eds.), *Structural Engineering in Vibrations, Dynamics, and Impacts*. CRC press, Taylor & Francis Group, Abingdon, UK, in press.

Rugarli, P., Vaccari, F. and Panza, G.F. (2019). Seismogenic nodes as a viable alternative to seismogenic zones and observed seismicity for the definition of seismic hazard at regional scale. *Vietnam Journal of Earth Sciences* , *41*: 289–304.

Wen, Z.P. and Wang, G.X. (2024). Earthquakes and Sustainable Infrastructure: Neodeterministic (NDSHA) Approach Guarantees Prevention Rather Than Cure. *Earthquake Science* , *37*: 494–497.

Zhang, Y., Rugarli, P., Ji Z., Jiang, C., Wu, Z. and Panza, G.F. (2025). Falsification of Seismic Hazard Assessment (SHA) by Real Earthquakes: The Design Magnitude M_{design} for Neo-deterministic Seismic Hazard Assessment (NDSHA). In: Y. Li et al. (Eds.), *Scientific Investigation of Continental Earthquakes and Relevant Studies*. Higher Education Press Ltd.

CHAPTER

5

How Many Sources Must A Man Write Down Before You Call Him A Seismologist?

Franco Vaccari[1*]

345, but 8648 is a much, much better guess.

1. Introduction

In seismic hazard studies, seismologists are grappling with the challenge of accurately characterizing seismic sources in tectonically active regions. This pioneering work presents an intriguing alternative: can seismic sources collaborate harmoniously to properly define a seismologist?

The author's idea of exploring the complex bidirectional relation between seismologists and seismic sources matured during his early years at Asilo di via Mamiani. After six decades of attempts and failures, he finally understood the phenomenon and solved the problem.

2. From Seismic Sources to the Optimal Seismologist

In the well-known NDSHA approach, the observed seismicity of the studied area is discretized in a regular grid. Seismic sources are then retained within a set of seismogenic zones (for more details, see Panza et al., 2012 and its references). In this work, a sophisticated recursive AI-based algorithm has been developed to identify the optimal set of seismic sources that collectively may define the ideal seismologist. The algorithm has been implemented in a user-friendly software. Figure 1 illustrates the results of 12 random executions of the program, where the complexity of the solution increases with each execution.

1 Olim Dipartimento di Ingegneria e Architettura, Univ. di Trieste.
Olim Dipartimento di Informatica, Matematica e Geoscienze, Univ. di Trieste.
Olim Dipartimento di Matematica e Geoscienze, Univ. di Trieste.
Olim Dipartimento di Scienze della Terra, Univ. di Trieste.
Olim Istituto di Geodesia e Geofisica, Univ. di Trieste.
Olim Liceo Scientifico G. Oberdan, Trieste.
Olim Scuola Media ai Campi Elisi, Trieste.
Olim Scuola Elementare E. de Amicis, Trieste.
Olim Asilo di via Mamiani, Trieste.
https://orcid.org/0000-0002-0056-5103
* Corresponding author: vaccari@units.it

DOI: 10.1201/9781003684121-6

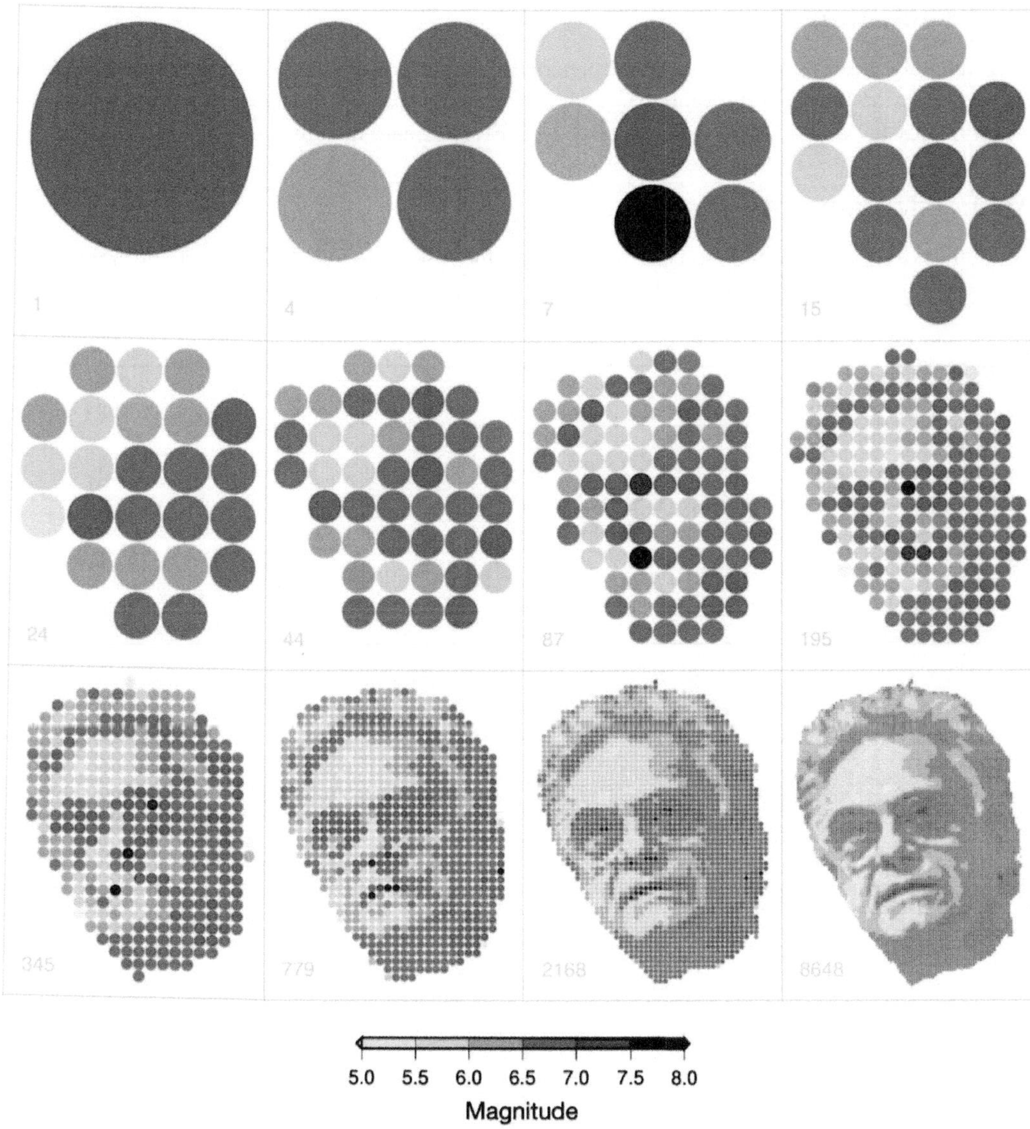

Fig. 1 Point sources of magnitude varying between 5 and 8 generated by 12 random executions of the algorithm aimed at the definition of the optimal seismologist. The requested level of complexity of the solution increases from top-left to bottom-right.

3. Discussion

Although Japanese seismologists are extremely satisfied with the simple single-source model depicted in the top-left corner of the figure, the majority of the scientific community agrees that 87 or even 195 sources may not be sufficient to accurately represent the optimal seismologist. In fact, an acceptable response to the title question could be: "at least 345 if you look from a distance, but 8648 is a much more accurate estimate".

4. Conclusion

Unfortunately, the optimal seismologist generated by the 8,648 sources immediately protested, claiming that the numbers in Figure 1 are barely visible. He also suggested increasing the source magnitude by 0.7 units. After his complaints, a software update is in the works.

Acknowledgements

The author expresses gratitude to B.D. for inspiring the title of this work. Despite their best efforts, no anonymous reviewers were able to improve the quality of this manuscript.

Reference

Panza, G.F., La Mura, C., Peresan, A., Romanelli, F. and Vaccari, F. (2012). Seismic hazard scenarios as preventive tools for a disaster resilient society. *Adv. Geophys.*, *53*, 93–165.

CHAPTER

6

Can The Profile of A High-profile Seismologist Perturb A Ground-shaking Scenario?

Franco Vaccari[1*]

Yes!

1. Introduction

We know that lateral heterogeneities and low-velocity sediments in the uppermost part of the medium traveled by seismic waves can generate complex amplification patterns at the free surface. Scientists have proposed various methodologies to model wave propagation in laterally heterogeneous media, including sophisticated numerical, analytical, and hybrid techniques. However, the interpretation of these results often depends on scientists' personal beliefs.

2. Can a Seismologist's Belief Perturb a Ground-shaking Scenario?

This question is challenging. To understand how a seismologist's mind can affect seismic waves, a laterally heterogeneous profile, 80 × 0.5 hm long, has been prepared. It includes the profile of a high-profile seismologist's head at its right edge, as shown in Figure 1. This will be referred to as "the local model".

3. Computations

The hybrid approach developed by Fäh and Panza (1994) has been employed to model wave propagation from the source to the mind of interest. The source has been positioned 80 hm to the left of the left edge of the local model, in a layered structure that reflects the average lithospheric

1 Olim Dipartimento di Ingegneria e Architettura, Univ. di Trieste.
Olim Dipartimento di Informatica, Matematica e Geoscienze, Univ. di Trieste.
Olim Dipartimento di Matematica e Geoscienze, Univ. di Trieste.
Olim Dipartimento di Scienze della Terra, Univ. di Trieste.
Olim Istituto di Geodesia e Geofisica, Univ. di Trieste.
Olim Liceo Scientifico G. Oberdan, Trieste.
Olim Scuola Media Al Campi Elisi, Trieste.
Olim Scuola Elementare E. de Amicis, Trieste.
Olim Asilo di via Mamiani, Trieste.
https://orcid.org/0000-0002-0056-5103.
* Corresponding author: vaccari@units.it

DOI: 10.1201/9781003684121-7

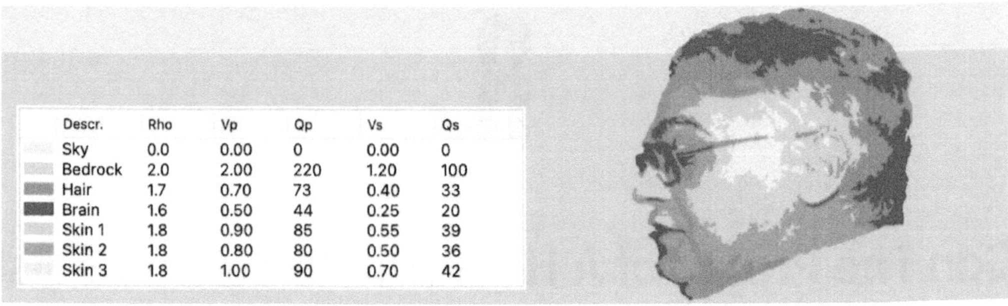

Fig. 1 A profile of a high-profile seismologist is positioned near the right edge of a bedrock profile to investigate wave propagation within his mind. Please note the absence of clouds in the sky.

properties of the region. The modal summation technique has been utilized to propagate the waves from the source to the local model. Subsequently, the incoming wavefield has been introduced into the local model, where wave propagation has been treated using the finite difference technique.

An event of magnitude $M = 5$ has been considered for the experiment, with the intention of causing just a slight damage to the mind of the high-profile seismologist. Table 1 presents the source properties. To simplify the waveforms, a size-scaled point source model has been employed to generate the seismic waves, thereby eliminating the complexities associated with the rupturing process of a fault of finite dimensions. This choice facilitates the identification of the effects induced by the presence of the profile of the high-profile seismologist along the profile.

Table 1 Properties of the source.

Distance to the profile (hm)	80
Depth (hm)	80
Strike/profile angle (°)	80
Dip (°)	80
Rake (°)	80

4. Results and Discussion

Approximately 100 three-component accelerograms have been generated along the profile, as depicted in Figure 2. It is evident that the profile of the high-profile seismologist significantly influences the amplitude and duration of ground shaking, particularly in correspondence of its mind, elevated with respect to the flat conditions that characterize the profile.

As expected, there are also evident reflections generated by the high-profile seismologist's mind, mostly visible in the vertical component of motion. Many seismologists were expecting remarkable amplifications also in the transverse component of motion. Regrettably, owing to the peculiar characteristics of the focal mechanism, coupled with the source position relative to the profile, nothing can be said about that.

A heated debate is currently underway between seismologists and neuropsychiatrists to unravel the underlying causes of the observed amplifications. While some members of the former group advocate for the involvement of brain waves, most representatives of the latter group insist on explaining everything in terms of head waves.

5. Conclusion

Interdisciplinary discussions can undoubtedly contribute to scientific progress, provided that researchers refrain from venturing beyond their areas of expertise.

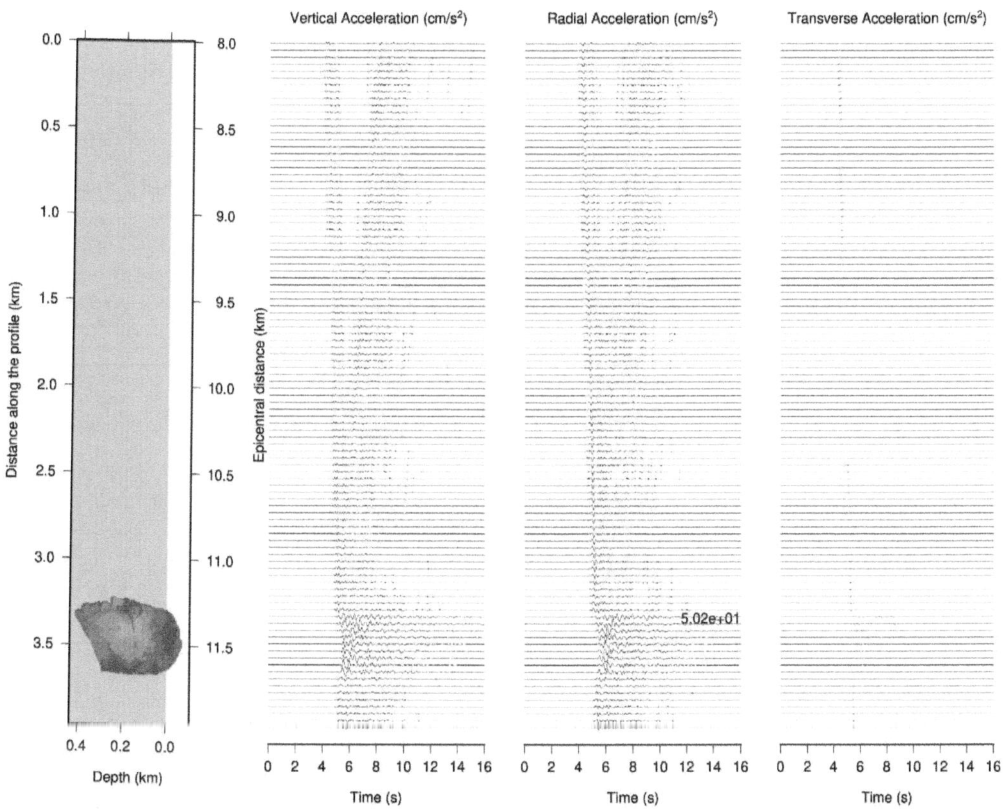

Fig. 2 The profile of the high-profile seismologist positioned near the end of the bedrock profile clearly influences the ground shaking. The largest peak of acceleration, about 80 × 0.67 cm/s^2, appears close to where the forehead emerges above the average ground level.

On the contrary, a remarkable example of fruitful interaction can be seen among the members of a Monty Python fan club, who posed an intriguing question: how the profiles of TWO high-profile seismologists positioned near the edge of a bedrock profile can impact the ground-shaking scenario. The model has already been prepared (Figure 3), and the analysis of its influence on wave propagation will be the subject of a future paper.

Fig. 3 The so-called "Kilimanjaro" model suggested by a Monty Python fan club member.

Reference

Fäh, D. and Panza, G.F. (1994). Realistic modelling of observed seismic motion in complex sedimentary basins. *Ann. di Geofisica*, *37*(6), 1771–1797.

Part II

Practical Implementations and Case Studies

CHAPTER

7

Recent Seismicity of the West African Region

Paulina Amponsah[1*], Mustapha Meghraoui[2], Bekoa Ateba[3], Prince Amoah[4] and Edmund Amartey[5]

The West African region is generally considered as a stable continental tectonic domain with few seismically active features. However, several earthquakes have struck the area in historic and recent times with damaging earthquakes dating as far back as 1615. Major events with magnitudes between 6 and 7 occurred in 1862, 1906, 1939, and 1983. The seismicity is not well known due to the sparse seismic network coverage, incomplete and inhomogeneous earthquake catalogues, and poor knowledge of the ground motion parameters. Most of the historical and recent events have not been recorded. However, in recent times, there have been significant improvements in the deployment of seismic equipment and networks for monitoring earthquake activity. Seismic events of magnitudes ranging from 3.5 to 6.5 have been observed and recorded in the region. These events are mostly associated with the Romanche, Chain, St. Paul transform faults, and the Cameroon volcanic line. Continuous strike-slip movement on the Romanche fracture zone adjacent to the West African continental margin may be the cause of the earthquakes in Ghana. The extension of the Akwapim fault zone south-westwards to the continental margin provides a link with the Ghana-Sierra Leone Ridge which is considered to be associated with the Romanche Fracture Zone. Evidence of seismic activity along the Akwapim fault zone through southern Ghana into Togo and Benin and its link with the Romanche fracture zone indicates a lineament of active tectonics extending about 1000 km into the African continental plate from the mid-Atlantic ridge. This paper is a study on the historical and recent seismicity in the West African region and is designed to raise awareness on the rising seismic activities in the region and awaken relevant authorities, organizations/institutions, and the public on the need for timely and sustainable proactive measures for mitigating the risks and disasters associated with earthquakes in agreement with the Sendai recommendations.

1. Introduction

The West African region is located relatively far from plate boundaries and major earthquake zones, and therefore mostly considered as a stable continental area. Though the region is considered as

[1] School of Nuclear and Allied Sciences, Ghana Atomic Energy Commission, Accra, Ghana.
[2] Institut Terre et Environnement de Strasbourg (ITES), UMR 7063, France.
[3] Institut de Recherches Geologiques et Minieres (IRGM), Yaounde, Cameroon.
[4] Nuclear Regulatory Authority, Ghana.
[5] National Nuclear Research Institute, Ghana Atomic Energy Commission.
E-mail: m.meghraoui@unistra.fr[2]; b.ateba@yahoo.com[3]; amprince77@gmail.com[4]; edalaska2000@yahoo.com[5]
* Corresponding author: pekua2@yahoo.com

DOI: 10.1201/9781003684121-9

a low to moderate seismic region, it experienced damaging earthquakes dating as far back as 1615 (Ambraseys & Adams, 1986). Major events which had local magnitude (M_L) and Moment magnitude (Mw) between 6 and 7 occurred in 1862, 1906, 1939, and 1983 (Table 1).

The instrumental seismicity is not well known due to the sparse seismic network. The historical earthquake records are incomplete as many past events may not have been recorded (Ambraseys & Adams, 1986). With the establishment lately of several seismic networks in the continent such as the Africa Array seismic network, the South African National seismograph network, Ghana National Digital seismic network among others, the monitoring and recording of seismic events in recent times has improved.

Earthquake magnitudes (Ms, M_L, Md, Mb, Mw) ranging from 3.5 to 6.5 have been observed recently in the region as shown in Table 1 (Meghraoui et al., 2016; Ambraseys & Adams, 1986). A more recent event recorded in the Republic of Guinea on 22 December 1983 was of body wave magnitude (Mb) 6.3 (Langer et al., 1985). 300 people were killed, about 1,500 were injured, more than 5,000 houses were damaged and close to 18,000 became homeless. The 22 June 1939 earthquake in Ghana recorded 17 deaths and many injuries. These are examples of infrequent, but very destructive earthquakes that occurred in the presumably stable intraplate regions such as West Africa (Meghraoui et al., 2016).

Seismic events recorded in south-western Nigeria in recent times is an indication that the country may not be aseismic as had been thought (Kadiri & Amponsah, 2022). On 27 June 1990, an earthquake of local magnitude (M_L) 3.7 was recorded in Nigeria. Another event of local magnitude (M_L) 4.2 was recorded on 7 July 1994, and on 7 March 2000 an earthquake of body-wave magnitude (Mb) 4.4 was recorded.

According to Adenkule et al. (2012), geologic and geodetic evidence suggest the existence of large fracture zones (Romanche and Charcot) in Nigeria. A possible originator of intraplate earthquakes in Nigeria can be associated with the regional stress generated with the West African Craton and the Congo Craton with an inferred principal stress trending WNW–ESE (Eze et al., 2012). Bondessen and Schmidt (1972), attributed the seismicity in Accra, Ghana to possible recent movements along the Togo, the Birimian, and the Dahomeyan thrust faults. Sykes (1978) emphasized the role of these faults to the high seismic activity in the Greater Accra Metropolitan Area as compared to those of its neighbouring regions and attributed the relative intense seismicity to the resurgence of movement along the ancient tectonic boundaries.

In the paper, we first present the geological and seismotectonic settings of West Africa, followed by the history of earthquakes in the region. A detailed seismicity of the region is presented as well. Finally, we discuss the implications on the seismic hazard and risk with a focus on the mitigation of disasters in urban areas in West Africa.

Table 1 Some historical and significant earthquakes in West Africa. Database is from Meghraoui et al. (2016); Ambraseys and Adams (1986). MSK: Medvedev Sponheuer Karnik scale.

Date	*Magnitude*	*Intensity (MSK)*	*Location*	*Effects*
1615	–	–	Elmina, Ghana	Fortress at Elmina was destroyed
1618	5.2 (Ms)	VI	Mali	Inhabitants panicked
18 Dec 1636	5.7 (M_L)	IX	Axim, Ghana	Workers buried in a Goldmine
1788	5.6 (Ms)	VIII	Togo	Strongly felt in Agonah, Abomey and Kana
20 May 1795	5.2 (Ms)	–	Sierra Leone	Strongly felt in Freetown
20 June 1796	5.2 (Ms)	–	Sierra Leone	Strongly felt in Freetown
1832	4.1(Ms)	–	Senegal	Felt in Dakar and St. Louis

Contd.

Table 1 *Contd.*

Date	*Magnitude*	*Intensity (MSK)*	*Location*	*Effects*
10 July 1862	6.5 (M_L)	IX	Accra, Ghana	Stone buildings in Accra were razed to the ground. The Christiansburg castle was uninhabitable
1872	4.9 (M_L)	VII	Accra, Ghana	–
11 February 1879	5.7 (Ms)	VIII	Ivory Coast	Felt in Baidu and Gregbeu; Old houses collapsed and trees toppled
20 November 1906	5.0 (M_L)	VIII	Ho, Ghana	Felt in the north-eastern part of Ghana. Government buildings were damaged
1927	4.0 (Ms)	–	Kakulima, Guinea	Strong shaking which made inhabitants to panic
1928	4.8 (Ms)	–	Kakulima, Guinea	Damage in local houses; inhabitants panicked
1914– 1933	–	–	Southern Ghana	Several shocks were recorded by the seismograph
1930	3.5 (Ms)	–	Togo	Felt in Palime and Ago
6 January 1933	4.0 (Ms)	VI	Togo	Felt in Palime and Misohoe
1935	4.0 (Ms)	–	Bofa, Western Guinea	Some damages in Bofa
22 June 1939	6.5 (M_L)	IX	Offshore near Nyanyano-Accra, Ghana	17 People died, 1,000 injured and over 1,377,870; US dollars' worth of property damage.
1945	5.6 (Ms)	VIII	Cameroon	Buildings were damaged
1945	5.6 (Ms)	–	Guinea	Strongly felt; caused damage to buildings
4 January 1957	5.7 (Ms)	V	Sierra Leone	–
1964	4.7 (M_L)	–	Akosombo, Ghana	Felt around Akosombo area
1969	4.6 (Ms)	IV	Cameroon	Felt in Fumban and Okola
1969	4.7 (M_L)	–	Tema, Ghana	Felt around Tema
22 December 1983	6.3 (Mw)		Republic of Guinea	300 killed,1500 injured, 5,000 houses damaged, 18,000 made homeless
27 June 1990	3.7 (M_L)	–	South-western Nigeria	Felt around Ibadan and its environs, there were violent vibrations of building foundations, braking of louvre blades. People panicked
7 July 1994	4.2 (M_L)	–	South-western Nigeria	
7 November 1994	4.2 (M_L)	–	Dan Gulbi near Gusau, Northern Nigeria	No vibration or damage was reported
6 March 1997	4.8 (M_L)	–	Accra, Ghana	Cracks in buildings in Accra
11 January 1999	4.2 (M_L)	–	Kolokain, Mali	Less stable structures collapsed
16 October 1999	4.4 (M_L)	–	Saint-Louis, Senegal	Unstable structures collapsed
7 March 2000	4.4 (Mb)	–	Okitipupa east of Lagos, Nigeria	No visible damage or casualty was reported

Contd.

Table 1 *Contd.*

Date	*Magnitude*	*Intensity (MSK)*	*Location*	*Effects*
17 January 2001	4.4 (M_L)	–	Buedu, Eastern province, Sierra-Leone	Less stable structures collapsed
2002	3.6 (M_L)	–	Kribi, Cameroon	Severe ground vibration and panic among the populace
18 May 2003	4.0(M_L)	–	Accra, Ghana	Slight cracks in some buildings
19 March 2005	4.6 (M_L)	–	Yaoundé, Cameroon	Felt in Monatele and Douala
10 January 2006	3.5 (M_L)	–	Accra, Ghana	No visible damage identified
14 June 2007	5.0 (M_L)	–	Bas, Sassandra, Ivory Coast	Objects fell from shelves and many buildings suffered moderate cracks
11 September 2009	4.4 (M_L)	–	Abomey–Calavi, Benin	Less stable structures collapsed
2009	4.4 (M_L)	–	Nigeria	Felt around Abeokuta, Ago-Iwoye, Ajambata, Ajegunle, Imeko, Ijebu-Ode, and Ilaro
24 March 2018	4.0 (M_L)	–	Accra, Ghana	No damage reported
13 January 2019	3.5 (M_L)	–	Accra, Ghana	These events were felt around Kasoa, Gbawe, McCharthy Hill, Weija, and its environs
2 March 2019	3.5 (M_L)	–		
24 June 2020,	4.2 (M_L)	–	Accra, Ghana	Felt within the southern part of Ghana. Residents panicked
12 December 2022,	3.5 (M_L)	–	Accra, Ghana	Felt mostly in parts of the Greater Accra Region
12 December 2023	3.6 (M_L)	–	Accra, Ghana	Felt in Accra

2. Geological and Tectonic Setting

By the early Cretaceous, the African continent had separated from the South American continent along the Mid-Atlantic spreading ridge (Campbell et al., 2018). This period of intense tectonic and thermal activity was contemporaneous with the formation of various basins such as the Bida and Volta Basins (Campbell et al., 2018). Following the last major period of thermal and tectonic activity, the African continent had formed, and the Atlantic margin of West Africa stabilized forming the passive continental margin (Campbell et al., 2018).

The West African craton composed of Archaean and Paleoproterozoic metamorphic and magmatic shields separated by sedimentary basins (Abouchami et al., 1990; Liégeois et al., 1991; Boher et al., 1992), comprises the Man Shield to the south, and the Reguibat Shield as well as the Anti-Atlas Belt both to the north.

The craton comprises a core of older Precambrian rocks which are surrounded by younger Precambrian and Phanerozoic units. The Paleoproterozoic Birimian rocks form a large portion of the West African craton, and due to their importance in the crustal evolution of the West African craton, extensive studies have been done in areas underlain by the Birimian terrane to better understand the geology.

In Ghana, the Paleoproterozoic Birimain Supergroup is mainly characterized by northeast-southwest trending volcanic belts with intervening basins. Both belts and basins were intruded by granitoids of Proterozoic age. Leube et al. (1990) suggested that magmatism and sedimentation were coeval and that these composite belts were predominantly volcano-plutonic ridges which generated coeval sediments through *in situ* strippings of the volcanic rocks.

Most Paleoproterozoic formations in West Africa can be assigned to the Birimian Supergroup, which is largely subdivided into a sedimentary succession, consisting of phyllites, tuffs, and greywackes, overlain by conglomerates, sandstones and shales, and a volcanic succession of tholeiitic affinity.

These two units were originally defined as Lower and Upper Birimian, respectively. It has, however, been demonstrated that the Lower and Upper Birimian rocks represent a coeval sequence, in which the volcanoclastic assemblage is emplaced as a series of evenly spaced volcanic belts. These volcanic belts are several hundred km long and consist mainly of low grade metamorphic tholeiitic lavas (Leube et al., 1990) and are separated by basins containing chiefly isoclinally folded dacitic, volcaniclastic, and argillitic sediments as well as granitoids.

The West African region consists of pre-existing structures that extend significantly (with great lengths) within the continent and the oceanic crust (Neev et al., 1982). Sections of deep external zones are made of dipping reflectors which could indicate high strain zones in the patchily deformed margin of the West. The Akwapim Fault zone marks the southeastern boundary of the West African craton and represents the Pan African thrust (Blundel & Banson, 1975). Fault plane solutions and geological observations of the Akwapim Fault zone suggest a strike-slip movement in response to a regional east southeast trending tensional stress (Bacon & Quaah, 1981). The fault zone is bounded by the Birimian rocks, the Togo and Dahomeyan Formations.

The coastal boundary fault defines the contact between Jurassic and Tertiary sedimentary units, and it is believed to have developed during the Jurassic and continued to be active during Cretaceous and Tertiary times. This fault intersects the Akwapim fault zone at Nyanyanu, a coastal village near Accra. The intersection is marked by intense shearing and fracturing visible at the coastal zone and beach outcrops (Blundell & Banson, 1975).

Fault systems in Nigeria include the Ifewara, Zungeru, Anka, and Kalangai fault systems. They are interpreted to have resulted from transcurrent movements (Riboulet et al., 2012). Adenkule et al. (2012) reported that geologic and geodetic evidence suggest the existence of large fracture zones, namely, Romanche and Charcot in Nigeria (Figure 1). A possible originator of the intraplate earthquakes in Nigeria can be attributed to regional stress created by the West African Craton and the Congo Craton with an inferred principal stress acting West northwest–East southeast (Eze et al., 2012). The tectonic history of West Africa and to an extent Africa dates as far back as the era of continental drift. As rifting separated Africa from South America, many developments occurred because of the tectonic forces and related kinematic properties. These episodes of crustal deformation acted as the initial shapers of the main topographic and tectonic makeup of the region (Campbell et al., 2018).

3. Seismotectonic Setting

The seismicity distribution in this intraplate region is complex and is likely due to the deformation of the oceanic floor and related transform faults combined with major shear zones and thrusts of the Archean and Pan African Orogenies. Seismic activity in the coastal areas may be related to the interaction of the continental margin with the offshore chain transform zone (Burke, 1971).

Although the Akwapim Fault zone of Ghana may seem an old tectonic structure, the present-day seismic activities in Accra and the southeastern parts of Ghana are influenced by the Romanche transform fracture zone which offsets the mid-Atlantic Ridge and extends until the Africa coastline. Meghraoui et al. (2019) establish that the St. Paul's fault moves along the Ivory Coast and reaches the coastline of Ghana (through Axim and Elmina, it travels to Nyanyano along the coast and branches inland) forming the Akwapim Fault zone. Axim and the western coastline of Ghana are influenced by the St. Paul's fracture zone. The mapping by Arens et al. (1971) and the work of

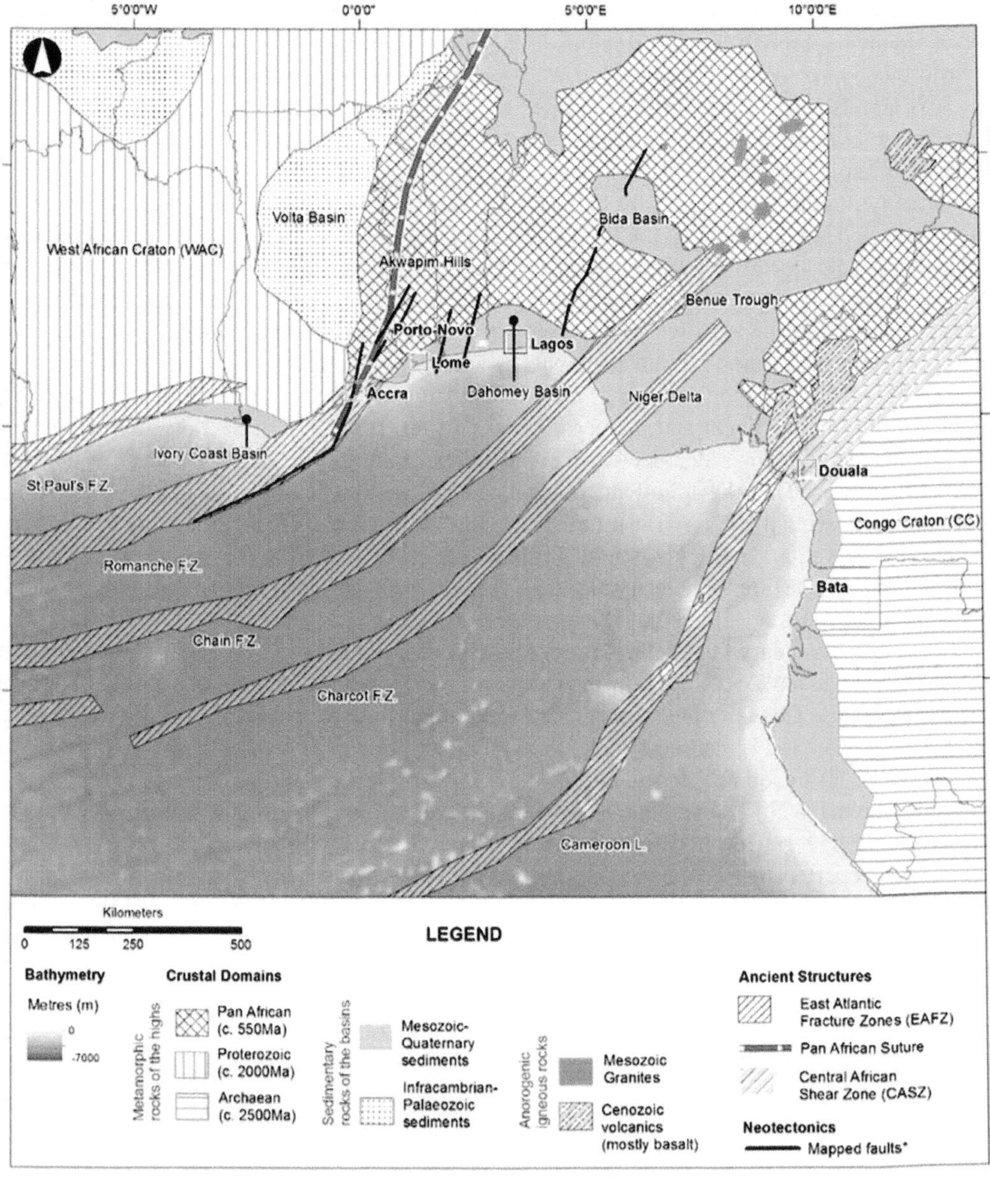

Fig. 1 Geology and mapped faults of West Africa (Campbell et al., 2018).

Mascle and Sibuet (1974) reveal that the Chain fracture zone enters the coast of Ghana further east (Quaah, 1980). Blundell and Banson (1975) and Blundell (1976) contend that continuing strike-slip movement on the Romanche fracture zone adjacent to the West African continental margin may be the cause of the earthquakes in Ghana. The extension of the Akwapim fault zone south-westwards to the continental margin provides a link with the Ghana-Sierra Leone Ridge which is considered to be associated with the Romanche Fracture zone.

Evidence of tectonic stresses obtained from focal mechanism solutions (Figure 2) along fracture zones in the mid-Atlantic outside the limits of ridge offsets is used to explain how stresses may be transmitted from the Romanche fracture zone to affect Accra and the Akwapim fault zone (Blundell & Banson,1975; Meghraoui et al., 2019). The seismic activity along the Akwapim fault zone through southern Ghana into Togo and Benin and its link with the Romanche fracture zone as reported by

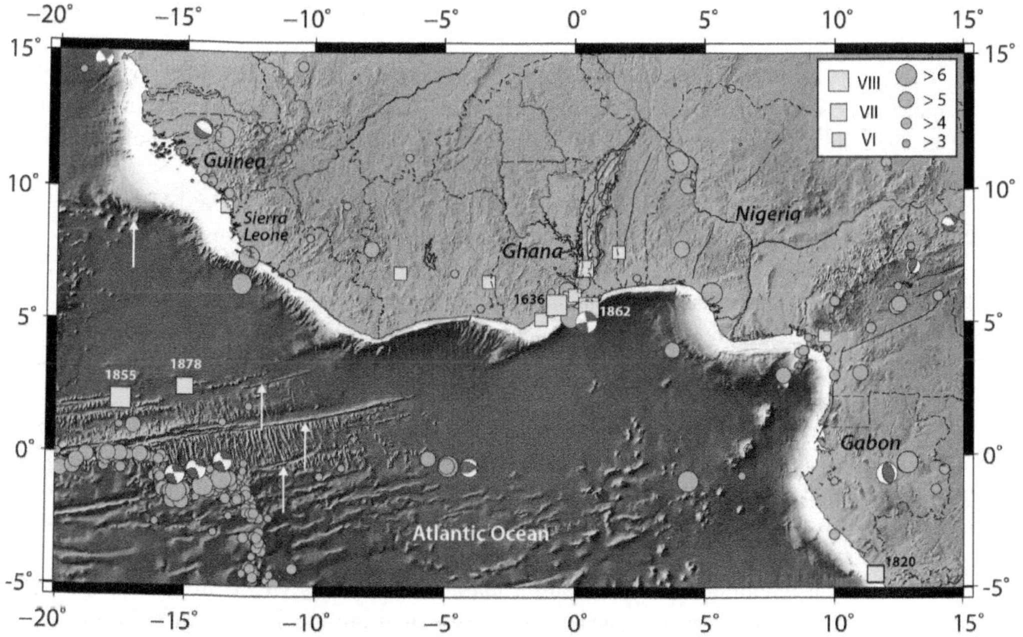

Fig. 2 Seismotectonic Map of West Africa with arrows indicating from north to south the Arkhangelsky, Saint Paul, Romanche, and Chain transform faults, as well as focal mechanisms of major seismic events (Mw > 5) from CMT-Harvard (Meghraoui et al., 2019).

Delteil et al. (1974) indicates a line of active tectonism with faulted Quaternary deposits extending about 1000 km into the African plate from the mid-Atlantic ridge (Blundel, 1976).

4. History of Earthquakes in West Africa

Over the years, some regions in West Africa have experienced moderate and smaller magnitude earthquakes. The most recent destructive earthquake in West Africa was the Guinea earthquake of 1983 (Langer et al., 1985; Ambraseys & Adams, 1986). In 1935, a surface wave magnitude (Ms) 4.0 shock rocked western Guinea and caused some damage in Bofa (Figure 3). This was followed by another seismic event of surface wave magnitude (Ms) 4.1 in the coastal area in 1939 (Ambraseys & Adams,1986). In 1945, an earthquake of surface wave magnitude (M_S) 5.6 struck Guinea. This event was strongly felt and caused damage to public buildings. A surface wave magnitude (M_S) 4.0 earthquake struck Guinea again in 1927. The Kakulima region experienced strong shaking and local inhabitants panicked. It is of interest to mention that about a year after the 1927 earthquake, another event of surface wave magnitude (Ms) 4.8 occurred in the Kakulima area and caused damage in local houses and panic among the population (Ambraseys & Adams,1986).

On 20 May 1795 and 20 June 1796 two earthquakes of estimated surface wave magnitude (Ms) 5.2 (inferred from isoseismal radii) struck Sierra Leone and were strongly felt in Freetown. Again, on January 4, 1957, an earthquake of magnitude 5.7 took place in the coastal area in Sierra Leone (Sykes & Landisman, 1964).

On 11 February 1879, an earthquake of surface wave magnitude (Ms) 5.7 and intensity VIII (MSK) occurred in the eastern part of Ivory Coast and caused considerable damage to buildings. In 1889, another earthquake of surface wave magnitude (Ms) 4.7 and intensity VIII(MSK) struck Ivory Coast and the mainshock was strongly felt in Baibu and Gregbeu where old houses collapsed and trees toppled (Ambraseys & Adams, 1986). Several shocks were reportedly felt in Ivory Coast in 1950.

In Togo, in 1788 an earthquake with estimated surface wave magnitude (Ms) 5.6 (inferred from isoseismal radii) was strongly felt in Agonah, Abomey, Kana, and several neighbouring villages. Again in 1930 several shocks were felt in Palime and Agu in Togo. The surface wave magnitude was 3.5. Another shock occurred in 1933 and was felt in Palime-Misahoe.

In 1832, Senegal experienced an earthquake of surface wave magnitude (Ms) 4.2 (inferred from isoseismal radii) which was felt in Dakar and St. Louis (Ambraseys & Adams, 1986).

In Burkina Faso, the then Upper Volta experienced an earthquake in 1831 which was felt in Baku but there was no information on the maximum intensity (Ambraseys & Adams, 1986).

On 18 February 1618 Mali experienced an earthquake of surface wave magnitude 5.2 (inferred from isoseismal radii). It caused panic among the populace and reptiles moved out from their hiding places (Ambraseys & Adams, 1986).

Destructive earthquakes have been reported in Ghana in the last two centuries in 1636, 1862, and 1939 (Table 1). Earthquakes of intensities ranging between VIII and IX and an estimated magnitude less than 7.0 occurred in the country with severe damage to buildings. Twenty lives were lost. Ambraseys and Adams (1986) estimated the surface and body wave magnitudes of the 1939 earthquake to be (Ms) 6.4 and (M_b) 6.8, respectively, and the maximum intensity as IX (uncertain). A local magnitude (M_L) 6.5 and a maximum intensity (MM) of IX were estimated for this earthquake (Junner, 1941). Quaah (1980) estimated the local magnitude (M_L) to be 6.4. Amponsah and Leydecker (2004) estimated the maximum intensity to be VIII and the depth as 18.0 km. Seventeen people were killed, 133 persons were injured, and an estimated US $ 103,77,870 worth of damage was done to buildings (Junner, 1941).

In recent times, several moderate earthquakes of local magnitude (M_L) less than 7.0 have been experienced in different regions in the subregion.

Seismic events recorded in South-western Nigeria attest to the fact that the country may not be aseismic as had hitherto thought (Akpan & Yakubu, 2010; Alaneme & Okotee, 2018). The first extensively reported occurrence of an earthquake in Nigeria was in 1933 (Akpan & Yakubu, 2010). Other events have been reported in recent times with intensities ranging from III to VI on the Modified Mercalli Intensity (MMI) Scale and local magnitude between (M_L) 3.7 and 4.2 (Akpan & Yakubu, 2010).

In Cameroon, earthquakes have been recorded with moment-magnitude Mw < 6. They are mostly linked to tectonics or the Cameroon Volcanic Line (CVL), a 1500 km long chain of volcanic islands and continental volcanoes (Eloumala et al., 2014; Ateba et al., 2009). Mount Cameroon, an active volcano, has the highest seismicity along the CVL. Its volcanic eruptions are usually accompanied by seismic swarms (Ateba et al., 2009; Ndibi et al., 2021). A large earthquake reaching Mw 6.0 and maximum observed intensity VI (MM) was strongly and widely felt in 1911 in the southern part of the country. Its epicentre was near Lolodorf town (Ambraseys & Adams, 1986). Two years later, in 1913, another earthquake of surface wave magnitude (Ms) 5.1 and maximum intensity VI, was located around Akonolinga. Another earthquake of surface wave magnitude (Ms) 5.6 and maximum intensity VIII, located around the Cameroon/Congo border was widely felt in 1945 in the southern part of Cameroon. In 1986, a surface wave magnitude (Ms) 4.6 earthquake occurred in Cameroon; it was intensely felt in Fumban and Tibati districts.

The seismic activity in West Africa is associated with faulting and an example is the Mw 6.3 earthquake of December 22, 1983, which occurred in Guinea. The main shock was accompanied by an ESE to EW trending surface faulting (Dorbath et al., 1984). The volcanic activity of the CVL and other moderate earthquakes within the continent have also occurred on known faults in the continent (Amponsah et al., 2008; Campbell et al., 2018).

Figure 4 shows a consolidated summary of seismic events in the West African region. Inserted in the figure is the map of Africa with the area of study squared.

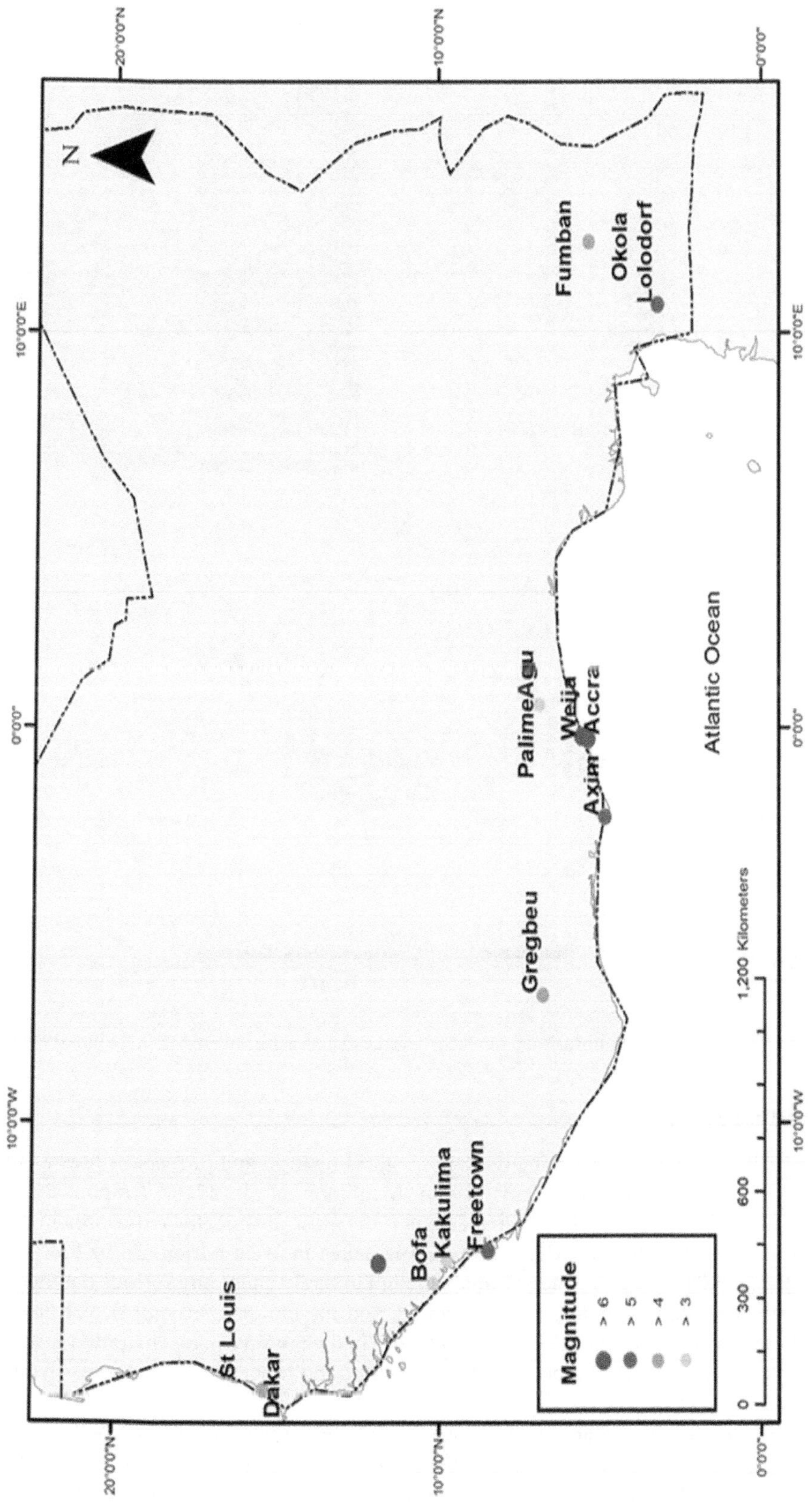

Fig. 3 Map of localities affected by historical earthquakes in West Africa.

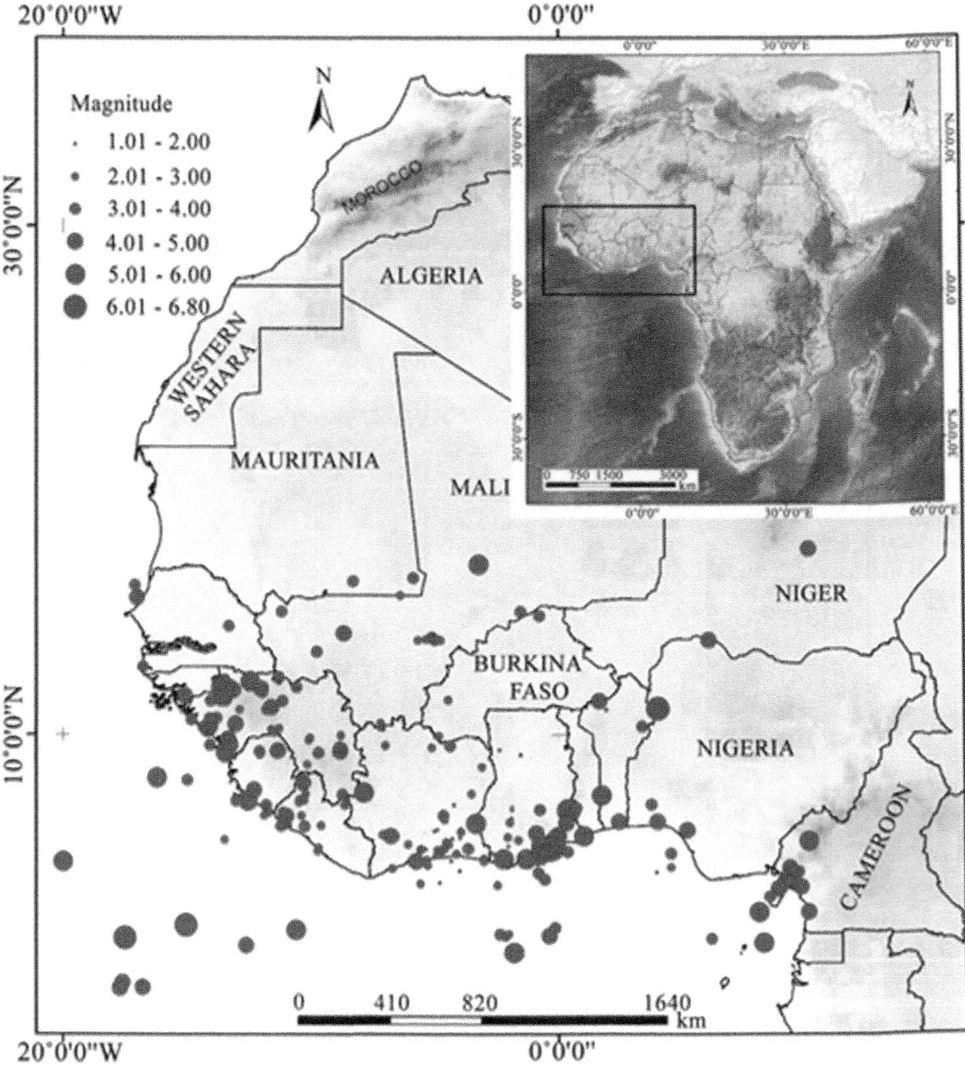

Fig. 4 Map showing both historical and recent seismic events across West Africa (Mbossi et al., 2021).

5. Discussion

The West African region is seismically active both in historic and in recent times as indicated by Quaah, (1980), Ambraseys and Adams (1986), Meghraoui et al. (2016), Amponsah et al. (2020), Mbossi et al. (2021). Research shows that Ghana is the most seismically active area in the subregion (Ambraseys & Adams, 1986). Most of the earthquakes in West Africa are in the range of local magnitude 3.0–5.0 (Figure 4) which is small enough to rarely cause any serious damage (depending on the earthquake depth, geological background, and the building structure), yet large enough to have been felt by many people. The seismic activity in the region can be attributed to and correlated with unique surface features, especially for the minor and moderate earthquakes that characterize the region.

One of the key problems faced in seismic hazard studies, however, has to do with the identification of seismic sources. It is important to understand the location and evolution of the major seismic sources and associated fault ruptures so that advance action is taken to address the risk and impact of earthquake disasters on life and property. This normally requires many organizations/institutions of diverse skills and expertise to implement specific activities such as the development of seismic hazard maps and seismic codes aimed at reducing the seismic risk (Amponsah, 2022; Amponsah et al., 2022).

The methodologies, strategic initiatives, and approach should be effectively coordinated by one body and in most cases, the National Disaster Mitigation Organizations have oversight of this.

The assessment of seismic hazards and the appropriate measures for risk reduction in the vulnerability of buildings and infrastructure is key for seismic hazard mitigation (Adetokunbo et al., 2025; Doku et al., 2023). Building codes could be improved based on the assessment of the hazard to address their inadequacies and enforced satisfactorily to reduce earthquake disasters.

6. Conclusion

Although West Africa's seismic activity is relatively low globally, it has historically been considerable, with destructive earthquakes recorded along major fault zones and the main Gulf of Guinea transform systems. The region is not seismically stable and still has the potential for catastrophic earthquakes (Mw > 7), due to the presence of major seismotectonic structures including the Cameroon Volcanic Line and the Romanche, St. Paul, Chain, and Charcot fracture zones. The need for thorough hazard classification and the inclusion of seismic parameters in regional development planning is reinforced by recent earthquake events that empirically confirm ongoing tectonic activity. A sustained investment in seismic infrastructure is essential. The establishment of a permanent, regional integrated seismic monitoring network would enhance earthquake detection threshold, improve hazard models, and support the production of reliable risk assessments. This will feed into periodic revision of seismic hazard maps, detailed micro-zoning studies of urban centres, and systematic public risk communication. By coupling scientific research with institutional capacity building, West Africa can advance towards evidence-based seismic risk management that mitigates the consequences of future large-magnitude earthquakes.

Acknowledgements

We acknowledge the efforts and support of the staff and National Service personnel of the National Data Center of the Ghana Atomic Energy Commission. We appreciate with gratitude the immense assistance from the Ghana Geological Survey Authority.

References

Abouchami, W., Boher, M., Michard, A. and Albarede, F. (1990). A major 2.1 Ga event of mafic magmatism in West Africa: An early stage of crustal accretion. *Journal of Geophysical Research: Solid Earth*, *95*(B11): 17605–17629.

Adetokunbo, P., Eluyemi, A.A., Jegede, E., Aguda, S., Omo seyin, T., Mohanty, D., Brassnavy, M., Amponsah, P. and Baruah, S. (2025). Geostatistical Analysis for Magnitude Scale Conversion and Gutenberg-Richter Parameter Estimation: Insights from Regional Seismicity of Nubian-Eurasian Plate Boundary Region. *Journal of Geoscience and Environment Protection*, *13*: 230–248. https://doi.org/10.4236/gep.2025.137014.

Adekunle, S.O., Desmond, O. and Nwakanma, A. (2012). Quantitative interpretation of petrophysical parameters for reservoir characterization in an Onshore Field of Niger Delta Basin. *Journal of Geography, Environmental Earth Science Inter.*, *33*(April), 14–36.

Akpan, U.O. and Yakubu T.A. (2010). A review of earthquake occurrences and observations in Nigeria. *Earthquake Science*: 289–294.

Alaneme, K.K. and Okotee, E.A. (2018). Critical evaluation of seismic activities in Africa and curtailment policies: A review. *Geoenvironmental Disasters*, *5*: 24.

Ambraseys, N.N. and Adams, R.D. (1986). Seismicity of West Africa. *Annales Geophysicae*, *4* (B 6): 679–702.

Amponsah, P. (2022). Natural hazards governance in Ghana. *Oxford Research Encyclopedia of Natural Hazard Science*. doi: https://doi.org/10.1093/acrefore/9780199389407.013.444.

Amponsah, P.E., Banoeng, B.K., Andam, A. and Asiedu, D. (2008). Soil radon concentration along fault systems in parts of south eastern Ghana. Journal of African Earth Sciences, volume 51, issue 1, pp. 39–48.

Amponsah, P., Midzi, V., Amoah, P. and Tetteh-Cofie Tetteh, A. (2022). The 24 June 2020, Earthquake in Southern Ghana. In: Unnikrishnan, A.S., Fredolin Tangang, and Raymond J. Durrheim (Eds.), *Extreme Natural Events-Sustainable Solutions for Developing Countries* (Chapter 14, pp. 381–390). https://doi.org/10.1007/978-981-19-2511-5. Springer Nature, published by Springer Nature Singapore Pte Ltd. ISBN 978-981-19-2510-8; ISBN 978-981-19-2511-5 (eBook).

Amponsah, P., Opoku-Ntim, I. and Nortey, G. (2020). Seismic risk in Ghana: Efforts and challenges. *Arabian Journal of Geosciences, 13*: 1–5.

Amponsah, P.E. and Leydecker, G. (2004). *Earthquake Catalogue for the Republic of Ghana and Adjacent Areas since the Year 1615* (pp. 1–11).

Arens, G., Delteil, J.R., Vallery, P., Damotte, B., Montadert, L. and Patriate, P. (1971). *The Continental Margin off the Ivory Coast and Ghana*. Institute of Geological Science Report, No. 70/16 (pp. 61–78).

Ateba, B., Dorbath, C., Dorbath, L., Ntepe, N., Frogneux, M., Aka, FT., Hell, J.V., Delmond, J.C. and Manguelle, D. (2009). Eruptive and earthquake activities related to the 2000 eruption of Mount Cameroon volcano (West Africa). *J. Volcanol. Geotherm. Res., 179*: 206–216. https://doi.org/10.1016/j.jvolgeores.2008.11.021.

Bacon, M. and Quaah, A.O. (1981). Earthquake Activity in Southern Ghana (1977–1980) *Bulletin of Seismological Society of America, 71* (3): 771–785.

Blundell, D.J. (1976). Active faults in West Africa. *Earth and Planetary Science Letters, 31*: 287–290. Elsevier Scientific Publishing Company, Amsterdam.

Blundell, D.J. and Banson, J.K.A. (1975). *Interpretation of Seismic Reflection Survey across the Continental Shelf of Accra and Its Bearing on Earthquakes in the Area*. Report Geological Survey No. 75/1 (pp 1–7).

Boher, M., Abouchami, W., Michard, A., Albarede, F. and Arndt, N.T. (1992). Crustal growth in west Africa at 2.1 Ga. *Journal of Geophysical Research: Solid Earth, 97*(B1): 345–369.

Bondessen, E. and Schmidt, A.F.J. (1972). Holocene tectonic activity in West Africa dated by archaeological methods: Discussion. Geological Society of America. *Bulletin, 83*: 1193–1196.

Burke, K. (1971). Recent faulting near the Volta dam, Nature *231*, 439–440.

Campbell, G., Lubkowski, Z., Villan, M. and Polidoro, B. (2018). A seismic source model for West Africa. In: *16th European Conference on Earthquake Thessaloniki Engineering*, Thessaloniki.

Delteil, J.R., Valery, P., Montadert, L., Fondeur, C., Patriat, P. and Mascle, J. (1974). Continental margin in the northern part of the Gulf of Guinea. *The Geology of Continental Margins*, 297–311.

Doku, M.S., Sam, F., Adams, S. J. and Amponsah, P.E. (2023). An investigation into the stress conditions of critical and lifeline infrastructure in South-eastern Ghana using the maximum likelihood estimation method. *Journal of Ghana Science Association, 21*(1): 69–73.

Dorbath, C., Dorbath, L., Gaulon, R., George, T., Mourgue, P., Ramdani, M., Robineau, B. and Tadili, B. (1984). Seismotectonics of the Guinean Earthquake of December 22, 1983. *Geophysical Research Letters, 11* (10): 971–974. Bibcode: 1984GeoRL..11..971D. doi:10.1029/GL011i010p00971.

Eloumala Onana, P.N., Mouzong, P.M. and Ateba, B. (2014). Crustal Structure and Seismogenic Zone of Cameroon: Integrated Seismic, Geological, and Geophysical Data. *Open Journal of Earthquake Research, 3*: 152–161. http://dx.doi.org/10.4236/ojer.2014.34015.

Eze, C.L., Sunday, V.N., Ugwu, S.A., Uko, E.D. and Ngah, S.A. (2012). Mechanical model for Nigerian intraplate earth tremors. *Advances in Science and Technology, 6*(2): 80–84.

Junner, N.R. (1941). The Accra Earthquake of 22nd June 1939. *Bulletin, 13*: 3–41. Gold Coast Geological Survey.

Kadiri, U.A. and Amponsah, P. (2021). Computation of area-characteristic seismicity parameters in Ghana, Nigeria, and immediate neighbors. *Arabian Journal of Geosciences, 14*(13).

Langer, C.J., Bonilla, M.G. and Bollinger, G.A. (1985). *The Guinea, West Africa, Earthquake of December 22, 1983: Reconnaissance Geologic and Seismologic Field Studies*. Open-File Report 85-282. United States Department of the Interior Geological Survey.

Leube, A., Hirdes, W., Mauer, R. and Kesse, G.O. (1990). The early Proterozoic Birimian Super group of Ghana and some aspects of its associated gold mineralization. *Precambrian Res., 46*: 139–165.

Liegeois, J.P., Claessens, W., Camara, D. and Klerkx, J. (1991). Short-lived Eburnian orogeny in southern Mali: Geology, tectonics, U-Pb, and Rb-Sr geochronology. *Precambrian Research, 50* (1–2): 111–136.

Mascle and Sibuet (1974). Continental Margin in the Northern Part of the Gulf of Guinea. *The Geology of Continental Margins* (pp. 297–311) SpringerLink..

Mbossi, E.F., Ndi bi, D.D.E., Nguet, P.W., Essi, J.M.A., Ntomb, E.O.B., Dili-Rake, J., Ateba, B. and Tabod, T.C. (2021). Preliminary Seismic Hazard Assessment in West Africa Based on Incomplete Seismic Catalogs. *Open Journal of Earthquake Research, 10*: 75–93. https://doi.org/10.4236/ojer.2021.102006.

Meghraoui, M., Amponsah, P., Bernard, P. and Ateba, B. (2019). Active Transform Faults in the Gulf of Guinea: Insights from geophysical data and implications for the seismic hazard assessment. *Canadian Journal of Earth Science, 56*: 1398–1408.

Meghraoui, M., Amponsah, P., Ayadi, A., Ayele, A., Ateba, B., Bensuleman, A., Delvaux, D., El Gabry, M., Fernandes, R., Midzi, V., Roos, M. and Timoulali, Y. (2016). The seismotectonic map of Africa. *Episodes, 39* (1): 9–18.

Ndibi, E.D.D., Mbossi, E.F., Wokwenmenda, N.P., Ateba, B. and Ndougsa-Mbarga. T. (2021). Probabilistic seismic hazard assessment for the main cities along the continental section of the Cameroon Volcanic Line. *Arabian Journal of Geosciences, 14*: 796. https://doi.org/10.1007/s12517-021-07033-2.

Neev, D., Hall, J.K. and Saul, J.M. (1982). The Pelusium megashear system across Africa and associated lineament swarms. *Journal of Geophysical Research: Solid Earth, 87*(B2): 1015–1030.

Quaah, A.O. (1980). *Microseismicity, Past Seismic Activity, and Seismic Risk in Southern Ghana*. Ph.D Theses (pp. 23–198). University of London.

Riboulot, V., Cattaneo, A., Berne, S., Schneider, R.R., Voisset, M., Imbert, P. and Grimaud, S. (2012). Geometry and chronology of late Quaternary depositional sequences in the Eastern Niger Submarine Delta. *Marine Geology*, (319–322): 1–20. doi: 10.1016/j.margeo.2012.03.002.

Sykes, L.R. and Landisman, M. (1964). The seismicity of east Africa, the Gulf of Aden, and the Arabian and Red Seas. *Seismological Society of America Bulletin, 54*: 1927, 23–198–1940.

Sykes, R.L. (1978). Intraplate Seismicity, Reactivation of Pre-existing Zones of Weakness, Alkaline Magmatism, and other Tectonism Postdating Continental Fragmentation. *Reviews of Geophysics and Space Physics, 16*(4): 621–682.

CHAPTER

8

Application of Bicharacteristic Curves of Body and Surface Waves in Geophysics

Georgi Boyadzhiev[1*]

In this chapter is presented an algorithm for 3D modelling of elastic piecewise homogeneous media, with application to Earth's crust and upper Mantle studies. The method is based on the principle of tomography. The source of the signal is an earthquake and by presumption there are at least three receiving stations on the surface.

Wave propagation in a solid media is described by a system of three strongly coupled hyperbolic equations with piece-wise constant coefficients. The 3D algorithm is based on one important feature of the hyperbolic PDEs, the singularities of the solution propagate on the bicharacteristic curves. In this chapter, the characteristic set and bicharacteristic curves are derived for the hyperbolic system in a homogeneous half-space with free boundary, as well as formulae of reflection and diffraction of the bicharacteristics on the internal boundaries of the media. Therefore, setting the presumed parameters of the source and approximated structure of the media, the propagation of the singularities can be computed and the set source/media to be verified to the real data from the receiver stations. The algorithm is applicable to the inverse problem in geophysics and Earth modeling.

Development of this 3D algorithm was inspired by cooperation of the author with Prof. G.F. Panza at the University of Trieste and ICTP Trieste.

1. Introduction

The aim of this chapter is the introduction of one geometric approach to 3D modelling of elastic piecewise homogeneous media, in particular of Lithosphere, the Earth's crust and upper Mantle. The method is based on classical tomography, and the main source of information are seismic waves, generated by a point source S and recorded by a set of seismic stations on the Earth surface. Irregularity of earthquakes is counterpoised by the density of seismic stations and plenty of data are available for geophysical surveys.

The structure of our planet and the strong ground motion have always been a subject of human curiosity and research. One main obstacle in exploring seismic mechanisms is the scale of the object in study. The planet is too large to be examined by probes. As a matter of fact, human best technology provides samples of Earth structure just a few kilometres deep. The cost of drilling is high as well.

[1] Institute of Mathematics and Informatics, Bulgarian Academy of Sciences, 8 Acad. G. Bonchev str, 1113 Sofia, Bulgaria. University of Architecture, Civil Engineering and Geodesy, 1 Hristo Smirnenski Blvd, 1046 Sofia, Bulgaria.

* Corresponding author: gpb@math.bas.bg

DOI: 10.1201/9781003684121-10

There are very few places like the San Andreas fault in California, where the seismic source is close to the surface and geologists examine the fault directly, providing best data for the features of the seismic source and the ground structure. For the rest of the planet there are applicable methods such as tomography. Several well-posed receivers record the data from a certain source. Then the data are processed with different methods like inverse tomography, for instance, and an approximation of the ground structure is obtained.

As a wave source in shallow and relatively small areas explosives are used. For larger areas data records of actual earthquakes are employed. Analysis of accurate records of the exact time and shape of the wave, provided by different seismic stations, allow determination of the depth and other features of the source with methods such as time-frequency analysis and others. Then tomography could be used to determine the ground structure.

Unfortunately, the tomography is not an exact method even when the source of the signal is known and under control. In seismology and geophysics, it is even more imprecise for neither the media, nor the source of waves is known. All we have is the seismogram as a data source. It is one set of data for determining two unknown quantities. That is why an iteration process is employed. Firstly, a model of the media is created over some assumptions to the source features. Then the media model is used to approximate the source features, etc.

For the purposes of geophysics, it is sufficient if the Earth is considered as an elastic body that is a continuum, in other words the matter is continuously distributed in space. Furthermore, since seismicity has relatively local effect, the planet can be approximated with no loss of generality by a half-space Ω with free surface $z = 0$ (in Cartesian coordinates) and axis z positive in downward direction. If the elastic parameters depend only on vertical coordinate z then the wave propagating in solid media satisfies the strongly coupled linear hyperbolic system:

$$\rho\frac{\partial^2 u_x}{\partial t^2} = X + (\lambda+2\mu)\frac{\partial^2 u_x}{\partial x^2} + \mu\frac{\partial^2 u_x}{\partial y^2} + \mu\frac{\partial^2 u_x}{\partial z^2} + (\lambda+\mu)\frac{\partial^2 u_y}{\partial x\partial y}$$

$$+ (\lambda+\mu)\frac{\partial^2 u_z}{\partial x\partial z} + \frac{\partial\mu}{\partial z}\frac{\partial u_x}{\partial z} + \frac{\partial\mu}{\partial z}\frac{\partial u_z}{\partial x}$$

$$\rho\frac{\partial^2 u_y}{\partial t^2} = Y + \mu\frac{\partial^2 u_y}{\partial x^2} + (\lambda+2\mu)\frac{\partial^2 u_y}{\partial y^2} + \mu\frac{\partial^2 u_y}{\partial z^2} + (\lambda+\mu)\frac{\partial^2 u_x}{\partial x\partial y}$$

$$+ (\lambda+\mu)\frac{\partial^2 u_z}{\partial y\partial z} + \frac{\partial\mu}{\partial z}\frac{\partial u_y}{\partial z} + \frac{\partial\mu}{\partial z}\frac{\partial u_z}{\partial y} \qquad \text{...(1)}$$

$$\rho\frac{\partial^2 u_z}{\partial t^2} = Z + \mu\frac{\partial^2 u_z}{\partial x^2} + \mu\frac{\partial^2 u_z}{\partial y^2} + (\lambda+2\mu)\frac{\partial^2 u_z}{\partial z^2} + (\lambda+\mu)\frac{\partial^2 u_x}{\partial y\partial z} + (\lambda+\mu)\frac{\partial^2 u_y}{\partial y\partial z}$$

$$+ \frac{\partial\lambda}{\partial z}\left(\frac{\partial u_x}{\partial x} + \frac{\partial u_y}{\partial y} + \frac{\partial u_z}{\partial z}\right) + 2\frac{\partial\mu}{\partial z}\frac{\partial\mu_z}{\partial z}$$

where, (u_x, u_y, u_z) is the displacement function. λ, μ and ρ are piece-wise continuous functions of z. Boundary conditions at the free surface $z = 0$ are given by:

$$\sigma_{zz} = (\lambda+2\mu)\frac{\partial u_z}{\partial z} + \lambda\left(\frac{\partial u_x}{\partial x} + \frac{\partial u_y}{\partial y}\right) = 0$$

$$\sigma_{zz} = \mu\left(\frac{\partial u_x}{\partial z} + \frac{\partial u_z}{\partial x}\right) = 0 \qquad \text{...(2)}$$

$$\sigma_{zy} = \mu\left(\frac{\partial u_y}{\partial z} + \frac{\partial u_z}{\partial y}\right) = 0$$

and in addition, the continuity of the functions u_x, u_y, u_z, σ_{zz}, σ_{zx} and σ_{zy} in Ω is required.

Coefficients ρ, λ, and μ depend on the geological properties of the rock. Though there is little information we have about exact ground structure, geological surveys near the surface show that the Earth crust is heterogeneous and consists of areas with homogeneous rock. That is why in any realistic model of the Earth crust and upper mantle the coefficients ρ, λ, and μ are piece-wise continuous functions and the results for the wave front set (Theorem 8.3.1, Hormander, v. I, p. 271) are not applicable. On the other hand, a reasonable approximation of the real Earth is a 3D structure of homogeneous blocks in welded contact $\{B_{i,j,k}\}$, where i and j are integers, and k is a natural number. Blocks $B_{i,j,k}$ and $B_{i+1,j,k}$ are neighbours along x axis, $B_{i,j,k}$ and $B_{i,j+1,k}$ along y, and $B_{i,j,k}$ and $B_{i,j,k+1}$ along z axis. Without loss of generality, we suppose the source S of the seismic signal is in block $B_{0,0,0}$.

Numerical methods such as finite differences, finite elements and hybrid one are applicable if the domain of system (1), (2) is relatively small. Otherwise, the grid is too large, and computational time is too costly or the approximation error is too high. There are also different analytical approaches to system (1), (2). One is based on the fundamental solution of (1) in integral form and the presumption that so called Rayleigh and Love modes give good approximation of the solution when the distance from the source is large enough compared to the wavelength (Levshin, 1973; or Aki & Richards, 2002). Some other methods are briefly mentioned in Section "Brief review of some 2D methods". All of them and others are well described in Panza et al. (2001) and are based on one standard approach in geophysics, namely, in case the body forces are neglected, the solution of (1) is considered as a plane harmonic wave propagating along the positive x axis:

$$u(x, t) = F(z) \cdot e^{i(\omega t - kx)} \qquad \text{...(3)}$$

where, ω is the angular frequency and k is the wavenumber corresponding to the phase velocity c, that is, $k = \omega/c$.

The main disadvantage of this approach is that the plane wave (3) is a 2D one, living in the plane $y = 0$ only. Roughly speaking, all information on y coordinate is lost and it is impossible to build a reasonable 3D model on a plane wave (3). Therefore, in order to build a realistic 3-D model of the ground structure, plane waves in the form (3) is a significant restriction. The information lost in splitting the system (1) is a serious obstacle to true-to-life modelling. This is the reason a new approach is suggested in this chapter, based on the information from system (1) itself.

Three major components of the ground motion modelling can be distinguished: the source of the motion, the media, or the ground structure, and the equation of motion. The last component is given by system (1). Let us focus on the first two.

There are different models corresponding to very different sources of seismic waves. In this chapter, we consider point sources that produce an impulse in a certain direction, namely, along-side vector $\xi_0 = (\xi_1^0, \xi_2^0, \xi_2^0)$. If the source is more complicated it can be represented as a vector field on a given curve, that is, the fault. In this case the procedure that is described below for a point source can be easily adapted.

The ground structure of the Earth is really complicated. That is why the standard approximation is used, the real ground is approximated by parallelepiped like a grid, mostly rectangular, i.e., a set of parallelepipeds with common sides which we call "blocks". In each block, the elasticity coefficients, i.e., the coefficients of system (1), are supposed to be constant ones. The method proposed in this chapter works as well if some of the sides of the blocks are not plane ones, but any smooth and convex surface which is not parallel to the bicharacteristics of system (1), i.e., if there are no gliding rays.

Considering the ground model above, the system (1) is just a strongly coupled system of three linear PDEs with piece-wise constant coefficients. Since the system (1) has constant coefficients within every block, the wavefront of the solution is a subset of the characteristic set of system (1). Furthermore, as the principal part is real with constant coefficients, the wavefront set is invariant under the bicharacteristic flow. Having in mind the source model described above, a point source with seismic impulse in some direction, actually the singularities of the solution carry all the information about the wave. Briefly speaking, the solution propagates over bicharacteristic curves within every homogeneous block. Then we can use so-called train solutions, i.e., the solution in the first block, the one containing the source, determine the boundary conditions for its neighbouring blocks, etc. According to geometrical optics and microlocal analysis, if a bicharacteristic curve reflects off the sides of every block the angle of incidence to the surface is equal to the angle of reflection. As for refraction at a surface, it is computed in the usual way, more details and exact computations are given in Chapter 5 below. Therefore, if we know the position of the source, the direction of the seismic impulse and media structure we can compute the point s_0 where the bicharacteristic curve has contact with the surface $z = 0$. The point s_0 is in fact the centre of the surface waves in the plane $z = 0$ generated by the section of the wave front and the plane $z = 0$. When actual measurement of the seismic waves is done, the coordinates of the point s_0 can be triangulated using the data of several stations. This way verification of the media model could be done. Exact coordinates of the epicenter of an earthquake and the centre of the surface waves r_0 can be computed using different and quite reliable techniques, like time-frequency analysis, based on the data from seismic stations. Given a certain 3 D media model, we can compute the point s_0. If the points s_0 and r_0 coincide within the error of the computations, then the media model is plausible.

For practical purposes, 3 D models to be tested with the procedure mentioned above can be generated using Monte Carlo or other well-known methods. Of course, like any other inverse problem, this algorithm has multiple solutions in the sense that many models can cover the requirement $s_0 \sim p_0$. Unfortunately, this is the best result we can hope for, given the complexity of the object to study and the information obtained by seismograms.

Another application of the bicharacteristics is in all 3 D models. Strictly speaking, approximating the Earth structure by a grid of rectangular parallelepipeds introduces artificial singularities at the vertices of the parallelepipeds as the boundary data there are not harmonized. These singularities "propagate" over the bicharacteristic curves and therefore the model should be chosen such that the bicharacteristics do not contain a vertex of the parallelepiped. In coupling coefficient models, the "coupling" of the coefficients should be done at one side of the bicharacteristic cone only since the solution is not smooth at the cone.

2. Brief Review of Some 2D Methods

All the following methods are well described in Panza et al. (2001).

If body forces are neglected, suppose solutions of (1) are harmonic waves (3) propagating along the positive x axis. Then system (1), (2) reduces to a single equation and 2D system:

$$\frac{\partial}{\partial z}\left[\mu\frac{\partial F_x}{\partial z} - ik\mu F_z\right] - ik\lambda\frac{\partial F_z}{\partial z} + F_x\,[\omega^2 \rho - k^2(\lambda + 2\mu)] = 0$$

$$\frac{\partial}{\partial z}\left[(\lambda + 2\mu)\frac{\partial F_z}{\partial z} - ik\lambda F_x\right] - ik\mu\frac{\partial F_x}{\partial z} + F_z[\omega^2 \rho - k^2\mu] = 0 \qquad ...(4)$$

where, ω is the angular frequency and k is the wavenumber corresponding to the phase velocity c, that is, $k = \omega/c$. Boundary conditions at surface $z = 0$ are:

$$\sigma_{zz} = (\lambda + 2\mu)\frac{\partial F_z}{\partial z} - ik\lambda F_x = 0$$

$$\sigma_{zx} = \mu\left(\frac{\partial F_x}{\partial z}\right) - ikF_z = 0$$

in which the solutions are so-called P-SV waves.

The equation is:

$$\frac{\partial}{\partial z}\left(\mu\frac{\partial F_y}{\partial z}\right) + F_y(\omega^2\rho - k^2\mu) = 0 \quad ...(5)$$

with boundary conditions at $z = 0$

$$\mu\frac{\partial F_y}{\partial z} = 0$$

in which the solutions are called SH waves.

If the ground structure is approximated by parallel to the surface $z = 0$ homogeneous layers, in other words, if parameter functions λ and μ are step - wise ones, i.e., constants in each layer, then one widely used and efficient method of solving system (4) and equation (5) is so called Multimodal method, or modal summation technique. The idea is to compute the solution of (4) or (5) in each layer, that is to solve the equation (or system) with constant coefficients, and then use the welded contact on the boundary, i.e., the boundary conditions on the common boundary are the same. For instance, if we consider Love modes in model with N layers, and $(u_y)_m$ is the solution of (5) in the m-th layer, then:

$$[k_m(u_y)_m\ (\sigma_{zy})_m] = a_m[k_{m-1}\ (u_y)_{m-1}\ (\sigma_{zy})_{m-1}]$$

where the constant 2×2 matrix a_m is dependent only on ρ_m, α_m, μ_m and density of the layer d_m. Denoting $A = a_{N-1}.a_{N-2}...\ a_1$ we have:

$$[k_m(u_y)_N\ (\sigma_{zy})_N] = a_m[k_{m-1}\ (u_y)_0\ (\sigma_{zy})_0].$$

Then the eigenfunctions, which are the basis of the space of solutions of equation (5), can be easily computed. Similar procedures can be applied to system (4).

For a ground model with lateral heterogeneous media, other approaches are employed as so-called Ray theory and various variants of mode coupling, like WKBJ method, etc. In Ray theory, for instance, the basic assumption is that the solution of the elastic equations of motion has the form $u(x, \omega) = A(x, x_0, \omega)\cdot e^{i\omega\theta(x,x_0)}/\sqrt{J(x, x_0)}$, where, $\theta(x, x_0)$ is the phase function, representing the time that wave takes to travel from point x_0 to point x, and J is the geometric decay of the wavefront (Babich, 1956). Please incorporate Further assuming that A has the form $A(x, x_0, \omega) = S(\omega)\cdot\sum A_i(x, x + 0)\omega^{-i}$, where $S(\omega)$ is the wavefront of the source time function. Then, function $A(x, x_0, \omega)$ is approximated by $S(\omega)\cdot A_0(x, x_0)$, and respectively, $u(x, \omega) \approx S(\omega)\cdot A_0(x, x_0)\cdot e^{i\omega\theta(x,x_0)}/\sqrt{J(x, x_0)}$, and the wavefront can be easily computed. As the wavefront propagates on the bicharacteristics, the so-called ray tracking and techniques from the geometrical optics allow the computation of rays and the following wave fronts. Unfortunately, this method is quite sensitive to small local perturbations of the velocity field and it works well if the dimension of the heterogeneities is much larger than the dominated wavelength of the considered waves.

Mode coupling uses ideas like the one Multimodal method described above. It is focused on the study of the surface waves, giving the fact that they, mostly fundamental, first and few higher modes (eigenfunctions), represent the longest and strongest part of a seismic signal generated by an earthquake. Different methods such as WKBJ, Invariant Embedding Technique, and Coupling Coefficients Method implement this idea in practice.

The main disadvantage of all the methods mentioned above is the limitation of the plane wave described in (3). It is in fact a 2D one, living in the plane $y = 0$ only. That is why the system (1) decomposes to system (10) and equation (11). Roughly speaking, the plane wave (3) is like a restriction operator on plane $z = 0$. The residue is equation (11), which is independent of x and z coordinates, but some information is lost. This assumption is not a problem at all if we suppose the source and the receiver of the signal to be situated in this plane and assume that everything meaningful happens within this plane. In other words, if we consider a 2D model then a plane wave in the form (3) is acceptable.

3. Characteristic Set and Bicharacteristic Strip in Homogeneous Block $B_{i,j,k}$

As it is well known, the characteristic set of a linear scalar operator $L(x, D) = \sum_{|\alpha| \leq 22} a_\alpha (x)D^\alpha$ is given by the zeroes of its principal symbol $p_L(x, \xi) = \sum_{|\alpha|=2} a_\alpha (x)\xi^\alpha$. In the case of a linear strongly coupled system, the characteristic set contains the zeroes of the determinant of the characteristic matrix of the system.

Each element of the characteristic matrix is the principal symbol of the corresponding equation with respect to the corresponding argument. For instance, if the system is $L_i(u_1, u_2, ...u_n) = 0$, $i = 1,\ldots n$, then the element (k, m) of the characteristic matrix is the principal part of L_k with respect to u_m. Then the characteristic set of system (1) in every block $B_{i,j,k}$ is given by the equation:

$$0 - p(x, \xi) -$$

$$|\rho\tau^2 - (\lambda + 2\mu)\xi_1^2 - \mu\xi_2^2 - \mu\xi_3^2 - (\lambda + \mu)\xi_1\, \xi_2 - (\lambda + \mu)\xi_1\, \xi_3 - (\lambda + \mu)\xi_1\, \xi_2\, \rho\tau^2 - \mu\xi_1^2 - (\lambda + 2\mu)$$
$$\xi_2^2 - \mu\xi_3^2 - (\lambda + \mu)\xi_2\, \xi_3 - (\lambda + \mu)\xi_1\, \xi_3 - (\lambda + \mu)\xi_2\, \xi_3\, \rho\tau^2 - \mu\xi_1^2 - \mu\xi_2^2 - (\lambda + 2\mu)\xi_3^2|.$$

If we denote by $\alpha = -(\lambda + \mu)^{-1} ([\rho\tau^2 - \mu(\xi_1^2 + \xi_2^2 + \xi_3^2)])]$ we have

$$0 = |\alpha + \xi_1^2\, \xi_1\, \xi_2\, \xi_1\, \xi_3\, \xi_1\, \xi_2\;\; \alpha + \xi_2^2\, \xi_2\, \xi_3\, \xi_1\, \xi_3\, \xi_2\, \xi_3\, \alpha + \xi_3^2|$$
$$= \alpha|\alpha + \xi_2^2\, \xi_2\, \xi_3\, \xi_2\, \xi_3\, \alpha + \xi_3^2| + |\xi_1^2\, \xi_1\, \xi_2\;\; \xi_1\, \xi_3\, \xi_1\, \xi_2\, \alpha + \xi_2^2\, \xi_2\, \xi_3\, \xi_1\, \xi_3\, \xi_2\, \xi_3\, \alpha + \xi_3^2|$$
$$= \alpha|\alpha + \xi_2^2\, \xi_2\, \xi_3\, \xi_2\, \xi_3\, \alpha + \xi_3^2\, | + \alpha|\xi_1^2\, \xi_1\, \xi_3\, \xi_1\, \xi_3\, \alpha + \xi_3^2\, |$$
$$+ |\xi_1^2\, \xi_1\, \xi_2\, \xi_1\, \xi_3\, \xi_1\, \xi_2\, \xi_2^2\, \xi_2\, \xi_3\, \xi_1\, \xi_3\;\; \xi_2\, \xi_3\, \alpha + \xi_3^2| =$$
$$= \alpha|\alpha + \xi_2^2\, \xi_2\, \xi_3\, \xi_2\, \xi_3\, \alpha + \xi_3^2| + \alpha|\xi_1^2\, \xi_1\, \xi_3\, \xi_1\, \xi_3\, \alpha + \xi_3^2| + \alpha|\xi_1^2\, \xi_1\, \xi_2\, \xi_1\, \xi_2\, \alpha + \xi_2^2|$$
$$+ |\xi_1^2\, \xi_1\, \xi_2\;\; \xi_1\, \xi_3\, \xi_1\, \xi_2\, \xi_2^2\, \xi_2\, \xi_3\, \xi_1\, \xi_3\, \xi_2\, \xi_3\, \xi_3^2| =$$
$$= \alpha^2(\alpha + \xi_1^2 + \xi_2^2 + \xi_3^2).$$

These simple calculations show that the characteristic set of system (1) consists of two subsets:

$$p_1\,(x, \xi) = \rho\tau^2 - \mu(\xi_1^2 + \xi_2^2 + \xi_3^2) = 0\; p_2(x, \xi) = \rho\tau^2 - (\lambda + 2\mu)(\xi_1^2 + \xi_2^2 + \xi_3^2) = 0 \qquad \text{...(6)}$$

since $(\lambda + \mu) > 0$.

Therefore, the wave propagating in homogeneous block $B_{i,j,k}$ is actually a composition of two waves. This result corresponds to the theory of P (primary) and S (secondary) body waves. P wave corresponds to the set defined by $p_2(x, \xi) = 0$, and S wave corresponds to the one defined by $p_1(x, \xi) = 0$.

Another important object in our study is the bicharacteristic strip of the linear strongly coupled system (1), since the characteristic set of an operator with real principal part $p(x, \xi)$ and constant coefficients is invariant under the bicharacteristic flow (Hormander, I Chapter 8). By definition, if $p(x^0, \xi^0) = 0$, then the bicharacteristic strip at point (x^0, ξ^0) is defined by the Hamilton equations:

$$\frac{dx}{ds} = \frac{\partial p(x, \xi)}{\partial \xi}, \frac{d\xi}{ds} = \frac{\partial p(x, \xi)}{\partial x}$$

with initial data $(x, \xi) = (x_0, \xi_0)$ for $t = 0$. The bicharacteristic strip l_1 generated by $p_1(x, \xi)$ through point (x^0, ξ^0) is:

$$|x_i = 2\xi_i^0 \cdot s + x_i^0,\ i = 1, 2, 3\ t = 2c\sqrt{(\xi_1^0)2 + (\xi_2^0)2 + (\xi_3^0)2} \cdot s + t^0,\ \xi_i = \xi_0,$$

$$i = 1, 2, 3,\ \tau = \tau^0 = c\sqrt{(\xi_1^0)2 + (\xi_2^0)2 + (\xi_3^0)2},\ i = 1, 2, 3 \quad ...(7)$$

as $p_1(x^0, \xi^0) = 0$. Constant $c = \sqrt{\mu/\rho}$ for bicharacteristics generated by $p_1(x, \xi)$ and $c = \sqrt{(\lambda + 2\mu)/\rho}$ for ones generated by $p_2(x, \xi)$.

The values of ξ_1^0, ξ_2^0, and ξ_3^0 are determined by the features of the seismic source. Without loss of generality, one can assume source of the seismic wave to be a point one with direction of the impulse $(\xi_1^0, \xi_2^0, \xi_3^0)$.

The restriction of the bicharacteristic strip into R4 is named the bicharacteristic curve. It is applicable to 3D modeling of the Earth, for, generally speaking, the singularities propagate over the bicharacteristic curves. In other words, singularity that is generated by an Earthquake in block $B_{0,0,0}$ propagates over the bicharacteristic curve in $B_{0,0,0}$ until it intersects at point (x_1, y_1, z_1) the boundary to the neighbouring block, $B_{1,0,0}$ for instance. Continuous boundary conditions mean that at point (x_1, y_1, z_1) system (1) in the block $B_{1,0,0}$ has singularity that propagates over the bicharacteristic curves in $B_{1,0,0}$, etc. For computational purposes it is convenient to write (5) in the form:

$$|x_1 = c^{-1}\ \xi_1^0 \cdot (t - t^0) + x_1^0\ x_2 = c^{-1}\ \xi_2^0 \cdot (t - t^0) + x_2^0\ x_3 = c^{-1}\ \xi_3^0 \cdot (t - t^0) + x_3^0 \quad ...(8)$$

since $t - t^0 = 2c\sqrt{(\xi_1^0)2 + (\xi_2^0)2 + (\xi_3^0)2} \cdot s = 2c|\xi^0|$ and without loss of generality we may assume $|\xi^0| = 1$.

4. Reflection and Refraction

All calculations above are made for system (1) with constant coefficients, i.e., for homogeneous block or half space. Geologically, the Earth crust and upper mantle consist of sub-domains $B_{i,j,k}$ containing the same material. In terms of our model, $\Omega = \cup B_{i,j,k}$ and the coefficients of (1) are constant in each $B_{i,j,k}$, or in other words realistic system (1) is the one with piecewise constant coefficients. Without loss of generality, we assume the boundary of $B_{i,j,k}$ to be piece-wise smooth:

$$\partial B_{i,j,k} = \cup\{F_{i,j,k,l}(x, y, z) = 0, 1 = 1, N_m\},$$

where, $F_{i,j,k,l}(x, y, z) = 0$ is a smooth surface in R^3 and N_m is a finite number.

Equation (5) describes the bicharacteristic curves of (1) in each $B_{i,j,k}$ and their behaviour on the boundary $\partial B_{i,j,k}$ is studied by geometrical optics and microlocal analysis. Meeting a surface $F_{i,j,k,l}(x, y, z) = 0$ at point p_b where the coefficients of (1) are not smooth, the bicharacteristic curve b can be reflected or refracted. In both cases there is singularity at boundary point p_b. It propagates over the bicharacteristics as well and this way the well-known formula for reflection and refraction from geometrical optics are obtained. When bicharacteristic curve is reflected the angle θ_{in} of incidence to the surface $F_{i,j,k,l}(x, y, z) = 0$ is equal to the angle of reflection θ_{rl}, since in the same block the equation (6) has the same coefficients. As for refraction at a surface, the match of the boundary conditions of the neighbouring blocks at the two sides of the boundary lead to the well-known formula from geometric optics $v_1 \sin \theta_r r = v_2 \sin \theta_{in}$, where, θ_{rr} is the angle of refraction, v_1 is the speed of the wave in the "incidence" block, and v_2 is the one in "refraction" block.

Let $\vec{n} = (n_1, n_2, n_3) = \left[\left(\frac{\partial F}{\partial x}\right)^2 + \left(\frac{\partial F}{\partial y}\right)^2 + \left(\frac{\partial F}{\partial z}\right)^2\right]^{-\frac{1}{2}} \left(\frac{\partial F}{\partial x}, \frac{\partial F}{\partial y}, \frac{\partial F}{\partial z}\right)(p_b)$ be the normal unit vector to surface $F_{i,j,k,l} = 0$ at the point of incidence p_b, $\xi^{in} = (\xi_1^{in}, \xi_2^{in}, \xi_3^{in})$ be the unit vector along the incidental bicharacteristic curve, $\xi^{rr} = (\xi_1^{rr}, \xi_2^{rr}, \xi_3^{rr})$ be the unit vector along refracted one, and $\xi^{rl} = (\xi_1^{rl}, \xi_2^{rl}, \xi_3^{rl})$ be the unit vector along the reflected one.

The speed of the wave is a physical feature of every material, and it is preliminary known. For instance, the velocity of the P-wave in homogeneous isotropic medium is $v_P = \sqrt{(\lambda + 2\mu)/\rho}$, while for the S-wave it is $\sqrt{\mu/\rho}$.

Quantities $\sin \theta_{in} = \sin \theta_{rl}$ and $\sin \theta_{rr}$ are easy to compute using scalar, or dot product $\cos \theta = \xi \cdot \vec{n}$ of unit vectors ξ and the normal unit vector $\vec{n}$, for instance,

$$\sin^2 \theta_{in} = 1 - (\xi_1^{in} n_1 + \xi_2^{in} n_2 + \xi_3^{in} n_3)^2.$$

Then equations of refraction and reflection from geometrical optics yield:

$$\xi_1^{rr} n_1 + \xi_2^{rr} n_2 + \xi_3^{rr} n_3 = \left[1 - \left(\frac{v_2}{v_1}\right)^2 (1 - [\xi_1^{in} n_1 + \xi_2^{in} n_2 + \xi_3^{in} n_3]^2)\right]^{1/2} \quad ...(9)$$

$$\xi_1^{rl} n_1 + \xi_2^{rl} n_2 + \xi_3^{rl} n_3 = |\xi_1^{in} n_1 + \xi_2^{in} n_2 + \xi_3^{in} n_3|$$

In addition, the incidental bicharacteristic curve, the refracted one, and the normal to the surface vector lie on the same plane, give us the relation:

$$|n_1\ n_2\ n_3\ \xi_1^{in}\ \xi_2^{in}\ \xi_3^{in}\ \xi_1^{rr}\ \xi_2^{rr}\ \xi_3^{rr}| = 0. \quad ...(10)$$

The same relation is valid for vector ξ^{rl}.

Finally, since we consider vectors ξ^{in}, ξ^{rr} and ξ^{rl} be unit ones, we obtain:

$$\begin{aligned} (\xi_1^{rr})^2 + (\xi_2^{rr})^2 + (\xi_3^{rr})^2 &= 1 \\ (\xi_1^{rl})^2 + (\xi_2^{rl})^2 + (\xi_3^{rl})^2 &= 1 \end{aligned} \quad ...(11)$$

Equations (9), (10), and (11) define uniquely vectors of refraction ξ^{rr} and reflection ξ^{rl} up to the sigh.

5. Characteristic Set and Bicharacteristic Strip on the Surface

On the surface $z = 0$, the solution of system (1) satisfies boundary conditions (2). Then the characteristic set and bicharacteristic strip on the surface are computed similarly to the ones in Chapter 2 applied to system (2) instead of system (1). The characteristic set is given by:

$$0 = \lfloor \lambda\xi_1\ \lambda\xi_2\ (\lambda + 2\mu)\xi_3\ \mu\xi_3\ 0\ \mu\xi_1\ 0\ \mu\xi_3\ \mu\xi_2 \rfloor = \lambda\mu^2 \lfloor \xi_1\ \xi_2\ \xi_3\ \xi_3\ 0\ \xi_1\ 0\ \xi_3\ \xi_2 \rfloor + \mu^2 \lfloor \lambda\xi_1\ \lambda\xi_2\ 2\mu\xi_3\ \xi_3\ 0\ \xi_1\ 0\ \xi_3\ \xi_2 \rfloor.$$

Simple calculations give:

$$0 = -\lambda\xi_3\,[\xi_1^2 + \xi_2^2 - \xi_3^2] - \xi_3[\xi_1^2 + (1 + \lambda)\xi_2^2 - (1 + 2\mu)\xi_3^2 + \lambda\xi_1]$$

since $\mu \neq 0$. Finally, the bi-characteristic set of (2) is:

$$0 = \xi_3[(1+\lambda)\xi_1^2 + (1+2\lambda)\xi_2^2 - (1+\lambda+2\mu)\xi_3^2 + \lambda\xi_1].$$

Then the characteristic set of system (2) contains two subsets:

$$q_1(x,\xi) = \xi_3 = 0 \qquad \text{...(12)}$$

$$q_2(x,\xi) = (1+\lambda)\xi_1^2 + (1+2\lambda)\xi_2^2 - (1+\lambda+2\mu)\xi_3^2 + \lambda\xi_1 = 0$$

System (2) is of first order and calculation of the bicharacteristic strip is much simpler than the one for system (1) that is of second order. By definition if $p(x^0, \xi^0) = 0$ then the bicharacteristic strip at point (x^0, ξ^0) is defined by the Hamilton equations:

$$\frac{dx}{ds} = \frac{\partial p(x,\xi)}{\partial \xi}, \quad \frac{d\xi}{ds} = \frac{\partial p(x,\xi)}{\partial x}$$

with initial data $(x, \xi) = (x_0, \xi_0)$ for $t = 0$.

The bicharacteristic strip k_1 generated by $q_1(x, \xi)$ through point (x^0, ξ^0) is:

$$x_1 = x_1^0, \quad x_2 = x_2^0, \quad x_3 = s + x^0, \quad t = t_3^0, \qquad \text{...(13)}$$

$$\xi_1 = \xi_1^0, \quad \xi_2 = \xi_2^0, \quad \xi_3 = \xi_0, \quad \tau = \tau^0.$$

The bicharacteristic strip k_1 generated by $q_2(x, \xi)$ through point (x^0, ξ^0) is:

$$|x_1 = 2(1+\lambda)s + \lambda + x_1^0, x_2 = 2(1+2\lambda)s + x_2^0, x_3 = -2(1+\lambda+2\mu)s + x_3^0, t$$

$$= t_0\ \xi_1 = \xi_1^0, \xi_2 = \xi_2^0, \xi_3 = \xi_3^0\ \tau = \tau_0 \qquad \text{...(14)}$$

as

$$q_2(x^0, \xi^0) = 0.$$

6. 3D Modelling of the Earth Crust and Upper Mantle

Suppose the source of the signal is point one and it is determined at point S with direction alongside vector ξ^0. 3D modelling by bicharacteristic curves is performed in following steps:

Step 1: Solving the inverse problem for system (2).

System (2) contains the boundary conditions to system (1). It describes the wave propagation on the free surface $z = 0$. Using data from several seismic stations, one solves inverse problems (13) and (14) and obtains the position of point E. Roughly speaking, E is the "epicentre" of the surface waves.

Step 2: Initial grid of blocks $\{B_{i,j,k}^k\}$ is imported.

Step 3: The set of bicharacteristic curves starting at point S with direction alongside vector ξ^0 are computed by (6) within a homogeneous block $\{B_{i,j,k}^k\}$. The reflection and refraction vectors between adjacent blocks are computed by (9), (10), and (11).

Step 4: Check.

A key point in every approximation is the error, or in terms of mathematical modelling, how accurate the model is compared to the real object. Using bicharacterstic curves, described in Section 2, it is possible to define the following criterion for accuracy:

Definition: *Let $\{B_{i,j,k}\}$ be a set of blocks and the source of seismic wave be a point one at the point S with direction alongside vector ξ^0. Let P is the point of the Earth surface belonging to the bicharacteristic curves generated by system (1), set of blocks $\{B_{i,j,k}\}$ and source S. Given set of blocks $\{B_{i,j,k}\}$ is plausible if the point P coincides with the epicenter E of the surface waves generated by the earthquake.*

Step 5: Decision.

If P is in preliminary defined vicinity of E, the model is $\{B^k_{i,j,k}\}$ is plausible. If not, a new set $\{B^{k+1}_{i,j,k}\}$ is generated by some probabilistic methods such as Monte Carlo method and algorithm turns to step 2.

Computing the bicharacteristic curves in all sets $\{B_{i,j,k}\}$ arises an important question. At the boundaries between two blocks, surfaces $F_{i,j,k,l}(x, y, z) = 0$, - is the bicharacteristic curve reflected, refracted, or both? The answer comes from the so-called reflection and refraction index.

It is a physical feature of the material that builds the block. How to compute refraction and reflection index is well described in Aki and Richards (2002), W.M. Ewing, W.S. Jardetzky, F. Press. (1957), or in W.L. Plant (1979).

Furthermore, the body waves records are useful to determine the block structure of the closest to the seismic stations block. The Wave front in a homogeneous block is a subset of the characteristic set of system (1), therefore it has constant velocity by (4).

Using bicharacteristic curves and the characteristic set we can compute arrival time for P- and S-waves. In combination with the criteria from the Definition, we can generate and test plausible 3-D models of the Earth crust and upper mantle.

Acknowledgements

This study is dedicated to Prof. G.F. Pansa. He opened the world of geophysics to the author many years ago and one of the results of this fruitful collaboration is the application of bicharacteristic curves of body and surface waves for modeling the Earth's crust and upper mantle. For the author, Prof. Pansa will always be a great scientist, a great teacher and a great man.

References

Aki K. and Paul G. Richards. (2002). *Quantitative Seismology* (2nd Edn.). University Science Books.

Babich, V.M. (1956). "Calculation of intensities of wavefronts by means of the ray method," Doklady AN SSSR, vol. 110, no. 3, pp. 353–357.

Babich, V.M. and Molotkov I.A. (1985). *Space-time Ray Method. Linear and Non-linear Waves.* Leningrad University Publishing House (in Russian).

Ewing W.Maureice and Wenceslas S. Jardetzky, Frank Press. (1957). *Elastic Waves in Layered Media.* McGraw-Hill series in the Geological Sciences.

Hormander L. (1983). *The Analysis of Linear Partial Diffetrential Operators I.* Springer-Verlag.

Levshin A. (1973). *Surface and Channel Seismic Waves.* Nauka, Moscow (in Russian).

Panza G.F., Fabio Romanelli and Franco Vaccari. (2001). Seismic wave propagation in laterally heterogeneous anelastic media: Theory and applications to seismic zonation. *Advances in Geo- Physics, 43,* 1–95.

Plant Walter L. (1979). Elastic waves in the Earth. *Developments in Solid Earth Geophysics, 11*. Elsevier.

CHAPTER

9

NDSHA (Fault- and Physics) Based Ground Motion Modelling for the City of Plovdiv, Bulgaria

Lyuba Dimova[1], Milen Tsekov[2], Fabio Romanelli[3] and Reneta Raykova[4*]

In this chapter we present the modelling of the kinematic rupture processes along several faults near the city of Plovdiv, Bulgaria. Specifically, we consider five seismogenic faults: three closest to Plovdiv faults from the European Database of Seismogenic Faults EDSF13 database and two faults associated with the 1928 M = 6.8 Chirpan and M = 7.1 Popovitsa (Plovdiv) earthquakes. We use the XeRiS web-app to compute response spectral acceleration, as well as a set of synthetic accelerograms, for a selected location in the city centre of Plovdiv, utilising a definition of an earthquake scenario for each fault and a local seismic velocity structure model for the region of interest. In the modelling, we used observed fault parameters for the 1928 earthquakes near Plovdiv and averaged parameters with a given maximum expected magnitude and size for the seismogenic EDSF13 faults. The maximum obtained response spectral acceleration for the $M_W = 6.8$ Chirpan earthquake in 1928 is about 200 cm/s^2 at periods between 1.0 and 2.0 s, while for the $M_W = 7.0$ Popovitsa earthquake in 1928 scenario, it reaches 1000 cm/s^2 for periods between 0.3 and 0.6 s, and a prominent extremum at about 1.6 s with an acceleration of 1200 cm/s^2 is observed. A hypothetical $M_W = 7.1$ event along the Asenovgrad Fault scenario generates response spectral acceleration peaks of about 1400 cm/s^2 at periods of 0.3-0.4 s, as the deepening fault surface lies beneath the city of Plovdiv. The modelled rupture along the Choba Fault, which can generate $M_W = 7.0$ seismic event, produces response spectral acceleration amplitudes above 1000 cm/s^2 at periods ranging from 0.3 s to 0.4 s and smaller maximums at 0.6 and 1.0 s. A hypothetical rupture along the 120 km-long Plovdiv Fault, with the capacity to generate $M_W = 7.0$ event (according to the EDSF13 database), may produce spectral accelerations of about 500 – 550 cm/s^2 at periods of 0.2 – 0.4 s. In an alternative $M_W = 7.0$ scenario along the Plovdiv Fault where only the western half of the fault, located close to Plovdiv, is activated (a scenario similar to the April 18th, 1928 event), the resulting

[1] Department of Meteorology and Geophysics, Sofia University. https://orcid.org/0000-0002-0970-0279.
[2] Department of Meteorology and Geophysics, Sofia University. https://orcid.org/0000-0003-1856-2573.
[3] Department of Mathematics, Informatics and Geosciences, University of Trieste. https://orcid.org/0000-0002-7478-8457.
[4] Department of Meteorology and Geophysics, Sofia University. https://orcid.org/0000-0001-5125-675X.
Email: lyuba_dimova@phys.uni-sofia.bg[1]; tsekov@phys.uni-sofia.bg[2]; romanel@units.it[3]
* Corresponding author: rraykova@phys.uni-sofia.bg

DOI: 10.1201/9781003684121-11

response spectral accelerations for low periods are even higher than the obtained values for the 1928 Popovitsa and the $M_W = 7.1$ Asenovgrad earthquake scenario. In conclusion, our results demonstrate that at the frequencies of interest to earthquake engineering the modeled ruptures on the seismogenic faults near Plovdiv can produce response spectral accelerations comparable to or greater than those associated with the 1928 earthquake, which caused severe damage to the city of Plovdiv. A more detailed (e.g., considering more sites with their geophysical conditions) seismic hazard study should be conducted for the city to identify the zones most exposed to seismic risk.

1. Introduction

The city of Plovdiv, with a population of 370 thousands, is the second largest city in Bulgaria, representing the major economic and cultural centre in the southern part of the country, where numerous seismogenic faults exist (Basili et al., 2013). On April 14th and April 18th 1928, two major earthquakes, with estimated magnitude of $M = 6.8$ and $M = 7.1$, respectively, struck near the city, destroying around 75 thousands buildings, killing 114 people, and injuring more than one thousand persons (Christoskov, 2005). According to other studies (e.g., Shebalin et al., 1998; Dimitrov et al., 2020) the magnitude of the stronger event is $M_W = 7.0$. The April 14th Chirpan earthquake did not cause significant damage to Plovdiv, whereas the April 18th earthquake devastated the city. For this reason, the event is known in Bulgaria as the 1928 Plovdiv earthquake, although its epicentre is near the village of Popovitsa, located about 25 km from Plovdiv. Both the April 1928 earthquake ruptures activated normal faults along the northern and southern borders of a near east-west striking graben, situated northeast of Plovdiv. These earthquakes are an early example of applying geodesy to study co-seismic tectonic movements (Kirov, 1935; Yankov, 1935; Richter, 1958). These unique geodetic measurements are still used in present-day studies, applying modern methodologies and techniques (e.g., Dimitrov et al., 2020; Dimova & Raykova, 2025).

The available historical information about seismic activity in the region of Plovdiv is scarce and incomplete due to the long occurrence intervals for strong earthquakes along the active faults in the region. Paleoseismic studies indicate that the occurrence interval for $M_W = 7.0$ earthquakes along the Chirpan 1928 rupture is 2350 ± 643 years (Vanneste et al., 2006), while along the Popovitsa (Plovdiv) 1928 rupture it is between 1000 and 3000 years (Yaneva et al., 2004). Given the low long-term slip rates in the region, ranging from 0.024 to 0.149 mm/year (Radulov, 2007), the occurrence intervals for strong earthquakes along the southern and northern borders of the Plovdiv Depression are likely to be in the order of thousands of years. A study of modelling block movements and synthetic seismicity (Dimitrova et al., 2022), shows long occurrence intervals, about one thousand years, for events along the southern boundary of the Upper Thrace Basin, part of which is the Plovdiv Depression, with events typically grouped together and often having magnitudes greater than 7.0. In contrast, synthetic seismicity along the northern boundary of the Upper Thrace Basin usually has magnitudes equal to or smaller than 7.0, with lower occurrence intervals, averaging around 250 years.

Seismic hazard studies at national scale based on the probabilistic (Orozova-Stanishkova & Slejko, 1994; Simeonova et al., 2006; Leydecker et al., 2008) and neo-deterministic approach (Orozova-Stanishkova et al., 1996) estimate the hazard in and around Plovdiv as relatively high. Recently, both probabilistic seismic hazard assessment and traditional deterministic earthquake scenarios, based on an event along the active fault to the north of the city, have been applied for the city of Plovdiv (Solakov et al., 2023). The authors obtained an expected peak ground acceleration of up to 0.44 g for the northern part of the city.

In this paper, we apply the neo-deterministic seismic hazard approach (NDSHA, e.g., see Panza et al., 2001, 2012; Panza & Bela, 2020) to calculate synthetic accelerograms and response spectral acceleration for a selected location in the city centre of Plovdiv, considering 5 possible earthquake sources: (1–2) the ruptures associated with the 1928 Chirpan and Popovitsa earthquakes;

(3–5) rupture along the EDSF13 faults nearest to Plovdiv (Basili et al., 2013): Plovdiv, Choba, and Asenovgrad faults (Figure 1). Our results show that $M_W = 7.1$ and $M_W = 7.0$ events along the Asenovgrad and the western segment of Plovdiv Faults may produce even higher effects than the devastating Plovdiv April 18th, 1928 earthquake.

2. Methodology

We use the so called Maximum Credible Seismic Input – Neo-Deterministic Seismic Hazard Assessment, MCSI-NDSHA, procedure (as described by Fasan et al. (2016) and Rugarli et al. (2019)), implemented in the web application XeRiS (see Vaccari & Magrin, 2022, and references therein), to calculate the seismic input (e.g., acceleration time histories and related response spectra) at a selected location in the city centre of Plovdiv considering 5 earthquake scenarios.

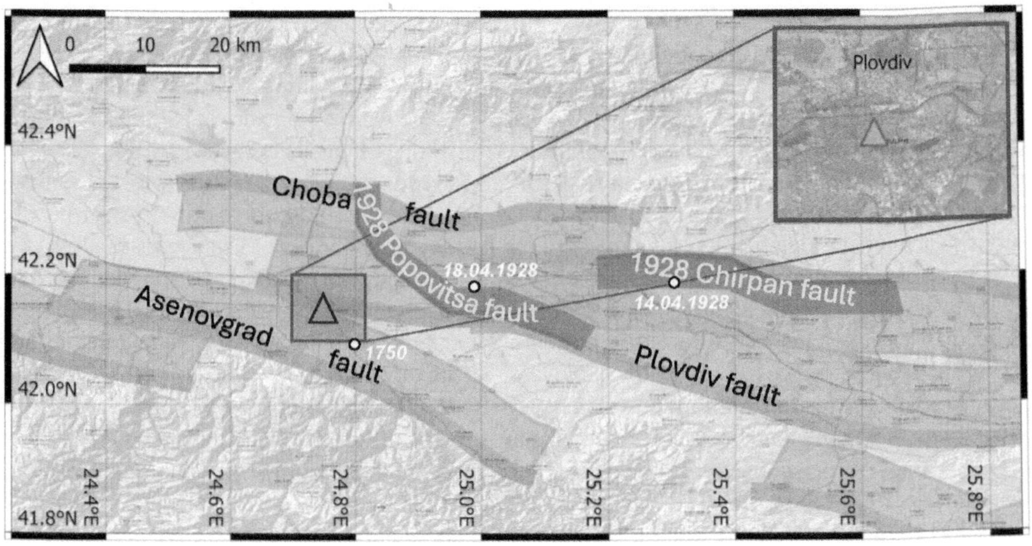

Fig. 1 The region of Plovdiv with nearby seismogenic faults. The triangle indicates the selected location in the city centre of Plovdiv (PL), for which ground acceleration is estimated. White dots represent the epicentres of the strong earthquakes in 1750 and 1928. Three active faults (according to EDSF13, Basili et al., 2013) nearest to Plovdiv (Plovdiv, Asenovgrad, and Choba faults – orange stripes), as well as the ruptures associated with the 1928 Popovitsa and Chirpan earthquakes (purple stripes) are also depicted. The thick lines specify the respective fault's top-end.

The procedure starts by defining earthquake scenarios for the selected seismogenic faults and establishing the corresponding geophysical structural models for seismic wave propagation. The seismogenic faults, characterized by their location, magnitude, and focal parameters, are modelled as extended sources (near-source domain) to account for the effects associated with the kinematic rupture process (i.e., directivity). The radiated seismic wavefield at the site of interest is computed using the Discrete Wave Number (DWN) technique, including near-field terms, following the theoretical framework proposed by Pavlov (2009). For each earthquake scenario multiple realisations of the kinematic rupture process are generated and synthetic accelerograms are calculated for every realization. Assuming a random variability of the nucleation point of the simulation, the methodology is based on the Monte-Carlo technique quantifying the distribution of the final slip, taking into account different (1, 5, 20, 100, or 1000) statistical realisations. The PULSYN06 algorithm, which is based on the theory presented by Gusev (2011), was used to simulate the source rupture process.

Regarding the MCSI response spectrum's procedure, defined and discussed in Fasan et al. (2016) and Rugarli et al. (2019), it performs the results of the propagation of all the considered scenarios for the selected site (in terms of magnitude, epicentral distance, and focal mechanism of the earthquake). The resulting spectral acceleration (SA) values are analysed and the variability values (in terms of percentiles) of the scenario providing the largest median spectral acceleration at each vibrational period are retained.

3. Data

We select a location in the city centre of Plovdiv where we estimate the ground motion based on physical modelling taking into account information on the seismogenic faults in the region. Figure 1 shows the chosen location (hereafter denoted as PL) along with the seismogenic faults potentially generating significant ground motion in the city of Plovdiv. To model the propagation of the seismic wavefield, we define the depth-velocity structure of the region around Plovdiv, obtained by combining sedimentary information from the geological map (Kozhuharov et al., 1999) with a local velocity model (Raykova & Panza, 2015; Raykova et al., 2019) for the lithosphere. The Earth structure model used in this study is summarised in Figure 2 and Table 1.

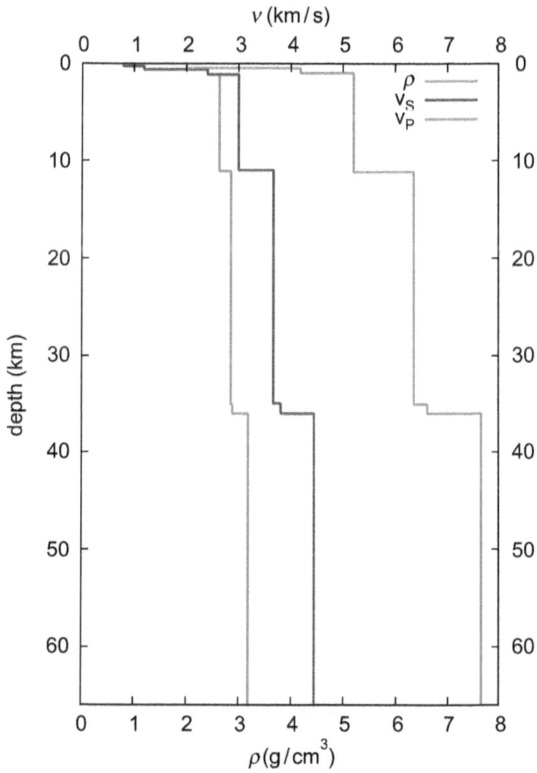

Fig. 2 Local crust and upper mantle structural model, with seismic velocities and density profiles.

For the selected location, we calculate synthetic accelerograms and response acceleration spectra for five earthquake scenarios. First, we consider earthquake scenarios corresponding to the strongest regional seismic events in recent history, the 1928 Chirpan and Popovitsa earthquakes. The source parameters for these events are obtained on the basis of the best approximation of the geodetic measurements carried out in the region before and after the earthquakes. Several studies have examined these faults, with slightly differing source parameters. Here we use the re-estimated fault parameters from Dimova and Raykova (2025), taking into account the faults location according to Bonchev and Bakalov (1928). Next, we consider earthquake scenarios involving the rupture of the entire length of the three seismogenic faults closest to the city of Plovdiv, as listed in the EDSF13 database (Basili et al., 2013): Plovdiv Fault ($M_W = 7.0$), Asenovgrad Fault ($M_W = 7.1$), and Choba Fault ($M_W = 7.0$). For these three earthquake scenarios we consider the maximum magnitude and maximum fault length as provided in the EDSF13, while for the faults width we use 20 km, as the seismicity in Bulgaria is concentrated mainly in the upper crust, down to the depths of about 20 km (Grigorova et al., 1978; Solakov et al., 2020; Tsekov et al., 2015). The depth of the fault top is considered to be 0 km for the 1928 Chirpan and Popovitsa earthquakes, as surface ruptures were observed. For the seismogenic EDSF13 faults, we consider a more conservative estimate of the fault's top depth of 2 km. The 1928 Popovitsa Fault partially coincides with a segment of the Plovdiv EDSF13 fault (eastern part of Popovitsa Fault). However, there is a significant difference in the geometry of the western parts of these faults. Table 2 summarizes the source parameters for the considered rupture scenarios.

Table 1 Parameters of the used Earth's structure model.

Thickness (km)	*Depth (km)*	*Density (g/cm^3)*	*Vp (km/s)*	*Vs (km/s)*	*Qp*	*Qs*
0.04	0.04	1.50	0.30	0.20	220	100
0.06	0.10	1.60	0.60	0.35	220	100
0.10	0.20	1.80	1.40	0.80	220	100
0.30	0.50	2.00	2.00	1.20	220	100
0.50	1.00	2.40	4.20	2.40	220	100
10.00	10.50	2.60	5.20	3.00	220	100
24.00	34.50	2.85	6.35	3.70	220	100
1.00	35.50	2.90	6.60	3.80	220	100
30.00	65.50	3.20	7.65	4.45	220	100

Table 2 Earthquake rupture scenarios considered in this study.

Name / *Parameter*	*1928CH*	*1928POP*	*ASENOVGRAD*	*CHOBA*	*PLOVDIV*
Longitude (°E)*	25.4	25.8	24.7	24.9	24.6
Latitude (°N)*	42.2	42.2	42.1	42.3	42.2
Depth (km)**	0.0	0.0	2.0	2.0	2.0
Magnitude (M_w)	6.8	7.0	7.1	7.0	7.0
Strike (°)	95	300	290	94	283
Dip (°)	60	60	65	60	60
Rake (°)	270	290	270	270	310
Length (km)	36.0	40.0	80.0	65.0	120.0
Width (km)	13.9	16.3	20.0	20.0	20.0

***Notes*:** *Longitude and Latitude refer to the top-centre point of the fault.
**The depth of the top of the fault plane.

4. Results and Discussion

We first calculate synthetic accelerograms and MCSI response spectrum for each of the five earthquake scenarios. For each scenario, we perform 100 computations using random sets of rupture process parameters, including various locations of the nucleation point, slip distribution and rupture duration. The results are shown in Figures 3–7. We present in each figure: (a) the location of the

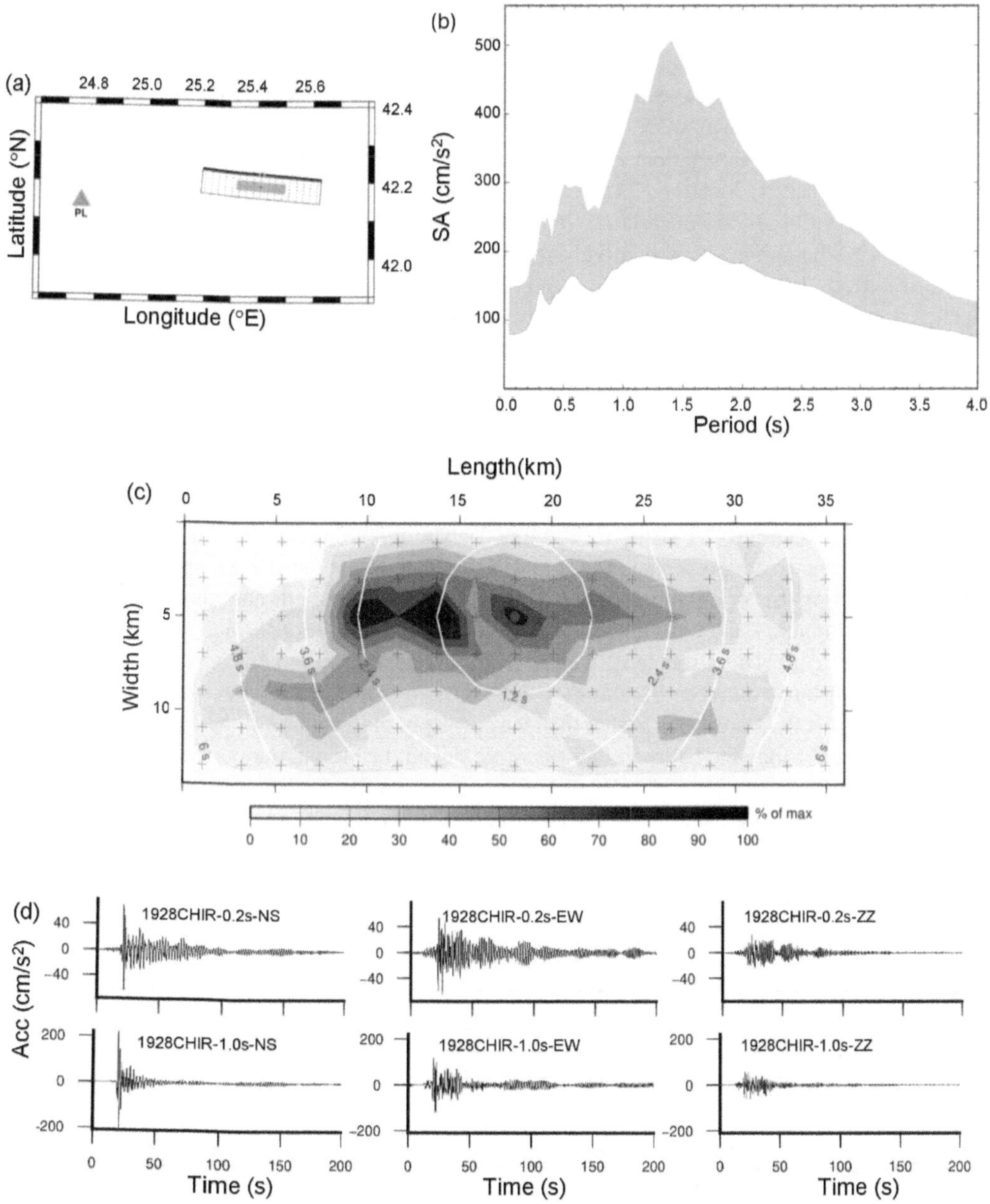

Fig. 3 Results for 1928 CH fault: (a) the blue rectangle indicates the surface projection of the fault rupture surface, while the orange rectangle indicates the area where the nucleation point is made varying; thick line indicates the fault top boundary; the triangle indicates the selected location PL; (b) MCSI spectrum; the line is the median curve for the 100 rupture realisations, while the shaded gray area extends from the median to the 95th percentile; (c) example of slip distribution over the modelled fault rupture area for one realisation; (d) the three-components of synthetic accelerograms at PL, for the illustrated rupture scenario and for conditional periods equal to 0.2 s (top) and 1.0 s (bottom).

modelled fault on the surface along with the point of interest; (b) the MCSI median acceleration spectrum and the 95% of uncertainty range. For illustration purposes, additional figures are presented only for the 1928 Chirpan scenario, showing one realization of the slip rupture process (Figure 3c) and the corresponding synthetic accelerograms (Figure 3d).

The results for the 1928 Chirpan rupture scenario are presented in Figure 3. This fault source is relatively far from the city of Plovdiv (the distance between PL and the centre of the fault surface is 54 km, as shown in Figure 3a), thus the computed MCSI response spectrum indicate low earthquake hazard (Figure 3b). One of 100 realisations created for the 1928CH source is shown in Figure 3c illustrating the slip rupture process. A 2D final slip map function is displayed from the simulated scenario, where $M_W = 6.8$, L = 36.0 km (length), and W = 13.9 km (width) are the fault parameters. The sub-source grid consists of X×Y crosses, locating double-couple forces, the white contours show the slip rupture propagation velocity of a non-stationary random process, and the red dot indicates the nucleation location. Examples of synthetic accelerograms at conditional periods of 0.2 s and 1.0 s with the 50% uncertainty range are shown in Figure 3d.

The MCSI median curve is relatively flat, with acceleration spectral amplitudes at about 180 cm/s^2 for periods 1.0 – 2.0 s. The peak ground accelerations (PGA) (Figure 3d) and the response spectral acceleration (Figure 3b) have relatively low values, indicating low seismic hazard for Plovdiv associated with the Chirpan 1928 rupture scenario. This result is consistent with the fact that the April 14th, 1928 earthquake caused negligible damage to the city of Plovdiv. However, this earthquake was destructive and deadly for the nearby city of Chirpan, located over the western part of the fault (Kirov, 1935).

We next model a rupture process corresponding to the April 18th, 1928 Popovitsa earthquake. A surface rupture of length between 40 and 50 km is observed following this quake, and its estimated magnitude is between 7.0–7.1 (Christoskov, 2000). The modelling results for the 1928 Popovitsa Fault rupture scenario are presented in Figure 4. The shape of the real 1928 Popovitsa Fault is quite complex, with NW–SE orientation for the western branch of the fault and almost W-E orientation for the eastern branch. We approximate the fault with a single rupture surface, averaging the fault's geometry (Figure 4a). The distance between PL and the centre of the fault is approximately 21 km, about half the distance between PL and the 1928 Chirpan Fault. The response spectral acceleration (Figure 4b) and the PGA are much higher than the corresponding values for the 1928 Chirpan rupture, which is expected, since the Chirpan Fault is more distant to PL and the corresponding earthquake has a lower magnitude than the 1928 Popovitsa earthquake. Again, the response acceleration spectrum is relatively flat for periods between 0.2 s and 2 s, however having much higher spectral

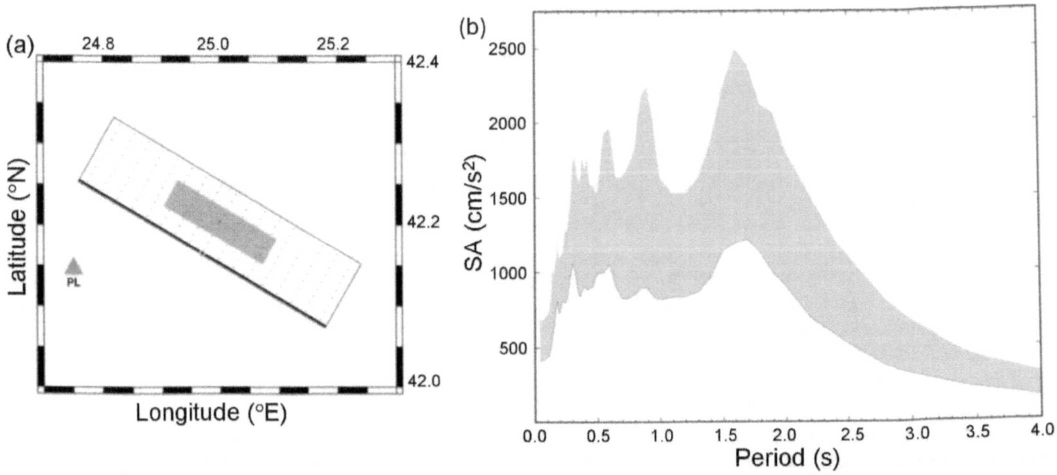

Fig. 4 Results for 1928 POP Fault (a) as MCSI spectrum at PL. (b) For a detailed explanation of the figure elements, see Figure 3.

acceleration values than for the 1928 Chirpan scenario. Response spectral acceleration values between 500 cm/s^2 and over 1000 cm/s^2 are observed over a wide range of periods (0.2 – 2.5 s), with peaks exceeding 1000 cm/s^2 for periods of 0.3 s, 0.6 s, and 1.7 s. PGA values for some of the realizations exceed 400 cm/s^2. The obtained high response spectral acceleration estimates for this scenario are consistent with the fact that the April 18th, 1928 earthquake with an epicenter near Popovitsa actually devastated the city of Plovdiv (Kirov, 1935).

The 1928 Chirpan and 1928 Popovitsa earthquakes are the only strong earthquakes near the city of Plovdiv for which surface rupture has been observed (Radulov, 2007). Thus, the available geologic information may allow us to properly model the rupture source. However, according to the EDSF13 database (Basili et al., 2013) there are three additional active faults in the vicinity of Plovdiv capable of generating earthquakes with Mw=7.0 or Mw=7.1: (1) the Plovdiv Fault; 2) the Asenovgrad Fault, and (3) the Choba Fault.

The April 18th, 1928 Popovitsa earthquake is associated with a segment of the so-called Plovdiv Fault, which is approximately 120 km long. It is the longest active fault with high seismogenic potential near to the city of Plovdiv. According to the EDSF13 database, an earthquake with a maximum expected magnitude of M_W = 7.0 can occur along the Plovdiv Fault. In fact, the 1928 Popovitsa Fault only partially coincided with the Plovdiv Fault (Figure 1). Due to the long length of the Plovdiv Fault, there are more than one plausible M_W = 7.0 earthquake scenarios involving a rupture of the entire fault or a part of it. First, we model a rupture along the entire length of the Plovdiv Fault, following the geometry presented in the EDSF13 database. Our results for the hypothetical rupture along the entire Plovdiv Fault with magnitude M_W = 7.0 are presented in Figure 5. The most western part of the Plovdiv Fault is located beneath the city of Plovdiv. However, since the fault's large length, the distance between the observed location (PL) and the centre of the modelled fault is relatively large, about 52 km (Figure 5a). The MCSI spectrum exhibits more spectral power at short periods (high frequencies) than at long periods (Figure 5b). Spectral acceleration exceeds 500 cm/s^2 for periods of 0.2 – 0.4 s, with additional peaks of about 400 cm/s^2 at 0.6 s and about 300 cm/s^2 at 1.0 s. For all the periods, the response spectral acceleration obtained for this scenario is significantly lower than for the Popovitsa Fault source scenario. At first sight, this result appears somewhat unexpected, since the Plovdiv Fault is closer to the city of Plovdiv than the 1928 Popovitsa Fault. However, it should be noted that this Mw = 7.0 model scenario (Figure 5) involves a relatively large rupture area of about 2400 km^2; consequently, the modeled slip along the rupture surface is relatively small.

According to empirical relationships between earthquake magnitude and rupture area (Wells & Coppersmith, 1994), the most likely rupture area for a M_W = 7.0 earthquake is about 1000 km^2. Thus, we consider two additional model scenarios along the Plovdiv Fault, generating M_W = 7.0 earthquakes, which involve rupture area close to the most expected value according to Wells and Coppersmith (1994). *Scenario (I)* involves a rupture along a 60 km-long segment in the most eastern part of the fault, i.e., the eastern half of the Plovdiv Fault. This scenario is plausible taking into consideration that the 1928 Popovitsa Fault is related to stress release over the western half of the Plovdiv Fault, and future strong seismic activity may occur along the eastern segment of the fault. The results show that this scenario does not pose a significant seismic hazard for the city of Plovdiv. The maximum observed median MCSI is less than 100 cm/s^2 for periods around 1.0 s (not shown). Such an estimate is expected since the distance between location PL and the centre of the rupturing model fault segment is about 80 km. While this rupture scenario is essentially harmless for Plovdiv, it may pose a significant seismic hazard for the city of Dimitrovgrad and other settlements in the region. *Scenario (II)* involves a rupture along a 60 km-long segment in the most western part of the fault, i.e. the western half of the Plovdiv Fault. This scenario is motivated by the assumption that the 1928 Popovitsa Fault is a ramification of the Plovdiv Fault, with the April 18th, 1928 event possibly increasing the stress in the neighbouring western part of the fault. The results show that such a scenario would be devastating for the city of Plovdiv: the modelled response spectral acceleration exceeds 1000 cm/s² for periods between 0.1 s and 0.8 s, with distinct maximums reaching nearly 1500 cm/s² (not shown). The obtained spectral acceleration estimates indicate that strong earthquakes along the western part of the Plovdiv Fault could produce even greater effects effects on the city of Plovdiv than the April 18th, 1928 seismic event.

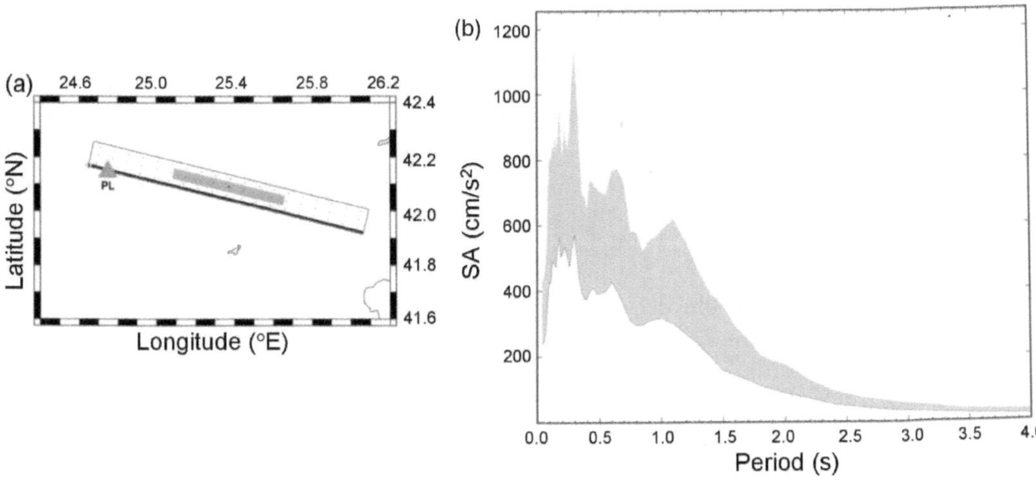

Fig. 5 Results for the entire PLOVDIV Fault (a) as MCSI spectrum at PL (b). For a detailed explanation of the figure elements, see Figure 3.

The Plovdiv Fault is not the only fault with significant seismic potential in the considered region. The city of Plovdiv is situated in the Plovdiv Depression, bordered to the south by the Rhodope Mountains and to the north by the Sredna Gora Mountains. According to geological evidence, both the northern and southern borders of the Plovdiv Depression are seismically active and capable of producing $M_W = 7.0$ earthquakes (Radulov, 2007; Basili et al., 2013). The southern end of the Plovdiv Depression is marked by the Asenovgrad Fault, which has seismogenic potential for an $M_W = 7.1$ earthquake according to the EDSF13 (Basili et al., 2013). Near the northern end of the Plovdiv Depression is located the Choba Fault, with a seismogenic potential $M_W = 7.0$, according to the same database. There is no historical evidence (Grigorova et al., 1978) of strong seismic events (M > 6.0) occurring along the northern border of the Plovdiv depression. However, there is one historical earthquake possibly located on the Asenovgrad Fault. It occurred in 1750, and although historical information about the event is scarce (Ambraseys, 2009), it was most likely a strong earthquake. Grigorova et al. (1978) give an estimated magnitude of 6.8 and epicenter very close to Plovdiv for this earthquake (Figure 1). Shebalin et al. (1998) give an even higher magnitude estimate of M7.2 and an epicentre location 42.2N, 24.5E, which puts it also on the Asenovgrad Fault.

The EDSF13 considers the Asenovgrad Fault as seismogenic, with the potential to generate events with $M_W = 7.1$ magnitude. The results of our computations for $M_W = 7.1$ rupture scenario along the entire Asenovgrad Fault are shown in Figure 6. The Asenovgrad Fault lies along the southern boundary of the Plovdiv depression, with the observed point PL situated over its rupture plane. The distance between location PL and the centre of the rupture plane is quite small, only about 5 km. Peak ground acceleration amplitudes in the synthetic accelerograms exceed 400 cm/s^2 (not shown), and median response spectral acceleration reaches a peak value of 1400 cm/s^2 for the engineeringly interesting period of 0.3 s (Figure 6b), exceeding the values obtained for the 1928 Popovitsa scenario corresponding to the seismic event that devastated the city of Plovdiv in 1928. Thus, our results indicate that the Asenovgrad Fault poses significant earthquake hazard for the city of Plovdiv.

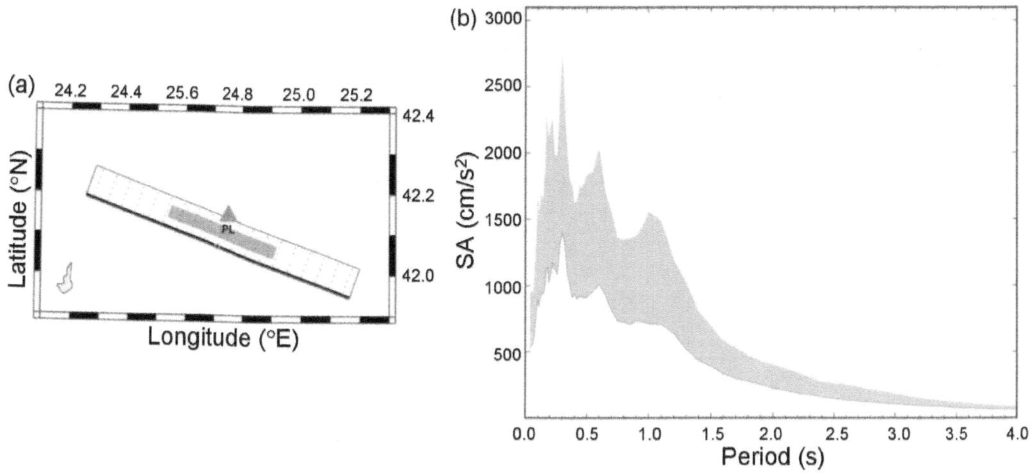

Fig. 6 Results for ASENOVGRAD Fault (a) as MCSI spectrum at PL (b). For a detailed explanation of the figure elements, see Figure 3.

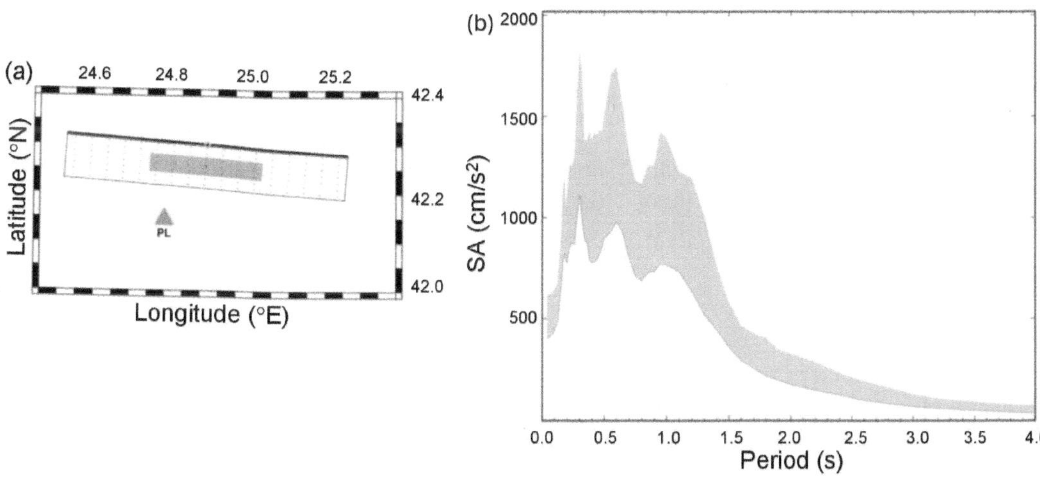

Fig. 7 Results for CHOBA Fault (a) as MCSI spectrum at PL (b). For a detailed explanation of the figure elements, see Figure 3.

Next, we consider another rupture scenario potentially affecting the city of Plovdiv, an $M_W = 7.0$ earthquake along the entire Choba Fault, which is situated near the northern boundary of the Plovdiv depression. Despite having $M_W = 7.0$ seismogenic potential according to the ESDF13 (Basili et al., 2013), no strong earthquakes ($M > 6.0$) have been documented along this fault in historical times (Grigorova et al., 1998; Shebalin et al., 1998). However, it should be considered that the recurrence intervals for strong earthquakes along active faults in the region are on the order of thousands of years (Radulov, 2007; Vanesste et al., 2006). Thus, the lack of historical information on strong earthquakes along the Choba Fault may simply reflect its long recurrence interval. The simulation results for an $M_W = 7.0$ seismic event along this fault are shown in Figure 7. The Choba Fault, with a length of 65 km, has an almost W-E orientation (Figure 7a), and the distance between PL and the centre of the fault is approximately 15 km. The median response spectral acceleration exceeds 600 cm/s^2 for periods between 0.2 – 1.3 s (Figure 7b) and has a peak value of more than 1000 cm/s^2 for period 0.4 s. The maximum synthetic PGA reaches values of about 400 cm/s^2. The simulations demonstrate that the Choba Fault poses a significant seismic hazard for the city of Plovdiv. At low periods, its effects are comparable to those of the April 18, 1928, event, but weaker than those associated with an $M_W = 7.0$ rupture along the western segment of the Plovdiv Fault or an $M_W = 7.1$ rupture along the Asenovgrad Fault.

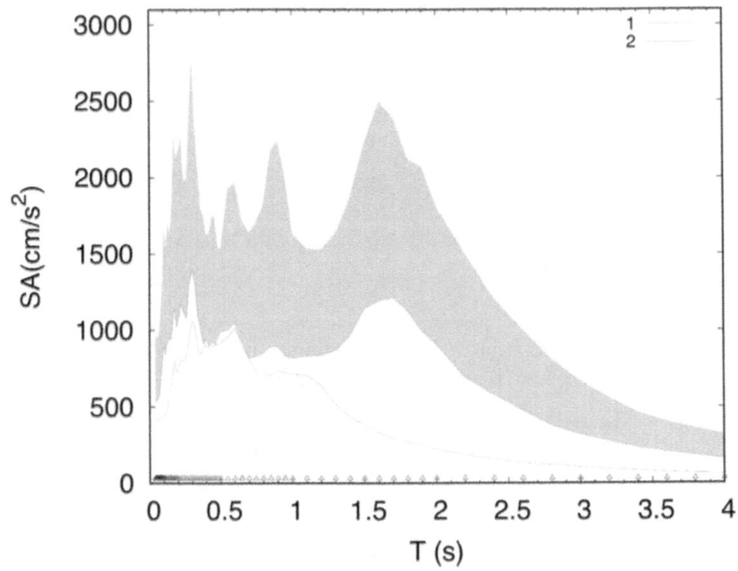

	source	Mw	edi (km)	dpth (km)	strike (°)	dip (°)	rake (°)	sre (°)	slon (°E)	slat (°N)
1	1928POP	7.0	21.6	8.6	300	60	290	47	25.0	43.2
2	ASENOVGRAD	7.1	4.8	11.1	290	65	290	367	24.7	42.1

Fig. 8 Multi-scenario MCSI spectrum at the location PL (24.75°E, 42.15°N) computed over the 5 considered earthquake scenarios: 1928 Chirpan, 1928 Popovitsa, $M_W = 7.0$ Plovdiv Fault, entire length, $M_W = 7.1$ Asenovgrad Fault, and $M_W = 7.0$ Choba Fault. Asenovgrad Fault (2) and 1928 Popovitsa Fault (1) scenarios dominate the response spectral acceleration at the different periods (see the coloured triangles).

Next, we calculate MCSI spectrum for a composite multi-scenario including the following individual ones: (1) 1928 Chirpan $M_W = 6.8$, (2) 1928 Popovitsa $M_W = 7.0$, (3) $M_W = 7.0$ on the Plovdiv Fault over its entire length, (4) $M_W = 7.1$ on the Asenovgrad Fault, and (5) $M_W = 7.0$ on the Choba Fault. For each period, response spectral acceleration values determined for each of the five scenarios are compared and the largest value is selected. Thus, at different periods the MCSI spectrum is obtained from different individual scenarios. The resulting composite MCSI spectrum is shown in Figure 8. Only two of the considered five scenarios determine the highest spectral acceleration at different periods. For periods up to 0.5 s the highest response spectral acceleration values are obtained for the $M_W = 7.1$ Asenovgrad Fault rupture scenario, while for periods over 0.7 s, the dominant contribution is that of the 1928 Popovitsa rupture scenario. At periods 0.5 – 0.7 s both scenarios produce comparable maximum response spectral accelerations. If an additional rupture scenario involving $M_W = 7.0$ event over the western half of the Plovdiv Fault is included, it dominates the response spectral acceleration at periods up to 0.8 s, while over 0.8 s again the response is dominated by the 1928 Popovitsa rupture scenario.

According to a recent study approximately 90% of all buildings in Plovdiv, representing however approximately 55% of total floor area, are up to six storeys buildings (Stefanov et al., 2023); they are the most vulnerable to $M_W = 7.1$ Asenovgrad Fault and $M_W = 7.0$ on western segment of the Plovdiv Fault rupture scenarios. Approximately 10% of Plovdiv buildings, representing however 45% of total floor area, have seven or more storeys; these may be strongly affected by a 1928 Popovitsa rupture scenario.

5. Conclusions

In this chapter, adopting a neo-deterministic, seismic-physics- and fault-based hazard approach, we estimate the ground-motion accelerations and response spectra at a location representative of the city of Plovdiv. This is based on available information about active faults with high seismogenic potential in its vicinity.

Our ground motion modelling for the earthquake scenarios related to fault ruptures corresponding to the April 1928 earthquakes, with epicentres near Chirpan and near Popovitsa, are consistent with the observed effects during these seismic events. We use the response acceleration spectrum obtained for the 1928 Popovitsa scenario as a reference scenario and compare it to possible $M_W = 7.0$ earthquake scenarios along active faults in the vicinity of Plovdiv for which no historical information about the rupture process is available.

Our results show that three faults in the vicinity of Plovdiv pose significant seismic hazard for the city: (1) Plovdiv Fault (western part), (2) the Asenovgrad Fault, and (3) the Choba Fault. Modelling the rupture process along these faults demonstrates that strong earthquakes along them could have effects comparable to, or even greater than, the effects of the devastating 1928 Popovitsa earthquake for periods up to 0.8 s. Our results indicate that at small periods, up to 0.8 s, the (1) $M_W = 7.1$ Asenovgrad Fault rupture scenario and (2) $M_W = 7.0$ rupture along the western half of Plovdiv Fault scenario show the highest response spectral acceleration values, i.e., pose greatest risk to low and medium height buildings (up to about 7– 8 storeys high) in the city of Plovdiv. For periods beyond 0.8 s, the 1928 Popovitsa scenario shows the highest response spectral acceleration among the considered scenarios. Thus, for the tallest buildings the 1928 Popovitsa scenario poses the highest risk.

Due to the high seismic hazard related to the Asenovgrad Fault and Plovdiv Faults, it is important to study carefully the seismic activity along these faults. For example, a recent study on the seismogenic nodes in Bulgaria capable of generating events with magnitudes greater than 6.0 (Gorshkov et al., 2024) identifies two such nodes along the studied faults: one located on the Plovdiv Fault and the other on the Asenovgrad Fault.

Since the last strong earthquake in the region of Plovdiv, April 18th, 1928 Popovitsa event, occurred almost 100 years ago, instrumental accelerograms are not available for $M_W = 7.0$ events in the region. One of the advantages of the neo-deterministic approach to seismic hazard assessment, followed here, is that it also provides realistic synthetic accelerograms, as well as synthetic ground motion displacement and velocity time histories, which can be used by the engineering community as a proxy for the missing instrumental records.

Acknowledgements

This study is financed by the European Union - NextGenerationEU, through the National Recovery and Resilience Plan of the Republic of Bulgaria, project SUMMIT BG-RRP-2.004-0008-C01, internal project number 70-123-265/19.02.2024.

References

Ambraseys, N. (2009). *Earthquakes in the Mediterranean and Middle East: A Multidisciplinary Study of Seismicity up to 1900* (1st Edn.). Cambridge University Press. https://doi.org/10.1017/CBO9781139195430.

Basili, R., Kastelic, V., Demircioglu Tumsa, M.B., Garcia Moreno, D., Nemser, E.S., Petricca, P., Sboras, S.P., Besana-Ostman, G.M., Cabral, J., Camelbeeck, T., Caputo, R., Danciu, L., Domaç, H., Fonseca, J.F. de B.D., García-Mayordomo, J., Giardini, D., Glavatovic, B., Gulen, L., Ince, Y., … Wössner, J. (2013). *European Database of Seismogenic Faults (EDSF)*. Istituto Nazionale di Geofisica e Vulcanologia (INGV). https://doi.org/10.6092/INGV.IT-SHARE-EDSF.

Bonchev, S. and Bakalov, P. (1928). Les tremblements de terre dans la Bulgarie du sud les 14 et 18 avril 1928. *Rev. Bulg. Geol. Soc., 1*(2), 51–63.

Christoskov, L. (2000). Energy and source parameters of the strong Bulgarian earthquakes after 1900. *Rep. Geod., Warsaw Univ. Technol., 3*(48), 15–20.

Christoskov, L. (2004). *Seismology* (1st Edn., Vol. 1). St. Kliment Ohridski University Press (in Bulgarian).

Dimitrov, D., Ruegg, J.-C., Meyer, B., Chabalier, J.-B., Botev, E. and Briole, P. (2020). The 1928 Plovdiv Sequence: Fault Model Constrained from Geodetic Data and Surface Breaks. In: *Proceedings SES 2020*, 250–261. http://space.bas.bg/SES/archive/SES%202020_DOKLADI/PROCEEDINGS%20SES%202020.pdf.

Dimitrova, S., Gorshkov, A., Vorobiova, I., Dimova, L., Novikova, O. and Raykova, R. (2022). Synthetic seismicity obtained by block movements modelling for the territory of Bulgaria: Preliminary results. *Review of the Bulgarian Geological Society, 83*(3), 299–302. https://doi.org/10.52215/rev.bgs.2022.83.3.299.

Dimova, L. and Raykova, R. (2025). Effects of the Fault Parameters on the Focal Mechanism Solutions: A Case Study of the 1928 Chirpan and Plovdiv Earthquakes. In: N. Dobrinkova and S. Fidanova (Eds.), *Environmental Protection and Disaster Risks (EnviroRisks 2024)* (Vol. 883, pp. 276–283). Springer Nature Switzerland. https://doi.org/10.1007/978-3-031-74707-6_30.

Fasan, M., Magrin, A., Amadio, C., Romanelli, F., Vaccari, F. and Panza, G.F. (2016). A seismological and engineering perspective on the 2016 Central Italy earthquakes. *International Journal of Earthquake and Impact Engineering, 1*(4), 395. https://doi.org/10.1504/IJEIE.2016.083253.

Gorshkov, A.I., Novikova, O.V., Dimitrova, S.Y., Dimova, L.D. and Raykova, R.B. (2024). Potential Locations of Strong Earthquakes in Bulgaria and the Neighbouring Regions. *International Journal of Geophysics, 2024*, 1–13. https://doi.org/10.1155/2024/8103337.

Grigorova, E., Christoskov, L., Sokerova, L., Rizhikova, S. and Roglinov, A. (1978). *Catalogue of Earthquakes in Bulgaria and the Nearby Territories during the Period 1st Cent. BC–1977* [Dataset]. Bulg. Acad. Sci.; Archives Bulletin of the Institute of Geophysics.

Gusev, A.A. (2011). Broadband Kinematic Stochastic Simulation of an Earthquake Source: A Refined Procedure for Application in Seismic Hazard Studies. *Pure and Applied Geophysics, 168*(1–2), 155–200. https://doi.org/10.1007/s00024-010-0156-3.

Kirov, K. (1935). *Contribution to the Study of South Bulgaria earthquakes of April 14th and April 18th 1928 (in Bulgarian). XXIX.* Institut de Physique du Globe de Paris.

Kozhuharov, D., Marinova, R. and Kozhuharova, E. (1999). *Geological Map of Bulgaria on the Scale 1:100 000. Sheet Plovdiv.* [Map]. Committee for Geology.

Leydecker, G., Busche, H., Bonjer, K.-P., Schmitt, T., Kaiser, D., Simeonova, S., Solakov, D. and Ardeleanu L. (2008). Probabilistic seismic hazard in terms of intensities for Bulgaria and Romania: Updated hazard maps. *Nat. Hazards Earth Syst. Sci., 8*, 1431–1439.

Orozova-Stanishkova, I. and Slejko, D. (1994). Seismic Hazard of Bulgaria. *Natural Hazards, 9*, 247–271.

Orozova-Stanishkova, I.M., Costa, G., Vaccari, F. and Suhadolc, P. (1996). Estimates of 1 Hz maximum acceleration in Bulgaria for seismic risk reduction purposes. *Tectonophysics, 258*(1–4), 263–274.

Panza, G.F. and Bela, J. 2020. NDSHA: A new paradigm for reliable seismic hazard assessment. *Eng. Geol., 275*, 105403.

Panza, G.F., Mura, C.L., Peresan, A., Romanelli, F. and Vaccari, F. (2012). Seismic Hazard Scenarios as Preventive Tools for a Disaster Resilient Society. In: *Advances in Geophysics* (Vol. 53, pp. 93–165). Elsevier. https://doi.org/10.1016/B978-0-12-380938-4.00003-3.

Panza, G.F., Romanelli, F. and Vaccari, F. (2001). Seismic wave propagation in laterally heterogeneous anelastic media: Theory and applications to seismic zonation. In: *Advances in Geophysics* (Vol. 43, pp. 1–95). Elsevier. https://doi.org/10.1016/S0065-2687(01)80002-9.

Pavlov, V.M. (2009). Matrix impedance in the problem of the calculation of synthetic seismograms for a layered-homogeneous isotropic elastic medium. *Izvestiya, Physics of the Solid Earth, 45*(10), 850–860. https://doi.org/10.1134/S1069351309100036.

Radulov, A. (2007). Active faults in the Plovdiv Depression and their long-term slip rate. *Geologica Balcanica, 36*(3–4), 51–56.

Raykova, R.B. and Panza, G.F. (2015). *VS structure of the crust and upper mantle in the Balkan Peninsula Region* (arXiv:1511.02852). arXiv. https://doi.org/10.48550/arXiv.1511.02852.

Raykova, R., Dimova, L. and Panza, G.F. (2019). Lithosphere-asthenosphere velocity structure along the Transmed VII section (Moesian platform–Aegean Sea). In: *Proceedings of the Geosciences Conference* at Sofia, Bulgaria. Review of the Bulgarian Geological Society, *80*, 105–107.

Richter, C. (1958). *Elementary Seismology*. W.H. Freeman and Co.

Rugarli, P., Amadio, C., Peresan, A., Fasan, M., Vaccari, F., Magrin, A., Romanelli, F. and Panza, G.F. (2018). Neo-Deterministic Scenario-Earthquake Accelerograms and Spectra: A NDSHA Approach to Seismic Analysis. In: J. Jia and J.K. Paik (Eds.), *Engineering Dynamics and Vibrations* (1st Edn., pp. 187–241). CRC Press. https://doi.org/10.1201/9781315119908-6.

Shebalin, N., Leydecker, G., Mokrushina, N., Tatevossian, R., Erteleva, O. and Vassiliev, V. (1998). *Earthquake Catalogue for Central and Southeastern Europe 342 BC–1990 AD*. Final Report to Contract ETNU - CT 93 - 0087.

Simeonova, S., Solakov, D., Leydecker, G., Busche, H., Schmitt, T. and Kaiser, D. (2006). Probabilistic seismic hazard map for Bulgaria as a basis for a new building code. *Nat. Hazards Earth Syst. Sci.*, *6*, 881–887.

Solakov, D., Simeonova, S., Raykova, P. and Aleksandrova, I. (2020). *Catalogue of the Earthquakes in Bulgaria and Surroundings since 1981* [Dataset]. National Institute of Geophysics, Geodesy and Geography, Bulgarian Academy of Sciences. https://doi.org/10.34975/CTLG-2020.V.1.

Solakov, D., Simeonova, S., Trifonova, P., Metodiev, M. and Raykova, P. (2023). Earthquake scenarios and population exposure for the city of Plovdiv. *Natural Hazards*, *118*(3), 2153–2171. https://doi.org/10.1007/s11069-023-06087-7.

Stefanov, D., Solakov, D. and Milkov, J. (2023). Assessment of the effects of strong earthquakes on the city of Plovdiv. *IOP Conf. Series: Materials Science and Engineering*, *1276*, 012028. https:// doi:10.1088/1757-899X/1276/1/012008.

Tsekov, M., Botev, E. and Borisov, R. (2015). Seismicity of the Rhodope zone over the period 1980–2014. In: *7th BgGS National Conference With International Participation "GEOPHYSICS 2015"*.

Vaccari, F. and Magrin, A. (2022). A user-friendly approach to NDSHA computations. In: *Earthquakes and Sustainable Infrastructure* (pp. 215–237). Elsevier. https://doi.org/10.1016/B978-0-12-823503-4.00016-6.

Vanneste, K., Radulov, A., De Martini, P., Nikolov, G., Petermans, T., Verbeeck, K., Camelbeeck, T., Pantosti, D., Dimitrov, D. and Shanov, S. (2006). Paleoseismologic investigation of the fault rupture of the 14 April 1928 Chirpan earthquake (*M* 6.8), southern Bulgaria. *Journal of Geophysical Research: Solid Earth*, *111*(B1), 2005JB003814. https://doi.org/10.1029/2005JB003814.

Wells, D. and Coppersmith, K. (1994). New Empirical Relationships among Magnitude, Rupture Length, Rupture Width, Rupture Area, and Surface Displacement. *Bulletin of the Seismological Society of America*, *84*(4), 974–1002.

Yaneva, M., Radulov, A., Nikolov, G., Shanov, S., Kamelbeeck, T., Nikolov, N., Mitev, A., Kostov, K., Vanneste, K. and Verbeeck, K. (2004, 17 Dec.). Geological records for paleoseismicity of Popovitsa fault in a trench near Popovitsa, Southern Bulgaria. In: *Annual Scientific Conference "Geology 2004"*.

Yankov, Y. (1935). *Regional Deformations Caused by the 14 and 18 April 1928 Earthquakes (in Bulgarian and French)*. Inst. Met. Sentr. De Bulgarie.

CHAPTER

10

Neo Deterministic Seismic Hazard Assessment (NDSHA) applied to Seismic Response Analyses: the case study of Chieti's Hill (Italy)

Chiara Faraone[1], Fabio Romanelli[2], Nicola Venisti[3] and Giovanna Vessia[4*]

Local seismic effects play a crucial role in urban planning and infrastructures design. These effects can amplify or attenuate ground shaking, significantly altering the impact of an earthquake in a given area due to non-linear and heterogeneous local seismic response (LSR) of subsurface formations (both soils and weathered rocks). One of the advanced methodologies that has been emerging in this field is Neo-Deterministic Seismic Hazard Assessment (NDSHA), which relies on physical principles to model earthquake scenarios. Contrary to the standard probabilistic approach, NDSHA provides credible estimation of seismic scenarios which can affect the target site. NDSHA has proved its ability to predict seismic hazard at national and regional scale worldwide in the last 30 years. In this study, we have used the NDSHA approach to investigate the impact of 7 tectonic structures on the city of Chieti (Abruzzo Region, Central Italy). Due to its varied geological structure, different areas within Chieti experience distinct seismic responses. Hereinafter, NDSHA have demonstrated that the lithology and stratigraphy of the sequences along a 2D subsurface vertical section plays a critical role in the wavefield amplification. Simulations indicate that the lower parts of Chieti, situated at the base of its hills, exhibit an amplification factor (FA) of approximately 4-5 times for a broadband period range (0.125-1 s) whereas the FA values are lower, about 2-3 at only at long periods (0.8-1 s) along the slopes and uphill. By combining NDSHA with LSR studies, local authorities can enhance earthquake preparedness and ensure safer construction practices for promoting resilience of urban communities.

[1] University "G. d'Annunzio" of Chieti-Pescara, Department of Engineering and Geology, https://orcid.org/0009-0006-9742-3553.

[2] Department of Mathematics and Geosciences, University of Trieste, https://orcid.org/0000-0002-7478-8457.

[3] Osservatorio Sismologico - Università degli Studi di Bari "Aldo Moro", Campus Universitario, https://orcid.org/0000-0001-9747-9617.

[4] University "G. d'Annunzio" of Chieti-Pescara, Department of Engineering and Geology, https://orcid.org/0000-0003-1733-7112.

Email: chiara.faraone@unich.it[1]; romanel@units.it[2]; nicola.venisti@uniba.it[3]

* Corresponding author: g.vessia@unich.it

DOI: 10.1201/9781003684121-12

1. Introduction

Neo-Deterministic Seismic Hazard Assessment (NDSHA) methodology has been used for about 30 years to estimate seismic hazard at regional and local scales both in Italy and abroad (e.g: Panza et al., 2001; Panza et al., 2012; Panza and Bela, 2020 and references therein; Panza et al., 2022). NDSHA methodology and its software implementation, developed in the context of a significant international collaboration (for a recent review see Panza et al. 2022), enables both first-order seismic zonation of a large area and site-specific modeling of the seismic wavefield in urban areas with relevant and cultural heritage sites, for the purpose of seismic microzoning. Actually, based on a region's seismotectonic, geologic and geophysical characteristics, NDSHA uses a multi-scenario and multi-scale (regional and local) procedure.

NDSHA naturally provides synthetic time series of ground motion comparable to the natural ones (i.e. Ricci et al. 2025), establishing a workable link among the engineering, geotechnical and seismological communities. Actually, in accordance with earthquake scenarios, at the regional scale, maps of peak values for bedrock-level ground displacement, velocity, and spectral accelerations can be extracted (e.g. Panza et al, 2012; Magrin et al., 2016; Fasan et al., 2016, and references therein). At the local scale, incorporating all available information collected in geological, seismotectonic and geotechnical databases of the sites of interest, NDSHA can provide groundshaking scenarios and broad-band time histories and any derived damage indicators, ready to be used as seismic input for further engineering design (e.g. Panza et al., 1996; Romanelli and Vaccari, 1999; Panza et al. 2001, 2012; Rugarli et al., 2019; Romanelli et al., 2022, and references therein) and seismic response analyses for microzonation studies (e.g. Vessia et al., 2021 and reference therein).

As a multi-scenario approach, NDSHA may incorporate seismic sources defined according to various methods, such as seismogenic zones, seismogenic nodes found by applying pattern recognition to a morpho structural analysis (MSZ) of the considered territory (e.g. Gorshkov et al., 2022, and references therein), and information on active faults that are currently available. In this work, the information available from the Italian Database of Individual Seismogenic Sources (DISS) has been used to define earthquake scenarios (DISS Working Group, 2021). A 2D vertical section of Chieti's town encompassing uphill and foothill areas (see Ricci et al., 2025) has been used to predict the various strong motion shaking induced by earthquake scenarios (of different magnitudes, distance and focal parameters) generated by different seismic sources along Chieti's vertical section.

In the following, Section 2 introduces and characterizes the seismic sources that may impact the Chieti site, detailing the geological and seismic-tectonic aspects of the Chieti region. Section 3 delineates the NDSHA local scale methodology and presents the profile analysis for several scenarios.

In Section 3, the NDSHA local scale approach is described and the profile analyses for the different scenarios are shown. Subsequently, Section 4 presents a discussion of the analytical results and offers concluding observations.

2. Seismotectonic and geologic features of Chieti (Abruzzo Region, Italy)

The city of Chieti is located in central Italy (Abruzzo Region), within the subsident basin of the Italian peninsula known as the Periadriatic Sector. This sector extends 30 km inland from the Adriatic coast, characterized by a hilly topography and subject to the high seismicity of the Central Apennine Chain. The active tectonics in the Apennine chain are dominated by seismogenic normal faults, which are responsible for the majority of seismic activity in central Italy (Boncio et al., 2004; D'Agostino, 2014; DISS Working Group, 2025; Galli, 2020; Faure Walker et al., 2021). The latest Italian historical earthquake catalogue, CPTI15 (Rovida et al., 2022; Rovida et al., 2020), indicates a significant decrease in seismic activity moving from the Adriatic coastline towards the eastern edge of the Apennine chain.

Chieti's urban development has spanned a large area encompassing both an isolated hill and the Pescara River valley. Chieti belongs to the seismic zones 918 (Meletti et al., 2004) which characterises its historical and current seismicity (Faraone et al., 2023). The city is located between the Apennines foothill belt and the Adriatic Sea. The Apennine area includes the Maiella massif which is experienced devastating earthquakes over the past three centuries (Table 1). Both near-field and far-field seismic events have caused damage to newly constructed buildings as well as historic structures (Faraone et al., 2025). Two significant earthquakes have impacted the area near Chieti: the 1706 event (M=6.8) and the 1933 event (M=5.9), both with epicenters in the Mount Maiella region, approximately 30 kilometers southwest of Chieti. The 1706 earthquake caused extensive damage across a vast area, reaching an intensity of VIII on the Mercalli-Cancani-Sieberg scale (IMCS) in Chieti (D'Amico et al., 1998).

Table 1 Historical earthquakes that produced damages at Chieti city of Intensity ≥ VI MCS DBMI15 Database (after Locati et al., 2022, modified). *Seismic intensity, **Epicentral intensity, ***Moment Magnitude.

SI (MCS)*	*Year*	*Month*	*Day*	*Epicentral Area*	*Lat (°)*	*Long (°)*	*EI***	*M****
VI	1456	12	5	Central-Southern Apennine	41.302	14.711	XI	**7.2**
VI	1688	6	5	Sannio	41.283	14.561	XI	**7.1**
VII	1703	1	14	Valnerina	42.708	13.071	XI	**7.0**
VIII	1706	11	3	Maiella	42.076	14.08	X	**6.8**
VI	1881	9	10	Chietino	42.237	14.335	VIII	**5.4**
VII	1882	2	12	Chietino	42.291	14.347	VII	**5.3**
VII	1915	1	13	Marsica	42.014	13.53	XI	**7.1**
VII	1933	9	26	Maiella	42.079	14.093	IX	**5.9**
VI	1984	5	7	Meta mountains	41.667	14.057	VIII	**5.9**

In this work, seven individual seismogenic sources (ISS), reported in the DISS database, were analyzed and discussed as potential seismic sources affecting Chieti's town. The composed seismic source which was responsible for the 1706 earthquake is not considered here. It has already been discussed in Ricci et al. (2025).

The ISS which will be hereinafter used as seismic sources are listed below and figured out in Figure 1 are responsible for the largest earthquakes that caused damages at Chieti's town (Ricci et al., 2025): 1. ITIS015 – Montereale Basin; 2. ITIS131 – Paganica; 3. ITIS132 – San Pio delle Camere; 4. ITIS002 – Fucino Basin; 5. ITIS003 – Aremogna-Cinque Miglia; 6. ITIS027 – Sulmona Basin; 7. ITIS094 – Tocco da Casauria.

They are located between 30 and 70 km far from Chieti and they all are characterized by a magnitude greater than 6. The principal features of these faults are described below:

- The Montereale Basin source (ITIS015) is an NNW-SSE striking fault with 6.5 M_w maximum magnitude is associated. It is considered responsible for the 1703 earthquake, as indicated in Table (2).
- The Paganica source (ITIS131) caused the 6 April 2009 earthquake (6.3 M_w). The fault parameters were constrained through the numerous seismological, geodetic and geological data collected after 2009 (Chiaraluce et al., 2011; Nocentini et al., 2018). This was the most recent earthquake affected Chieti. It was a far-field seismic event causing differentiated damages on

buildings set along the borders of the hilltop, most of which were damaged at the first two stories (Faraone et al., 2025).

- The San Pio delle Camere source (ITIS132) was recognized as active by several independent groups (e.g. Bosi, 1975; Bagnaia et al., 1989; Vezzani and Ghisetti, 1998; Morewood and Roberts, 2000; Pizzi et al., 2002; Roberts and Michetti, 2004; Servizio Geologico d'Italia, 2006). It is located with a left en-echelon step at the south-eastern edge of Paganica seismogenic source. Its location and geometry was constrained by field work and geomorphic analysis on the basin running from Barisciano to Civitaretenga – Navelli, where converging morphotectonic, geological and structural evidence point to the presence of an active large SW-dipping normal fault (Di Bucci et al., 2011a).
- The Fucino Basin source (ITIS002) is defined as a surface-breaking fault characterized by pure normal slip. This definition agrees with direct observations of the coseismic fault scarps formed during the 13 January, 1915 earthquake, (M_w = 7.0) with the inversion of 1915 coseismic elevation changes, and with paleoseismological analyses of the main fault scarps observed in the Fucino Plain (DISS Working Group, 2025). Chieti suffered a VII I_{MCS} in the 1915 earthquake (Locati et al., 2022).
- The Aremogna – Cinque Miglia source (ITIS003) is an NNW-SSE striking fault with associated moment magnitude of 6.4. Two historical earthquakes may have been generated by this fault occurred in 801 A.D. and 1349 A.D. If none of these events ruptured this fault, and the true recurrence interval is in the lower part of the estimated range, this source could hold a significant potential for producing a large earthquake in the near future (DISS Working Group, 2025).
- The Sulmona Basin source (ITIS027 in DISS) is a well-known normal fault that dips SW (Galli et al., 2015). The most significant earthquakes potentially associated with this source are those of November 3, 1706 (M_w 6.8) and September 26, 1933 (M_w 5.9).
- The Tocco da Casauria source (ITIS094) is an E-W striking, normal-oblique fault with a potential of 6.0 M_w that is believed to be the origin of one shock of the extensive 1456 seismic sequence (Fracassi and Valensise 2007).

In Table 2 the technical characteristics of each of the seven individual seismogenic sources considered in this study are summarized.

The specific values of Strike, Dip, and Rake for each structure listed in the DISS and represented in Table 2 have enabled the construction of the corresponding focal mechanisms, illustrated in Figure 1. The analysis of these mechanisms reveals that structures 1, 2, 3, 4, 6, and 7 exhibit extensional focal mechanisms, which can be correlated with the flexural deformation of the Apulian foreland and the development of the associated foredeep. In contrast, structure 5 is associated with a right-lateral strike-slip focal mechanism (E-W orientation). This characteristic is justified by the fact that ITIS094 (no. 5 in Figure 1) is located near the transition zone between the central and southern Apennines, an area well known for focal mechanisms of this type.

From a geological point of view, Chieti is characterized by the sub-horizontal layers of the terraced Mutignano formation (FMT) ranging from the Upper Pliocene to Lower Pleistocene age. FMT consists on marine clays, that are the bedrock of the entire territory. Within the geological sequence, the marine clays transit upwards into sands and conglomerates. Hence, FMT can be divided into several associations, arranged from bottom to top, according to their litho-technical properties. In Chieti, the following units are recognized from the surface in depth: (1) FMT_a pelitic–sandy association (clays with thinly stratified sandy and sandy silty layers interbed in over consolidated grey-marly clays and clays), (2) FMT_c sandy–pelitic association (yellow silty sands alternate with sands of varying cementation levels, as well as thinly laminated, over-consolidated greyish silty clays and clays), and (3) FMT_d sandy conglomeratic association (lenses and layers of gravels and conglomerates made of cm-sized clasts alternate with extremely dense yellowish sands and sandstones in medium to thick beds) (ISPRA, 2010).

Table 2 Parametric information about each individual seismogenic sources (DISS Working Group, 2025). **Based on paleosismology*

DISS-ID	*Location (Lat/Long)*	*Length (km)*	*Width (km)*	*Min depth (km)*	*Max depth (km)*	*Strike (deg)*	*Dip (deg)*	*Rake (deg)*	*Mw*	*Lastes Earthquake (dd/mm/yy)*
ITIS015	42.42/13.26	23.4	13.6	3	13.4	132	50	270	6.5	02/02/1703
ITIS131	42.32/13.45	14	9.5	3	9.5	133	43	275	6.3	06/04/2009
ITIS132	42.27/13.62	16.2	10.5	0.7	8.7	127	50	270	6.2	Unknown
ITIS027	42.09/13.88	20	12.2	1	11.6	135	60	270	6.4	Unknown
ITIS094	42.21/13.91	12	8	11	18.5	89	70	230	6.0	30/12/1456
ITIS002	41.96/13.61	28	15.5	1.5	14.9	135	60	270	6.7	13/01/1915
ITIS003	41.84/13.97	20	12.2	3	12.3	144	50	270	6.4	*800 B.C-1030 A.D.

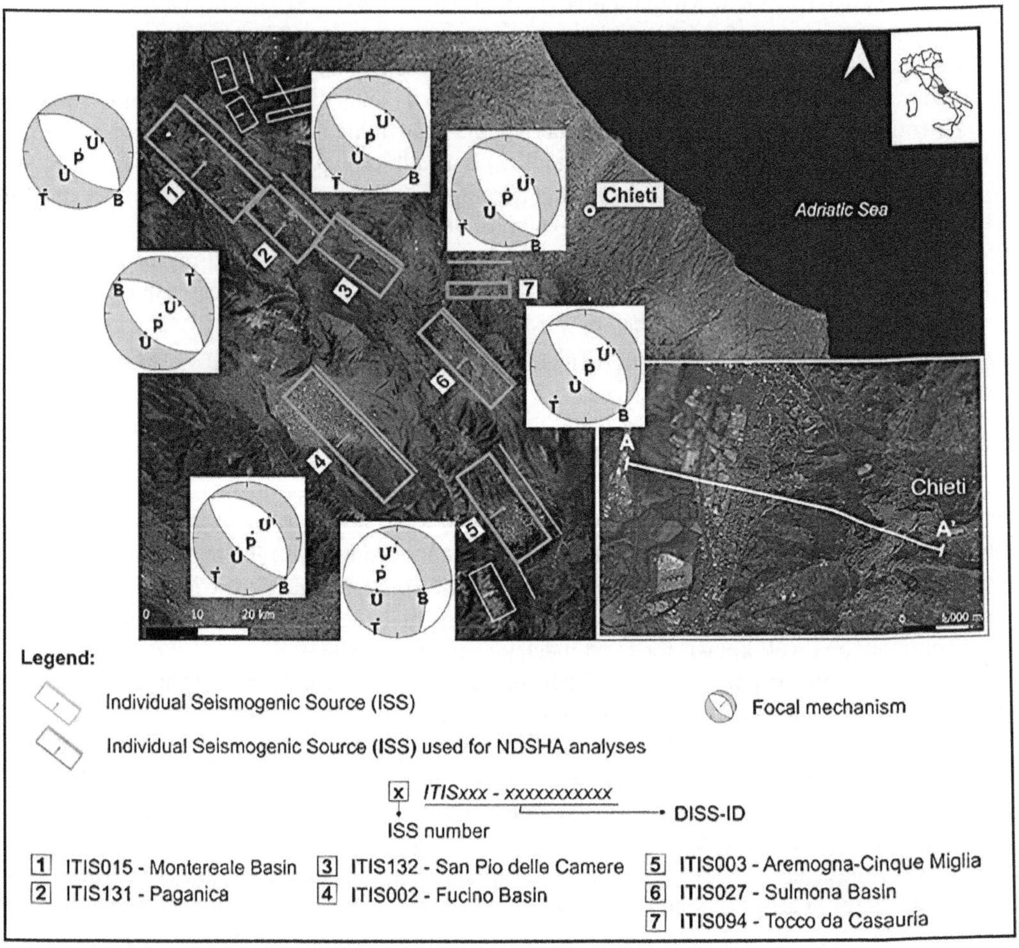

Fig. 1 Individual Seismogenic Sources – ISS (DISS Working Group, 2025) considered in this study. The inset shows the trace A – A' in white color of the geological section used for NDSHA analyses. Grey/white beach balls represent the focal mechanism of ISS: compressive (P-axis), distensive (T-axis), fault plane (U-line) and nodal point (B).

The slopes of the Chieti hill are often covered of landslides deposits, anthropogenic material deposits and Holocene eluvial-colluvial deposits. The Pescara River cuts the FMT formation and alluvial deposits in several orders of fluvial terraces, whose age ranges from Late Pleistocene (“AVM” units in Figure 2A) to Holocene (“olob” units in Figure 2A).

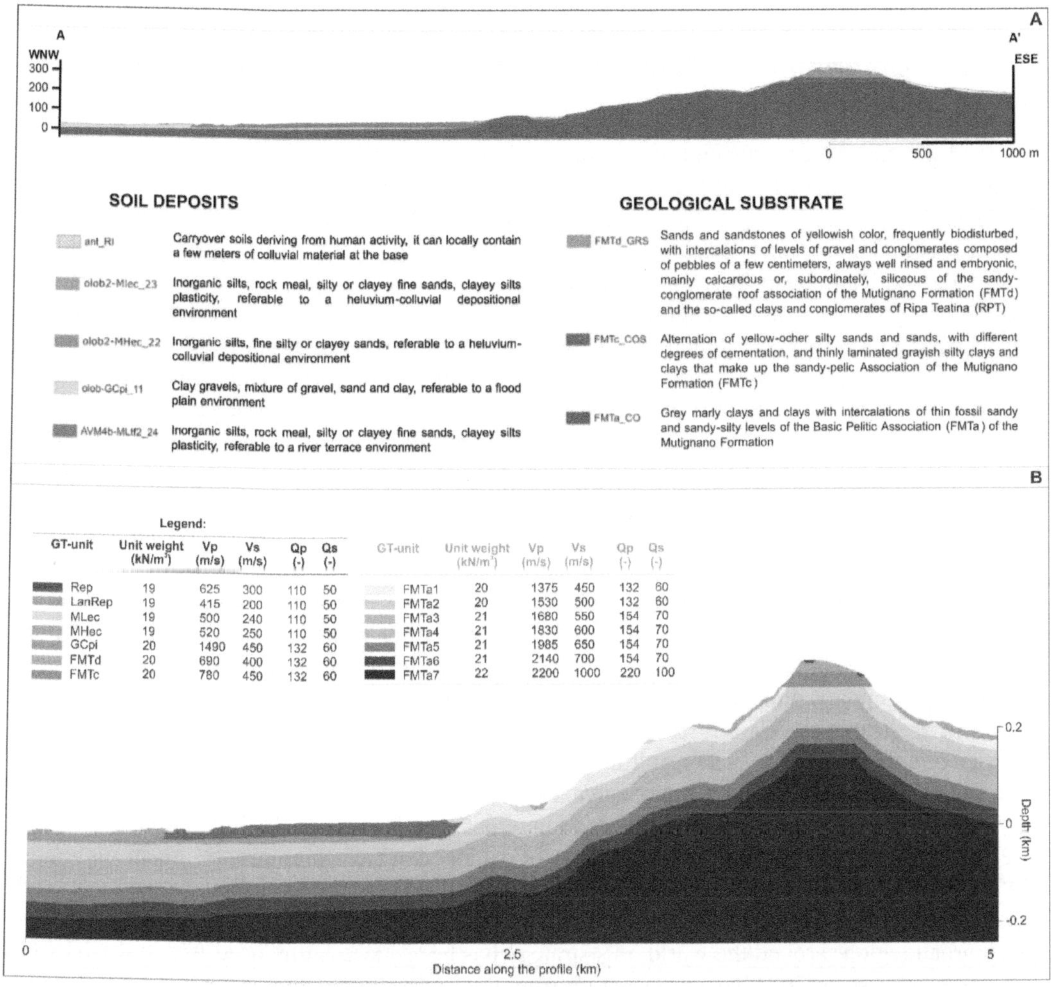

Fig. 2 (A) Chieti’s 2D geological subsoil model. Modified from Ricci et al., 2025. (B) Chieti’s 2D subsoil model used in the seismic modelling with its geophysical properties.

The Holocene Quaternary deposits, listed in legend of Figure 2A, are categorized according to the following rule: the first four lowercase characters refer to the age of the deposit, the second two uppercase characters are the lithotype units according to the Seismic Microzonation standards (SM Working Group, 2008), and then two lowercase characters refer to the depositional environment (tf - alluvial terrace; pi - alluvial plain; ec - eluvial – colluvial). Finally, the first number refers to the order of the river terraces and the second one indicates the consistence of each lithotype (the lower the number the lower the consistence). For further lithological details of these formations refer to Ricci et al. (2025).

The shallow subsurface geometry of the lithological series that makes up the Chieti hill and the neighbouring Pescara River valley is depicted in the geo-lithological section (Figure 2A). Details on the physical-mechanical and seismic characteristics of each lithotype used for the numerical simulations of local seismic response (LSR) and the model, are shown in Figure 2B: each color represents different values of shear wave velocity V_S, longitudinal wave velocity V_P, the constant damping for both wave types Q_S and Q_P according to the viscous-elastic behavior assumed for both rocky and soil formations.

3. Neo-Deterministic Seismic Hazard Assessment – NDSHA – at local scale

A hybrid approach, integrating multimodal summation and the finite difference technique, effectively models the influence of lateral heterogeneities along seismic wave propagation paths (Fäh et al., 1993). Recent applications of this methodology in Italy can be found in Panza et al. (2015), Hassan et al. (2016), Nunziata and Costanzo (2022), and Romanelli et al. (2022). Leveraging expertise in seismic verification and fostering close collaboration between seismologists and engineers, the NDSHA program suite for generating realistic synthetic seismograms has undergone significant advancements. The primary objective of these developments is to simplify its application for seismic input estimation, structural verification of existing engineering works, and the design of new structures. To facilitate this, a web application, XeRiS, has been introduced (Vaccari, 2016; Vaccari and Magrin, 2019; Vaccari and Magrin, 2022), providing a user-friendly graphical interface that abstracts the complexity of the computational engine. For defining earthquake scenarios in this study, we utilized data from the Italian Database of Individual Seismogenic Sources (DISS Working Group, 2025), including geometrical and focal parameters as well as maximum reported magnitudes, as detailed in Section 2. To model seismic wavefields radiated from an extended source, the kinematic rupture process is simulated using the PULSYN algorithm, based on the theoretical framework presented by Gusev (2011). The fault plane, where the primary rupture occurs, is discretized into a grid of point sub-sources, with slip distribution described stochastically and activated upon rupture front arrival. This approach enables PULSYN to generate multiple stochastic realizations of the source model—incorporating slip distribution and rupture velocity—for each source-to-site path. The resulting spectra (amplitude and phase) of the equivalent source time functions account for both rupture dynamics and directivity effects, making them ready for convolution with the corresponding Green's Functions (see Hassan et al., 2020; Romanelli et al., 2022 for further details and applications).The hybrid method, which combines the modal summation (MS) and finite difference (FD) techniques is used to simulate the ground motion caused by laterally heterogeneous structure models at the sites of interest (e.g. Fäh et al., 1993; Panza et al., 2001). In many urban regions around the world, this strategy has been successfully used within the NDSHA (e.g., Hassan et al., 2020). Wave propagation from the source to the studied section (with Chieti's uphill and foothill) is simulated using the modal summation (MS) technique for the Rayleigh (P-SV motion) and Love (SH motion) wavefields (Panza, 1985; Florsch, 1991). Since this is a pure analytical technique, there is no time penalty for the size of the model in this part of the modeling. The incoming wavefield is then propagated using the Finite Difference (FD) method in the laterally heterogeneous region of the structural model (e.g. see Fäh et al., (1993)). In this way, each of the two modeling techniques, MS and FD, is applied to the region of the structural model where it is most effective, and the two techniques are coupled by incorporating the time histories that are obtained using MS into the FD computations (Figure 3).

The local structure is discretized into a grid mesh with the purpose of using the FD approach, and the grid mesh size must be small enough to have at least 10 points per wavelength. As a bedrock model, we used the pertaining lithospheric structure between those adopted for calculations at the national scale by Panza et al. (2012, see Model 6); the computed wavefield is then introduced in the laterally heterogeneous model discussed in Section 2 (see Figure 2A) and shown in Figure 3.

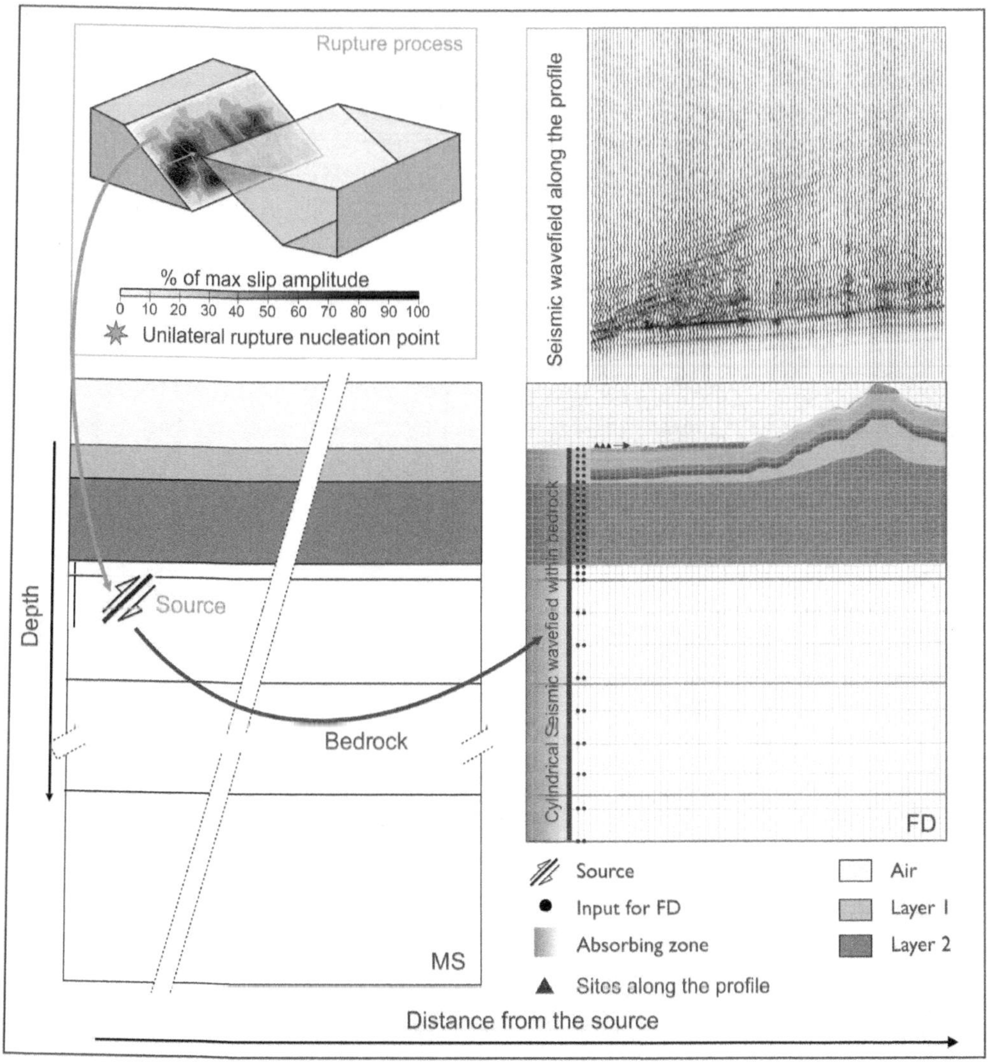

Fig. 3 Scheme of the hybrid, Modal Summation (MS) – Finite Differences (FD), method.

4. Results and Discussion

The wavefields associated with the seven seismic sources, along with the amplifications calculated as the ratio between the surface 2D and 1D response spectra, have been simulated. The acceleration time series computed along the analysed section for the seven scenarios, described in Section 3, are presented through three components: vertical, radial, and transverse. From results, the interaction between the seismic wavefield and the geophysical-geotechnical properties of the materials (up to the surface) generates highly variable ground shaking scenarios depending on the seismic source considered. Moreover, the amplitude and duration of the time series at different sites exhibit significant variability due to the combined effects of mechanical and geometrical factors, such as the differential generation of basin-edge-induced waves. The seven seismic sources show the same extensional focal mechanisms (normal faulting directed along NW-SE) although the Tocco da Casauria fault direction is tilted towards W-E. Nonetheless, the three wavefields of Tocco da Casauria seismic source do not show special features compared to the others even accounting for its shortest distance to Chieti. Due of its vicinity, the peak of the transverse wavefield has the highest peak among all.

All seven seismic sources exhibit high amplitudes and extended signal durations along the sediment layers (approximately 40 m thick) and the hill slopes (where slope deposits are as thin as 10 m), across different period ranges and with varying extents.

For discussion purpose, only the case of Sulmona (Figures 4-5) and Tocco da Casauria (Figures 6-7) seismic sources have been presented hereinafter. The remainder results have been illustrated in Appendix 1.

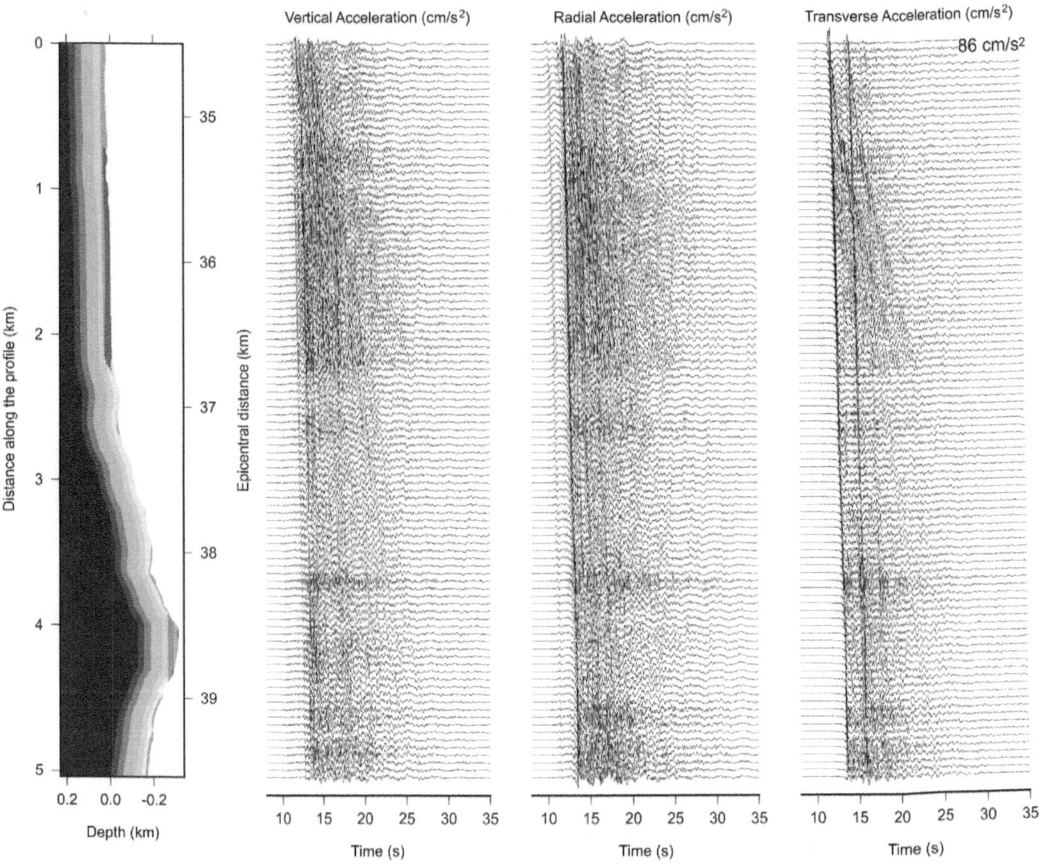

Fig. 4 Sulmona Basin (ITIS027) source. Acceleration time histories along the 2D section at Chieti town site, for P-SV (Vertical and Radial components) and SH (Transverse component is wavefields. The plotted amplitude of the signals is scaled to the maximum value (marked by the label).

The Sulmona source (Figure 4), located only 35 km from Chieti, demonstrates high amplification across all acceleration components throughout the section, with a maximum peak acceleration of 86 cm/s² along the transverse component (Figure 4). Figure 5 highlights strong amplifications of the vertical acceleration component (3–4) at the foothill within the period range of 0.13–0.5 s (2–8 Hz). On the hilltop, amplifications reach values of 4–5 at 2 s (0.5 Hz) and within the 0.2–0.3 s (3.5–5 Hz) range. The radial and transverse components exhibit localized amplifications (3–4) within a smaller foothill area at a period range of 0.25–1 s (1–4 Hz). Along the hill slope, an amplification of 3 is observed within the 0.14–0.33 s (3–7 Hz) range.

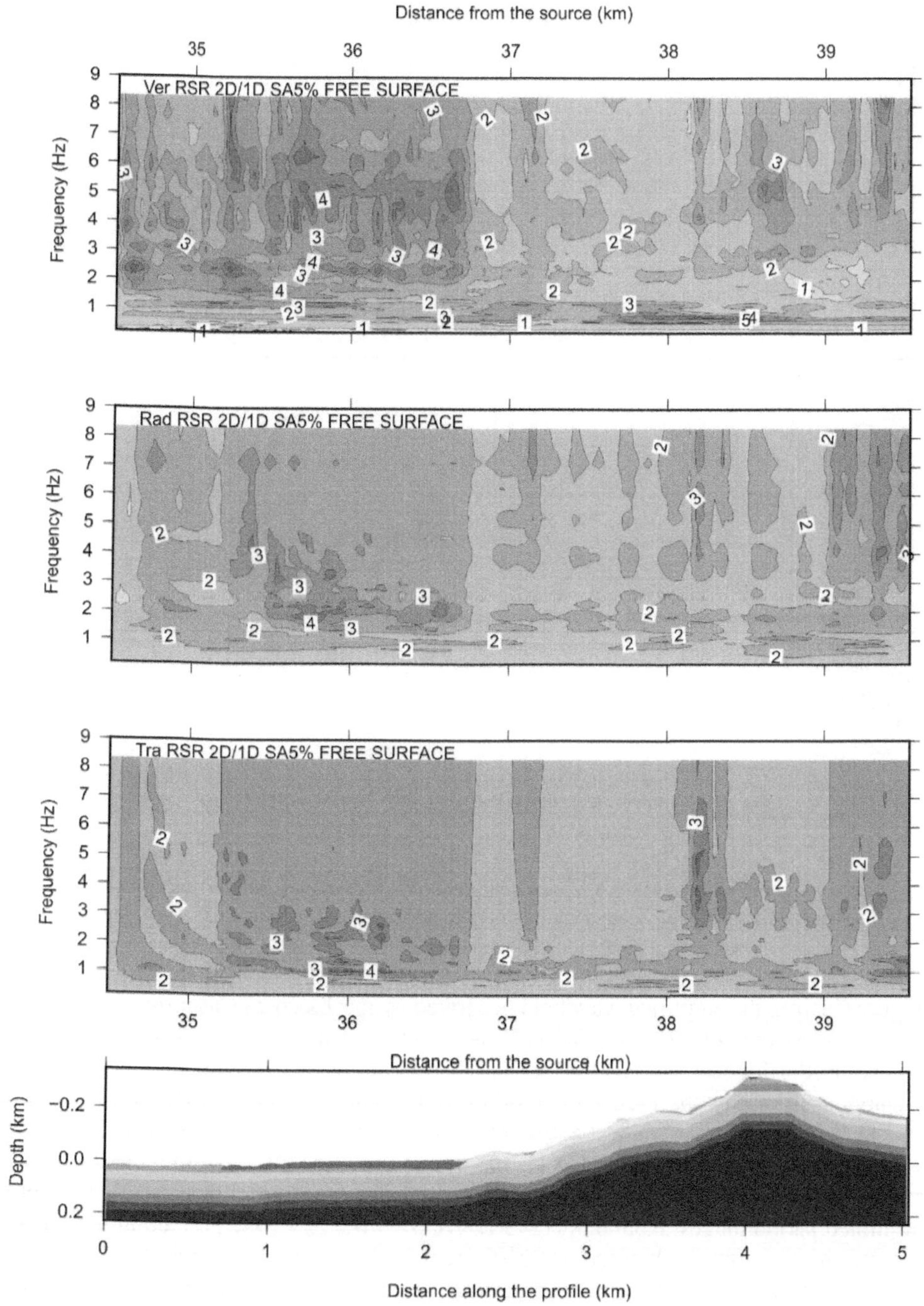

Fig. 5 Sulmona Basin (ITIS027) source. Spectral amplifications obtained as the ratio between the response spectra of the synthetic seismograms computed along the profile and the synthetic signals computed at the same sites considering the reference bedrock model. From top to bottom: vertical component, radial component, transverse component.

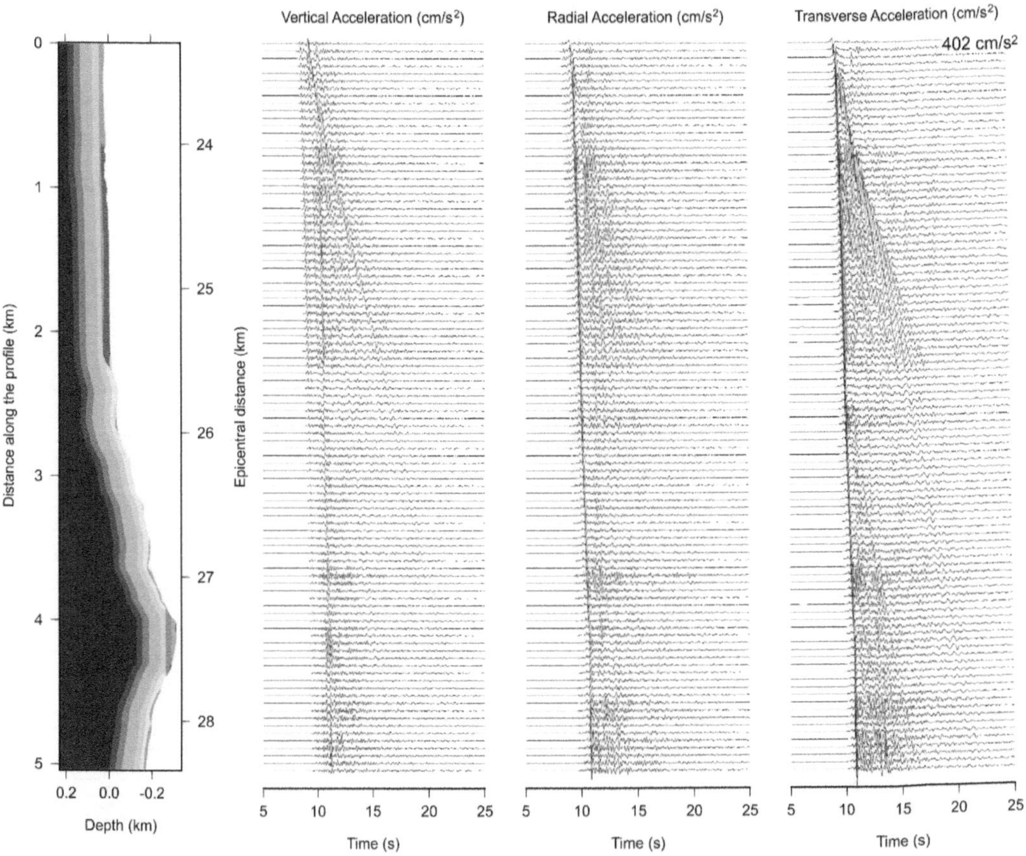

Fig. 6 Tocco da Casauria (ITIS094) source. Acceleration time histories along the 2D section at Chieti town site, for P-SV (Vertical and Radial components) and SH (Transverse component) wavefields. The plotted amplitude of the signals is scaled to the maximum value (marked by the label).

Figure 6 shows the amplified wavefield generated by the Tocco da Casauria source, the closest to Chieti at approximately 25 km. The peak acceleration value calculated along the transverse component is the highest among the seven seismic sources, reaching 402 cm/s^2. However, amplification is primarily observed at the foothill, where soil deposits are thicker, and along the hill slopes. Figure 7 confirms the highest amplifications along the vertical component (3–4) across the entire considered frequency range of 1–8 Hz (0.13–1 s) and at the hilltop within the 0.14–0.2 s (5–7 Hz) range. Regarding the radial and transverse components, lower amplifications of 3 are observed within limited period ranges: 0.33–0.5 s (2–3 Hz) and 0.14–0.25 s (4–7 Hz) along the hill slopes.

Finally, these results can be used in Chieti town for both (1) seismic structural design of new buildings and retrofitting of historical and modern structures (Romanelli et al., 2022) through the response spectra calculated at the surface and (2) microzoning mapping of seismic amplification factors according to Italian rules (SM working group, 2008).

4. Concluding Remarks

The wavefields generated by seven seismic sources have been computed along the vertical section of Chieti's urban area. The results of these LSR simulations, conducted using the NDSHA method, provide insights into the distribution of seismic amplifications not only at the surface but also within

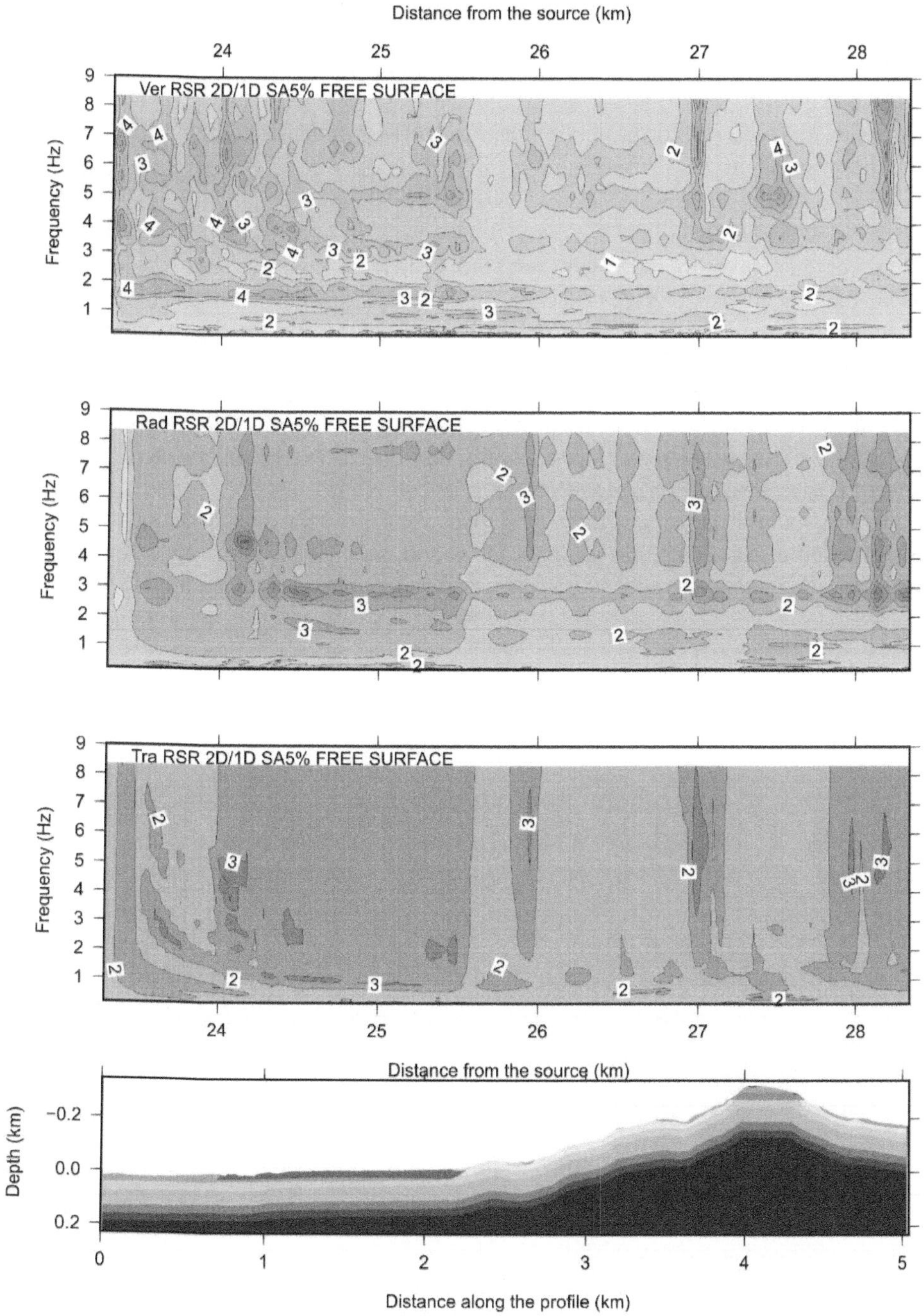

Fig. 7 Tocco da Casauria (ITIS094) source. Spectral amplifications obtained as the ratio between the response spectra of the synthetic seismograms computed along the profile and the synthetic signals computed at the same sites considering the reference bedrock model. From top to bottom: vertical component, radial component, transverse component.

the section. The wavefields have been calculated for the three components of the synthetic signals recorded at the site. The key findings are summarized as follows:

1. All seismic sources exhibit amplified accelerations in areas with thicker deposits.
2. At the foothill, the vertical component is typically highly amplified (3–5), whereas along the hill slopes and at the hilltop, accelerations reach values of 3–4, varying according to the seismic source.
3. Amplifications do not correspond to identical periods or frequencies but fluctuate depending on the seismic source.
4. The long period of 1 s (1 Hz) is generally associated with the strongest amplifications, particularly along the hillslope.

These results underscore the capability of the NDSHA method to analyse the spatial variability of seismic amplifications along a 2D section for each seismic signal component, considering both the incoming wave direction and the geomorphological characteristics of the site at the surface and within the section.

However, some limitations of this method rely on simplifications such as: 2D analyses for approximated 3D site effects; generation of synthetic seismic signals simulating the natural ones; disregard the seismic wave reflection and refraction due to heterogeneity of lithospheric rocks; simplified 2D viscous-elastic behaviour of rocks and soils.

Thus, this study demonstrates that NDSHA is well-suited for microzonation studies, enabling the investigation of source-site path contributions and the directional effects of incoming seismic waves, factors that cannot be adequately addressed using conventional decoupled methods.

Appendix 1. Simulated seismic wavefields and amplification patterns in Chieti

In this section five out of seven seismic scenarios of wavefields (Figures A2-A4-A6-A8-A10-A12) and surface amplifications (Figures A1-A3-A5-A7-A9-A11-A13) have been discussed.

Beginning with the Montereale fault, located approximately 70 km from Chieti (Figures A1-A2), the vertical component is observed to be less amplified than the radial and transverse components, both at the foothill and along the hill. All components are scaled to the peak acceleration value, which for the Montereale fault is 50 cm/s^2. As illustrated in Figure A2, the vertical component exhibits the highest amplification (3–4) across the entire frequency range of 1–8 Hz (0.13–1 s) at the foothill, whereas lower amplifications (2–3) are observed along the slope at a 1 s period and between 0.12–0.3 s (3–8 Hz). The radial and transverse components, which lie in the horizontal plane, show broadband amplifications (2–3) across the soil deposits at the foothill and along the slope deposits. The thickness of the deposits appears to play a dominant role in surface shaking amplification, while the hill morphology influences the selection of amplified periods, particularly in the transverse and radial directions.

Figures A3-A4 pertain to the Paganica fault, which is closer to Chieti, at approximately 50 km. Similar to the Montereale fault, the transverse and radial components exhibit greater amplification

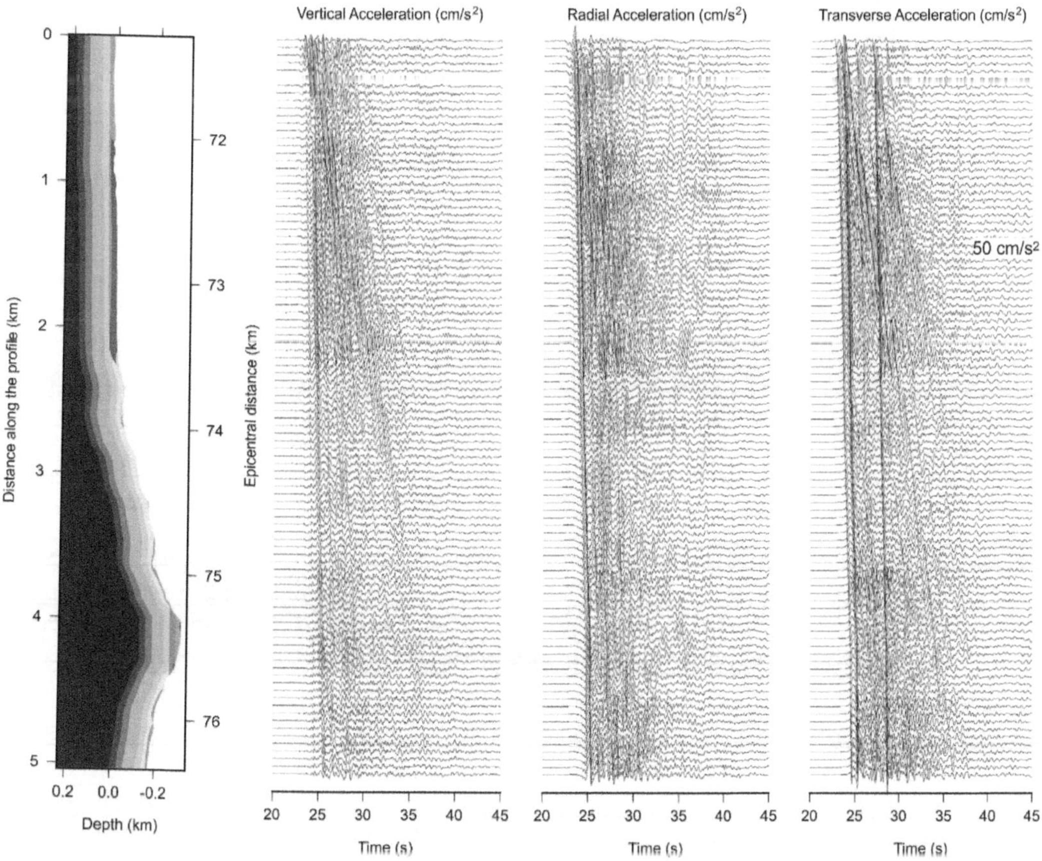

Fig. A1 Montereale Basin (ITIS015) source. Acceleration time histories along the 2D section at Chieti town site, for P-SV (Vertical and Radial components) and SH (Transverse component) wavefields. The plotted amplitude of the signals is scaled to the maximum value (marked by the label).

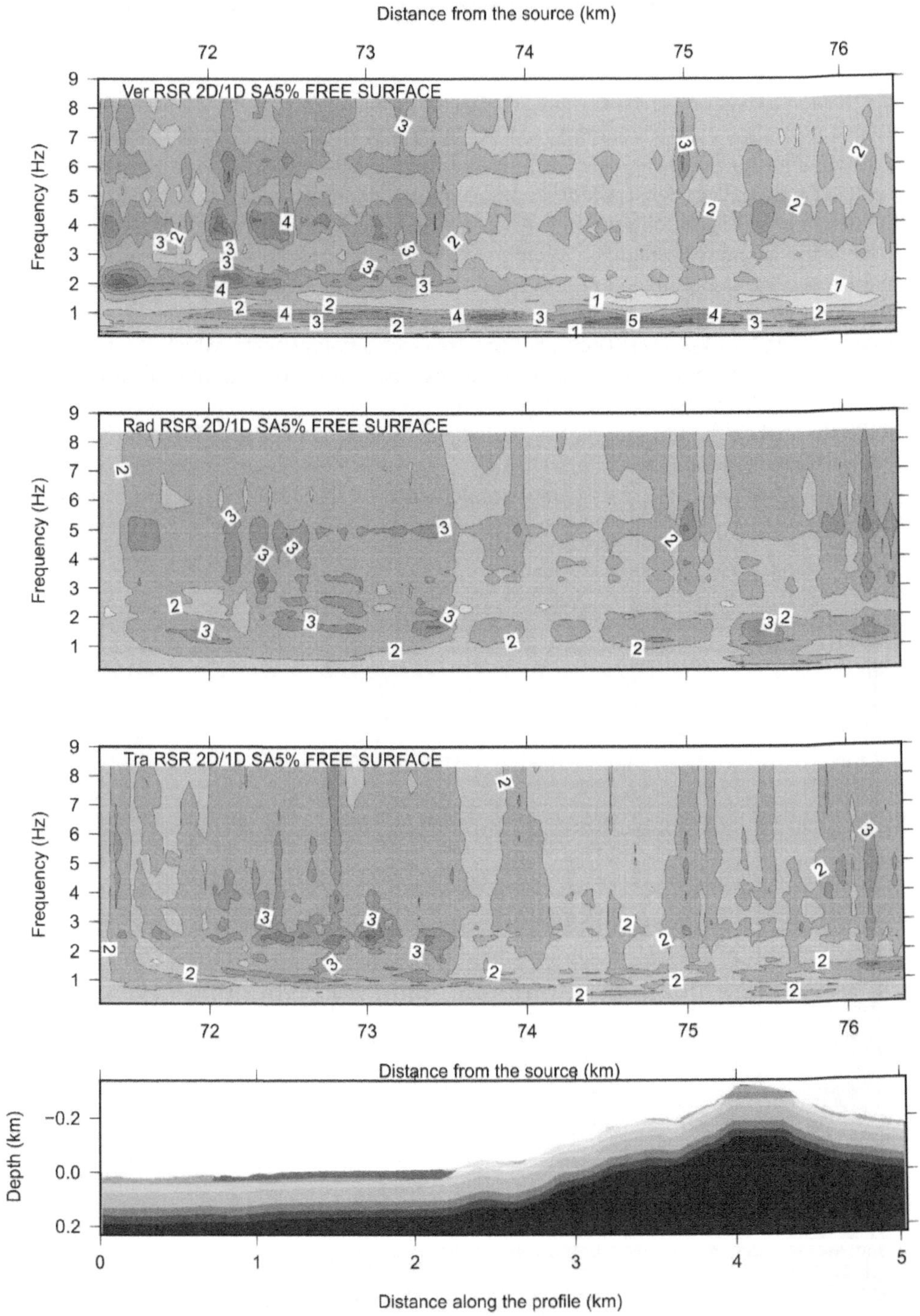

Fig. A2 Montereale Basin (ITIS015) source. Spectral amplifications obtained as the ratio between the response spectra of the synthetic seismograms computed along the profile and the synthetic signals computed at the same sites considering the reference bedrock model. From top to bottom: vertical component, radial component, transverse component.

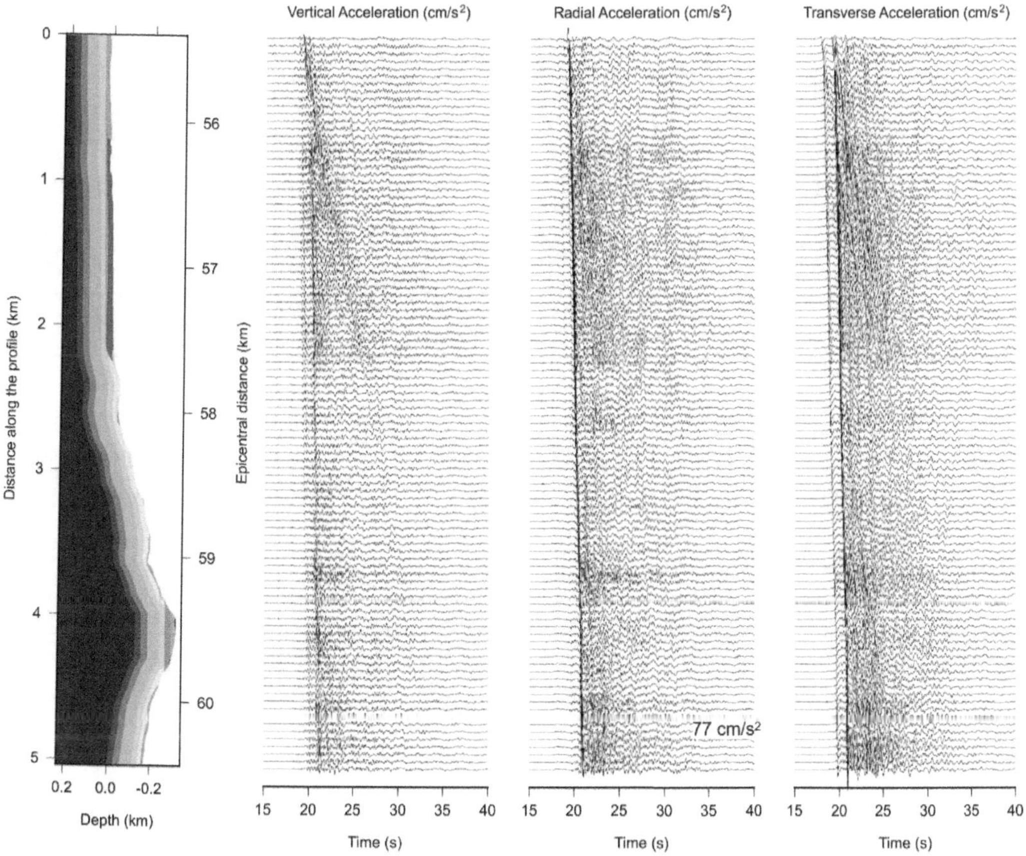

Fig. A3 Paganica (ITIS131) source. Acceleration time histories along the 2D section at Chieti town site, for P-SV (Vertical and Radial components) and SH (Transverse component) wavefields. The plotted amplitude of the signals is scaled to the maximum value (marked by the label).

than the vertical component. The peak acceleration value for this fault is 70 cm/s^2, likely due to its proximity to Chieti compared to the Montereale fault. As shown in Figure A4, the vertical component displays significant broadband amplification along the foothill (3–4) and very high amplification (4–5) over the hilltop at long periods (1 s, 1 Hz). Large amplification values (3) are observed on the hilltop at lower periods (0.14–0.25 s, corresponding to higher frequencies of 4–7 Hz). Conversely, the radial and transverse components indicate high amplifications (3) along the foothill deposits and slopes, concentrated within frequency ranges of 2–5 Hz (0.17–0.5 s) at the foothill and 2–8 Hz along the slopes (1.13–0.5 s). However, amplification levels vary across the different components.

Figures A5-A6 correspond to the San Pio delle Camere source, located approximately 40 km from Chieti. Figure A5 highlights the amplified wavefield across the entire Chieti section, despite the peak acceleration value reaching only 57 cm/s^2. This observation aligns with distinct amplification patterns across the three wavefield components (Figure A6): the vertical acceleration component reaches an amplification of 5 at a long period (1 s), while amplifications of 4 occur at 1 Hz (1 s), 2.5 Hz (0.4 s), and 4–5 Hz (0.2–0.25 s). The radial and transverse components exhibit amplifications

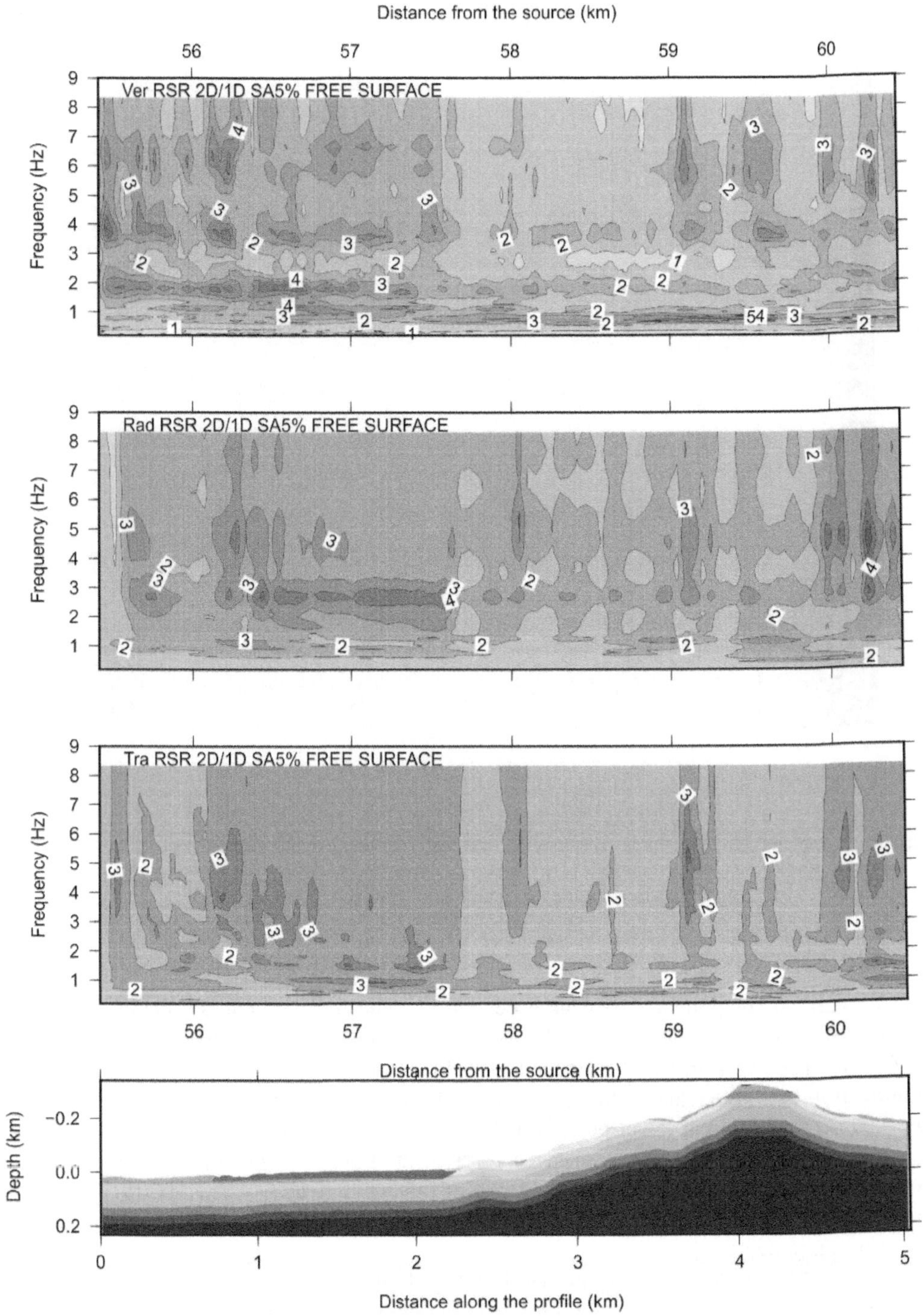

Fig. A4 Paganica source (ITIS131). Spectral amplifications obtained as the ratio between the response spectra of the synthetic seismograms computed along the profile and the synthetic signals computed at the same sites considering the reference bedrock model. From top to bottom: vertical component, radial component, transverse component.

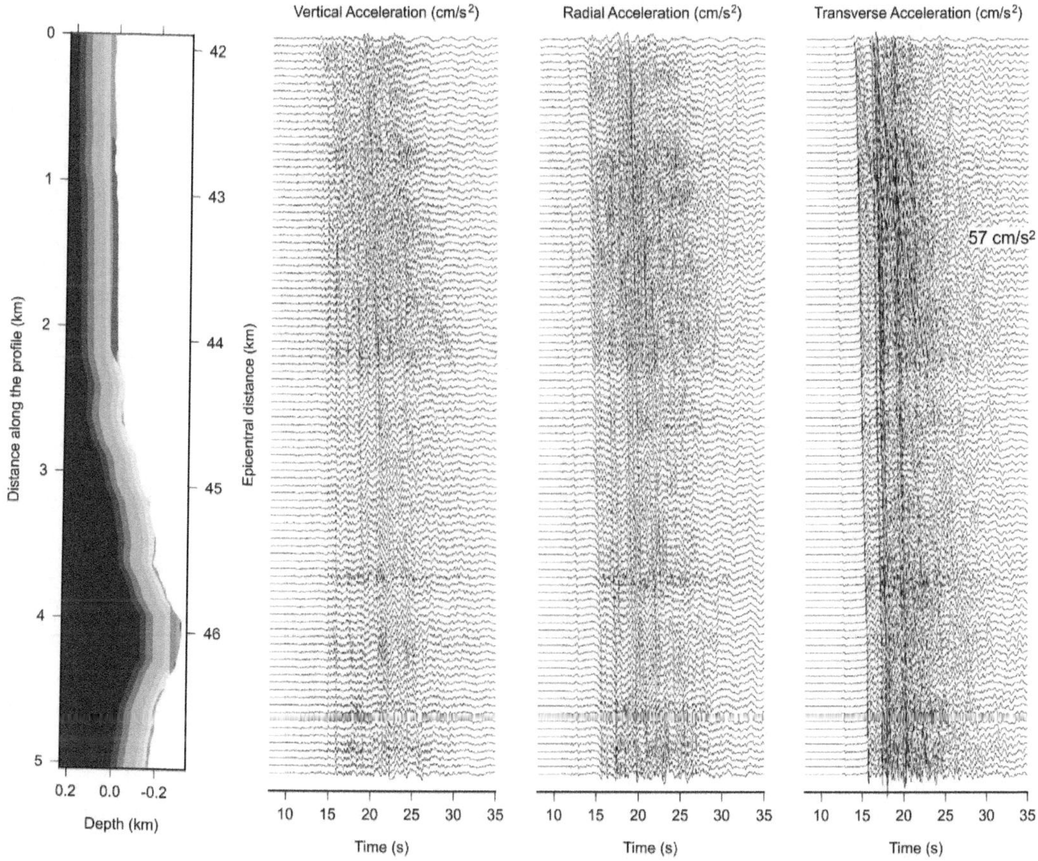

Fig. A5 San Pio delle Camere (ITIS132) source. Acceleration time histories along the 2D section at Chieti town site, for P-SV (Vertical and Radial components) and SH (Transverse component) wavefields generated by a rupture process. The plotted amplitude of the signals is scaled to the maximum value (marked by the label).

of 3–4 along the foothill at 1–4 Hz (0.25–1 s) and along the hill slope at 1 Hz (1 s) and 4–8 Hz (0.13–0.25 s).

Figures A7-A8 illustrate the case of the Fucino basin source, located approximately 60 km from Chieti. Figure A8 reveals a very high amplification throughout the section in the radial acceleration component, with a peak value of 170 cm/s^2. Conversely, the transverse component exhibits the lowest amplification, although it remains higher at the foothill. As shown in Figure A8, the vertical component reaches an amplification value of 4 at the foothill within the period range of 0.14–0.25 s (4–6.5 Hz) and along the hill slope at 1 s and 0.13–0.25 s (4–8 Hz). The radial component gets an amplification value of 3 at 1 s and at 0.33 s (3 Hz) at the foothill, as well as at 1 s at the hill crest. Meanwhile, the transverse component exhibits amplification values of 3 at the foothill but does not maintain consistent amplification across the entire soil deposit area. The amplification pattern varies across the different acceleration components.

Finally, Figures A9-A10 illustrate the acceleration wavefield and its amplifications caused by the Aremogna-Cinque Miglia seismic source, which is located at a distance from Chieti of 60 km. Figure A9 shows that the entire section amplifies the wavefield, with a peak acceleration of 97 cm/s^2 observed along the radial component. As depicted in Figure A10, significant amplifications are

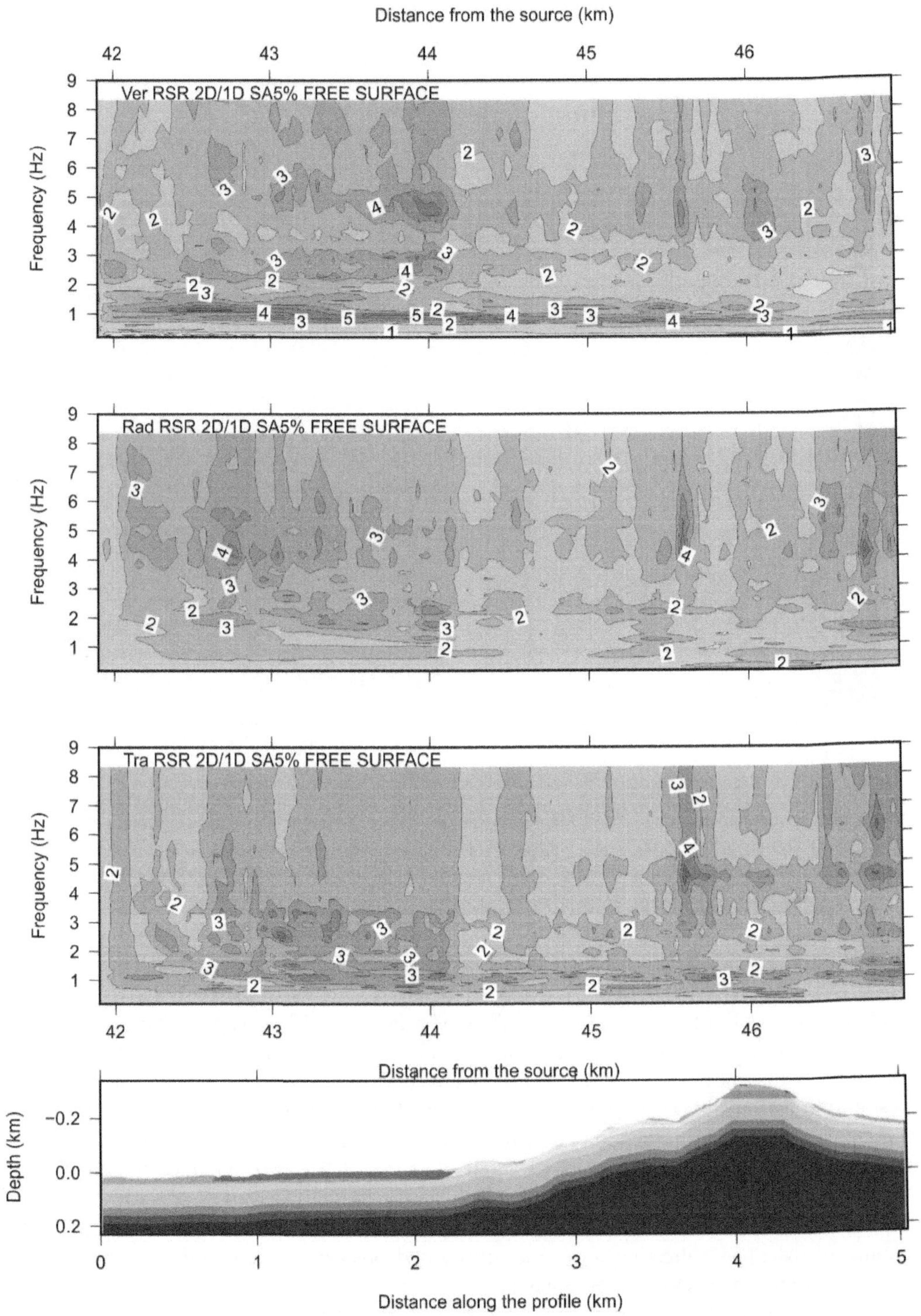

Fig. A6 San Pio delle Camere (ITIS132). Spectral amplifications obtained as the ratio between the response spectra of the synthetic seismograms computed along the profile and the synthetic signals computed at the same sites considering the reference bedrock model. From top to bottom: vertical component, radial component, transverse component.

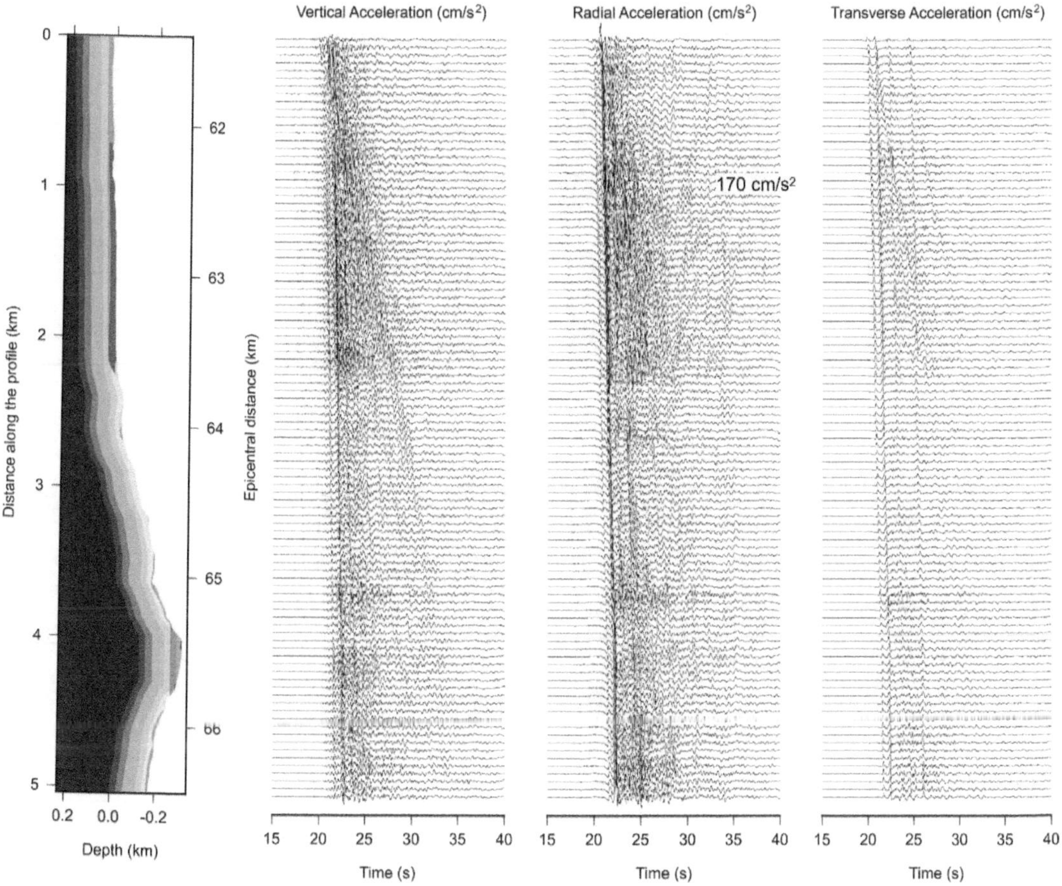

Fig. A7 Fucino Basin (ITIS002) source. Acceleration time histories along the 2D section at Chieti town site, for P-SV (Vertical and Radial components) and SH (Transverse component) wavefields. The plotted amplitude of the signals is scaled to the maximum value (marked by the label).

again evident along the vertical component (3–4) at the foothill, corresponding to the period range of 0.13–0.3 s (3.5–8 Hz). At the hilltop, amplifications as high as 3–4 are reached at 2 s (0.5 Hz), while along the hill slope, similar amplifications occur within the 0.13–0.3 s (3.5–8 Hz) range. The radial and transverse components exhibit amplification values of 3 at 1 s and within the 0.2–0.5 s (2–5 Hz) range at the foothill, whereas along the hill slopes, an amplification of 3 is observed at 0.14–0.25 s (4–7 Hz).

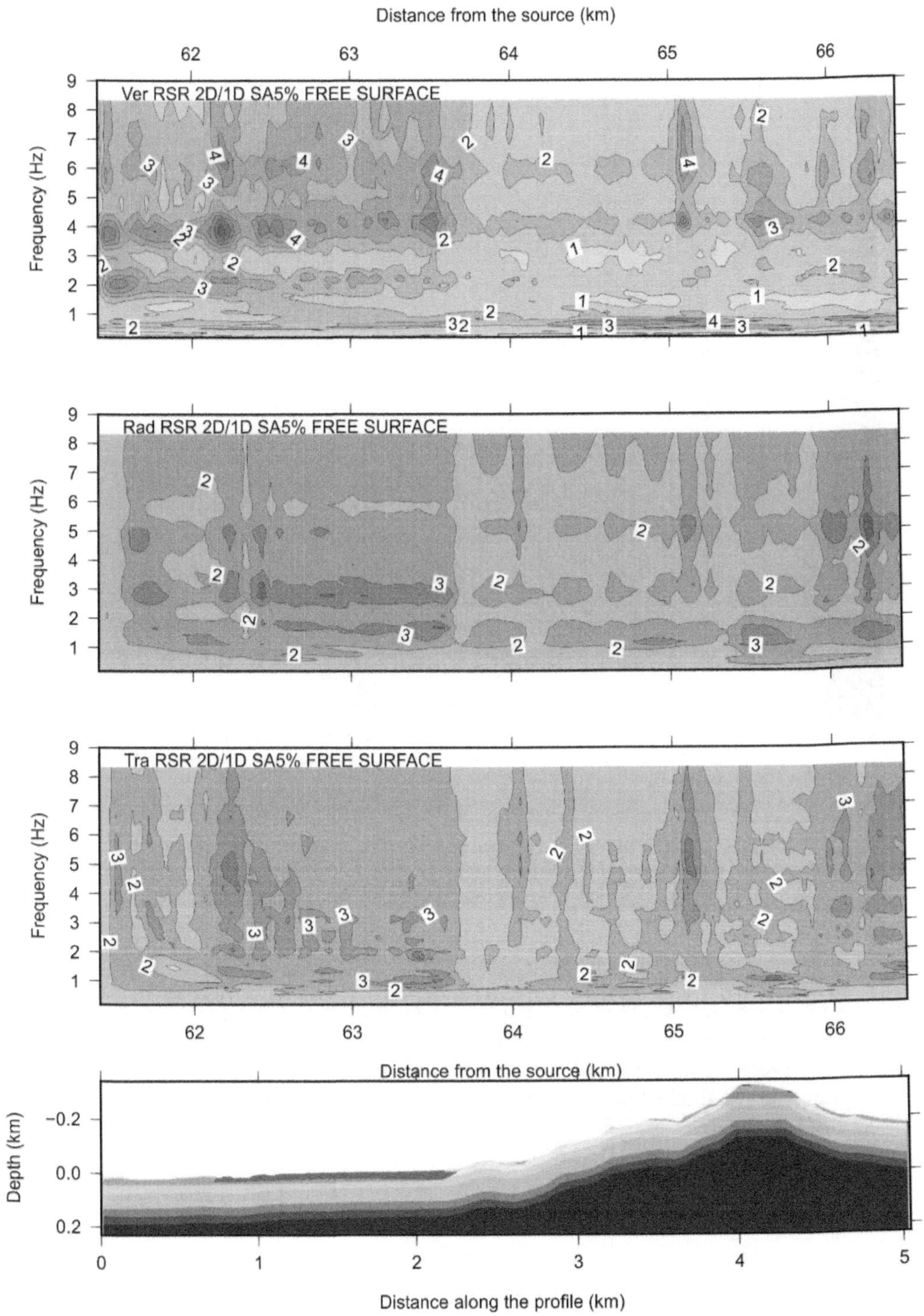

Fig. A8 Fucino Basin (ITIS002) source. Spectral amplifications obtained as the ratio between the response spectra of the synthetic seismograms computed along the profile and the synthetic signals computed at the same sites considering the reference bedrock model. From top to bottom: vertical component, radial component, transverse component.

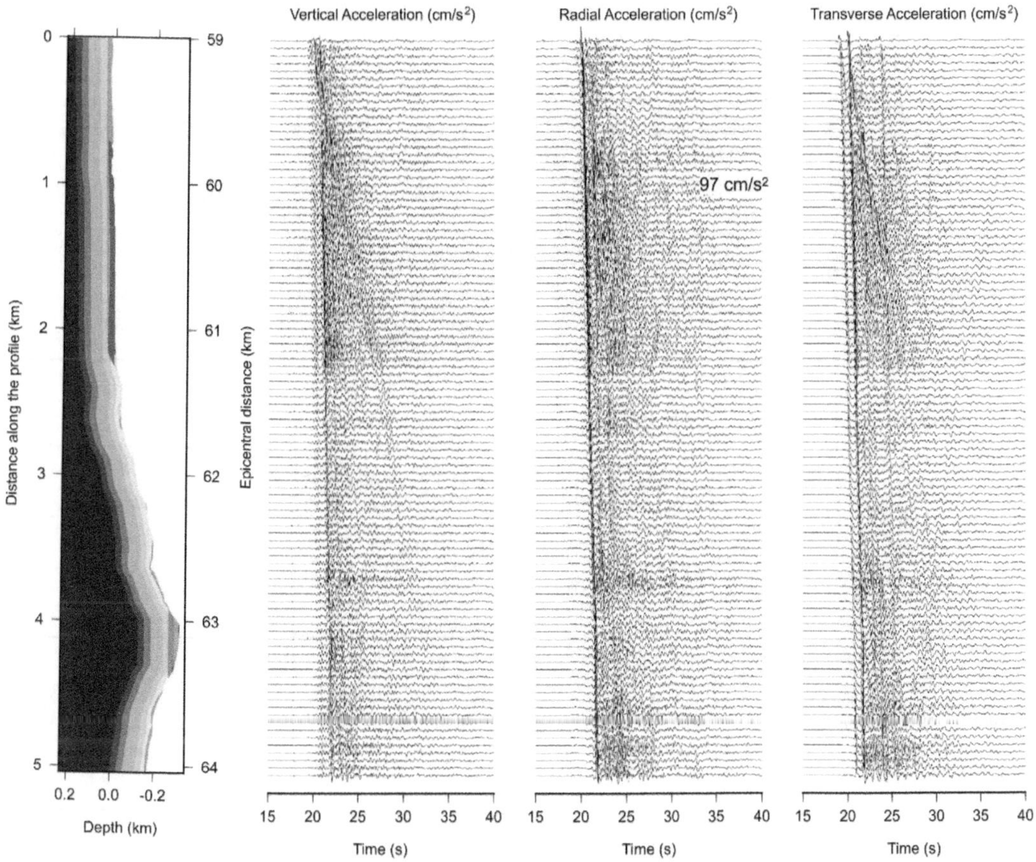

Fig. A9 Aremogna-Cinque Miglia (ITIS003) source. Acceleration time histories along the 2D section at Chieti town site, for P-SV (Vertical and Radial components) and SH (Transverse component) wavefields. The plotted amplitude of the signals is scaled to the maximum value (marked by the label).

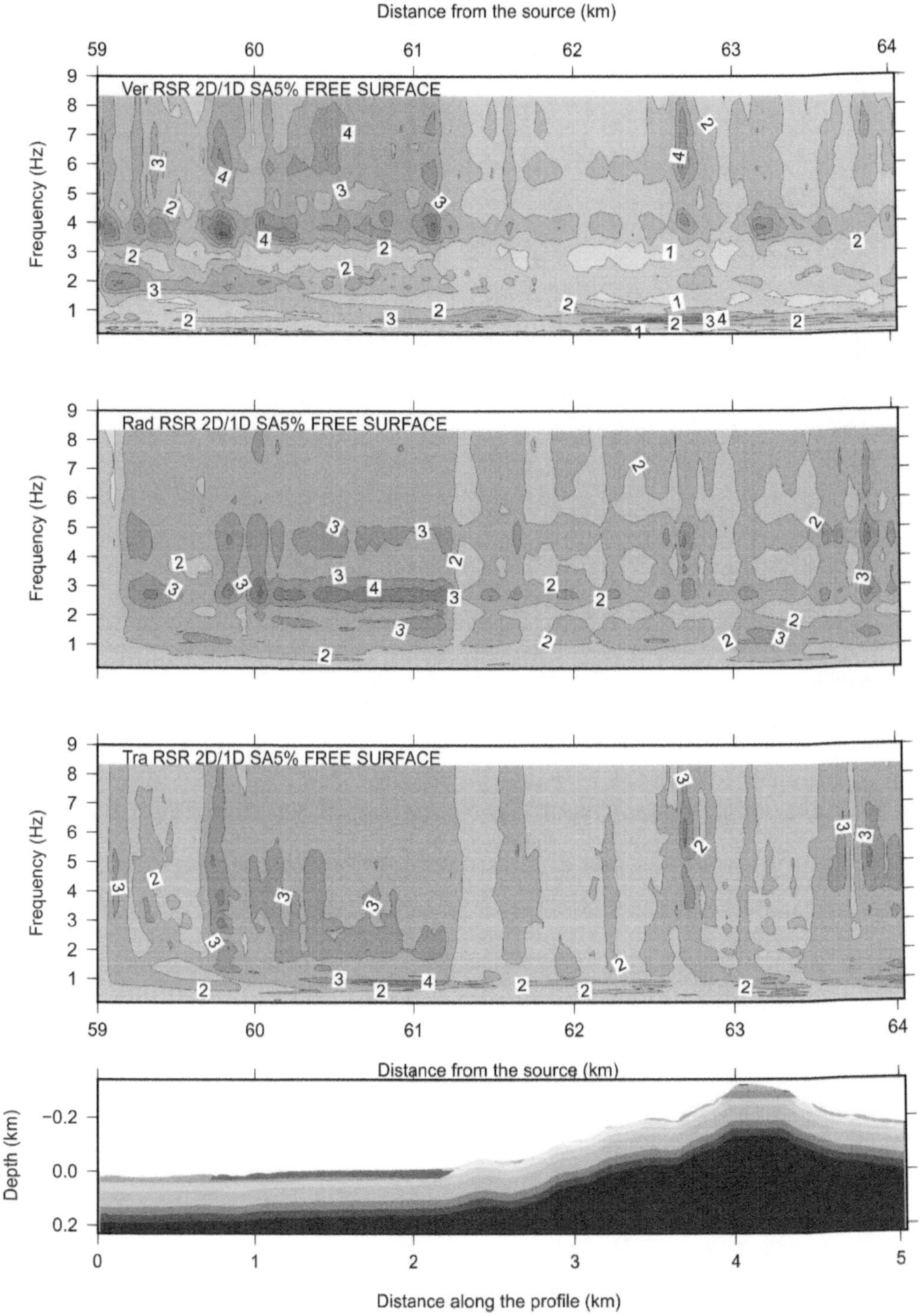

Fig. A10 Aremogna-Cinque Miglia (ITIS003) source. Spectral amplifications obtained as the ratio between the response spectra of the synthetic seismograms computed along the profile and the synthetic signals computed at the same sites considering the reference bedrock model. From top to bottom: vertical component, radial component, transverse component.

References

Bagnaia, R., A.D' Epifanio and S. Sylos Labini, 1989. Aquila and Subequan basins: an example of Quaternary evolution in central apennines, Italy. Quaternaria Nova, 1, 1–23.

Boncio, P., Lavecchia, G., Pace, B., 2004. Defining a model of 3Dseismogenic sources for Seismic Hazard Assessment applications: the case of central Apennines Italy. J. Seismol. 8(3), 407–425.

Bosi, C., 1975. Osservazioni preliminari su faglie probabilmente attive nell'Appennino Centrale. Boll. Soc. Geol. It., 94, 827–859.

Chiaraluce L., Valoroso L., Piccinini D., Di Stefano R., De Gori P. 2011. The anatomy of the 2009 L'Aquila normal fault system (central Italy) imaged by high resolution foreshock and aftershock locations. Solid earth. DOI: 10.1029/2011JB008352.

D'Agostino, N., 2014. Complete seismic release of tectonic strain and earthquake recurrence in the Apennines (Italy). Geophys. Res.Lett.41, 1155–1162. https://doi. org/10.1002/2014GL059230.

D'Amico, V., Albarello, D., Mantovani, E., 1998. A distribution-free analysis of the magnitude-intensity relationship: an application to the Mediterranean region. Ann. Geophys. 16 (Suppl IV), C1192.

DISS Working Group, 2025. Database of Individual Seismogenic Sources (DISS), Version 3.3.1: A compilation of potential sources for earthquakes larger than M 5.5 in Italy and surrounding areas. Istituto Nazionale di Geofisica e Vulcanologia (INGV). https://doi.org/10.13127/diss3.3.1.

Fäh, D., Iodice, C., Suhadolc, P. and Panza, G.F., 1993. A new method for the realistic estimation of seismic ground motion in megacities: The case of Rome. Earthquake Spectra, 9(4), 643–668. https://doi.org/10.1193/1.1585735.

Faraone, C., Colantonio, F., Vessia, G., 2025. Local Seismic Effects Responsible for Differentiated Damages in Historical City Centers: The Case Study of San Giustino's Square (Chieti, Italy). Geosciences 2025, 15, 76. https://doi.org/10.3390/geosciences15030076.

Faraone, C., Colantonio, F., Vessia, G. Seismic Amplification Effects Induced by Ancient Shallow Cavities underneath the Urban Area of the Historical City Center of Chieti, Italy. Eng. Geol. 2023, 324, 107259.

Fasan, M., Magrin, A., Amadio, C., Romanelli, F., Vaccari, F., Panza G.F., 2016. A seismological and engineering perspective on the 2016 Central Italy earthquakes. International Journal of Earthquake and Impact Engineering, 11 (4), 395–420

Faure Walker, J., Boncio, P., Pace, B., Roberts, G., Benedetti, L., Scotti, O., Visini, F., Peruzza, L., 2021. Fault2SHA Central Apennines database and structuring active fault data for seismic hazard assessment. Sci. Data 8,1–20.https://doi.org/10.1038/ s41597-021-00868-0.

Florsch, N., Fäh, D., Suhadolc, P. and Panza, G.F., 1991. Complete synthetic seismograms for high-frequency multimode SH-waves. PAGEOPH, 136, 529–560. https://doi.org/10.1007/BF008 78586.

Fracassi, U., and G. Valensise, 2007. Unveiling the sources of the catastrophic 1456 multiple earthquake: Hints to an unexplored tectonic mechanism in Southern Italy. Bull. Seismol. Soc. Am., 97, 3, 725–748, 10.1785/0120050250.

Galli, P., Giaccio, B., Peronace, E., Messina, P., 2015. Holocene paleoearthquakes and early-late Pleistocene slip-rate ontheSulmona fault (central apeninnes, Italy). Bull. Seismol. Soc.Am.105(1), 1–13.https://doi.org/10.1785/0120140029.

Galli, P., 2020. Recurrence times of central southern Apennine faults (Italy): Hints from palaeoseismology. Terra Nova 32, 399–407. https://doi.org/10.1111/ter.12470.

Gorshkov, A.I., Soloviev, A.A., 2022. Morphostructural zoning for identifying earthquake-prone areas. In: Panza, G.F., Kossobokov, V.G., Laor, E., De Vivo, B., (Eds). Earthquakes and sustainable infrastructure (1st Edition): Neodeterministic (NDSHA) approach guarantees prevention rather than cure. Elsevier, pp. 135–149.

Gusev, A.A., 2011. Broadband kinematic stochastic simulation of an earthquake source: A refined procedure for application in seismic hazard studies, Pure Appl. Geophys. 168, 155–200.

Hassan, H., Fasan, M., Sayed, M., Romanelli, F., Elgabry, M., Vaccari, F., Hamed, A., 2020. Site-specific ground motion modeling for a historical Cairo site as a step towards computation of seismic input at cultural heritage sites. Eng. Geol. 268, 105524.

ISPRA, 2010. Cartografia geologica ufficiale Foglio CARG 1:50.000 n.361-Chieti. Istituto Superiore per la Protezione e la Ricerca Ambientale, Roma, Italy available at https://www.isprambiente.gov.it/Media/carg/361_CHIETI/Foglio.html (last access March 2023).

Locati M., Camassi R., Rovida A., Ercolani E., Bernardini F., Castelli V., Caracciolo C.H., Tertulliani A., Rossi A., Azzaro R., D'Amico S., Antonucci A. (2022). Database Macrosismico Italiano (DBMI15), versione 4.0 [Data set]. Istituto Nazionale di Geofisica e Vulcanologia (INGV). https://doi.org/10.13127/dbmi/dbmi15.4.

Magrin, A., Gusev, A.A., Romanelli, F., Vaccari, F., Panza, G.F., 2016. Broadband NDSHA computations and earthquake ground motion observations for the Italian territory. Int. J. Earthq. Impact Eng. 1 (1–2), 131–158.

Meletti C., Galadini F., Valensise G., Stucchi M., Basili R., Barba S., Vannucci G., Boschi E. 2004. Zonazione sismogenetica ZS9 [Data set]. Istituto Nazionale di Geofisica e Vulcanologia (INGV). https://doi.org/10.13127/sh/zs9.

Morewood, N.C., and G.P. Roberts, 2000. The geometry, kinematics and rates of deformation within an en échelon normal fault segment boundary, central Italy. J. Struct. Geol., 22, 8, 1027–1047, 10.1016/S0191-8141(00)00030-4.

Nocentini M., Cosentino D., Spadi M., Tallini M. 2018. Plio-Quaternary geology of the Paganica-San Demetrio-Castelnuovo Basin (Central Italy). Journal of Maps, 14(2): 411–420. DOI: 10.1080/17445647.2018.1481774.

Nunziata, C. and Costanzo, M.R., 2022. Application of NDSHA to historical urban areas. In *Earthquakes and Sustainable Infrastructure* (pp. 397–414). Elsevier.

Panza, G.F., 1985. Synthetic seismograms: the Rayleigh waves modal summation. J. Geophys., 58, pp. 125–145.

Panza, G.F., Vaccari, F., Costa, G., Suhadolc, P. and Fäh, D., 1996. Seismic input modelling for zoning and microzoning, *Earthquake Spectra* 12, 529–566.

Panza, G.F., Romanelli, F., Vaccari, F., 2001. Seismic wave propagation in laterally heterogeneous anelastic media: Theory and applications to seismic zonation. Advances in Geophysics, vol. 43; p. 1–95.

Panza, G.F., La Mura, C, Peresan, A., Romanelli, F, Vaccari, F., 2012. Seismic Hazard Scenarios as preventive tools for a disaster resilient society. Advances in Geophysics, vol. 53; p. 93–165.

Panza, G.F., Bela, J. 2020. NDSHA: A new paradigm for reliable seismic hazard assessment. Engineering Geology, Volume 275, 105403 (2023 Best Paper Award).

Panza, G.F., Kossobokov, V.G., Laor, E., De Vivo, B. (Eds), 2022. Earthquakes and sustainable infrastructure (1st Edition): Neodeterministic (NDSHA) approach guarantees prevention rather than cure. Elsevier, pp. 672.

Pizzi, A., F. Calamita, M. Coltorti and P. Pieruccini, 2002. Quaternary normal faults, intramontane basins and seismicity in the Umbria-Marche-Abruzzi Apennine Ridge (Italy): contribution of neotectonic analysis to seismic hazard assessment. Boll. Soc. Geol. It., vol. spec. 1, II, 923–929.

Ricci, A., Romanelli, F., Vaccari, F., Boncio, P., Venisti, N., Faraone, C., Vessia, G., Panza, G.F. Comparison between the NeoDeterministic Seismic Hazard and FEM Approach to Assessing 2D Local Seismic Response at Chieti's City Site (Abruzzo, Italy). Eng. Geol. 2025, 347, 107891.

Roberts, G.P., and A.M. Michetti, 2004. Spatial and temporal variations in growth rates along active normal fault systems: an example from The Lazio–Abruzzo Apennines, central Italy. J. Struct. Geol., 26, 2, 339–376.

Romanelli, F., Altin, G., Indirli, M., 2022. Spreading NDSHA application from Italy to other areas. In: Panza, G.F., Kossobokov, V.G., Laor, E., De Vivo, B. (Eds). Earthquakes and sustainable infrastructure (1st Edition): Neodeterministic (NDSHA) approach guarantees prevention rather than cure. Elsevier, pp. 175–194.

Romanelli, F., Vaccari F., 1999. Site response estimation and ground motion spectrum scenario in the Catania area. Journal of Seismology, 3 (3), 311–326. https://doi.org/10.1023/A:1009847213825.

Rovida A., Locati M., Camassi R., Lolli B., Gasperini P., Antonucci A., 2022. Catalogo Parametrico dei Terremoti Italiani (CPTI15), versione 4.0 [Data set]. Istituto Nazionale di Geofisica e Vulcanologia (INGV). https://doi.org/10.13127/cpti/cpti15.4.

Rovida A., Locati M., Camassi R., Lolli B., Gasperini P., 2020. The Italian earthquake catalogue CPTI15. Bulletin of Earthquake Engineering, 18(7), 2953–2984. https://doi.org/10.1007/s10518-020-00818-y.

Rugarli, P., Amadio, C., Peresan, A., Fasan, M., Vaccari, F., Magrin, A., Romanelli, F., Panza, G.F., 2019. Neo- deterministic scenario-earthquake accelerograms and spectra: a NDSHA approach to seismic analysis. In: Jia, J., Paik, J.K. (Eds.), Chapter 6 in "Engineering Dynamics and Vibrations: Recent Developments". CRC Press, Boca Raton, Florida, USA, ISBN 978-1-4987-1926-1, pp. 187e241. https://doi.org/10.1201/9781315119908-6.

Servizio Geologico d'Italia 2006, Foglio 359 L'Aquila della Carta Geologica d'Italia alla scala 1:50.000. Poligrafico dello Stato, Roma, 1:50.000.

SM Working Group, 2008. Guidelines for Seismic Microzonation. Conference of Regions and Autonomous Provinces of Italy, Civil Protection Department, Rome, (Original Italian Edition: Gruppo di lavoro MS, Indirizzi e criteri per la microzonazione sismica, Conferenza delle Regioni e delle Province autonome–Dipartimento della protezione civile, Roma, 2008), 3 vol e Dvd.

Vaccari, F., 2016. A web application prototype for the multi scale modelling of seismic input. In: S. D'Amico (ed.), Earthquakes and Their Impact on Society. pp. 563–584. Springer International Publishing.

Vaccari, F., Magrin, A. 2019. NDSHA computational aspects of the neo-deterministic seismic hazard assessment. In F. Dobran (Ed.), Resilience and sustainability of cities in hazardous environments (pp. 202–212). GVES.

Vaccari, F., Magrin, A., 2022. A user-friendly approach to NDSHA computation. In: Panza, G.F., Kossobokov, V.G., Laor, E., De Vivo, B. (Eds). Earthquakes and sustainable infrastructure (1st Edition): Neodeterministic (NDSHA) approach guarantees prevention rather than cure. Elsevier, pp. 215–237.

Vessia, G., Laurenzano, G., Pagliaroli, A., Pilz, M., 2021. Preface to the Special Issue: "Seismic site response estimation for microzonation studies promoting the resilience of urban centers". Engineering Geology, 284, 106031. DOI: 10.1016/j.enggeo.2021.106031.

Vezzani, L, and F. Ghisetti, 1998. Carta Geologica dell'Abruzzo. in: Vezzani, L., and F. Ghisetti (eds), Carta Geologica dell'Abruzzo, scale 1:100,000, S.E.L.C.A., Firenze.

CHAPTER

11

Lithospheric Structure below Seismic Stations in the East of Cuba, from the Cross-correlation Dispersion Analysis and Its Joint Inversion with Receiver Functions

O. González[1,2,3], B. Moreno[1,2], V. Clouard[2], E. Arango[1,2], Y. Rojas-Agramonte[4] and E. Calais[5]

New dispersion relations and tomographic maps of Rayleigh wave group velocities were calculated from seismic noise cross-correlations between pairs of seismic stations in eastern Cuba. The resulting tomographic maps for shorter periods (6–12 s) are consistent with known surface geological features such as the Cauto sedimentary basin, the Sierra Maestra mountain range, and the Moa-Baracoa ophiolitic massif. The joint inversion procedure of receiver functions (RF) and surface wave dispersion data, enabled the generation of S-wave velocity (Vs) models to characterize the crust and upper mantle structure. In the northeastern sector, the results reveal a thick accretionary crust, from 27 km to 31 km thick, featuring low-velocity channels likely associated with past compressional deformation and obduction processes. Conversely, stations in the central southern coast are closer to oceanic settings with a thin crust and shallow lithospheric mantle that becomes thicker to the east. For both regions, high seismic velocity layers (LID) below the upper lithosphere are present, which may belong to potential subducted or remnant slab, at depths ranging from 40 to 100 km, possibly associated with past regional geodynamic processes of Cretaceous or Paleogene age.

1. Introduction

Regional studies of the crust and upper mantle in the Caribbean region by seismic waves processing, have been done by Gaite et al. (2015); González et al. (2007, 2012); Moreno et al. (2023); Van Der

1 Centro Nacional de Investigaciones Sismológicas (CENAIS). Ministerio de Ciencia, Tecnología y Medio Ambiente, Cuba.
2 Géosciences Environnement Toulouse (GET), Observatoire Midi-Pyrénées, Université Toulouse III Paul Sabatier, France.
3 Observatoire Volcanologique et Sismologique de Martinique. Institut de Physique du Globe de Paris, France.
4 Institut für Geowissenschaften, Ruprecht-Karls-Universität Heidelberg, Germany.
5 École normale supérieure, Département de Géosciences, Paris, France.
* Corresponding author: oleary@cenais.cu

DOI: 10.1201/9781003684121-13

Hilst (1990); Van Benthem et al. (2013); Vdovin et al. (1999). These studies include some general characterization of the lithospheric structure of Cuba. The characterization of the physical properties of the crust in eastern Cuba was, until 2003, based on seismic refraction methods, gravimetric data, deep drilling and remote sensing (Shcherbakova et al., 1978; Bovenko et al., 1982; Bush & Shcherbakova, 1986). Moreno (2003) was the first to employ teleseismic recordings to compute receiver functions (RFs) and estimate the crustal structure beneath seven Cuban stations. Later on, González et al. (2012) combined RF analysis with regional surface wave dispersion data to provide information on the crust and for the first time, local information about the upper mantle, below eight stations located along Cuba. Arango et al. (2014) calculated a three-dimensional crustal model using gravity inversion for eastern Cuba, and Possee et al. (2021) applied S-to-P receiver functions to map lithospheric seismic velocity discontinuities around the North American-Caribbean plate boundary, in particular below 9 seismic stations in the eastern of Cuba. While Moreno (2003) and Arango et al. (2014) illustrate the main crustal features, Gonzalez et al. (2012) and Posee et al. (2021) also mapped some lithospheric characteristics, including the lithosphere-asthenosphere boundary (LAB), seismic low velocity zones (LVZ) and high seismic velocity layers (LID) below the lithosphere.

To improve upon these earlier studies in eastern Cuba, five permanent new broad band (BB) seismic stations were installed and five other temporary BB stations during the period 02/2018–12/2019 (Clouard, 2018) (Figure 1). This was done in the framework of the Interreg Caraïbe V PREST project (https://www.interreg-caraibes.com/prest). Combined with the six BB stations operating since 1996 (Moreno, 2002), the resulting network of 16 stations enabled us to employ ambient noise cross-correlation techniques for surface wave dispersion analysis (Bensen et al., 2007).

The ambient noise cross-correlation method, which retrieves Empirical Green's Functions, has been successfully applied worldwide (Shapiro et al., 2005; Moschetti et al., 2007; Bensen et al., 2008; Lin et al., 2009) and provides robust measurements of Rayleigh wave dispersion. Although these data yield valuable information on the absolute S-wave velocities (Vs), they are relatively insensitive to sharp vertical velocity contrasts. Receiver functions, in contrast, are sensitive to S-wave velocity contrasts (Ammon, 1991), which gives rise to converted phases, but allow for substantial trade-off between the depth and velocity above an impedance change (Ammon, 1991). The combination of both kinds of information in a joint inversion of RFs and dispersion curves minimizes the resolution gaps inherent in each individual dataset (Ozalaybey et al., 1997; Julià et al., 2000, 2003, 2005; Gök et al., 2008).

This joint inversion procedure is formulated as a nonlinear inversion problem that produces a set of possible solutions below each station, depending mainly on the initial models. By dividing the study region into equally-sized and regularly distributed cells, each one containing one station, the use of optimization procedures described by Boyadzhiev et al. (2008), allows us to determine a unique solution for each cell. Given the following:

(i) availability of surface wave dispersion data in the Caribbean region in the period range from 10 s to 40 s (González et al., 2007, 2012) and from Vdovin et al. (1999) from 60 s to 150 s;
(ii) possible use of the ambient noise cross-correlations procedure between stations in Eastern Cuba to determine more precise surface waves dispersion data at lower periods;
(iii) recent teleseismic records of the improved seismic network in Cuba;
(iv) efficiency of the Local Smoothness Optimization procedures, to identify an optimum 1D model of Vs below each station, after a joint inversion analysis of RFs and surfaces wave dispersion for cells containing seismic stations over a study area (González et al., 2018).

We determined new velocity models to constrain the distribution of shear wave velocity versus depth in the lithosphere-asthenosphere system. Then, where possible, we compare the new results with the ones available in literature.

2. Tectonic Settings

Eastern Cuba (Figure 1) is bounded to the north by the Bahamas Platform and to the south by the Gônave microplate. The collision between the Cuban segment and the Bahamas Platform occurred around 45 Ma (Itturalde-Vinente et al., 2008), about 5 Myrs after the initiation of the Cayman spreading ridge (Leroy et al., 2000). This collision was associated with an eastward shift in Caribbean plate motion (Mann et al., 1995) and the creation of a new plate boundary now expressed by the Septentrional-Oriente Fault Zone (SOFZ). The Oriente Fault (OF), part of the SOFZ, currently accommodates 10 mm/yr of left-lateral strike-slip motion between the Gonâve microplate and the Cuban segment (Benford et al., 2012; Calais et al., 2023). Approximately 95% of the earthquakes in Cuba, occur along its southeastern margin, associated with this plate boundary (Alvarez et al., 1999), making the region a high seismic hazard area (Garcia et al., 2003; Alvarez et al., 2016; Pagani et al., 2018).

The geology of Cuba results from Cretaceous Greater Antilles Arc volcanism initiated shortly before 135 Ma (Rojas-Agramonte et al., 2005). The Cretaceous volcanic arc (CVA) was originally built over a Pacific-derived lithosphere (Iturralde-Vinent, 1998; Pindell et al., 2006, 2011) and remained active until the Campanian (Hall et al., 2004). The CVA occupies large parts of the Cuban orogen, including the basement of the Sierra Maestra Paleogene arc in the southern part of our study area (Figure 1) with ages ranging between ~60 Ma–48 Ma (Rojas-Agramonte et al., 2004). The collision between the Caribbean and North American plates arrested the Cuban volcanism and a fold and thrust belt was formed along the length of the island. The fold and thrust belt include North American passive continental paleo-margin sequences, ophiolitic units obducted onto the passive margin and arc units tectonically imbricated with the ophiolites (Iturralde-Vinent, 1996, 1998). The Eastern Cuban Ophiolites (ECO) includes the Mayarí-Cristal and Moa-Baracoa units. The Cauto Basin and the Holguin Plain contain thick Neogene – Quaternary sedimentary sequences. The Cauto Fault is interpreted by some as a remnant of the early activity of the Cayman spreading center at 49 Ma that was abandoned in the early Oligocene when the oblique movement of the Caribbean plate shifted to a purely eastward motion (Leroy et al., 2000).

Previous studies, mostly based on gravimetry and seismic profiles, proposed three different types of crust in Eastern Cuba: a continental or a thick transitional crust in the north, an oceanic crust in the south, and in between a thin transitional crust (Bush & Shcherbakova, 1986; Otero et al., 1998). Four main geological provinces have been described by Iturralde-Vinent (1998). To the North (latitude > 20.75°) the rocks of the Cretaceous volcanic arc are predominant, while in its northern portion they are tectonically imbricated with the ophiolites and covered by the carbonate sequences of the Bahamas platform. In the central region, which comprises the Cauto Basin and the Holguin Plains (Figure 1), thick layers of Neogene-Quaternary sediments are present. In the south-eastern region, to the East of the Guantanamo Bay and up to the zone coinciding with the Moa-Baracoa Ophiolitic Massif Structure, rocks of the Paleogene Volcanic Arc predominate, covered in some places by younger sedimentary sequences.

The southeastern part of Cuba and its geology are juxtaposed, to the south, to the extended Mesozoic island arc crust of the eastern Cayman Trough. This crust is assumed to be the result of east-west rifting pre-dating the late Eocene to Present opening of the Cayman Trough as a pull-apart oceanic basin (Leroy et al., 1996). The boundary between these contrasted domains occurs along the OF, which is the locus of about 500 km of eastward displacement of Hispaniola with respect to southeastern Cuba since the late Eocene (Calais et al., 1991).

3. Data Processing

For the receiver function calculation, a set of 38 strong teleseismic events in the period from 2017 to 2021 (Table 1 and Figure 1) were recorded by the 16 BB stations in Eastern Cuba (Table 2 and Figure 1). For the six BB stations operating in 2011 (CCCC, LMGC, RCC, GTBY, MOA, and MASC)

Table 1 Earthquakes parameters for the new teleseismic records used in the receiver function analysis. Additional teleseismic data from González et al. (2012) were also incorporated.

Date (dd/mm/yyyy)	*Time (hh:mm:ss)*	*Latitude*	*Longitude*	*Depth (km)*	*Mag.*
30/11/2017	06:32:50	01.08°S	23.43°W	10.0	6.5
14/01/2018	09:18:45	15.76°S	74.709°W	39.0	7.1
02/04/2018	13:40:34	20.66°S	63.01°W	559.0	6.8
04/05/2018	22:32:54	19.31°N	155.00°W	2.06	6.9
15/08/2018	21:56:54	51.42°N	178.05°W	20.0	6.6
24/08/2018	09:04:06	11.04°S	70.82°W	609.5	7.1
25/10/2018	22:54:52	37.51°N	20.56°E	14.0	6.8
22/10/2018	05:39:40	49.25°N	129.4°W	11.0	6.6
01/11/2018	22:19:51	19.59°S	69.29°W	102.0	6.2
09/11/2018	01:49:40	71.63°N	11.25°W	10.0	6.7
30/11/2018	17:29:29	61.35°N	149.95°W	46.7	7
11/12/2018	02:26:32	58.60°S	26.47°W	164.7	7.1
05/01/2019	19:25:38	8.15°S	71.58°W	571.2	6.8
20/01/2019	01:32:51	30.07°S	71.42°W	54.8	6.7
01/03/2019	08:50:42	14.70°S	70.13°W	267	7
15/03/2019	05:03:50	17.87°S	65.91°W	359	6.3
26/05/2019	07:41:15	5.81°S	75.28°W	122.4	8
14/06/2019	00:19:12	30.06°S	72.08°W	11.0	6.4
04/07/2019	17:33:49	35.71°N	117.51°W	10.7	6.4
06/07/2019	03:19:53	35.77°N	117.60°W	8.0	7.1
01/08/2019	18:28:07	34.24°S	72.31°W	25.0	6.8
29/09/2019	15:57:53	35.48°S	73.16°W	11.0	6.7
25/12/2019	03:36:01	50.61°N	129.97°W	6.38	6.3
31/03/2020	23:52:31	44.46°N	115.14°W	14.53	6.8
15/05/2020	11:03:27	38.17°N	117.85°W	2.7	6.5
03/06/2020	07:35:34	23.30°S	68.42°W	96.84	6.8
22/07/2020	06:12:44	55.03°N	158.52°W	28.0	7.8
30/08/2020	21:20:29	0.78°N	29.87°W	10.0	6.5
18/09/2020	21:43:58	0.92°N	26.84°W	10.0	6.9
19/10/2020	20:54:39	54.61°N	159.65°W	31.1	7.6
30/11/2020	22:55:03	24.33°S	67.02°W	189.0	6.3
27/12/2020	21:39:14	39.34°S	74.99°W	10.0	6.7
19/01/2021	02:46:22	31.83°S	68.81°W	20.0	6.4
19/05/2021	00:42:19	33.06°S	109.39°W	10.0	6.7
29/07/2021	06:15:49	55.36°N	157.89°W	35.0	8.2
14/08/2021	11:57:43	55.17°N	157.65°W	20.7	6.9
21/09/2021	13:14:32	36.78°S	73.93°W	18.8	6.4
11/10/2021	09:10:24	56.26°N	156.55°W	69.1	6.9

Table 2 List of the seismic stations used in this study. Network codes: CU, Caribbean USGS; CW, Cuban Seismological Service; and XH, PREST, temporary seismic network installed in the framework of the PREST-INTEREG project.

Station Name	*Lat.N (°)*	*Lon.W (°)*	*Instruments*	*Network*
LTUP	20.96	– 77.00	CMG40T - Taurus	XH
NICP	20.69	– 75.54	CMG40T - Taurus	XH
YARP	20.36	– 76.36	Trillium 120P - Taurus	XH
CCCC	21.19	– 77.42	Trillium 120P - Centaur	CW
NMDO	20.56	– 74.98	Trillium 120P - Centaur	CW
MASC	20.18	– 74.23	BBVS, 60 s - EDAS-24IP	CW
QMBU	20.20	– 74.81	Trillium Compact, 120 s - Centaur	CW
RCC	19.99	– 75.70	Trillium 120P - Centaur	CW
MARV	20.01	– 75.91	Trillium Compact, 120 s - Centaur	CW
CHIV	19.98	– 76.42	Trillium Compact, 120 s - Centaur	CW
LMGC	20.06	– 77.00	BBVS, 60 s - EDAS-24IP	CW
PILO	19.91	– 77.41	Trillium Compact, 120 s - Centaur	CW
GTBY	19.93	– 75.11	Streckeisen STS - 2.5	CU
PINP	20.48	– 75.79	Trillium 120P - Taurus	XH
SABP	20.36	– 75.31	Trillium 120P - Taurus	XH
HLG	20.92	– 76.24	Trillium 120P - Centaur	CW

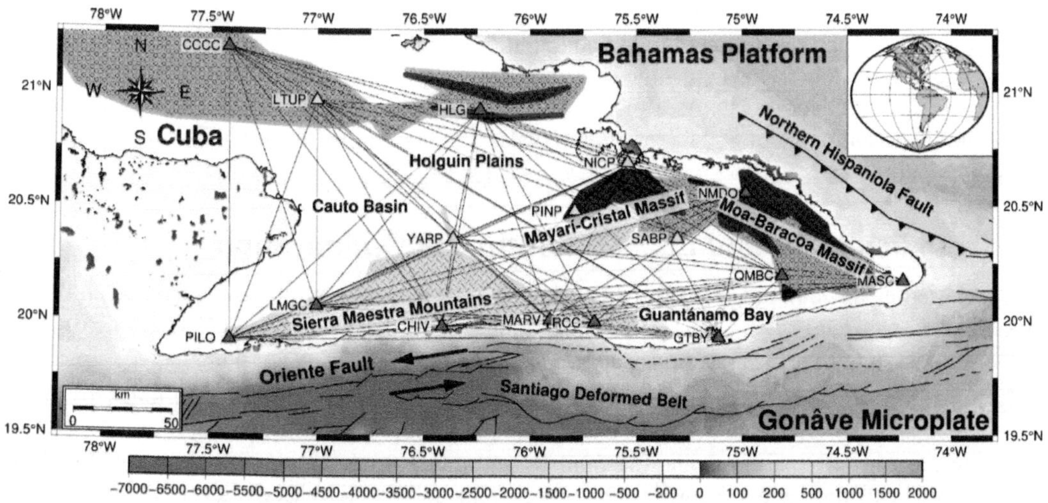

Fig. 1 Simplified geologic map of eastern Cuba (modified from Iturralde-Vinent et al. (2016)). Different geological units are distinguished by distinct map patterns or tonal fills: the Cretaceous volcanic arc, ophiolites and serpentinite mélanges, upper Eocene to recent deposits, and the Paleogene volcanic arc. Permanent broadband stations and temporary PREST stations are indicated by triangles with different symbols (see legend). Black lines represent inter-station paths used for dispersion analysis. The upper map shows the location of the teleseismic earthquakes and its corresponding paths (red lines) to the study region (earthquakes parameters in Table #1).

the teleseismic events used by González et al. (2012) were included. Only seismic events with an epicentral distance in the range 30°–90° and with a high signal-to-noise ratio were considered. Data preparation, including picking, decimation, filtering, rotation, and deconvolution of the vertical component of the selected earthquake signal from the corresponding horizontal component, was performed using the most updated version of the SAC software (Goldstein, 1999). The first 38 s of the RFs were computed, using Gaussian factors of 0.5 ($f < 0.25$ Hz), 1.0 ($f < 0.5$ Hz) and 2.5 ($f < 1.25$ Hz), with the Ligorría and Ammon (1999) procedure, which uses the iterative time-domain deconvolution technique. Because of the large background noise in the frequency range < 1.25 Hz, the corresponding Gaussian factor provided the smallest number of valid signals.

To compute the group velocity dispersion relations, the traditional method of ambient noise cross-correlation was used (Bensen et al., 2007). Data preparation for a single station, cross-correlation, and temporal stacking were performed using the SAC codes (Goldstein, 1999). The latest version of frequency-time analysis (FTAN) (Levshin et al., 1972, 1992) was applied to determine the surface wave dispersion curves at periods ranging from 6–24 s. Quality control of the dispersion relations was done using different approaches:

(i) For a phase speed c, 4 km/s was used, and at least 3 wavelengths (L), between stations at distance D, the dispersion measurement at maximum period (T) required $D > 3cT$.
(ii) To test the robustness of the dispersion curves, the set of correlated data was divided in two subsets for each pair of stations and stacked separately. Dispersion relation for each set was calculated and compared. Those dispersion curves showing inconsistencies (differences of group velocities > 0.1 at the same period between the two data sets from the same pair of stations) were removed from the analysis. The dispersion measurement error for each period was calculated on the basis of the average of these differences.
(iii) For the calculated Green functions, only those with signal to noise ratio SNR > 7 were selected.

Using the tomographic procedure described by Ditmar and Yanovskaya (1987), Yanovskaya and Ditmar (1990), Wu and Levshin (1994), and Yanovskaya (1997), new tomographic maps of group velocity in Eastern Cuba were obtained at periods ranging from 6 to 24 s, with 2 s intervals (Figure 2). The lateral resolution of the tomographic study (Figure 3), determined by the density of paths, is approximately 60 km across the entire region, except for stations located in the westernmost part, such as CCCC and PILO. As a result of the tomographic maps, a group velocity dispersion curve in the period range of 6–24 s (at 2 s intervals) was determined for each 50 km × 50 km cell in the studied region.

4. Join Inversion and Local Smoothness Optimization

For each station, the joint inversion procedure and the layering criteria follows the method and codes described by Gonzalez et al. (2012, 2018), including in the software package Computer Programs in Seismology (Herrmann & Ammon, 2002). We chose as input dispersion data:

(i) For the 6 s–24 s period range, we used the results from the tomographic maps from the dispersion analysis using the cross correlation between two seismic stations.
(ii) For the 25 s–40 s period range, the fundamental-mode Rayleigh wave group velocities have been taken from the pertinent 2° × 2° tomographic cell given by Gonzalez et al. (2012).
(iii) For longer periods from 40 s–150 s, the results of Vdovin et al. (1999) were used.

The set of models resulting from the fully non-linear inversion of the cellular dispersion curves in the Caribbean region (González et al., 2012) were used as the starting models. These models, derived from a broad search of possible solutions, are consistent with the dispersion data and their associated experimental errors, and represent the average physical properties of the cells in which the stations are located.

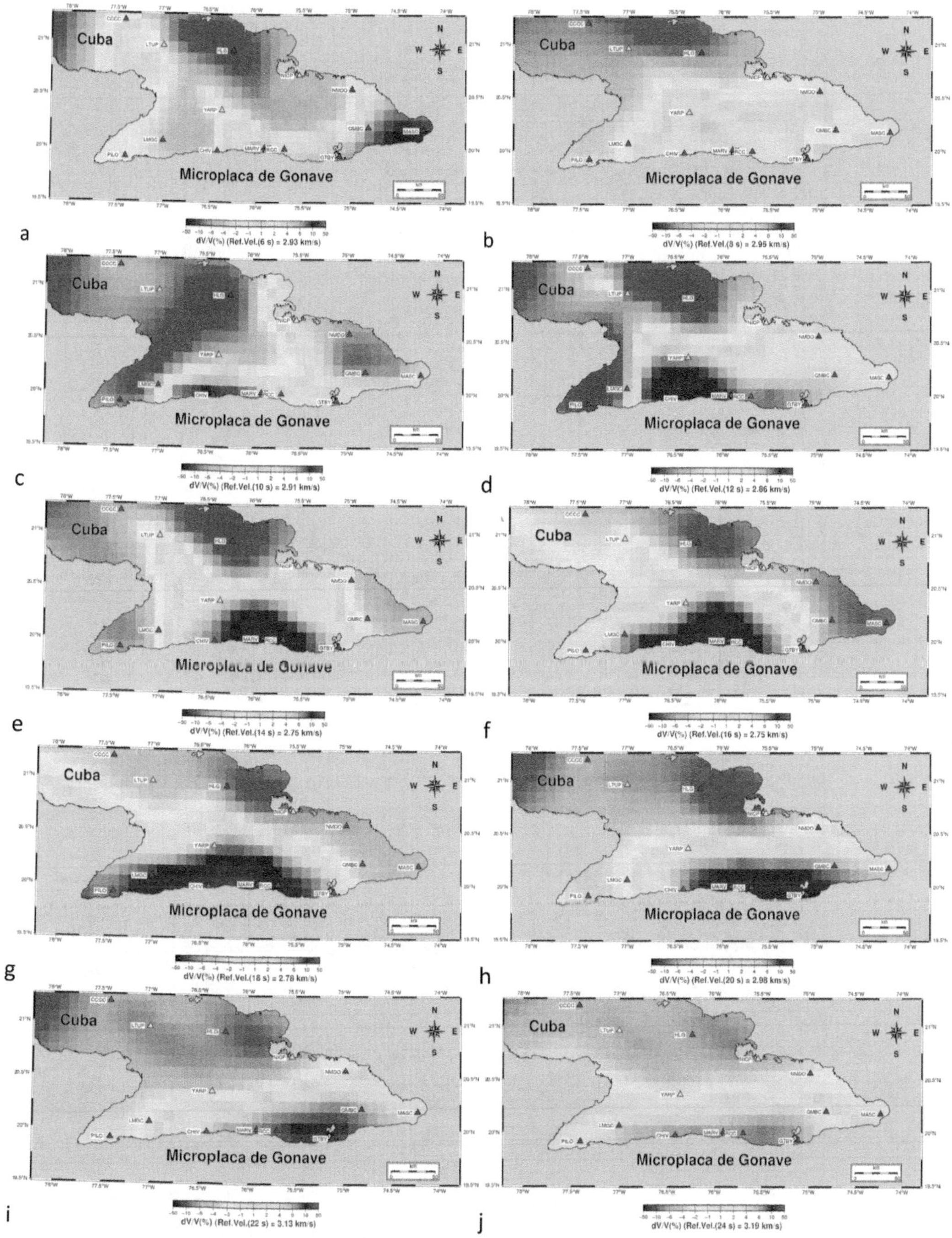

Fig. 2 Rayleigh waves group velocity tomographic maps at different periods (6–24s) shown as percent deviation from the average reference velocity at each period (Ref. Vel.)

The influence parameter, p, was selected by trial and error within the range of 0.1–0.5, ensuring that the resulting models achieve both: (i) the best possible percentage fit to the receiver functions (RFs), and (ii) dispersion curves that are consistent, within the error margins, with the experimental data. The damping factor, which balances the trade-off between resolution and stability (Julià et al., 2000), was fixed at $\sigma = 0.5$.

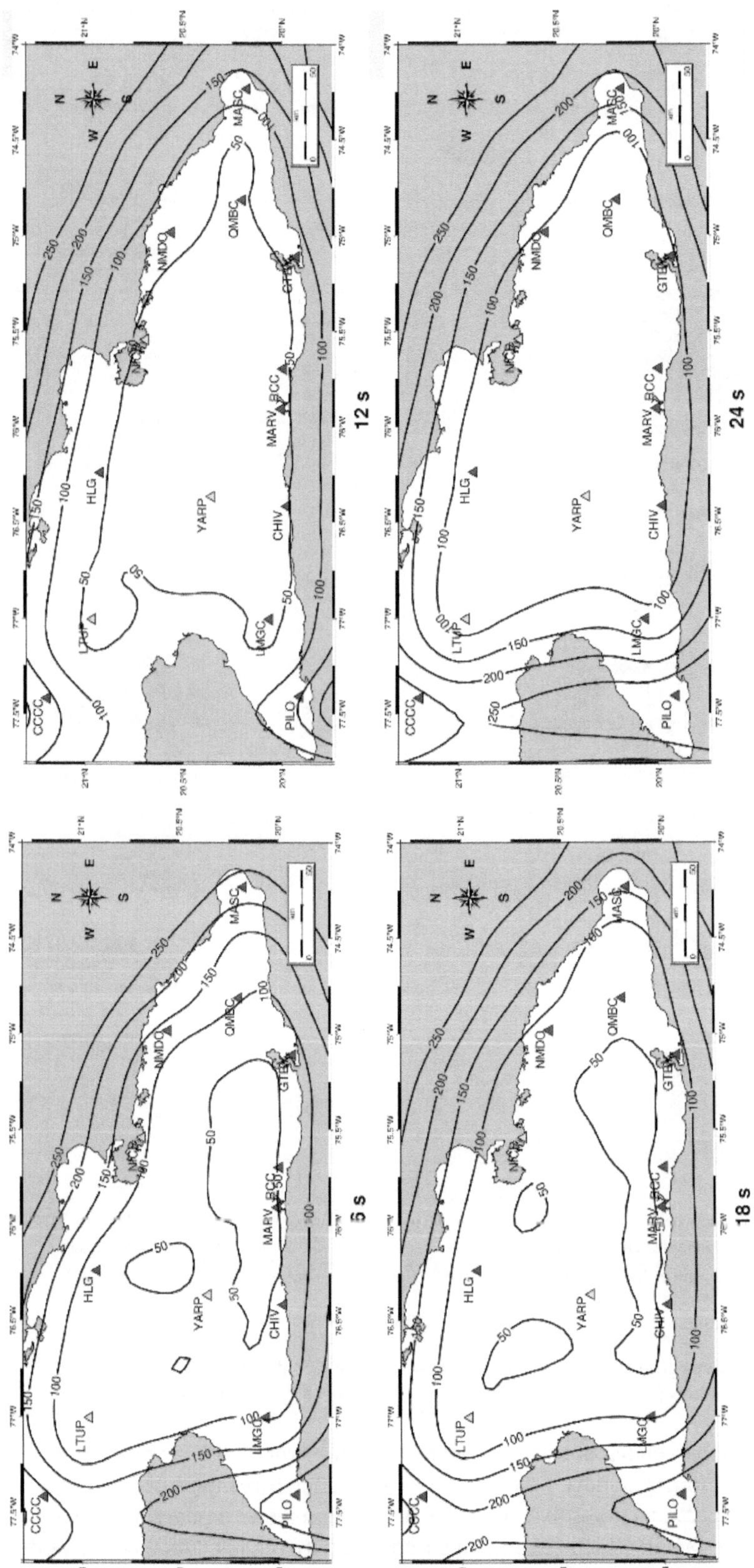

Fig. 3 Map of lateral resolution (km) at different periods (6–24 s).

As described by González et al. (2012, 2018), the inversion is controlled by a misfit function that quantifies the average percentage difference between the experimental and theoretical RFs. Additionally, all resulting Vs models were smoothed at 0.05 km/s to mitigate potential numerical artifacts. Since these smoothed models may not be direct solutions of the joint inversion, they are subsequently used as input for a second inversion. For the final models at each station, the results indicated that:

(i) the mean values of the fit between the experimental and the theoretical RFs associated to the selected earthquakes are over 75%;
(ii) the percentage fit for the RFs for the same station but from different starting models are relatively close to each other;
(iii) differences between the experimental surface wave dispersion curves and those generated during the inversion are consistent within the error of the measurements.

The depth resolution, mostly determined by the dispersion curves, is consistent with that described by González et al. (2012), considering: (i) the use of the same dataset for longer periods, and (ii) the influence of upper-layer information on the resolving power of the dispersion data (Chimera et al., 2003; Panza, 1981; Pontevivo & Panza, 2006). Finally, following the procedure described by González et al. (2018) for selecting the optimum model for each station, and considering both the relative geographic positions of the stations and the limitations of relying exclusively on statistical criteria, the Local Smoothness Optimization (LSO) procedure (Boyadzhiev et al., 2008) was employed. To adapt the LSO procedure to the geographic conditions of eastern Cuba, the cell dimensions based on the lateral resolution of the receiver functions were fixed, using the geometric criteria of multiple rays (Tkalčić et al., 2006).

Considering the geographic distribution of the stations, the study region was divided into regularly distributed cells, with each cell containing a single station (see Figure 4). In this region, only one distribution of cells was feasible; therefore, stability tests were performed based on subgroups (Figure 4). These subgroups were established according to regional geological settings (Stanek et al., 2009).

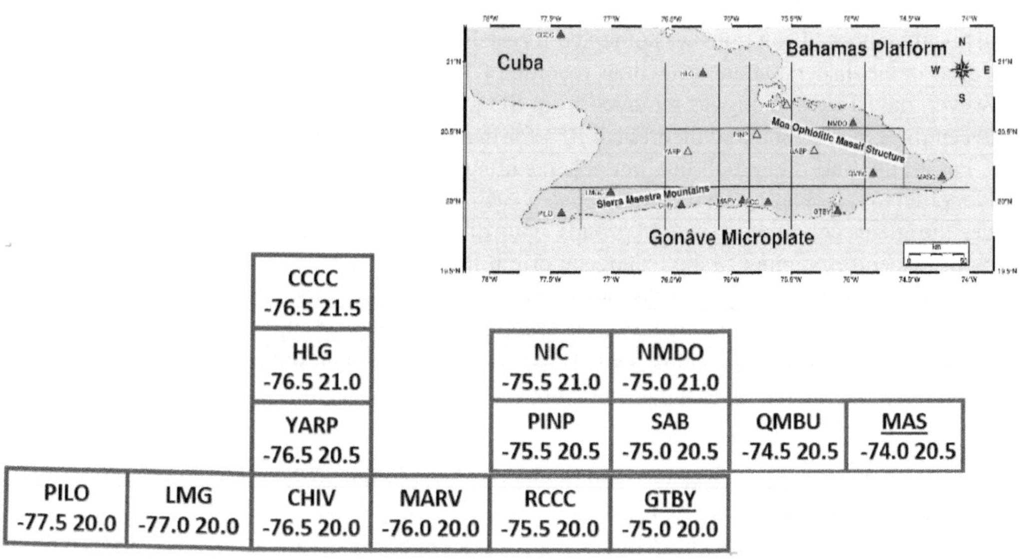

Fig. 4 Diagram of the distribution of the cells used during the LSO procedure, according to the relative position of the stations (see scheme in the map on the top). Rectangles identify two subgroups of cells used in the Local Smoothness Optimization (LSO) procedure, distinguished by different outline styles. Underlined names of stations denote the starting cell during the LSO procedure (see Boyadzhiev et al. (2008) for details). GTBY station was the starting cell for the LSO for all the cells and for its subgroup.

For the resulting models, the Moho depth was chosen for a layer with Vs > 4.0 Km/s. For the lithosphere-asthenosphere boundary, the criteria from Fischer et al. (2010) was used, which considers the negative gradient of the shear velocity due to the presence of a small amount of partial melt in the asthenosphere. Seismic high-velocity layer (LID) (Anderson, 1995), below the lithosphere, is identified as a possible slab starting at depths where a new positive gradient of the S-wave velocities is present.

5. Results

Dispersion curves of Rayleigh wave group velocities were calculated from 72 pairs of seismic BB stations using ambient noise and cross-correlation analysis. FTAN analysis for each pair was performed over the period range from 6 s–24 s, the upper limit being mainly determined by the inter-station distances and the bandwidth of some stations, while the lower limit depends on the signal-to-noise ratio. For longer periods, the upper limit was adjusted so that the distance between each pair of stations corresponds to at least three wavelengths. The experimental error at different periods ranges from 0.04 km/s to 0.08 km/s. As described in the data processing section, due to inconsistencies between the two sets of stacked signals, the inter-station distance, and the amount of data available for stacking, stations PINP and SABP were excluded from the dispersion analysis.

Ten new tomographic maps of Rayleigh wave group velocity dispersion were calculated, covering the southeastern part of Cuba at resolutions not previously studied for the period range of 6–24 s (Figure 2). Corresponding resolution maps were generated for each period (Figure 3), with most of the region having a spatial resolution of less than 50 km. The observed group velocity dispersion values are relatively lower than those reported in previous regional studies, particularly compared to the data used in earlier joint inversion procedures by González et al. (2012). Such difference is probably related to the influence, for broader areas considered in past studies, of the surrounding zones like the Gonave microplate or the Bahamas Platform, with different geological properties and origin.

For each station, the number of initial models derived from the cellular (2°×2°) models defined by González et al. (2011) ranged from 13–18. After applying the joint inversion procedure, an equivalent number of solutions was obtained for each one. Although stations PINP and SABP were not used in the dispersion analysis, their recorded RFs were included in the inversion procedures. However, the LTUP station was removed from the joint inversion analysis because the average fit between the experimental and theoretical RFs for the selected earthquakes was below 75%.

For all the stations and all final models resulting from the joint inversion, the LSO procedure was applied to select an optimal model beneath each station. To test the stability of this procedure, under the given conditions, two subgroups with different predominant geological origins were chosen: one corresponding to the Paleogene origin like Sierra Maestra Mountains (SMM) and the other to the Cretaceous origin in the northeast of the study area (Sommer et al., 2011). The Vs vs. depth models obtained via the LSO, for each subgroup, were identical to those obtained when applying the procedure to the entire area. These results suggest that the selection procedure, based on smoothing, is stable despite the local geological conditions.

6. Discussion

Due to the nonlinear relationship between the surface wave dispersion and the seismic velocity structure of the Earth, a direct comparison of tomographic maps with physical and geological properties is complex. However, the heterogeneity of the tomographic maps calculated at different periods shows complexity and depth variability. In general, group velocity maps for shorter periods correlate well with the known surface geological features of the study region (Iturralde, 1996, 1998; Sommer et al., 2011). For lower periods (6–12 s; Figure 2a, b, c, and d), group velocities are mainly influenced by the shallowest features, such as sedimentary basins, mountains, and upper crustal structures. Low-velocity anomalies at these periods are consistent with the presence of sedimentary layers in the Cauto Basin and the Holguín Plain (Figure 1). This effect is particularly evident at 10 s, suggesting significant sedimentary deposit thickness.

High Rayleigh surface wave group velocity anomalies at areas at shorter periods, spatially coincide with the presence of the main mountain ranges. At periods ~10 s, high group velocities were detected in the Moa-Baracoa ophiolitic massif structures, while in the Sierra Maestra mountain range, where rocks of the Paleogene Volcanic Arc are present, high velocities are more clearly observed from 10 to 18 s. This observation, suggests also the presence of high velocity lower crust and lithospheric mantle beneath this area.

From 12 s, as seen from Figure 2d onwards, high group velocities are observed at CHIV, MARV, RCC, and up to GTBY stations, where the crustal and lithospheric thicknesses may be indicative of an oceanic setting (Moreno et al., 2003; González et al., 2012). In contrast, lower velocities are found at stations HLG, LTUP, and CCCC, likely associated with a thicker crust. This difference is consistent with Moho depth estimates from receiver function analysis (González et al., 2012), which indicate a Moho depth of ~17 km at GTBY and RCC, whereas at CCCC, it reaches ~30 km. At periods greater than 12 s, lower velocity values were found at NMDO, LMGC, and MASC, when compared to CHIV, MARV, RCC, and GTBY, (Figure 2d onwards). It aligns with previous studies indicating crustal thicknesses ranging from 24–30 km (González et al., 2012; Possee et al., 2021).

For longer periods (>20 s), influenced by deeper structures, group velocities decrease from south to north, likely related to the thickening of the upper lithosphere or presence of a deeper seismic high-velocity layer (LID) (Anderson, 1995) beneath the southern region. This feature may be associated with a remnant slab involved in the formation of the Paleogene volcanic arc. The new dispersion values have better resolution than previous studies (González et al., 2012) and could be used in future inversion procedures to characterize crustal Vs velocities, as well as in combination with other data, such as receiver functions and gravimetric measurements.

The nonlinear inversion of Rayleigh wave dispersion relations and receiver functions results in a set of models beneath each station (ranging 13 to 18), consistent with both datasets, as well as with prior information on the uppermost layers. This set of models provides valuable insights into the crust and upper mantle structure of eastern Cuba. They can serve as initial models for other inversion procedures (e.g., Arango et al., 2014) or contribute to obtain an optimal solution based on statistical approaches (e.g., González et al., 2012), by employing alternative techniques such as LSO (Boyadzhiev et al., 2008) or by integrating new information.

In most cases, the resulting models from LSO procedure (Figure 5 and Supplementary material #1) evidence crustal thickness (Table 3) consistent with past similar studies (see Table 4 for comparison) and new information below the new installed stations.

The results for the Cascorro station (CCCC), located in the extreme west of the study region at latitudes greater than 21° within the Cretaceous Volcanic arc, indicate a crustal thickness of approximately 30 km, consistent with previous studies (e.g., Bovenko et al., 1982; Bush and Shcherbakova, 1986; González et al., 2012; Moreno, 2003; Otero et al., 1998). Below the Moho, high velocity lithospheric layers up to 75 km depth are followed by high mantle seismic velocities (~4.5 km/s) up to 150 km. This deep high velocity lithosphere was also observed by P-wave tomography profiles to the West of this region (van Benthem et al., 2013).

Even though HLG, NICP, NMDO, and MASC belong to different LSO subgroups, their crustal thicknesses range was 27–31 km. This relatively thick crust, characterized by the presence of crustal layers with relatively low-velocity channels (LVC), is consistent with an accretionary type possibly formed under compressional deformation between the Bahamas Platform and the allochthonous complex in northeastern Cuba (Moreno et al., 2002). Incorporating additional and more precise information for NMDO and MASC, has allowed us to identify an LID beneath these stations, that was not detected in previous studies (e.g., González et al., 2012). Beneath the HLG, NICP, and

Table 3 Moho depth (rounded to 0.5 km), lithospheric thicknesses (rounded to 5 km) and ranges of depth for the possible subducted slab (where present) below the seismic stations in eastern Cuba.

Station	*NICP*	*YARP*	*CCCC*	*NMDO*	*MASC*	*QMBU*	*RCC*	*MARV*	*CHIV*	*LMGC*	*PILO*	*GTBY*	*PINP*	*SABP*	*HLG*
Crustal thickness(km)	31.5	10	29.5	27.5	28.5	17.5	18.5	9.5	8	22	21.0	18.5	18	8.0	27
Lithospheric thickness(km)	75	110	75	65	40	165	25	20	20	65	60	70	95	75	75
Depth range for subducted slab(km)	80 165	–	–	80 130	50 175	–	50 100	45 110	35 115	95 155	95 135	85 150	–	80 165	80 165

Table 4 Comparison of the crustal and lithospheric thickness, and the depth range for the subducted slab at the seismic stations in eastern Cuba. Considering its short relative distance, the results for NMDO are compared with MOAC station in González et al. (2012).

Station	*Moreno (2003)*	*González et al. (2012)*	*This study*	*Contents by Lines*
CCCC	31 km	29.5 km 75 km 110 km 140 km	29.5 km 75 km	–Crustal thickness –Lithospheric thickness –Depth range for subducted slab
NMDO	29 km	26 km 130 km	27.5 km 65 km 80 km 130 km	
MASC	24 km	30 km 150 km	28.5 km 40 km 50 km 175 km	
RCC	18 km	16.5 km 95 km 130 km 190 km	18.5 km 25 km 50 km 100 km	
LMGC	19 km	24.5 km 65 km 105 km 155 km	22 km 65 km 95 km 155 km	
GTBY		19 km 75 km 85 km 150 km	18.5 km 70 km 85 km 150 km	

NMDO stations, an LVZ followed by a new LID is observed at approximately 80 km depth. This LVZ may be attributable to a mantle wedge located between northeastern Cuba and the Bahamas Platform. Beneath MASC, this potential subducted slab is observed at a shallower depth, beginning at 50 km, owing to its closer proximity to the Bahamas Platform. The presence of a lithospheric slab, possibly inherited from the frontal subduction of the North American plate beneath the Cretaceous Greater Antilles arc, has previously been hypothesized below Hispaniola from P-wave tomography (van Bethem et al., 2013) and wide-angle seismic imaging (Núñez et al., 2019). It is consistent with the interpretation of Corbeau et al. (2019), who attribute the intermediate-depth seismicity below northern and central Haiti to the North American lithospheric slab beneath Haiti, as also schematically proposed by Calais et al. (2023, their Figure 5) on the basis of a new seismic catalog for Haiti.

Based on the Moho depth and the Vs values, the stations closest to the center of the southern coast (CHIV, MARV, and RCC) exhibit below a thin crust (ranging from 10 to 20 km) and a lithospheric thickness from 20–25 km (Figures 5 and 6), with an oceanic-type character (e.g., Panza, 1980). These values are consistent with the gradual southward thinning of the crust observed in seismic profiles (Bush & Shcherbakova, 1986) and with the classification of the crust described by Otero et al. (1998).

Their lithosphere overlies a shallow low-velocity zone (LVZ) and a nearly aseismic LID extending from ~50 km to ~110 km deep. This LID may be interpreted as a remnant slab from the Paleogene volcanic arc (Sommer et al., 2011). To the east of RCC, the thickening of the lithosphere is consistent with the deepening of the seismicity in the Santiago Deformed Belt (Figure 6).

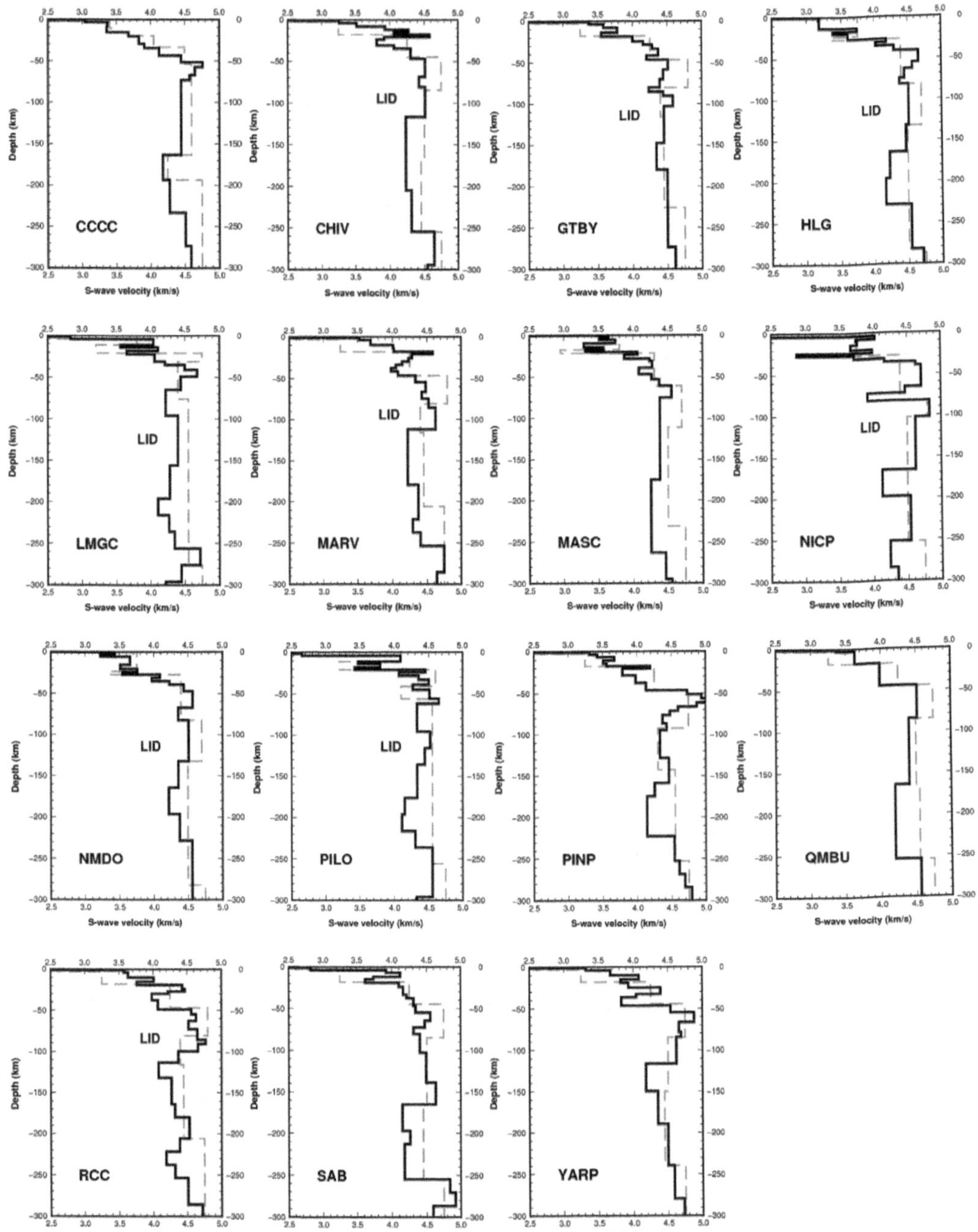

Fig. 5 Chosen solution (thick solid black line) and initial model (grey dashed line) for broadband seismic stations in the eastern region of Cuba, results of the joint inversion of RFs (using Gaussian factors at 0.5, 1.0, and 2.5) and dispersion data. LID represents the zone where high seismic velocities are present and the codes at 250 km depth are the name of the stations.

To the west, the LMGC and PILO stations, located on the northern and western margins of the Sierra Maestra Mountains (SMM), exhibit a thicker crust (~22 km), with low-velocity layers that suggest an accretionary-type crust, inherited from the volcanic origin of the SMM (Rojas-Agramonte et al., 2006). The lithosphere, approximately 65 km thick, is consistent with an oceanic-

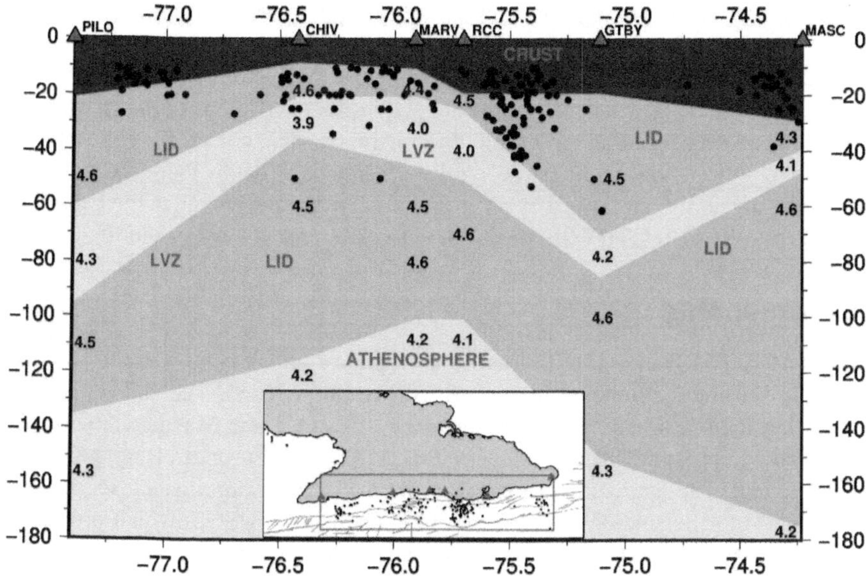

Fig. 6 Schematic interpretative cartoon of profiles Vs vs. depth along the stations close to the southern coast of Cuba. LID, seismic high velocity zone; LVZ, low velocity zone; numbers are Vs velocities values in km/s; seismic stations are indicated by triangular symbols, while earthquake hypocenters are shown as dots, registered by more than 4 digital stations, M > 3.0, depth below 10 km in a period from 2000 to 2024 (https://www.cenais.gob.cu/navegador/). In the interpretation some representative values are given that fall within the range of velocity for the layers. The bottom map shows the relative positions of the earthquakes and stations, with a framed rectangle indicating the profile area.

type structure (e.g., Panza, 1980). Overlaying by an LVZ, a new and aseismic LID is present ranging from 95 km–140 km. Hints of this LID to the west of PILO have been also observed by P-wave tomography (van Benthem, 2013).

YARP station, at the end of the northern margin of SMM, is situated on a huge basaltic dyke trending north-south (Gyarmati, 1983), of Cretaceous age (Núñez et al., 2003). It exhibits a thin crust (~10 km thick) overlain a lithosphere approximately 110 km thick, with several LVC in the uppermost mantle.

SABP station is located in the southern margin of the Moa-Baracoa Masiff, in a sequence of carbonate rocks of Upper Oligocene age in a basin over an ophiolitic complex from the Cretaceous age, while PINP is located over Mayarí-Cristal Massif ophiolitic complex (Núñez et al., 2003). For both stations, the presence of LVC below the shallow Moho, as well as a thick lithosphere, is consistent with the obduction process (Coleman, 1971), where ophiolitic slices of the oceanic domain are buckled and squeezed on top of the continental lithosphere.

7. Conclusions

The uppermost layers of the Vs vs. depth new models, and the tomographic maps of Rayleigh waves group velocities at shorter periods evidence large variability of the values which are consistent with the complex geological features of southeastern Cuba (Iturralde, 1996, 1998; Núñez et al., 2003; Sommer et al., 2011). The thick sediment deposits in the Cauto Basin, the ophiolitic complexes in accretionary crust, and the more recently formed Sierra Maestra Mountains, where oceanic crustal settings are present, are rephrased by such models and maps.

For deeper structures, the models also reveal the presence of lithospheric LVCs and the presence of high velocity mantle possibly associated with the past subduction and collision of the greater

Arc of the Antilles with the North American passive margin (Iturralde-Vinent & Gahagan, 2002; Rojas-Agramonte et al., 2006).

Thick accretionary crust with LVCs is observed below the stations near the northeastern coast (NICP, NMDO, and MASC) consistent with the obduction process (Coleman, 1971), where ophiolitic slices are buckled and squeezed on top of the lithosphere. Oceanic crustal and lithospheric settings are predominant below stations in the part central southern coast (below the stations CHIV, MARV, and RCC) overlaying an LVZ followed by an aseismic LID. The deepening to the East of the nearly aseismic LVZ, from RCC to GTBY, is consistent with earthquakes locations in the last 25 years.

Acknowledgements

NICP, PINP, SABP, YARP, and TUNP broadband seismic stations belong to the French national pool of portable seismic instruments SisMob-RESIF (INSU-CNRS). The authors are grateful to S. Roussel for her logistical help. Their installation by VC, OG, and Manuel Cutie (CENAIS), in 2018 was funded by the INTERREG Caraïbe V PREST project (European Regional Development Fund FEDER funds). The authors are grateful to the National Seismological Service of Cuba of the National Centre for Seismological Research of Cuba and the USGS, which provided the data used in this study. OG, BM, and EA thank the staff of Géosciences Environnement Toulouse (GET)/Observatoire Midi-Pyrénées (OMP) of the University of Toulouse III Paul Sabatier, for their support and facilities during their stay in Toulouse. The present work was done with the financial support of the INTERREG Caraïbe V PREST (Plateforme REgionale de Surveillance Tellurique) project and the "DISA-Desarrollo de Investigaciones Sismológicas Aplicadas en la República de Cuba" program, through "Analysis of the structure and dynamics of the crust in the arc of the Antilles", project Code PS211SC001-004. EC acknowledges funding from the French National Research Agency, "CAST project", grant number ANR-22-CE01-0019. OG sincerely appreciates Prof. Giuliano Francesco Panza for his exceptional guidance on the rigorous application of the employed tools and the interpretation of the results.

References

Álvarez, J.L., Chuy, T., García, J., Moreno, B., Álvarez, H., Blanco, M., Expósito, O., González, O. and Fernández, A.I. (1999). *An Earthquake Catalogue of Cuba and Neighboring Areas*. Internal report of the Abdus Salam International Centre for Theoretical Physics. IC/IR/99/1 (ICTP).

Álvarez, L., Lindholm, C. and Villalón, M. (2016). Seismic Hazard for Cuba: A New Approach. *Bulletin of the Seismological Society of America, 107* (1), 229–239. doi: https://doi.org/10.1785/0120160074.

Ammon, C.J. (1991). The isolation of receiver effects from teleseismic P waveforms. *Bull. Seism. Soc. Am.. 81*, 2504–2510.

Anderson, D.L. (1995), Lithosphere, asthenosphere, and perisphere. *Rev. Geophys., 33*(1), 125–149. doi:10.1029/94RG02785.

Arango, E.D., Pérez M.A. and Batista J.A. (2014, July–September). Crustal structure of eastern Cuba, derived by constrained 3D gravity inversion. *Geofísica Internacional, 53*(3), 259–275 Elsevier. doi: 10.1016/S0016-7169(14)71504-1.

Benford, B., DeMets, C. and Calais, E. (2012). GPS estimates of microplate motions, northern Caribbean: Evidence for a Hispaniola microplate and implications for earthquake hazard. *Geophys. J. Int., 191*, 481–490.

Bensen, G.D., Ritzwoller, M.H., Barmin, M.P., Levshin A.L., Lin F., Moschetti M.P., Shapiro, N.M. and Yang, Y. (2007). Processing seismic ambient noise data to obtain reliable broad-band surface wave dispersion measurements. *Geophys. J. Int., 169*, 1239–1260. https://doi.org/10.1111/j.1365-246X.2007.03374.x.

Bensen, G.D., Ritzwoller, M.H. and Shapiro, N.M. (2008). Broadband ambient noise surface wave tomography across the United States. *J. Geophys.*, Res., *113*, B05306, doi:10.1029/2007JB005248.

Bovenko, V.G., Shcherbakova, B.E. and Hernández, H. (1982). Novyye geofizicheskiye dannyye o glubinnour stroyenii vostochnoy kuby (New geophysical data on the deep structure of eastern Cuba). *Sovetskaya Geologiya, 9*, 101–109; translation in *International Geology Review, 24*, 1155–1162.

Boyadzhiev, G., Brandmayr, E., Pinat, T. and Panza, G.F. (2008). Optimization for nonlinear inverse problem. *Rendiconti Lincei: Scienze Fisiche e Naturali, 19*, 17–43.

Bush, V.A. and Shcherbakova, I.N. (1986). New Data on the Deep Tectonics of Cuba. *Geotectonics, 20*, 192–203.

Calais, Eric & de Lépinay, Bernard. (1991). From transtension to transpression along the northern Caribbean plate boundary off Cuba: Implications for the Recent motion of the Caribbean plate. *Tectonophysics, 186*, 329–350. doi:10.1016/0040-1951(91)90367-2.

Calais, E., Gonzalez, O.F., Arango-Arias, E.D., Moreno, B., Palau, R., Cutie, M., Diez, E., Montenegro, C., Rodriguez Roche, E., Garcia, J., Castellanos, E. and Symithe, S. (2023). Current deformation along the northern Caribbean plate boundary from GNSS measurements in Cuba. *Tectonophysics, 868*, 230068.

Calais, E., Symithe, S.J. and de Lépinay, B. (2023). Strain Partitioning within the Caribbean-North America Transform Plate Boundary in Southern Haiti: Tectonic and Hazard Implications. *Bulletin of the Seismological Society of America, 113* (1), 131–142. doi:10.1785/0120220121.

Chimera, G., Aoudia, A., Saraó, A. and Panza, G.F. (2003). Active tectonics in central Italy: constraint from surface wave tomography and source moment tensor inversion. *Phys. Earth Planet. Inter., 138*, 241–262.

Clouard, V. (2018). PREST in Cuba temporary experiment installed in the southeast of Cuba to better understand the dynamics of the lithosphere at the plate boundary, Cuba (RESIF-SISMOB) [Data set]. *RESIF: Réseau Sismologique et Géodésique Français*. https://doi.org/10.15778/RESIF.XH2018.

Coleman, R.G. (1971). Plate tectonic emplacement of upper mantle peridotites along continental edges, *J. Geophys. Res., 76*, 1212–1222.

Corbeau, J., Gonzalez, O.L., Clouard, V., Rolandone, F., Leroy, S., Keir, D., Stuart, G., Momplaisir, R., Boisson, D. and C. Prépetit. (2019). Is the local seismicity in western Hispaniola (Haiti) capable of imaging northern Caribbean subduction?. *Geosphere, 15* (6), 1738–1750. doi: https://doi.org/10.1130/GES02083.1.

Ditmar, P.G. and Yanovskaya, T.B. (1987). A generalization of the Backus-Gilbert method for estimation of lateral variations of surface wave velocity. *Izv. Akad. Nauk SSSR, Fizika Zemli* (6), 30–60.

Fischer, K.M., Ford, H.A., Abt, D.L. and Rychert, C.A. (2010). The lithosphere-asthenosphere boundary. *Annu. Rev. Earth Planet. Sci., 38* (1), 551–575. http://dx.doi.org/10.1146/annurev-earth-040809-152438.

Gaite, B., Villaseñor, A., Iglesias, A., Herraiz, M. and Jiménez-Munt, I. (2015). Shear velocity model of the southern North American and Caribbean plates from ambient noise and earthquake tomography. *Solid Earth, 6*, 271–284, https://doi.org/10.5194/se-6-271-2015

García, J., Slejko, D., Alvarez, L., Peruzza, L. and Rebez, A. (2003). Seismic Hazard Maps for Cuba and Surrounding Areas. *Bulletin of the Seismological Society of America, 93* (6), 2563–2590. doi: https://doi.org/10.1785/0120020144.

Gök, R., Mahdi, H., Al-Shukri, H. and Rodgers, A.J. (2008). Crustal structure of Iraq from receiver functions and surface wave dispersion: Implications for understanding the deformation history of the Arabian-Eurasian collision. *Geophys. J. Int., 172*, 1179–1187

Goldstein, P. (1999). *SAC User's Manual*. Lawrence Livermore Laboratory, University of California.

Gonzalez, O., Alvarez L., Guidarelli M. and Panza G.F. (2007). Crust and Upper Mantle Structure in the Caribbean Region by Group Velocity Tomography and Regionalization. *Pure Appl. Geophys. 164*, 1985–2007.

González, O., Moreno, B., Romanelli, F. and Panza, G.F.. (2012). Lithospheric structure below seismic stations in Cuba from the joint inversion of Rayleigh surface waves dispersion and receiver functions. *Geophys. Journal Int., 189*, 1047–1059. doi:10.1111/j.1365-246X.2012.05410.x.

González, O., Alvarez, J.L., Moreno, B. and Panza, G.F. (2012) S-Wave velocities of the lithosphere-asthenosphere system in the Caribbean region. *Pure Appl. Geophys., 169*(1–2), 101–122. Springer Basel AG. doi:10.1007/s00024-011-0321-3.

González, O., Clouard, V., Tait, S. and Panza, G. (2018). S-wave velocities of the lithosphere- asthenosphere system in the Lesser Antilles from the joint inversion of surface wave dispersion and receiver function analysis. *Tectonophysics*, 734–735, 1–15. https://doi.org/10.1016/j.tecto. 2018.03.021.

Gyarmati, P. (1983). Las rocas intrusivas intermedias de Cuba oriental. En: *Contribución a la Geología de Cuba*. (Editorial Científico-Técnica, pp. 99–111). La Habana. http://www.redciencia.cu/geobiblio/paper/1983_Gyarmati_rocas%20 intrusivas_CGPO.pdf.

Hall, C.M., Kesler, S.E., Russell, N., Pinero, E., Sánchez C,R., Pérez, R.M., Moreira, J.A. and Borges, M. (2004). Age and tectonic setting of the Camagüey volcanic-intrusive arc, Cuba: Late Cretaceous extension and uplift in the western Greater Antilles. *The Journal of Geology, 112*(5), 521–542.

Herrmann, R.B. and Ammon, C.J. (2002). *Computer Programs in Seismology: Surface waves, Receiver functions and Crustal Structure, Version 3.20* (110 pp).. Saint Louis University, St. Louis.

Iturralde-Vinent, M.A. (1996). Geología de las ofiolitas de Cuba. In: M.A. Iturralde - Vinent (Ed.), *Ofiolitas y arcos volcánicos de Cuba, IGCP Project 364*. Special Contribution n.1 pp. (83–120) Miami, USA.

Iturralde-Vinent, M. (1998). Sinópsis de la constitución geológica de Cuba. *Acta Geológica Hispánica, 33*(1–4), 9–56.

Iturralde-Vinent, M. and Gahagan, L. (2002). Late Eocene to Middle Miocene Tectonic Evolution of the Caribbean: Some principles and their Implications for Plate Tectonic Modeling. In: T.A. Jackson (Ed.), *Caribbean Geology into the Third Millennium. Transactions of the Fifteenth Caribbean Geological Conference* (pp. 47–62). Pear Tree Press Ltd., Jamaica.

Iturralde-Vinent, M.A., Otero, C.D., Garcia-Casco, A. and van Hinsbergen, D.J.J. (2008). Paleogene foredeep basin deposits of north-central Cuba: A record of arc–continent collision between the Caribbean and North American plates. *Int. Geol. Rev.*, *50*, 863–884.

Iturralde-Vinent, M., Garcia-Casco, A., Rojas-Agramonte, Y., Proenza, J., Murphy, J.B. and Stern, B. (2016). The geology of Cuba: A brief overview and synthesis. *GSA Today*, *26*, 4–10. 10.1130/GSATG296A.1.

Julia, J., Ammon, C.J., Herrmann, R.B. and Correig, A.M. (2000). Joint inversion of receiver functions and surface-wave dispersion observations. *Geophys. J. Int.*, *143*, 99–112.

Julia, J., Ammon, C.J. and Herrmann R.B. (2003). Lithospheric structure of the Arabian Shield from the joint inversion of receiver functions and surface-wave group velocities. *Tectonophysics*, *371*, 1–21.

Julia J., Ammon, C.J. and Nyblade, A.A. (2005). Evidence for mafic lower crust in Tanzania, East Africa, from joint inversion of receiver functions and Rayleigh wave dispersion velocities. *Geophys. J. Int.*, *162*, 555–569.

Leroy, Sylvie, Mauffret, A. and Pubellier, Manuel. (1996). Structural and Tectonic Evolution of the Eastern Cayman Trough (Caribbean Sea) From Seismic Reflection Data. *AAPG BULL.*, *80*, 222–247. doi:10.1306/64ED8796-1724-11D7-8645000102C1865D.

Leroy, S., Mauffret, A., Patriat, P. and Mercier de Lépinay, B. (2000). An alternative interpreta- tion of the Cayman trough evolution from a reidentification of magnetic anomalies. *Geophys. J. Int.*, *141*, 539–557.

Levshin, A.L., Pisarenko, V.F. and Pogrebinsky, G.A. (1972). On a frequency-time analysis of oscillations. *Ann. Geophys.*, *28*(2), 211–218.

Levshin, A.L., Radnikova, L.I. and Berger, J. (1992). Peculiarities of surface wave propagation across Central Eurasia. *Bull. Seismol. Soc. Am.*, *82*, 2464–2493.

Ligorría, J.P. and Ammon, C.J. (1999). Iterative deconvolution of teleseismic seismograms and receiver function estimation. *Bull. Seism. Soc. Am.*, *89*, 1395–1400.

Lin, F.C., Ritzwoller, M.H. and Snieder, R. (2009). Eikonal tomography: Surface wave tomography by phase front tracking across a regional broad-band seismic array. *Geophysical Journal International*, *177*(3), 1091–1110.

Mann, P., Taylor, F.W., Edwards, R.L. and Ku, T.L. (1995). Actively evolving microplate formation by oblique collision and sideways motion along strike-slip faults: An example from the northeastern Caribbean plate margin. *Tectonophysics*, *246*, 1–69.

Moreno, B. (2002). The new Cuban Seismograph Network, *Seism. Res. Lett.*, *73*, 505–518.

Moreno, B. (2003). The crustal structure of Cuba derived from Receiver Function Analysis. *Journal of Seismology*, *7*, 359–375.

Moreno, B., Aoudia, A., Manu-Marfo, D., Kherchouche, R. and Pachhai. (2023). S. Crust–Uppermost Mantle Structure beneath the Caribbean Region from Seismic Ambient Noise Tomography. *Bulletin of the Seismological Society of America*, *113*(3), 1064–1076. doi: 10.1785/012022006.

Moschetti, M.P., Ritzwoller, M.H. and Shapiro, N.M. (2007). Surface wave tomography of the western United States from ambient seismic noise: Rayleigh wave group velocity maps. *Geochemistry, Geophysics, Geosystems*, *8*(8).

Núñez, K., Castellanos, E., Alfonso, W. and Rosa, A. (2003). Cartografía Geológica Digital. Simbolización. IGP-DOC-SIGEOL-05-01-2003. En: Castellanos, E., Núñez, K., Carrillo, D., Pantaleón, G. Triff, J. (Eds.), *Diseño del Sistema de Información Geológica de Cuba*. CNDIG, IGP, La Habana (139 pp). Diseño del Sistema de Información Geológica de Cuba, ISBN 959-7117-13-4.

Nuñez, Diana & Córdoba, Diego & Kissling, Edi. (2019). Seismic structure of the crust in the western Dominican Republic. *Tectonophysics*, *773*, 228224. doi: 10.1016/j.tecto.2019.228224.

Otero, R., Prol, J.L., Tenreyro, R. and Arriaza, G.L. (1998). Caracteristicas de la corteza terrestre de Cuba y su plataforma marina (Characteristics of the Earth's crust in Cuba and its marine platform). *Mineria y Geologia*, *15*, 31–35.

Ozalaybey, S., Savage, M.K., Sheehan, A.F., Louie, J.N. and Brune, J.N. (1997). Shear-wave velocity structure in the northern basin and range province from the combined analysis of receiver functions and surface waves. *Bull. Seism. Soc. Am.*, *87*, 183–189.

Pagani, M., Garcia, J., Gee, F., Johnson, K., Poggi, V., Styron, R., Weatherill, G., Simionato, M., Viganò, D., Danciu, L. and Monelli D. (2018). Global Earthquake Model (GEM) Seismic Hazard Map (version 2018.1–December 2018). doi: 10.13117/GEM-GLOBAL-SEISMIC-HAZARD-MAP-2018.1.

Panza, G.F. (1980). *Evolution of the Earth's lithosphere*. NATO Adv. Stud. Inst. Newcastle, 1979. In: Davies, P.A., and Runcorn, S.K. (Eds.), *Mechanisms of Continental Drift and Plate Tectonics* (pp. 75–87). Academic Press.

Panza, G.F. (1981). The resolving power of seismic surface waves with respect to crust and upper mantle structural models. In: Cassinis, R. (Ed.), *The Solution of the Inverse Problem in Geophysical Interpretation* (pp. 39–77). Plenum Publ. Corp., New York, NY.

Pindell, J., Kennan, L., Stanek; K.P., Maresch, W.V. and Draper, G. (2006). Foundations of Gulf of Mexico and Caribbean evolution: Eight controversies resolved. In: Ituralde-Vinent, M., and Lidiak, E.G. (Eds.), *Caribbean Plate Tectonics:*

Stratigraphic, Magmatic, Metamorphic and Tectonics Events (UNESCO/IUGS IGCP Project 433). *Geol. Acta, 4*, 303–341.

Pindell, J., Maresch, W.V., Martens, U. and Stanek, K. (2011). The Greater Antillean Arc: Early Cretaceous origin and proposed relationship to Central American subduction me´langes: Implications for models of Caribbean evolution. *Int. Geol. Rev*. doi:10.1080/00206814.2010.510008.

Pontevivo, A. and Panza, G. (2006). The lithosphere–asthenosphere system in the Calabrian Arc and surrounding seas—southern Italy. *Pure and Applied Geophysics, 163*, 1617–1659.

Possee, D., Rychert, C., Harmon, N. and Keir, D. (2021). Seismic discontinuities across the North American Caribbean plate boundary from S-to-P receiver functions. *Geochemistry, Geophysics, Geosystems, 22*, e2021GC009723. https://doi.org/10.1029/2021GC009723.

Rojas-Agramonte, Y., Neubauer, F., Handler, R., Garcia-Delgado, D.E., Friedl, G. and Delgado Damas, R. (2005): Variation of palaeostress patterns along the Oriente transform wrench corridor, Cuba: Significance for Neogene–Quaternary tectonics of the Caribbean realm. *Tectonophysics, 396*, 161–180.

Rojas-Agramonte, Y., Neubauer, F., Kr.Ner, A., Wan, Y.S., Liu, D.Y., Garcia-Delgado, D.E. and Handler, R. (2004): Geochemistry and early Palaeogene SHRIMP zircon ages for island arc granitoids of the Sierra Maestra, southeastern Cuba. *Chemical Geology, 213* (4), 307–324.

Rojas-Agramonte, Y., Neubauer, F., Bojar, AV., Hejl, E., Handler, R. and García-Delgado, D.E. (2006). Geology, age, and tectonic evolution of the Sierra Maestra Mountains, southern Cuba. *Geologica Acta,* 123–150. ISSN 1695-6133.

Shapiro N., Campillo M., Stehly, L. and Ritzwoller, M. (2005, 11th Mar.). High-resolution surface-wave tomography from ambient seismic noise. *Science, 307*(5715), 1615-8. doi: 10.1126/science.1108339. PMID: 15761151.

Shcherbakova, B.E., Bovenko, V.G. and Hernández, H. (1978). Stroyeniye zemnoy kory Zapadnoy Kuby (Crustal structure in West Cuba). *Sovetskaya Geologya, 8*, 138–143; translation in *International Geology Review, 20*, 1125–1130.

Sommer, M., Hüneke, H., Meschede, M. and Cobiella-Reguera, J. (2011). Geodynamic model of the northwestern Caribbean: Scaled reconstruction of Late Cretaceous to Late Eocene plate boundary relocation in Cuba. *Neues Jahrbuch für Geologie und Paläontologie - Abhandlungen* (Band 259, Heft 2), *259*(3), 299–312.

Stanek, K.P., Maresch, W.V. and Pindell, J. (2009). The geotectonic story of the northwestern branch of the Caribbean Arc: Implications from structural and geochronological data of Cuba. *Geological Society, London, Special Publications, 328*, 361–398, https://doi.org/10.1144/SP328.15.

Tkalčić, H., Pasyanos, M.E., Rodgers, A.J., Gök, R., Walter, W.R. and Al-Amri, A. (2006). A multistep approach for joint modeling of surface wave dispersion and teleseismic receiver functions: Implications for lithospheric structure of the Arabian Peninsula. *J. Geophys. Res., 111*, B11311. doi:10.1029/2005JB004130.

Van der Hilst, R.D. (1990). *Tomography with P, PP, and pP Delay-time Data and the Three-dimensional Mantle Structure below the Caribbean Region*, Ph.D. Thesis, University of Utrecht, Holland.

Van Benthem, S., Govers, R., Spakman, W. and Wortel, R. (2013). Tectonic evolution and mantle structure of the Caribbean. *J. Geophys. Res. Solid Earth, 118*, 3019–3036. doi:10.1002/jgrb.50235.

Vdovin, O., Rial, J.A., Levshin, A.L. and Ritzwoller, M.H. (1999). Group-velocity Tomography of South America and the Surrounding Oceans. *Geophys. J. Int., 136*, 324–340.

Wu, F.T. and Levshin, A. (1994). Surface-wave group velocity tomography of east Asia. *Phys. Earth Planet Int., 84*, 59–77.

Yanovskaya, T.B. and Ditmar, P.G. (1990). Smoothness criteria in surface wave tomography. *Geophys. J. Int., 102*, 63–72.

Yanovskaya, T.B. (1997). Resolution estimation in the problems of seismic ray tomography. *Izvestiya, Physics of the Solid Earth, 33*(9), 762–765.

Supplementary material 1: Vs models derived from the joint inversion of dispersion relations and RFs and chosen using the LSO procedure.

CCCC	
Vs (km/s)	*Depth (km)*
2.41	0
3.04	1
3.37	3.5
3.36	8.5
3.68	15.5
3.83	20.5
3.91	29.5
4.13	34.5
4.45	43.5
4.76	51.5
4.65	57.5
4.57	67.5
4.45	73.5
4.18	163.5
4.28	193.5
4.51	233.5
4.59	273.5

CHIV	
Vs (km/s)	*Depth (km)*
2.01	0
3.26	1
3.52	3.7
3.93	7.7
4.27	11.7
4.06	14.7
4.58	17.7
3.92	20.7
3.81	23.7
4.06	30.7
4.30	34.7
4.51	46.7
4.42	68.7
4.51	80.7
4.23	116.7
4.31	204.7
4.64	254.7
4.54	294.7

GTBY	
Vs (km/s)	*Depth (km)*
1.88	0
3.37	1.0
3.58	3.7
3.78	7.7
3.55	13.7
4	18.7
4.15	24.7
4.28	28.7
4.37	33.7
4.2	41.7
4.51	46.7
4.44	58.7
4.39	70.7
4.23	80.7
4.44	85.7

HLG	
Vs (km/s)	*Depth (km)*
1.57	0
3.17	0.6
3.18	3.6
3.74	13.6
3.38	17.6
3.6	20.6
4.19	27.3
4.03	31.3
4.3	35.3
4.66	41.3
4.57	55.3
4.45	63.3
4.38	75.3
4.51	82.3
4.47	132.3

4.58	90.7
4.45	103.2
4.34	147.7
4.5	179.7
4.61	273.7

4.23	164.3
4.17	196.3
4.54	228.3
4.71	282.3

LMGC	
Vs (km/s)	*Depth (km)*
1.59	0
2.83	1.5
4.04	4.1
3.56	11.1
4.11	14.1
3.65	18.1
4.06	22.1
4.21	31.1
4.5	35.1
4.68	41.1
4.44	49.1
4.22	65.1
4.4	96.1
4.28	156.1
4.11	196.1
4.27	216.1
4.35	236.1
4.72	256.1
4.45	276.1
4.22	296.1

MARV	
Vs (km/s)	*Depth (km)*
2.04	0
3.53	1
3.69	3.7
4.01	9.7
4.02	13.7
4.58	17.7
4.28	20.7
4.23	25.7
4.14	29.7
4.05	33.7
3.98	37.7
4.08	41.7
4.33	46.7
4.48	54.7
4.42	66.7
4.52	74.7
4.62	85.7
4.22	111.9
4.37	179.7
4.29	221.7
4.4	237.7
4.75	253.7
4.64	285.7

MASC	
Vs (km/s)	*Depth (km)*
1.99	0
3.64	0.6
3.51	2.6
3.74	5.6
3.29	9.6
3.57	14.6

NICP	
Vs (km/s)	*Depth (km)*
0.85	0
2.44	0.6
4.03	3.6
3.74	7.6
3.76	10.6
3.66	14.6

3.31	17.3
4.06	20.3
3.87	23.3
4.23	28.3
4.27	31.3
4.08	39.3
4.26	47.3
4.37	53.3
4.55	61.3
4.38	75.3
4.25	175.3
4.46	263.3
4.55	295.3

3.99	20.6
2.87	24.3
3.71	27.3
4.17	31.3
4.63	35.3
4.71	39.3
4.47	65.3
3.91	73.3
4.83	82.3
4.62	102.3
4.13	166.3
4.54	198.3
4.24	252.3
4.35	284.3

NMDO	
Vs (km/s)	*Depth (km)*
1.83	0
3.43	0.6
3.22	2.6
3.65	5.6
3.51	15.6
3.76	20.6
3.54	24.3
4.09	27.3
3.97	31.3
4.23	35.3
4.44	39.3
4.57	47.3
4.36	67.3
4.51	82.3
4.36	132.3
4.22	164.3
4.38	196.3
4.56	228.3

PILO	
Vs (km/s)	*Depth (km)*
1.39	0
2.65	1.5
4.09	4.1
3.48	11.1
3.8	14.1
3.43	18.1
4.45	21.1
4.07	24.1
4.35	28.1
4.5	33.1
4.27	39.1
4.51	45.1
4.65	56.1
4.33	62.1
4.52	96.1
4.44	116.1
4.33	136.1
4.15	176.1
4.11	196.1
4.3	216.1
4.55	236.1
4.32	296.1

PINP	
Vs (km/s)	*Depth (km)*
1.81	0
3.31	1.0
3.42	3.7
3.67	6.7
3.51	10.7
3.56	14.7
4.19	17.7
3.79	20.7
3.98	28.7
4.13	37.7
4.73	46.7
4.94	51.7
4.95	56.7
4.87	61.7
4.6	66.7
4.48	71.7
4.37	76.7
4.43	86.7
4.33	95.2
4.46	128.7
4.25	158.7
4.14	174.7
4.54	222.7
4.61	252.7
4.69	268.7
4.79	284.7

RCC	
Vs (km/s)	*Depth (km)*
2.08	0
3.57	1.0
3.63	3.7
4.01	9.7
3.76	14.7
4.43	18.7
4.47	23.7
4.22	26.7
3.98	29.7

QMBU	
Vs (km/s)	*Depth (km)*
1.87	0
3.36	1
3.63	2.7
3.99	17.7
4.52	44.7
4.4	84.7
4.2	164.7
4.56	254.7
4.55	350
4.64	367

SABP	
Vs (km/s)	*Depth (km)*
1.32	0
2.82	1
3.92	3.7
4.12	6.7
3.74	11.7
3.62	14.7
4.1	18.7
4.16	23.7
4.21	33.7
4.31	37.7
4.34	46.7
4.55	54.7
4.47	64.7
4.31	72.7
4.4	80.7
4.49	102.7
4.63	138.7
4.15	164.7
4.26	196.7
4.18	212.7
4.84	254.7
4.92	270.7

4.07	37.7
4.56	48.7
4.63	54.7
4.52	62.7
4.65	72.7
4.77	85.7
4.66	90.7
4.37	99.5
4.08	113.3
4.27	131.8
4.32	163.8
4.53	179.8
4.39	205.8
4.19	221.8
4.32	237.8
4.51	253.8
4.72	285.8

YARP	
Vs (km/s)	*Depth (km)*
1.96	0
3.32	1
3.67	3.7
4.08	9.7
3.82	14.7
3.93	18.7
4.39	24.7
4.04	32.7
3.83	36.7
4.54	46.7
4.88	54.7
4.66	66.7
4.69	78.7
4.62	84.7
4.18	116.7
4.35	149.7
4.5	189.7
4.59	239.7
4.73	279.7

CHAPTER

12

Seismogenic Nodes in South Europe

Gorshkov A.[1], Panza G.F.[2] and Novikova O.[3]

Southern Europe is one of the most seismically active regions in the world, requiring continuous improvement of seismic hazard assessment research. In such studies, information about potential sources of earthquakes plays a key role. The chapter presents the results of seismogenic nodes' identification in the regions of Southern Europe obtained using the method of pattern recognition of earthquake prone areas (PREPA). The methodology assumes the occurrence of strong earthquakes in morphostructural nodes formed at the intersections of morphostructural lineaments, which are identified by morphostructural zoning. The nodes are characterized by the pertinent geophysical and morphological parameters, using which the pattern recognition algorithms identify potential nodes capable of generating earthquakes of the target magnitude. Information on capable nodes is of particular relevance for seismic hazard assessment methods including the Neo-Deterministic Seismic Hazard Assessment (NDSHA) that has evolved in recent decades.

1. Introduction

Southern Europe and especially the Mediterranean are among the most seismically active regions in the world (Boshi et al., 1993). Over the past decades, many studies have been conducted to assess seismic hazard for the entire Mediterranean region and its individual areas (e.g., Giardini et al., 2003). Most of these works implemented the probabilistic seismic assessment approach (PSHA), the reliability of which has come under considerable criticism in recent decades (Kossobokov & Nekrasova, 2012).

The methodology used in this work for recognizing potential seismogenic nodes is based on the empirical observations on the relationship between strong earthquakes and morphostructural nodes. As a hypothesis, it was first formulated by Gelfand et al. (1972) in the Pamir-Tien Shan region of Central Asia, where the morphostructural zoning (MSZ) method (Rantsman, 1979; Gorshkov et al., 2003) in combination with the pattern recognition (PR) technique (Bongard, 1967; Gelfand et al.,

[1,2] Institute of Earthquake Prediction Theory and Mathematical Geophysics, Russian Academy of Sciences, Moscow, Russian Federation.

[2] Accademia Nazionale dei Lincei, Rome, Italy. OR Earthquake Administration, Beijing, China. OR International Seismic Safety Organization (ISSO), Arsita, Italy.

[3] Beijing University of Civil Engineering and Architecture (BUCEA), Beijing, China.

https://orcid.org/0000-0001-5843-8468[1]; https://orcid.org/0000-0003-3504-3038[2]; giulianof.panza@lincei.it; https://orcid.org/0000-0001-7293-9314[3]

Email: onovikov@mitp.ru[2]; onovikov@mitp.ru[3]

* Corresponding author: gorshkov@mitp.ru

DOI: 10.1201/9781003684121-14

1972; Gorshkov et al., 2003) was implemented for pattern recognition of earthquake prone areas (PREPA).

Further studies using PREPA confirmed the occurrence of epicentres of strong earthquakes in the vicinities of intersections of morphostructural lineaments (nodes) in many seismically hazardous regions of the world (Gorshkov & Soloviev, 2021; Gorshkov & Novikova, 2018). Over time, targeted events have occurred in most of these regions. The consistency of the methodology is confirmed by post-publication target earthquakes that occurred in previously studied seismic regions: 87% of such events occurred at nodes previously identified as seismic for target earthquakes (Soloviev et al., 2014; Gorshkov & Novikova, 2018; Kossobokov & Soloviev, 2018). More recently, it has been shown that the destructive earthquakes in Turkey on February 6, 2023 occurred in a node that in 1974 was considered by Gelfand et al. (1974) to be potential for earthquakes of magnitude 6.5+ (Gorshkov et al., 2024a). All these data indicate satisfactory reliability of PREPA for identifying seismogenic nodes.

The hypothesis on the nucleating strong earthquakes at the nodes initially is based on observations (Gelfand et al., 1972). Later theoretical studies suggested a physical explanation for the concentration of stresses at fault intersections. The role of intersecting faults in generating earthquakes was first proposed by Aki (1979) by introducing geometric barriers, including both the intersections of large faults dividing blocks and bends in the stretch of the single fault. These geometric barriers were also discussed in the global scale as the intersection of transform faults (King, 1985). Talwani (1988) proposed that fault intersections can accumulate large crystals, thereby generating earthquakes. Increased fracturing and crustal disintegration characterize fault intersections, thereby facilitating the migration of liquid or gaseous fluids (Sibson, 1988), which can reduce the friction of the fault, leading to sudden movement along the fault causing earthquakes (Barenblatt, 1993). Recently, Martinelli et al. (2023) demonstrated the important role of geofluids in generating large earthquakes on a global scale and proposed to include geofluids to the list of fundamental geological parameters to be considered in hazard assessment research. Consequently, fault intersections or bends are generally considered to be seismogenic in nature.

In the period 1995–2015, studies on MSZ and identifying potential earthquake locations were actively conducted within the SAND group at the International Centre for Theoretical Physics in Trieste, Italy, under the leadership of G. Panza. At that time, the main focus of these tasks was on the Mediterranean regions. The aim of this work is to present the results of identifying seismogenic nodes in regions located in the European part of the Mediterranean.

Information on potential earthquake sources provided by PREPA is an integral part of seismic hazard assessment methodologies including NDSHA that was developed in the same period by G. Panza and his team (Panza et al., 2001, 2012, 2020; Peresan et al., 2011; Rugarli et al., 2019).

2. Methodology

The PREPA approach is based on two main stages and is described in detail in the works (Gorshkov et al., 2003; Gorshkov & Soloviev, 2021).The first one essentially consists of the analyzed objects definition – the morphostructural nodes – by MSZ. In the second stage, PR is used to classify all nodes identified by MSZ into two groups: nodes capable of generating earthquakes with a target magnitude of M_0, and nodes where seismic events cannot exceed M_0. MSZ is employed to define a system of areas that are hierarchically ordered. Each area is characterized by homogeneous topography and uniform tectonic structure. MSZ delineates lineament-and-blocks structure that includes the following three elements: (1) block areas (blocks) of specific rank (1st, 2nd, and 3rd); (2) morphostructural lineaments that block's boundary zones with different rank; (3) nodes that are intersections of lineaments. MSZ ignores information about the past seismicity of the region so as not to influence the definition of nodes. The nodes are formed at the intersections of the boundaries of blocks of different orientations and are characterized by particularly intense fracturing and contrasting neotectonic deformations, which leads to a mosaic structure and relief.

Nodes are areas of a certain extent (Rantsman, 1979; Gvisiani et al., 1988; Gorshkov et al., 2003). Their limits can be traced at sites where a complex topography pattern becomes more regular. This can be done through field work or by analyzing large-scale topographic maps, as was done for the junction zone of the Alps and Dinarides (Gorshkov et al., 2009a). However, mapping the limits of the nodes is a cumbersome task and it requires the detailed MSZ analysis of lineament intersection areas based on field studies. For pattern recognition purposes, the nodes of the areal shape can be approximated by circles of a certain radius, which is determined by the size of the target M_0 following the empirical relation between magnitude and dimension of earthquake source (Wells & Coppersmith, 1994; D'Amico et al., 1999).

The PREPA approach assumes that nodes hosting the epicenters of target earthquakes have similar geological and geophysical characteristic features. PR allows us to select such features and, based on them, to identify potential nodes that have not yet been marked by target events. PR separates all nodes into two classes: (1) class D ("dangerous" nodes) where earthquakes with magnitude $M \geq M_0$ may occur; (2) class N ("non-dangerous" nodes) where only smaller events of $M < M_0$ may occur.

Discrimination of nodes on D and N is carried out using a set of topographic, geological and geophysical parameters presented as scalar components of vectors, which are a unique characteristic of each node and are not associated with seismicity data in the studied region. These parameters include the geomorphic and morphometric information, the characteristics of the block-and-lineament geometry, the gravity, and magnetic data. Morphometric parameters characterize the contrast and intensity of deformations associated with the period of the relief building. Parameters of the lineament-and-block geometry reflect the complexity and fragmentation of the crust in the node vicinities. Gravity and magnetic data are indicators of the density heterogeneity in the upper crust.

Pattern recognition algorithms, among which Cora-3 (Bongard, 1967; Gelfand et al., 1972; Gorshkov et al., 2003) was commonly exploited in the studies presented in this chapter, use a set of vectors of all nodes as input. To apply recognition algorithms with training, it is necessary to specify training samples D_0 and N_0 in advance. For this purpose, information about target events known from earthquake catalogues is used. The sample set D_0 includes nodes already accommodating earthquakes with $M \geq M_0$. The subset N_0 consists of the remaining nodes, some of which may contain smaller events with $M < M_0$. It is practically impossible to select an "uncontaminated" N_0, since there is no guarantee that the node assigned to N_0 is definitely aseismic.

During the training phase, the algorithm selects characteristic features of classes *D* and *N* classes based on the parameters used. At the classification stage, the algorithm counts the numbers (n) of *D*- and *N*-traits (n_D and n_N, respectively) that each node possesses and appoints it to class *D* if $n_D - n_N \geq \Delta$ or to class *N* if $n_D - n_N < \Delta$, where Δ is the initially specified parameter of the algorithm.

Some of the nodes initially assigned to N_0 during the recognition procedure become *D*. Global validation studies showed that over time, 27% of target events occurred in some of these *D* nodes (Gorshkov & Novikova, 2018; Kosobokov & Soloviev, 2018; Gorshkov & Soloviev, 2021).

3. Seismogenic Nodes in the Mediterranean Regions

In the period 2000–2024, most seismically active regions of the Mediterranean were studied using PREPA (Gorshkov et al., 2000, 2002, 2004, 2009a, 2009b, 2010, 2019, 2021, 2024b; Peresan et al., 2015). This section presents the results of recognition of seismogenic nodes in each of the studied regions. As a rule, the study region encompasses a major tectonic structure such as the Alps, Apennines, Carpathians and others. Within the MSZ, these structures are considered as first-rank territorial units. In the case of Iberia, the entire lithospheric Iberian plate was studied. Recognition was carried out individually for each region. Below for each of them are the MSZ maps showing the lineament-and-block structure and recognized seismogenic nodes for the target magnitude (M_0) specified for the given region. We focus on the presentation of recognition results. MSZ map descriptions are available in the articles dedicated to each individual region (Gorshkov et al., 2000, 2002, 2004, 2009a, 2009b, 2010, 2019, 2020, 2021; Peresan et al., 2015).

The synoptic map shown in Figure 1 displays seismogenic nodes (D) defined in the Apennines and Sicily (Gorshkov et al., 2002) Alps, Dinarides (Gorshkov et al., 2004), and Carpathians-Balkanides (Gorshkov & Soloviev, 2009b). In all these regions, the target magnitude was specified as M_0 = 6.0 and the node size was defined as a circle with a radius of 25 km surrounding each lineament intersection point.

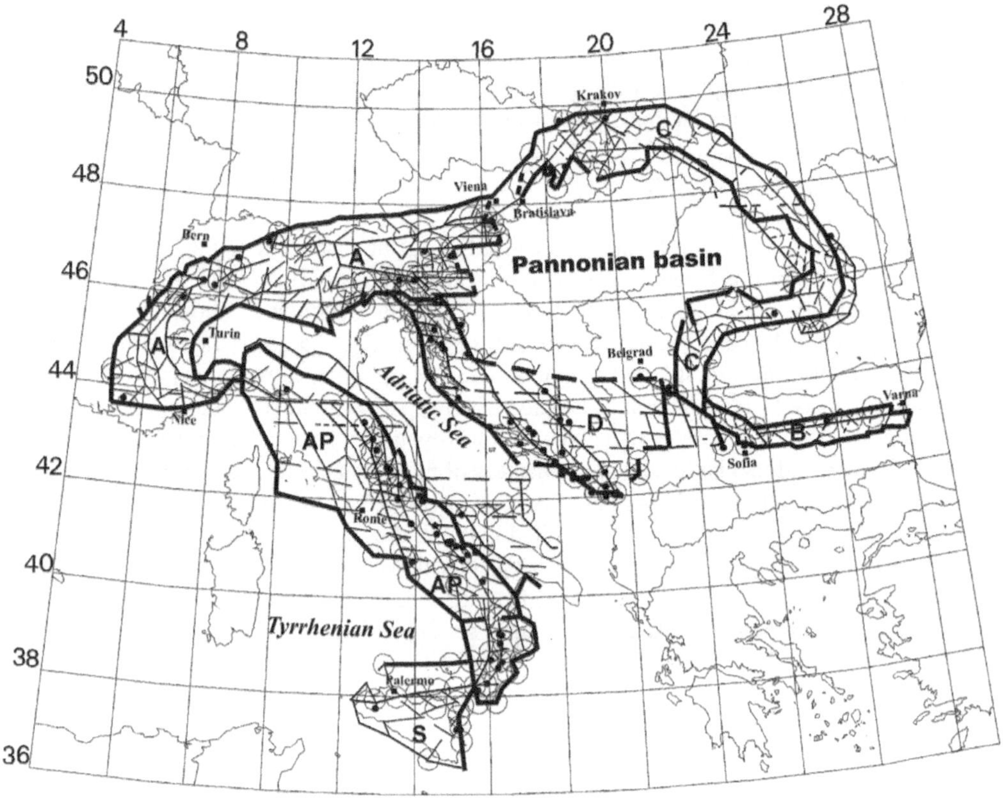

Fig. 1 Apennines and Sicily, Alps, Dinarides, and Carpathians-Balkanides: MSZ map and recognized seismogenic nodes (modified from Gorshkov & Soloviev, 2009b).

Continuous lines mark the longitudinal lineaments, dashed lines depict the transverse lineaments. Bold lines show the lineaments of the 1st rank, middle lines mark the lineaments of the 2nd rank, hairlines depict the lineaments of the 3rd rank. Circles mark recognized D nodes for M6+. Black dots show epicenters of M6+. A-Alps, AP-Apennines, B-Balkanides, C-Carpathians, D-Dinarides, S-Sicily.

Apennines and Sicily. In Apennines and Sicily, 146 nodes have been delineated by MSZ (Gorshkov et al., 2002). Within 43 nodes, earthquakes with M^3 6.0 already took place according to the catalogues used at the moment of recognition (Camassi & Stucchi 1997; Peresan et al., 1997). Nodes hosting these events composed the training set D_0. A total of 81 nodes were recognized as *D* for earthquakes with M6+ (Figure. 1). Forty-four of them were not marked by strong events at the time, but some of them are associated with paleoearthquakes, in particular, the nodes located in southern Italy on the Pollino and Palinuro faults. Since 2002, three events with magnitudes greater than the target have occurred in the region: the 2009 magnitude 6.3, 2016 magnitude 6.2, and 2016 magnitude 6.6 earthquakes. All of them fall into nodes *D* (Gorshkov & Novikova, 2018). The D nodes identified in the region by Gorshkov et al. (2002) were subsequently used to assess seismic hazard employing the NDSHA approach (Peresan et al., 2011).

Alps. In total, MSZ identified 142 lineament intersections, each of which was considered as a node defined as a circle with a radius of 25 km around each lineament intersection point (Gorshkov et al., 2004). 16 earthquakes with M6+ are known in the Alps, including historical ones (Camassi & Stucchi, 1996). The nodes hosting these events were assigned to the training set D_0. Of 142 nodes, 54 and 88 were classified *D* and *N*, respectively (Figure 1). All nodes hosting earthquakes with M6+ were recognized as *D*. At 31 out of the 54 *D* nodes, earthquakes with M6+ are not recorded till present. In the Alps, *D* nodes are concentrated mainly in the Western Alps and in the easternmost and southernmost sectors of the Eastern Alps indicating a higher seismic potential of these areas (Figure 1).

Dinarides. In addition to the Dinarides, this region also includes the northern part of the Serbo-Macedonian Massif. The MSZ outlined 81 nodes (Gorshkov et al., 2004). Earthquake catalogues report 46 earthquakes M6+ in the Dinarides (Herak et al., 1996; Živčič et al., 2000). In many cases their epicenters overlap (Figure 1). 21 nodes have been put in D_0. As a result of the recognition of 81 nodes, 46 and 35 were assigned to classes *D* and *N*, respectively. In 15 of the 46 *D* nodes, earthquakes with M6+ have not been recorded to date. The recognized *D* nodes spatially form several compact clusters elongated along the higher rank lineaments.

Carpathians and Balkanides. Seismogenic nodes for these regions were identified in the work (Gorshkov & Soloviev, 2009b). According to Shebalin et al. (1998), only a few M6+ earthquakes occurred in the Carpathians and the Dinarides. The *in corpore* recognition procedure is not applicable to this region because there are insufficient data to form the reliable D_0 training set. Therefore, the identification of nodes susceptible to M6+ earthquakes was performed using the *D* and *N* node features selected by CORA-3 for the Dinarides (Gorshkov et al., 2004). For this purpose, the number of *D* and *N* features identified for the nodes in the Dinarides was calculated for each node in the Carpathians and the Balkanides, and then the classification of nodes into *D* and *N* was established. The circles in Figure 1 show the *D* nodes susceptible to M6+ events. As a result, a total of 135 nodes in the Carpathians were 45 nodes were assigned to *D*. In the Balkanides, of the 29 nodes 13 ones were defined *D*. In both of these regions, all known M6+ earthquakes are situated at recognized *D* nodes (Figure 1).

Iberia. MSZ outlined 405 nodes in Iberia shown in Figure 2 (Gorshkov et al., 2010). The objective was to recognize the nodes prone to produce shallow earthquakes with $M \geq 5.0$. The largest observed magnitudes and macroseismic intensities in the Iberian Peninsula are mainly related to historical events (Martínez Solares, & Mezcua, 2002; IGN 2007). The events with $M \geq 5.0$ or epicentral intensity $I_0 \geq$ VII are plotted in Figure 2. These data have been used to select sample nodes for the learning step of the recognition.

Iberia is characterized by very heterogeneous topography and the complex tectonic structure. Because of this, the nodes are situated in very different morphological and tectonic environments. This leads to considerable variations in the values of the parameters describing the nodes. Therefore, for the recognition task, the nodes were assembled in four groups. Each group included nodes located within more or less similar morphological and tectonic settings. The nodes situated in mountain environments, low elevated mountains and plateaus, in basins, and those sitting on the continental slope were assigned to different groups.

The recognition has been performed individually for each of the four groups. The recognized nodes, capable of generating earthquakes with $M \geq 5.0$ or $I_0 \geq$ VII, are plotted in Figure 2.

The recognized *D* nodes are mainly concentrated at the periphery of the Iberian Peninsula.

Almost all nodes in the Betics and Pyrenees are identified as *D*, including those where there is no record of a target earthquake. The nodes located in the interior of the peninsula, apart from several

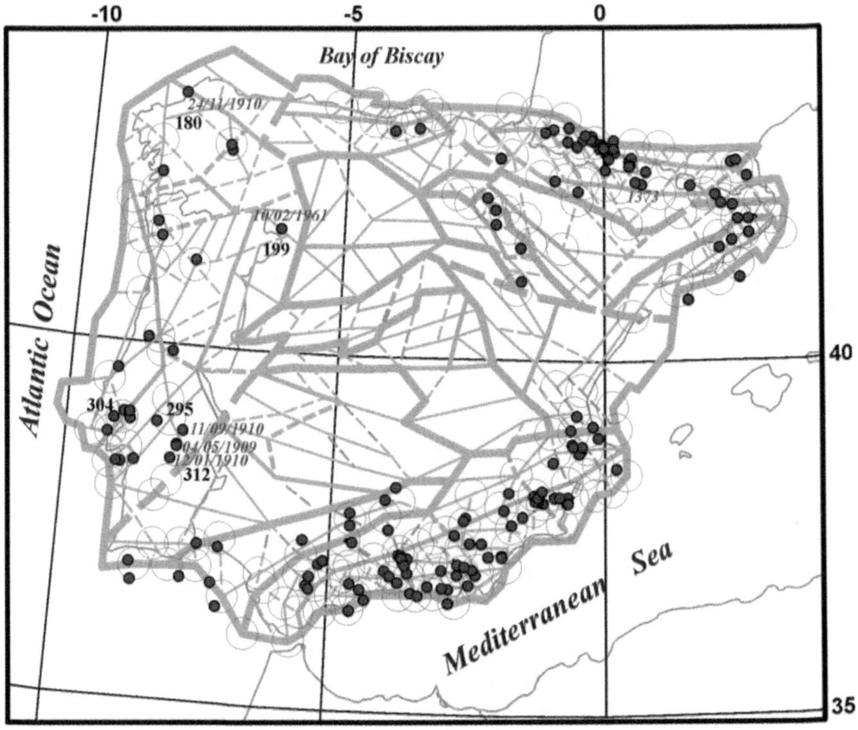

Fig. 2 Nodes prone to earthquakes with $M \geq 5.0$ or $I_0 \geq$ VII in Iberia (after Gorshkov et al., 2010).
Note: Circles denote nodes prone to earthquakes with $M \geq 5.0$ or $I_0 \geq$ VII. Dots are epicentres of the earthquakes with $M \geq 5.0$ or $I_0 \geq$ VII. Lines are the same as in Figure 1.

D nodes in the Iberian Chain, have been recognized as non-capable of $M \geq 5.0$. In topographically low areas, potential nodes were identified in the Portuguese, Lower Tagus, and Ebro basins, as well as inland of the Gulf of Valencia and the Cape St.Vicent. Few potential nodes were also recognized in the Cantabrian Mountains and on the continental slope.

Some of the recognized D nodes give more knowledge about seismic risk affecting special sites like nuclear and water power plants and large metropolitan areas.

French Massif Central. The goal was to identify seismogenic nodes capable of generating I_0^3 VI earthquakes in the French Massif Central, an intraplate region of low-to-moderate seismicity in the Mediterranean. The MSZ map of the region presented in Figure 3 is borrowed from (Gorshkov & Gaudemer, 2019; Gorshkov et al., 2022).

The Massif Central is characterized by low to rare moderate seismicity. Target earthquakes for this region were defined as earthquakes with $I_0 \geq$ VI. To select the sample nodes for the learning stage, the historical and instrumental earthquakes with $I_0 \geq$ V reported by the SisFrance earthquake database (SisFrance, 2008) were considered. The training set D_0 included 17 nodes for the selected events (Figure 3).

Figure 3 shows that all 17 nodes from D_0 hosting the recorded earthquakes with $I_0 \geq$ VI were properly recognized as *D*. Additionally, 28 nodes were assigned to *D*. Most of the *D* nodes are located on the higher - first and second rank - lineaments, i.e., on the boundaries of the larger blocks.

The spatial distribution of *D* nodes over the Massif interiors reveals the remarkable difference between its western and eastern parts separated by the lineament from 5 to 99 that corresponds to the Sillon Houiller fault. In the western nonvolcanic section of the Massif, only three nodes (7, 10, and 37 in Figure 3) have been assigned to *D*, while the overwhelming majority of *D* nodes is situated

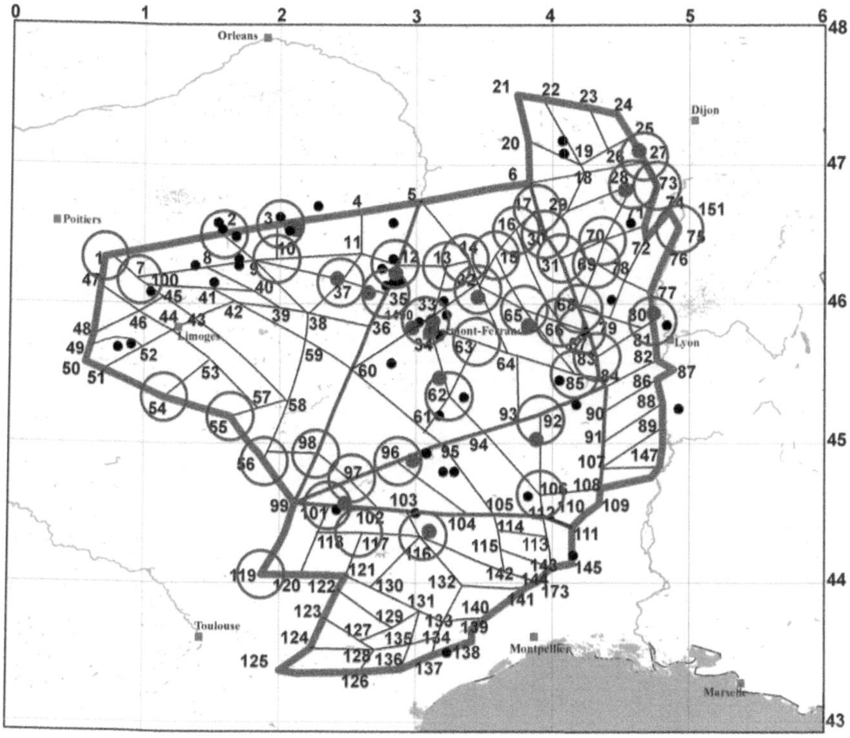

Fig. 3 MSZ map of the French Massif Central and recognized seismogenic nodes (after Gorshkov & Gaudemer, 2018; Gorshkov et al., 2022).

Note: Circles denote D nodes prone to earthquakes with $I_0 \geq$ VI, while the non-circled nodes are classified N. Roman numerals refer to the numbers of the megablocks. Arabic figures mark the numbers of nodes. Black dots mark earthquakes with $V \leq I_0 <$ VI, while the gray ones depict earthquakes with $I_0 \geq$ VI. Lines as in Figure 1.

in the eastern volcanic part of the Massif. Thus, the results of the recognition indicate the higher seismic potential in the eastern part of the Massif with respect to its western domain.

The recognition pinpoints a number of D nodes where the events of the target size have not been recorded up to now. This information on the potential earthquake sources is important for studies devoted to long-term seismic hazard assessment in the French Massif Central.

Po Basin. The nodes have been defined based on the MSZ map shown in Figure 4. In total, 102 intersections of lineaments are identified and each intersection is treated as a node. Formally the node is defined as a circle of radius $R = 20$ km surrounding each point of intersection of lineaments. The Po plain is generally characterized by moderate size earthquakes, with magnitudes not exceeding 5 for the majority of events. Nevertheless there are some important exceptions. In 2012, the south east part of the Po plain, region Emilia-Romagna, was struck by the $M = 6.1$ earthquake, which occurred at the node 54 (Figure 4).

To select the sample nodes for the learning stage, the information on the recorded events with $M \geq \varepsilon 5.0$ is taken into account, considering the available catalogs (Camassi & Stucchi, 1997; Peresan & Panza, 2002; Rovida et al., 2011). Specifically, 71 earthquakes have been considered to select the nodes for the learning stage. Eighteen (18) earthquakes are located within the Po Plain, and 53 are associated with the boundaries between the Po Plain and the surrounding mountains.

Out of the 102 nodes identified within the study region, 60 are located in lowland environments within the Po plain and the remaining 42 sit on the first rank lineaments that divide the Po plain from the surrounding mountain chains. The parameter values of the nodes located in the Po basin and the nodes at the mountain/plain contact differ greatly. Therefore, recognition was performed separately for these two groups of nodes, using the Cora-3 recognition algorithm for each of them.

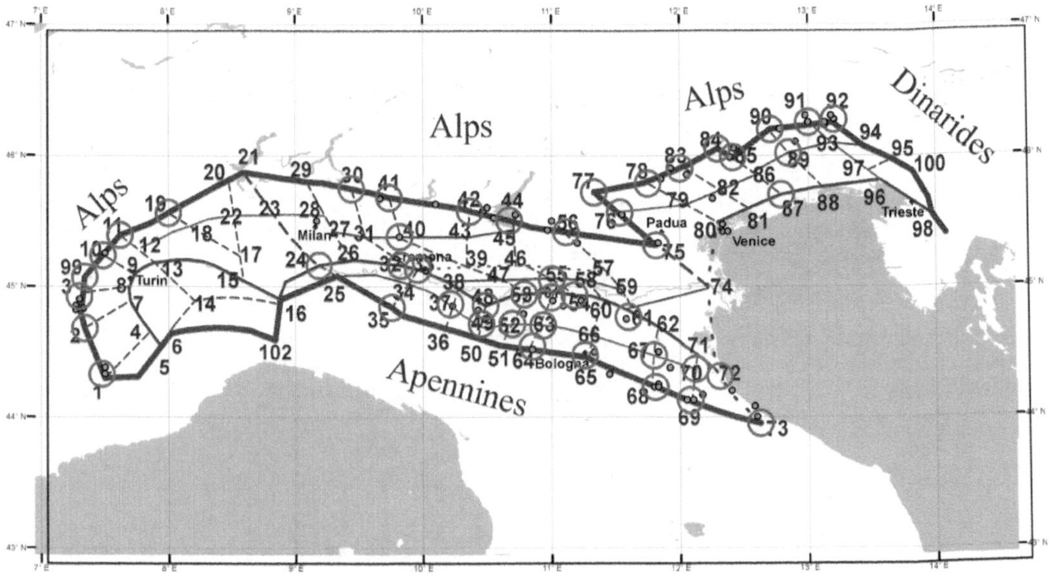

Fig. 4 MSZ map of the Po plain and *D* nodes for M5+.
Note: Lines are the same as in Fig. X1. Dots mark epicenters M5+. Red circles show recognized nodes prone to M5+ earthquakes. The figures indicate the node numbers.

The recognized *D* nodes, identified by PR both within the Po plain and on the first-rank lineaments separating the plain from the Alps, Dinarides, and Apennines, are shown in Figure 4. Nodes 28 and 80, 82 (Figure 4), which host ancient earthquakes whose location and magnitude are poorly defined, were misclassified as N.

The Corsica-Sardinia Lithospheric Block. The region is traditionally viewed as a region of notoriously low seismicity. Earthquake catalogs report only three moderate earthquakes in the range of magnitudes 5.0– 5.1. In such a situation, the use of PR is impossible due to the insufficient number of target earthquakes for the training stage. Therefore, an attempt was made to divide the Corsica-Sardinia nodes into seismogenic nodes (*D*) prone to M5+ and non-seismogenic ones (*N*), exploiting the existing characteristic traits (criteria of seismicity) of *D* and *N* nodes selected by means of the algorithm Cora-3 in Iberia (Gorshkov et al., 2010) and in the Western Alps (Cisternas et al. 1985).

MSZ identified 81 nodes in the region (Figure 5). The following voting procedure was used to classify them into *D* and *N*. The numbers of both Iberian and Alpine *D*-features and *N*-features were determined for each node in the Corsica-Sardinia region. The difference (Δ) between the numbers of *D*-features and *N*-features was then calculated for each node in the Corsica-Sardinia region. The classification of nodes obtained using the Iberian seismicity criteria more adequately identified the *D* nodes compared to the nodes defined by the Alpine criteria. Of the 81 nodes identified in the Corsica-Sardinia region, 24 for which $\Delta \geq 1$ were classified as seismogenic nodes for M5+. The identified D-nodes are plotted in Figure 5.

Figure 5 shows that three nodes hosting M5+ earthquakes are identified as *D*. Most of the nodes hosting smaller events with $4.0 \leq M < 5.0$ were also assigned to *D*. Information on these nodes was incorporated into the NDSHA for seismic hazard assessment in the Corsica-Sardinia block (Brandmayr et al., 2022).

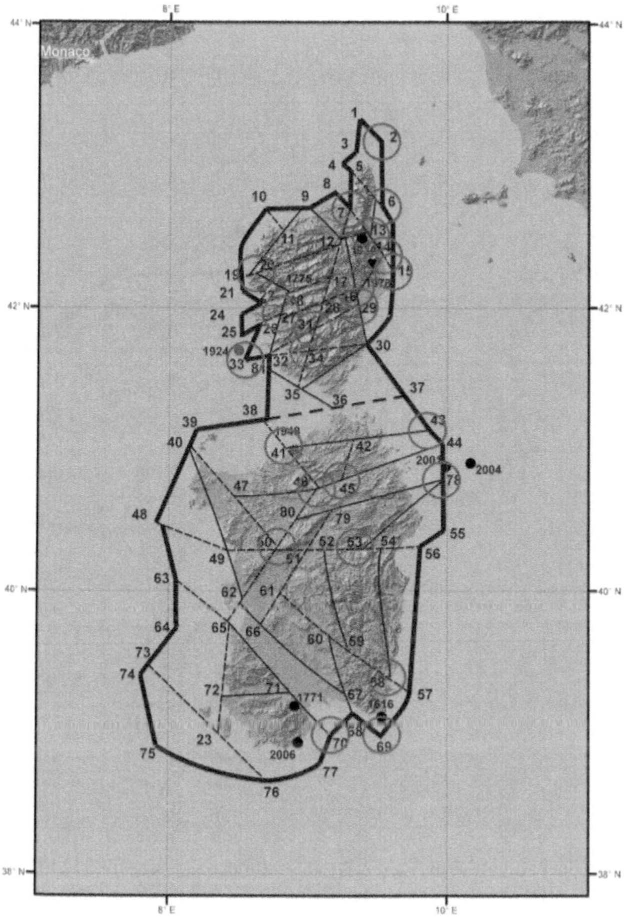

Fig. 5 MSZ map of the Corsica-Sardinia block and seismogenic nodes identified using Iberian criteria of seismicity (Gorshkov et al., 2021a).

Note: Lines are the same as in Fig. X.1. Circles are *D* nodes prone to $M \geq 5.0$ events; non-circled intersections of the lineaments are **N** nodes. Magenta dots denote epicenters of earthquakes with $M \geq 5.0$, while black ones mark epicenters of earthquakes with $4.0 \leq M < 5.0$.

Hellenides. The high level of seismicity in this region is associated with a complex geodynamic environment caused by the interaction of the Eurasian, African, Adriatic, Aegean, and Anatolian continental plates. According to (Papazachos et al., 2000), 19 events with $M \geq 7.0$ occurred there since 1444. The maximum magnitudes were observed during the earthquake of 1867 ($Mw = 7.4$) on the island of Kefalonia in the Ionian Sea and during the Peloponnesian earthquake of 1886 ($Mw = 7.3$) in the southwestern Greece (Papazachos et al., 2000).

MSZ delineated 139 nodes shown in Figure 6. The training set D_0 for class *D* comprised 19 nodes hosting earthquakes with M_w7+. Fifty-six (56) nodes marked by earthquakes with $6.0 \leq M_w < 6.9$ were attributed to the testing subset X whose objects are correlated to the events with magnitudes close to the target one. These objects were not used by the Cora-3 for determining the decision rule at the training stage; however, they were divided into classes *D* and *N* at the classification stage. The remaining 64 nodes composed the training set N_0 for class *N*.

The recognition results are displayed in Figure 6. All the nodes hosting the earthquakes with M_w7+ are attributed by the Cora-3 to the class *D*. Additionally, eleven nodes from the training set

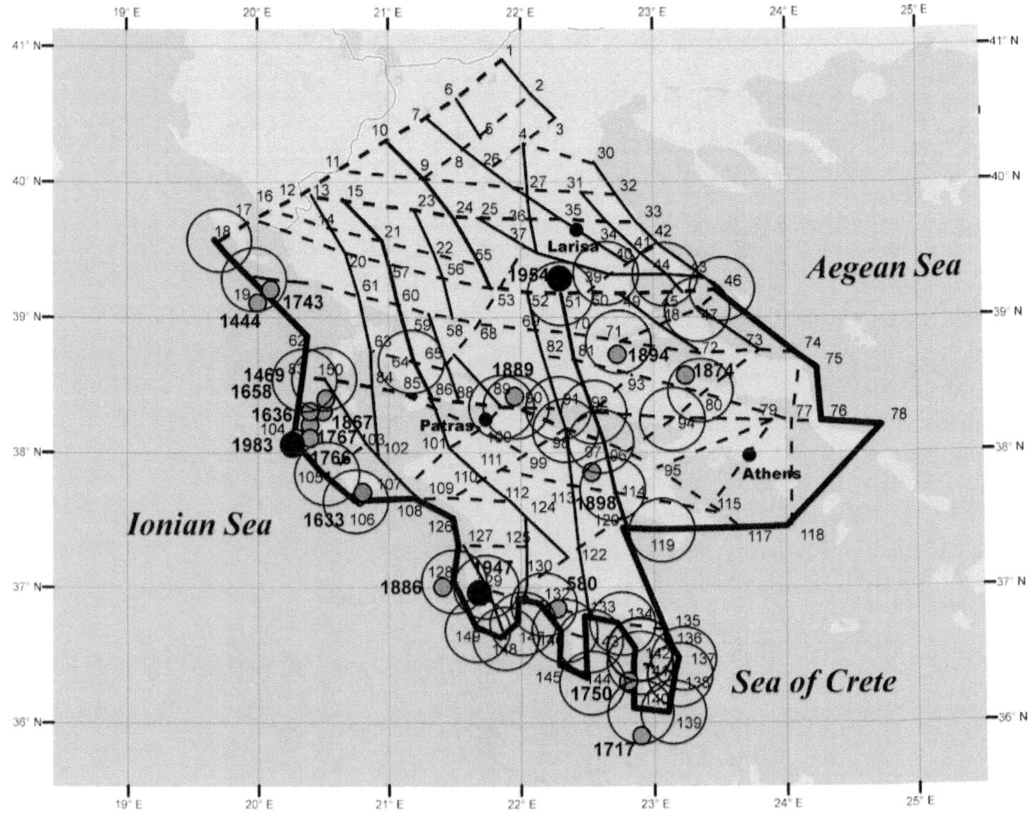

Fig. 6 MSZ map of the Hellenides and recognized seismogenic nodes for M7+ (Gorshkov et al., 2020). *Note:* Circles mark seismogenic nodes *D* for M 7+. Lines are the same as in Figure 1.

N_0 and 13 nodes from the testing set X which contains nodes marked by events with $6.0 \leq M_w < 6.9$ are also referred to class *D*.

The recognized *D* nodes outline several highly seismic zones in most of which the earthquakes with M_w7+ have already occurred. In particular, this is true for the clusters of *D* nodes recognized on the continental slope of the Ionian Sea and southern coast of the Peloponnese Peninsula. Some potential nodes were identified in areas where the earthquakes with M_w7+ have not been observed before. *D* nodes were recognized on the lineaments bordering the rift zone of the Gulf of Corinth that is characterized by the active tectonics (e.g., Briole et al., 2000). Another group of *D* nodes (Nos. 44, 46, and 47) was recognized in the region of Euboea Island where the western branch of the North Anatolian Fault approaches the Greek coast.

MSZ map (Figure 6) and the PR approach have been also exploited for identifying potential tsunamigenic sources (Novikova & Gorshkov, 2022).

Bulgarian Region. Over the centuries, the Bulgarian region has been exposed to a high seismic activity (Grigorova et al., 1978; Solakov et al., 2020). 178 nodes were identified by MSZ for the territory of Bulgaria and neighboring regions shown in Figure 7 (Gorshkov et al., 2024b).

At the training stage, all 178 nodes were assigned to three training sets. The subset D_0 consisted of 16 nodes, the subset N_0 included 142 nodes and the subset X was formed by 20 nodes.

As a result of recognition, 59 nodes out of 178 identified by MSZ were assigned to class *D*. These nodes are shown as circles in Figure 7.

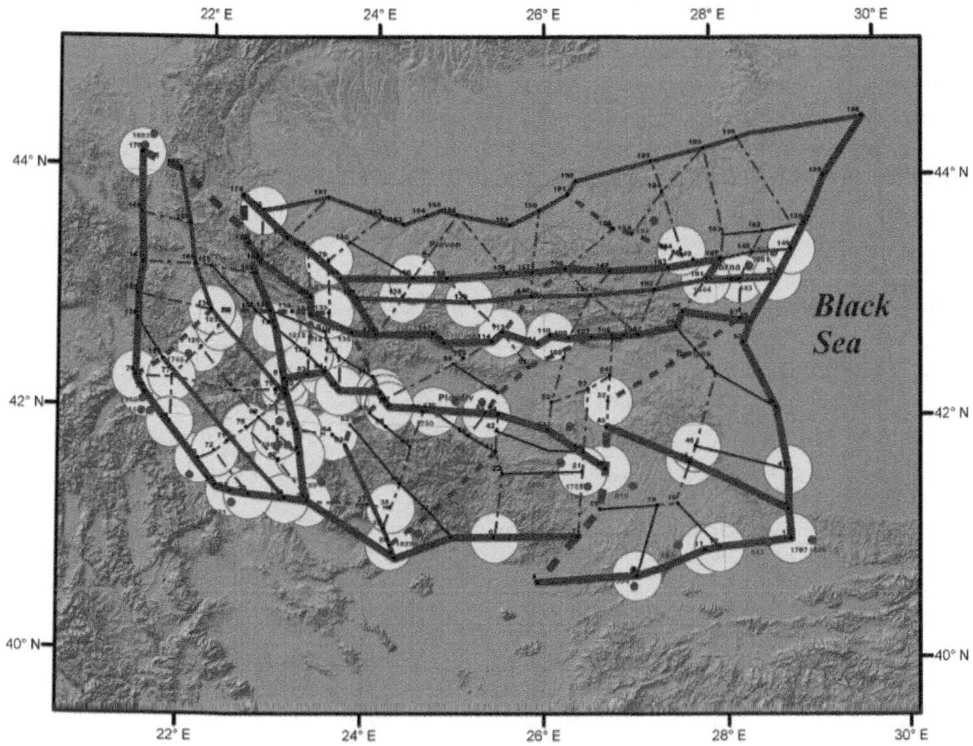

Fig. 7 MSZ map of the Bulgarian region and seismogenic nodes, capable of generating events with M6+. *Note:* *D* nodes are evidenced by circles, while all non-circled nodes are *N*. Lines as in Figure 1 (Gorshkov et al., 2024b).

The most of the identified earthquake-prone nodes are located on the lineaments of the highest (Ist and IInd) ranks, which correspond to the prominent faults and divide the largest blocks of the Earth's crust in the Bulgarian region. The majority of *D* nodes are located in the mountainous areas, i.e., in Rila, Pirin, and Rhodope massifs, as well as in Stara Planina Mountain range. In general, the spatial distribution of seismogenic nodes corresponds to the dispersion of the observed seismicity in the Bulgarian region. Most of the *D* nodes are identified in the western part of Bulgaria, where the observed seismicity is concentrated. The obtained results also confirm the high seismic potential of Shabla region (northeast Bulgaria). Nodes 101 and 100 are also identified as seismogenic in the Shabla area, including nodes 99 and 149 located in the Black Sea (see Figure 7). Potential nodes 46 and 15 were identified on the boundary between Strandzha area and the Thrace basin, where earthquakes M6+ have not yet been observed. In the most northwestern part of the region, node 170, in which two historical events of 1739 and 1895 took place, is not recognized as *D* type. This node is located in the Serbo-Macedonian massif, which differs in relief and structure from those in the Bulgarian region. This node is recognized as capable of generating earthquakes of M6+ in the similar study for the Dinarides (Gorshkov et al., 2004). Figure 7 displays node 170 as *D*.

The performed recognition technique identifies a number of *D* nodes where events M6+ have not been documented up to present. This is a significant amount of additional information in order to improve seismogenic source models and contribute to long-term seismic hazard assessment in the Bulgarian region. The information on the possible locations of strong earthquakes provided by Gorshkov et al. (2024) can be directly incorporated in the neo-deterministic procedure for seismic hazard assessment, thus, filling in possible gaps in known seismicity in the Bulgarian territory.

4. Concluding Remarks

Long-term application of PREPA in Southern Europe shows that it is sufficiently effective in identifying seismically hazardous areas for target events of different size. The results presented in this chapter can be considered positive, since in each of the studied regions the nodes hosting target events were recognized almost without errors. Several cases where nodes, mostly marked by historical target events, were misclassified as non-seismogenic can be explained by the low accuracy of magnitude determination of historical earthquakes.

The number of potential nodes that have not yet been marked with target events is relatively small in each region. This is consistent with the level of seismicity in the considered regions.

One of the main advantages of the PREPA is that it does not require prior knowledge of the regional seismicity of the studied region since the MSZ is based on the analysis of the available topographic, tectonic, and geological data, whereas paleoseismological studies need information about the past seismic activity. The methodology appears to be particularly useful in areas where information on historical seismicity is scarce, such as in the Corsica-Sardinia region discussed in this chapter. Thus, special attention should be paid to seismogenic nodes that are not yet marked by target earthquakes.

Therefore, the seismogenic nodes defined by PREPA provide first-order systematic information that may significantly contribute to reliable seismic hazard assessment. Information on potential earthquake source locations provided by PREPA can be directly incorporated into the NDSHA process, thereby filling possible gaps in seismic history (Panza et al., 2012; Panza & Bela, 2020). Specifically, the importance of seismogenic nodes in NDSHA applications has been demonstrated for the Italian region (Peresan et al., 2011; Zuccolo et al., 2011), for northeast Egypt (Gorshkov et al., 2019), and for the Corsica-Sardinia block (Brandmayr et al., 2022).

References

Aki, K. (1979). Characterization of barriers on an earthquake fault, *J. Geophys. Res.*, *84*, 6140– 6148.

Barenblatt, G.I. (1993). Micromechanics of fracture. In: E.R. Bodner, J. Singer, A. Solan and Z. Hashin, (Eds.), *Theoretical and Applied Mechanics* (pp. 25–52). Elsevier, Amsterdam.

Bongard, M.M. (1967). *Problema uznavaniya* (*Problem of Recognition*). Moscow: Nauka, (in Russian).

Boschi, E., Mantovani, E. and Morelli, A. (1993). *Recent Evolution and Seismicity of the Mediterranean Region*. Kluwer Academic Publishers, Dordrecht, Netherlands.

Brandmayr, E., Vaccari, F. and Panza, G.F. (2022). Neo-deterministic seismic hazard assessment of Corsica-Sardinia block. *Rend. Fis. Acc. Lincei*, *33*, 47–56. https://doi.org/10.1007/s12210-021-01033-w.

Briole, P., Rigo, A., Lyon-Caen, H., et al. (2000).. Active deformation of the Corinth rift, Greece: Results from repeated Global Positioning System surveys between 1990 and 1995, *J. Geophys. Res.*, *105*(11), 25606–25625. https://doi.org/10.1029/2000JB900148.

Camassi, R. and Stucchi, M. (1997). NT 4.1, *Un catalogo parametrico di terremoti di area Italiana al di Sopra della soglia di danno, Open Data File*. Consiglio Nazionale delle Ricerche GNDT.

Cisternas, A., Godefroy, P., Gvishiani, A., et al. (1985). A Dual Approach to Recognition of Earthquake Prone-Areas in the Western Alps. *Annales Geophysicae*, *3*(2), 249–270.

D'Amico, V., Albarello, D. and Mantovani, E. (1999). A distribution-Free Analysis of Magnitude-Intensity Relationships: An Application to the Mediterranean Region. *Physics and Chemistry of Earth*, *24*, 517–521.

Gelfand, I.M., Guberman, Sh.A., Izvekova, M., et al. (1972). Criteria of high seismicity determined by pattern recognition. *Tectonophysics*, *13*(1–4), 415–422.

Gelfand, I.M., Guberman, Sh.A., Zhidkov, M.P., et al. (1974). Recognition of earthquake-prone areas. II. Four regions in Asia Minor and South East Europe, in *Vychislitel'naya seis_mologiya, Vyp. 7: Mashinnyi analiz tsifrovykh seismicheskikh dannykh* (*Computational Seismology* (Vol. 7): *Computerized Analysis of Digital Seismic DataI* (pp. 3–40)). Keilis-Borok, V.I., Ed., Moscow: Nauka.

Giardini, D., Grünthal, G., Shedlock, K.M. and Zhang, P. (2003). The GSHAP global seismic hazard map. In: W. Lee, H. Kanamori, P. Jennings and C. Kisslinger (Eds.), *Handbook of Earthquake and Engineering Seismology* (pp. 1233–1239). Amsterdam, Netherlands: Academic Press.

Gorshkov, A., Kuznetsov, I., Panza, G. and Soloviev, A. (2000). Identification of future earthquake sources in the Carpatho-Balkan orogenic belt using morphostructural criteria. *PAGEOPH, 157*, 79–95.

Gorshkov, A.I., Panza, G.F., Soloviev, A.A. and Aoudia, A. (2002). Morphostructural zonation and preliminary recognition of seismogenic nodes around the Adria margin in peninsular Italy and Sicily, *JSEE*, Spring, *4*(1), 1–24.

Gorshkov, A.I., Kossobokov, V. and Soloviev, AA. (2003). Recognition of earthquake prone areas. In: Keilis-Bork, V., Soloviev, A.A. (Eds.), *Nonlinear Dynamics of the Lithosphere and Earthquake Prediction* (pp. 235–320). Springer, Heidelberg.

Gorshkov, A.I., Panza, G.F., Soloviev, A.A. and Aoudia, A. (2004). Identification of seismogenic nodes in the Alps and Dinarides. *Bolletino della Societa Geologica Italiana, 123*, 3–18.

Gorshkov, A.I., Panza, G.F., Soloviev, A.A., et al. (2009a). Delineation of the geometry of the nodes in the Alps-Dinarides hinge zone and recognition of seismogenic nodes ($M \geq 6.0$). Terra Nova. doi: 10.1111/j.1365-3121.2009.00879.x

Gorshkov, A.I. and Soloviev, A.A. (2009b). Recognition of Possible Locations of Future M ≥ 6.0 Earthquakes: The Mediterranean Mountain Belts 2009. *Journal of Volcanology and Seismology*, *3*(3), 210–219. doi:10.1134/S0742046309030075.

Gorshkov, A.I., Soloviev, A.A., Jiménez, M.J., et al. (2010). Recognition of earthquake-prone areas (M^3 5.0) in the Iberian Peninsula. *Rendiconti Lincei –Scienze Fisiche e Naturali, 21*(2), 131–162. doi:10.1007/s12210-010-0075-3.

Gorshkov, A. and Novikova, O. (2018). Estimating the validity of the recognition results of earthquake prone areas using the ArcMap. *Acta Geophysica, 66*(5), 843–853. doi: 10.1007/s11600-018-0177.

Gorshkov, A. and Gaudemer, Y. (2019). Seismogenic nodes defined with pattern recognition in the French Massif Central. *Journal of Iberian Geology, 45*(1), 63–72. https://doi.org/10.1007/s41513-018-0087-x.

Gorshkov, A.I., Hassan, H.M. and Novikova, O.V. (2019). Seismogenic Nodes (M≥5.0) in Northeast Egypt and Implications for Seismic Hazard Assessment. *Pure Appl. Geophys., 176*, 593–610. https://doi.org/10.1007/s00024-018-2012-9.

Gorshkov, A.I., Novikova, O.V. and Gaudemer, Y. (2020). Strong (M ≥ 7.0) Earthquake-Prone Areas in Hellenides, Greece. *Izvestiya, Physics of the Solid Earth, 56*(1), 45–52. doi: 10.1134/S1069351320010036.

Gorshkov, A., Panza, G.F., Soloviev, A., et al. (2021a). On the seismic potential of the Corsica–Sardinia block. *Rend. Fis. Acc. Lincei, 32*, 715–728. https://doi.org/10.1007/s12210-021-01014-z

Gorshkov, A. and Soloviev, A. (2021b). Morphostructural zoning for identifying earthquake-prone areas. In: Panza, G.F., Kossobokov, V.G., Laor, E. and De Vivo, B. (Eds.), *Earthquakes and Sustainable Infrastructure: Neodeterministic (NDSHA) Approach Guarantees Prevention Rather than Cure* (pp. 135–149). Elsevier. ISBN: 978-0-12-823503-4.

Gorshkov, A.I., Hassan, H.M., Mandal, P. and Novikova, O. (2022). Identifying Potential Earthquake Sources in Continental Environments. *Surv. Geophys.* https://doi.org/10.1007/s10712-021-09683-z.

Gorshkov, A.I., Kossobokov, V.G. and Novikova, O.V. (2024a). Prediction Results for the Strongest Earthquakes of February 6, 2023 in Southern Turkey. *Izvestiya, Physics of the Solid Earth, 60*(3), 339–345. doi: 10.1134/S1069351324700423.

Gorshkov, A.I., Novikova, O.V., Dimitrova, S.Y., et al. (2024b). Potential Locations of Strong Earthquakes in Bulgaria and the Neighbouring Regions. *International Journal of Geophysics, 11*, 1–13. doi: 10.1155/2024/8103337.

Grigorova, E., Christoskov, L., Sokerova, L., et al. (1978). Catalogue of earthquakes in Bulgaria and the nearby territories during the period 1st cent. BC–1977. *Bulg. Acad. Sci.*

Gvishiani, A., Gorshkov, A., Rantsman, E., et al. (1988). *Identification of Earthquake-prone-areas in the Regions of Moderate Seismicity*. Nauka, Moscow, 175 pp. (in Russian).

Herak, M., Herak, D. and Markusic, S. (1996). Revision of the earthquake catalogue and seismicity of Croatia, 1908–1992. *Terra Nova, 8*, 86–94.

IGN (Instituto Geográfico Nacional) (2007). *Catálogo Sísmico* (Earthquake catalogue) (in Spanish).

King, G. (1986). Speculations on the geometry of the initiation: A termination process of earthquake rupture and its relation to morphology and geological structure. *Pure Appl. Geophys., 124*, 567–583.

Kossobokov, V.G. and Nekrasova, A.K. (2012). Global seismic hazard assessment Program maps are erroneous. *Seism. Instrum., 48* (2), 162–170. doi:10.3103/ s0747923912020065.

Kossobokov, V. G. and Soloviev, A.A. (2018). Pattern recognition in problems of seismic hazard assessment. *Chebyshevskii Sbornik, 19*(4), 55–90. doi: 10.22405/2226-8383-2018-19-4-55-90.

Martinelli, G., Pierotti, L., Facca, G. and Gherardi, F. (2023). Geofluids as a possible unconventional tool for seismic hazard assessment. *Front. Earth Sci., 11*, 1286817. doi: 10.3389/feart.2023.1286817.

Martínez-Solares, J.M. and Mezcua, J. (2002). Catálogo sísmico de la Península Ibérica (880a.C.–1900) (Earthquake catalogue of the Iberian peninsula (AD 880–1900), *18*, 253 pp. Ministerio de Fomento, I.G.N., Madrid Monografia. (in Spanish).

Novikova, O.V. and Gorshkov, A.I. (2022). Local tsunamigenic sources in Greece, identified by pattern recognition. *Nat. Hazards, 113*, 1335–1348. https://doi.org/10.1007/s11069-022-05349-0.

Panza, G.F., Vaccari, F., Costa, G., Suhadolc, P. and Faeh, D. (1996, Aug.). Seismic input modeling for zoning and microzoning. *Earthquake Spectra, 12*(3), 529–566. https://doi.org/10. 1193/1.1585896.

Panza, G.F., Romanelli, F. and Vaccari, F. (2001). Seismic wave propagation in laterally heterogeneous anelastic media: theory and applications to seismic zonation. *Adv. Geophys.*, *43*, 1–95.

Panza, G.F., La Mura, C., Peresan, A., Romanelli, F. and Vaccari, F. (2012). Seismic hazard scenarios as preventive tools for a disaster resilient society. *Adv. Geophys. Chpt.*, *3*(53), 93–165. https://doi.org/10.1016/B978-0-12-380938-4.00003-3.

Panza, G.F. and Bela, J. (2020). NDSHA: A new paradigm for reliable seismic hazard assessment. *Engineering Geology*, *275*, 105403.

Papazachos, B.C., Comninakis, P.E., Karakaisis, G.F., et al. (2000). *A Catalogue of Earthquakes in Greece and Surrounding Area for the Period 550 BC–1999.* http://geohazards.cr.usgs.gov/iaspei/europe/greece/the/ catalog.htm.

Peresan, A., Zuccolo, E., Vaccari, F., Gorshkov, A. and Panza, G. (2011). Neo-Deterministic Seismic Hazard and Pattern Recognition Techniques: Time-Dependent Scenarios for North-Eastern Italy. *Pure Appl. Geophys.*, *168*, 583–607. doi:10.1007/s00024-010-0166-1.

Peresan, A., Costa, G. and Vaccari, F. (1997, April). CCI1996: *The Current Catalogue of Italy*. Internal Report IC/IR/97/9. Miramare-Trieste.

Peresan, A. and Panza, G.F. (2002). *UCI2001: The Updated Catalog of Italy*. The Abdus Salam International Centre for Theoretical Physics. ICTP, Miramare, Trieste. Italy. Internal report IC/IR/2002/3.

Peresan, A., Gorshkov, A., Soloviev, A. and Panza, G.F. (2015). The contribution of pattern recognition of seismic and morphostructural data to seismic hazard assessment. *Bollettino di Geofisica Teorica ed Applicata*, *56*(2), 295–328. doi: 10.4430/bgta0141.

Rantsman, E.Ya. (1979). *Morphostructure of Mountain Regions and Sites of Earthquakes*. Moscow: Nauka (in Russian).

Rovida, R. Camassi, P. Gasperini and M. Stucchi (Eds.) (2011). *CPTI11, the 2011 Version of the Parametric Catalogue of Italian Earthquakes*. Milano, Bologna, http://emidius.mi.ingv.it/CPTI. doi: 10.6092/INGV.IT-CPTI11.

Rugarli, P., Vaccari, F. and Panza, G. (2019). Seismogenic nodes as a viable alternative to seismogenic zones and observed seismicity for the definition of seismic hazard at regional scale. *Vietnam Journal of Earth Sciences*. https://doi.org/10.15625/0866-7187/41/4/14233.

Sibson, R.H. (1988). Earthquake faulting, induced fluid flow, and fault-hosted gold-quartz mineralization. In: Bartholomew, M.J., Hyndman, D.W., Mogk, D.W., and Mason, R. (Eds.), *Basement Tectonics: Characterization and Comparison of Ancient and Mesozoic Continental Margins* (pp. 603–614). *Proc. 8th Int. Conf. Basement Tectonics*. Butte, MT.

Shebalin, N., Leydecker, G., Mokrushina, N. et al. (1998). *Earthquake catalogue for Central and Southeastern Europe 342 BC–AD 1990.* European Commission, Report No. ETNU CT 93-0087, Brussels.

SisFrance (2016). *Base données de sismicité historique française*. BRGM, EDF. IRSN . http://www.sisfrance.net.

Solakov, D. Simeonova, S., Raykova, P. and Aleksandrova, I. (2020). *Catalogue of the Earthquakes in Bulgaria and Surroundings since 1981*. National Institute of Geophysics, Geodesy and Geography, Bulgarian Academy of Sciences. doi: 10.34975/CTLG-2020.V.1.

Soloviev, A.A., Gvishiani, A.D., Gorshkov, A.I. et al. (2014). Recognition of Earthquake-Prone Areas: Methodology and Analysis of the Results. *Izvestiya. Physics of the Solid Earth*, *50*(2), 151–168. DOI: 10.1134/S1069351314020116.

Talwani, P. (1988). The intersection model for intraplate earthquakes. *Seismological Research Letters*, *59*, 305–310.

Wells, D.L. and Coppersmith, K.J. (1994). New empirical relationships among magnitude, rupture length, rupture width, and surface displacement. *Bull. Seism. Soc. Am.*, *84*, 974–1002.

Živčič M., Suhadolc P. and Vaccari F. (2000). Seismic zoning of Slovenia based on deterministic hazard computations. PAGEOPH., 157, 171–184.

Zuccolo, E., Vaccari, F., Peresan, A. and Panza, G.F. (2011). Neo-deterministic (NDSHA) and probabilistic seismic hazard (PSHA) assessments: A comparison over the Italian territory. *Pure Appl. Geophys.*, *168*, 69–83. doi:10.1007/s00024-010-0151-8.

CHAPTER

13

Deterministic Seismic Hazard Computations (Hybrid Method) Applied for Two Cities in the Territory of Hungary

Katalin Gribovszki[1,2], Franco Vaccari[3], Sándor Szalai[1], Lili Czirok[4], Péter Varga[1] and Giuliano F. Panza[5,6,7,8]

Between 2002 and 2008, K. Gribovszki had the opportunity to stay in Trieste for several months and to participate in the research of the Seismological Group of the Department of Earth Sciences, University of Trieste, using data taken with her from Hungary. In this summary paper, we briefly report the results of the deterministic seismic hazard computations carried out during these years.

The primary objective of our work in Trieste was to prepare seismic hazard and risk maps for the cities of Debrecen and Budapest, capital of Hungary. In order to achieve this goal, synthetic seismograms were computed by the so-called hybrid technique by using different earthquake scenarios along several profiles.

The response spectra ratio values (RSR) along the investigated 2D profiles (at the predefined set of points) versus frequency and the epicentral distance have been automatically provided by the hybrid technique for the vertical, transversal, and radial components. Similarly to the RSR, curves of ratios of peak ground accelerations (PGA(2D)/PGA(1D)) were automatically provided as well at the predefined set of points at the surface along the investigated profiles. Besides these maps and curves, unique earthquake hazard curves have been produced too, such as Effective Peak Acceleration (EPA) curves for the inner town of the city of Debrecen. EPA curves made it possible to compare our results using our scenario No. 1 to the macroseismological intensity map of a devastating historical earthquake in 1834, thus it was able to validate our computations.

As the results of the computations peaked, ground acceleration grid maps for the territory of the downtown of Budapest and Debrecen for the three different components came into existence as well. Furthermore, for both cities, special seismic risk maps have been prepared. These special seismic risk maps were created based on the combination between the maximal amplitude frequencies of

[1] Institute of Earth Physics and Space Science, Hungarian Research Network.
[2] Institute of Geomatics and Civil Engineering, Faculty of Forestry, University of Sopron.
[3] Dipartimento di Matematica, Informatica e Geoscienze, Università di Trieste, Trieste, Italy.
[4] Hungarian Meteorological Service.
[5] Accademia Nazionale dei Lincei, Rome, Italy.
[6] Earthquake Administration, Beijing, China.
[7] International Seismic Safety Organization (ISSO), Arsita, Italy.
[8] Beijing University of Civil Engineering and Architecture (BUCEA), Beijing, China.
* Corresponding author: gribovszki.katalin@epss.hun-ren.hu

DOI: 10.1201/9781003684121-15

spectral acceleration of synthetic seismograms and the building's eigenfrequencies at every 0,1 km^2 or even less areas at the downtown of Budapest and Debrecen.

1. Introduction and Objectives

Deterministic seismic hazard assessment (DSHA) computations have been done for the inner town of two Hungarian cities: Debrecen and Budapest. Synthetic seismograms were computed by the so-called "hybrid technique".

The computations for the area of Debrecen were motivated by the fact that Debrecen, the second most populated city in Hungary (its population is about 200, 000), is located in the eastern part of the country and is close to the Érmellék seismo-active region, where one of the largest earthquakes ever occurred in Hungary in the first half of the 19th century (Szeidovitz et al., 2002). The seismically active fault system, the Gálospetri graben, together with its continuation in Hungary: the Mobile zone, principally determines the seismic hazard of Debrecen.

In our initial computations in 2001 (Gribovszki & Vaccari, 2004), 2-scenarios (2 different computations along 2 different profiles, scenarios No. 1 and No. 2, Figure 1) have been used, assuming that the earthquakes occurred in the Gálospetri graben (Gálospetri village), which is part of the main quake zone of the Érmellék region, or in the Mobile zone (Hosszúpályi village).

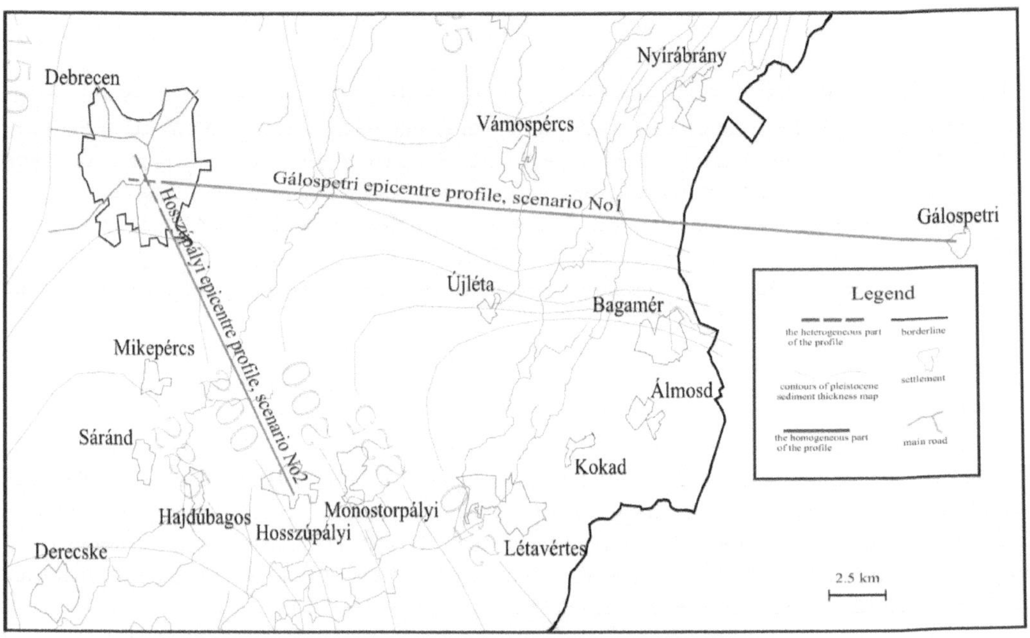

Fig. 1 Profiles between the city of Debrecen and the assumed epicentres in Gálospetri and Hosszúpályi villages, scenarios No. 1 and 2 (Figure 2 in Gribovszki and Vaccari, 2004).

We had the opportunity to compare our computations (scenario No. 1) with the macroseismic effects observed at the site of Debrecen in 1834, when the most devastating earthquake occurred in Érmellék seismo-active region (Zsíros, 1985).

One-and-a-half years later in 2003, as the continuation of our initial computations, a detailed one, was carried out using the same scenario as scenario No. 2 (Gribovszki & Panza, 2004). In that case, the initial computation using scenario No. 2 has been completed with developed input data (using more borehole data than before) at the 2D laterally heterogeneous profiles (Figure 2). Furthermore the computations were extended to the whole territory of Debrecen along 11 different laterally heterogeneous profiles, in order to prepare the synthetic scaled accelerograms along the profiles and the curves and maps derived from them (the ratio of maximum acceleration values: PGA(2D)/PGA(1D), response spectra ratio: RSR). These synthetic scaled accelerograms could be expected for the entire territory of Debrecen in the case of the highest earthquake hazard (assuming the nearest epicentre in the Mobil zone fault system). Another aim of the computations was to generate a special seismic risk map that, beyond the synthetic seismograms, takes into account the number of floors of the city's building in Debrecen as well (Gribovszki & Panza, 2004).

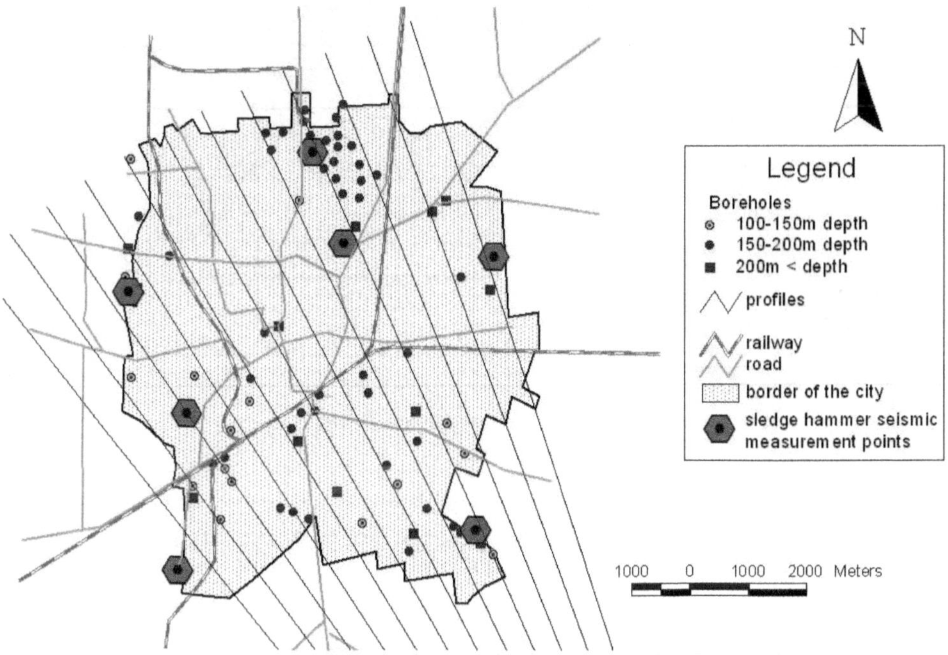

Fig. 2 Boreholes and 2D profiles inside the city of Debrecen (Figure 3 in Gribovszki and Panza, 2004).

In 2008, DSHA computations were performed by the hybrid technique along 4 different laterally heterogeneous profiles across the downtown of Budapest. Budapest, the capital city of Hungary, considering its population, cultural and economic values deserves accentuated attention from the point of view of natural hazards. The computations used the same seismic source (scenario No. 3) but different laterally heterogeneous profiles. The parameters of the seismic source were adopted from the parameters of the well-known 1956 Dunaharaszti earthquake. This Dunaharaszti earthquake was the strongest event (M5.6) that had ever occurred in the vicinity of Budapest, only 15 km far from the inner town of the capital (Figure 3).

Earlier than our computations, simpler bedrock based deterministic seismic hazard assessment computations were carried out with the use of modal summation for the whole territory of Hungary (Bus et al., 2000). In the case of the area of Budapest according to Bus et al. (2000) the designed ground acceleration values between 0.02*–0.04g can be expected for the bedrock. The same values were computed for Debrecen as well.

Szeidovitz et al. (2001) also computed the designed ground acceleration values for Budapest by modal summation method taking into account the thick loose sediments at the surface as well applying two different scenarios of the Dunaharaszti earthquake that occurred in 1956. They found 7° intensity on the MKS-64 scale (4.9 cm/sec horizontal component) for the southern Pest part of the inner town of the capital, when the epicentre was put into Dunaharaszti village.

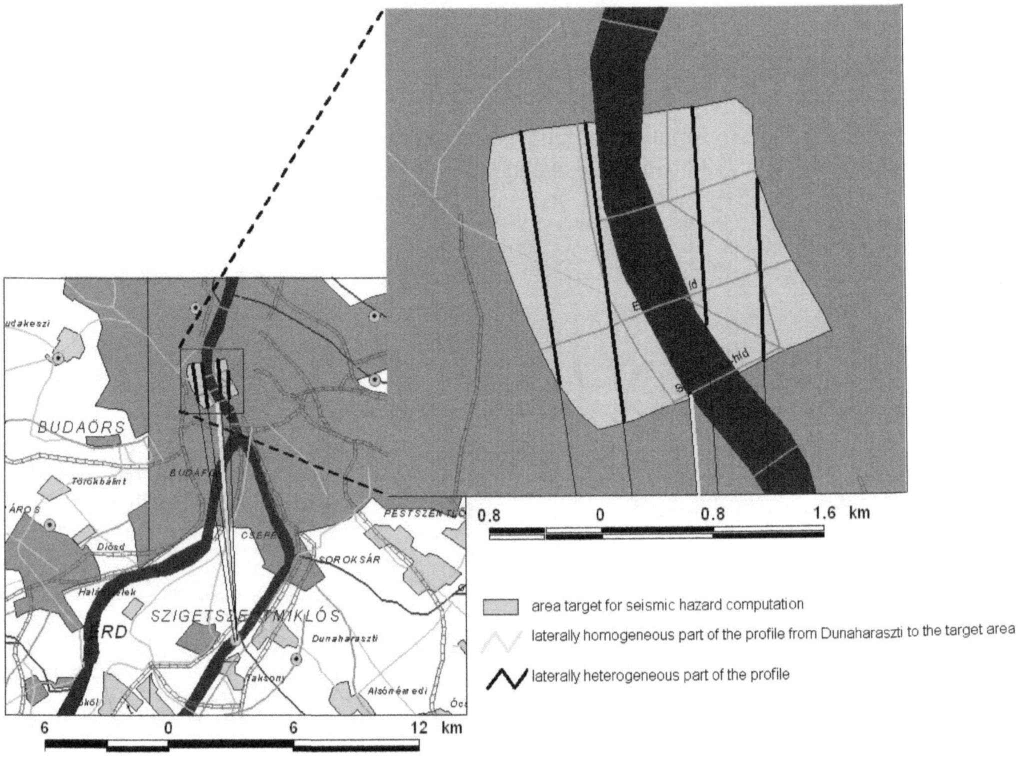

Fig. 3 Profiles between the downtown of Budapest, scenario No. 3 (Figure 2 in Gribovszki et al., 2010).

Panza et al. (2008) determined by the Neo-Deterministic Seismic Hazard Assessment (NDSHA) for Budapest 0.08–0.15g values, and for Debrecen 0.15–0.30g values.

This summary of our DSHA computations were preceded by the probabilistic seismic hazard assessment (PSHA) [Tóth et al., 2008, 2006]. According to Tóth et al. (2008, 2006) in the downtown of Budapest for a 50-year time-interval with 10% of no-exceedance the expected peak ground bedrock acceleration is 0.154g, and for Debrecen this value is between 0.1–0.105g.

According to results obtained by Bus et al. (2000) and Tóth et al. (2008, 2006) the bedrock related DSHA and PSHA earthquake hazard of Budapest and Debrecen in comparison with other cities of the world can be classable as medium.

The effects of local ground condition for the amplification of seismic waves was described by site response analysis in Hungary in Kegyes-Brassai (2017) as well. Earthquake hazard analysis and building vulnerability assessment has been done to determine the seismic risk of existing buildings in an urban area (city of Győr, Hungary) (Kegyes-Brassai 2015). Seismic microzonation in Budapest has been done as well by Győri et al. (2021) by joint analysis of active and passive surface wave methods.

2. Method

In the computations described in this paper synthetic seismograms are created by the hybrid method. A detailed modelling of the ground motion by hybrid method can take into account the local geophysical, geological, and geotechnical conditions at a site of interest (Panza et al., 2002). The hybrid method combines the modal summation method, valid for laterally homogeneous inelastic media (bedrock model) (Panza 1985; Panza & Suhadolc 1987; Florsch et al., 1991), with the finite difference method, which allows the modelling of the laterally inhomogeneous inelastic media (Fäh et al., 1990, 1994; Panza et al., 2000).

Wave propagation is treated by means of the modal summation technique from the source to the vicinity of the local heterogeneous inelastic structure that we want to model in detail. This laterally homogeneous inelastic structural model represents the principal crustal-properties of the propagation path from the focal source to the laterally heterogeneous area. This laterally homogeneous structure will hereafter be named "the bedrock model". It consists of several flat layers each of which is described by thickness and density values, P-wave and S-wave velocities and independent intrinsic attenuation values (Q).

The generated wavefield is then introduced into a grid that defines the heterogeneous area (2D profile) and it propagates according to the finite difference scheme. The laterally varying part of the applied hybrid model is constructed on the basis of the local geology.

Using this approach, the focal source, path and site effects are all taken into account and therefore it is possible to carry out a detailed study of the propagating wave field at even large distances from the epicentre.

3. Input Data

The target areas of DSHA computations are situated in a single profile (scenario No. 1), in the whole territory of Debrecen (scenario No. 2.) and in the downtown of Budapest. In case of Budapest the target area extends in N-S direction along the section of Danube from Batthyány square to Gellért square for a one km broad zone on both sides of the river (Figure 3).

The following initial data are requested for computation of synthetic seismograms:

- focal depths, focal mechanisms of the quakes, and the coordinates of epicentres;
- bedrock models between the hypocentres and the target areas of seismogram computations;
- 2D laterally heterogeneous models of near surface geological formations within the target areas.

3.1 Hypocentres and Focal Mechanisms of the Earthquakes

The main task was to define the scenarios corresponding to the earthquakes of the same size as the destructive events that occurred on 15th of October 1834 (in case of Debrecen), and the event occurred on 12th of January 1956 (in case of Budapest). For determining the parameters of the calculations of the Debrecen target area we relied on the work of Szeidovitz (2000) and for the calculations of the Budapest target area on the work of Szeidovitz (1986).

From the analysis of the felt intensity (up to IX) it has been possible to estimate a magnitude ranging from 5.5 to 7.1 (Szeidovitz, 2000) for scenarios No. 1 and 2. The supposed parameters of the source mechanism are dip: $\delta n = 70°$; rake: $\lambda = 0°$ (strike-slip); focal depth = 10 km; M = 6.0. Keeping fixed the dip and rake, we have considered two sources, one coinciding with the original epicenter of the earthquake of 1834 (Gálospetri, strike-receiver angle of 15°, scenario No. 1), the other close to Hosszúpályi village (strike-receiver angle of 165°, scenario No. 2). The parameters of the different scenarios' computations were summed up in Table 1.

The location of the epicentre at scenario No. 2 was justified by the fact that Hosszúpályi village is situated the closest to the city of Debrecen along the Mobile Zone fault system, so this could cause the highest seismic hazard for Debrecen. The explanation of this choice is the following. It can be possible to assume Hosszúpályi as the location for the earthquake's epicentre, since according to the generally accepted deterministic earthquake hazard principles, if an earthquake with a given magnitude has already occurred once at a given fault line, then another earthquake of the same magnitude can occur anywhere else along the same fault line.

When the epicenter is in Gálospetri (scenario No. 1), the length of the whole profile between the epicenter and the city of Debrecen is 46.8 km, of which the heterogeneous (local model) part is 2.3 km long. When the epicenter is in Hosszúpályi (scenario No. 2), the length of the whole profile between the epicenter and the city of Debrecen differs from 21 to 23 km long, of which the heterogeneous part is between 4.5–8.5 km long.

Only one earthquake scenario was used for the calculation of synthetic seismograms along the 4 different profiles in the case of the inner town of Budapest (scenario No. 3). The data of earthquake scenario No. 3 were the same as the data of the earthquake that occurred on 12th of January 1956 (Szeidovitz, 1986). The supposed parameters of the source mechanism are dip: $\delta = 90°$; rake: $\lambda = 0°$ (pure strike-slip left lateral); focal depth: h = 10 km; magnitude: M = 6.0; strike-receiver angle: $\varphi_0 = 40°$. The length of the wave path between the epicentre and the inner city of Budapest was 15 km; the length of the laterally heterogeneous paths (2D profiles) differs from 2.2 km to 3.4 km in the inner city. The location of the epicenter was in the centre of the triangle of Dunaharaszti, Szigetszentmiklós, Taksony villages. Figure 3 shows the 2D profiles on the surface. The detailed explanation of the focal mechanism can be found in Chapter 4/a-b of Varga et al. (2009) Final Report.

Table 3.1 Parameters of the different computations.

Parameters of the Computations	*Scenario No. 1*	*Scenario No. 2*	*Scenario No. 3*
Epicentre	Gálospetri	Hosszúpályi	Centre of the triangle of Dunaharaszti, Taksony, and Szigetszentmiklós
dip, δ	70°	70°	90°
rake, λ	0°	0°	0°
focal depth	10 km	10 km	10 km
magnitude, M	6.0	6.0	6.0
strike-receiver angle	15°	165°	40°
length of the whole profile	46.8 km	21–23 km	17.2–18.4 km
length of the laterally heterogeneous part of the profile	2.3 km	4.5–8.5 km	2.2–3.4 km

3.2 Bedrock Models

At the 1D part of the computation the seismic waves propagate along the laterally homogenous bedrock model (from the epicentre up to the beginning of the 2D part of the computation) for the model summation part of hybrid method synthetic seismogram computation.

In case of scenarios No. 1 and No. 2, P-wave velocity values measured at boreholes, and the results of some special research connection with the velocities, densities, and Q values of the upper 4000 m of the crust were taken into account in order to improve the previous parameters of the bedrock model. The PGT-1 seismic reflection section velocity parameters (Hegedűs and Takács, 1998) were used, whenever the layer depth has been larger than the well logging hole bottom data. The No. VI. structural unit of the Pannonian Basin model was used as a reference model (Bus et al., 2000) in the modal summation part of the computations at the deepest part of the bedrock model. In the above mentioned publication the Pannonian Basin has been divided into six regional structural units defined on the basis of geophysical and geological data. Parameters of the deepest layers have been adapted from the IASPE91 global model (Kennett and Engdahl, 1991) and from the velocity model of Mónus (1995) computed for the Pannonian Basin. In more details, see Gribovszki and Vaccari (2004) for scenario No. 1, and Gribovszki and Panza (2004) for scenario No. 2.

In case of scenario No. 3, the detailed data of the upper layer of this laterally homogeneous bedrock model can be found in Dövényi et al. (2008) Final report profile 5, V5-2 (XII. appendix). For the deeper part than 2000 m of the section, Bus et al. (2000) No. IV. model for the Pannonian basin was used as a reference model.

3.3 2D Laterally Heterogeneous Profiles

The destructive effects of seismic waves are strongly influenced by the physical parameters of the soil, within a few hundred metres from the topological surface. In order to characterize the laterally varying parts of the models, from the point of view of a heterogeneous inelastic structure, it was compiled on the basis of local geology, available geotechnical measurements, especially the knowledge of many shallow well loggings in Debrecen (Figure 2), in case of scenarios No. 1 and No. 2.

In case of scenario No. 3, the detailed data of the four 2D profiles can be found in "Geological structure of the inner town of Budapest and the geological-geophysical model of the northern part of Csepel island" Dövényi et al. (2008). Final Report, Appendix VII–XI. and XVI–XVIII.

3.4 Computation Parameters

The detailed explanation of computation parameters can be found in Gribovszki and Vaccari (2004) for scenario No. 1 (Gálospetri), in Gribovszki and Panza (2004) for scenario No. 2 (Hosszúpályi), and in Chapter 5 of Varga et al. (2009) Final Report for scenario No. 3 (Budapest).

4. Results

As the results of the computations P-SV (radial and vertical components) and SH (transversal component) synthetic displacement time series, velocity time series and accelerograms were computed along the local models,

- at 97 sites along the 1 laterally heterogeneous profile in Debrecen when the source was in Gálospetri (scenario No. 1);
- at about 100 sites at each 11 laterally heterogeneous profiles in Debrecen when the source was in Hosszúpályi (scenario No. 2);
- at about 90–95 sites in case of the all 4 laterally heterogeneous profiles at the inner town of Budapest (scenario No. 3).

All the computations two other kinds of figures were prepared too as results:

- Amplifications of peak ground accelerations along the profile (PGA(2D)/PGA(1D));
- Response spectra ratio along the profile (Sa(2D)/Sa(1D), RSR).

The 2nd and 3rd two figures were prepared in order to assess the site effects along the laterally varying part of the profiles. The response spectra ratio or spectral amplification represents the influence that local soil conditions introduce into the response spectra with respect to the bedrock model.

4.1 Results, Scenario No. 1

The influences in Debrecen of the Érmellék earthquake that occurred in 1834 were modelled by scenario No. 1. When the source is in Gálospetri (the original epicentre of the earthquake in fact occurred in 1834, at the Gálospetri graben, Érmellék seismo-active zone) the RSR (5% damping) along the profile are shown in Figure 4. The horizontal components of the ground acceleration caused the most significant damages, therefore we concentrate on the values of these components. The RSR maximum values (2.5) are seen in the horizontal components, at the second half of the profile and for frequencies below 1 Hz. This means that the multi-storey buildings at that part of Debrecen suffer a higher seismic hazard than the lower ones. Because of this, if an earthquake similar to the event of 1834 would occur, then it could cause much larger damage (larger macroseismic intensities) in a residential or industrial area with tall buildings than the damage was caused by the 1834 event, since in 1834 only single-storey buildings (and tall churches) existed in Debrecen.

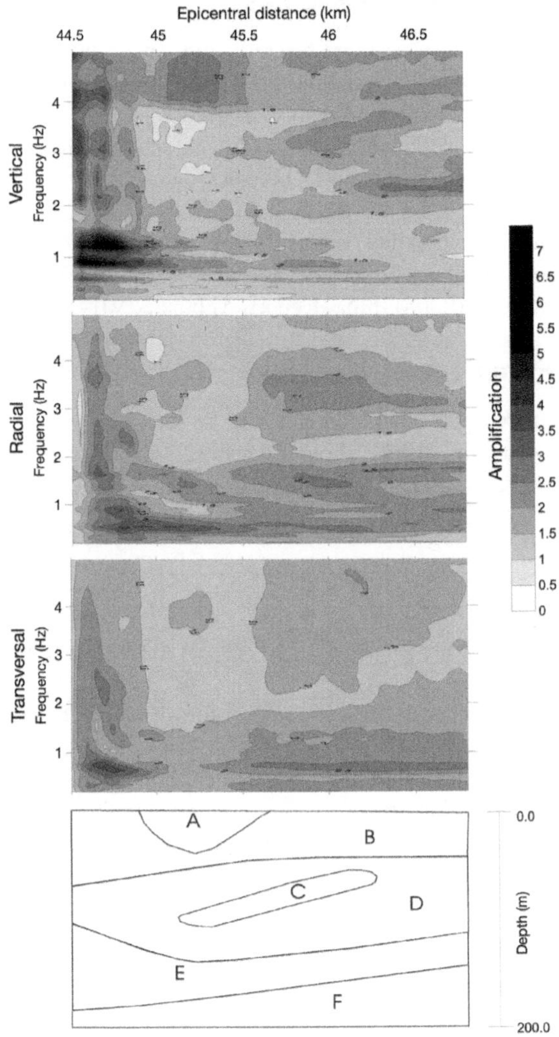

Fig. 4 Relative response spectra ratio with 5% damping versus frequency and versus the laterally varying profile in Debrecen, scenario No. 1 (Gálospetri epicentre).

The transversal component of PGA values, in case of the computations in scenario No. 1, can be converted into intensities and compared to the macroseismic intensity values estimated from the contemporary damage assessment reports in 1834. These PGA values extracted from synthetic signals are between 97–90 m/s^2 at the last 800 m of the profile (Figure 4). These values are approximately equal to 7° and 8° MSK-64 scale intensities (Bisztricsány, 1974).

In the majority of cases the application of the peak instrumental values (PGA, PGV, PGD) does not furnish sufficient information about the devastating effect of a given earthquake. Realizing this limitation, the Applied Technology Council (ATC, 1978) introduced the concept of the Effective Peak Acceleration (EPA). It is defined as the average spectral acceleration over the period range of 0.1–0.5 s divided by 2.5 (standard amplification factor for a 5% damping spectrum). EPA represents the acceleration which is most closely related to the damage potential of an earthquake (Newmark & Hall, 1982).

EPA values were calculated for three components in the case of scenario No. 1 (transversal component EPA values are shown in Figure 5). The curve shape of EPA values of transversal components is similar to the PGA curve but it represents significantly smaller values (50–57 m/s^2). By transforming EPA values into intensity values along the last 800 m part of the profile (where we have macroseismic intensities) 6° and 7° MSK-64 scale intensity values were obtained. The obtained results are in agreement with the observed macroseismic intensities due to the devastating 1834 earthquake, which occurred in the Érmellék seismo-active region (Figure 3 in Gribovszki & Vaccari, 2004).

These EPA and PGA values computed by the hybrid method (taking into account the effect of the subsoil) are an order of magnitude larger than the PGA values computed by Bus et al. (2000) for bedrock (0.020–0.040 g). Therefore, it can be concluded that the calculated values are in good agreement with each other, since in Debrecen loose, low velocity layered subsoil has been found below the surface (in the laterally 2D profile).

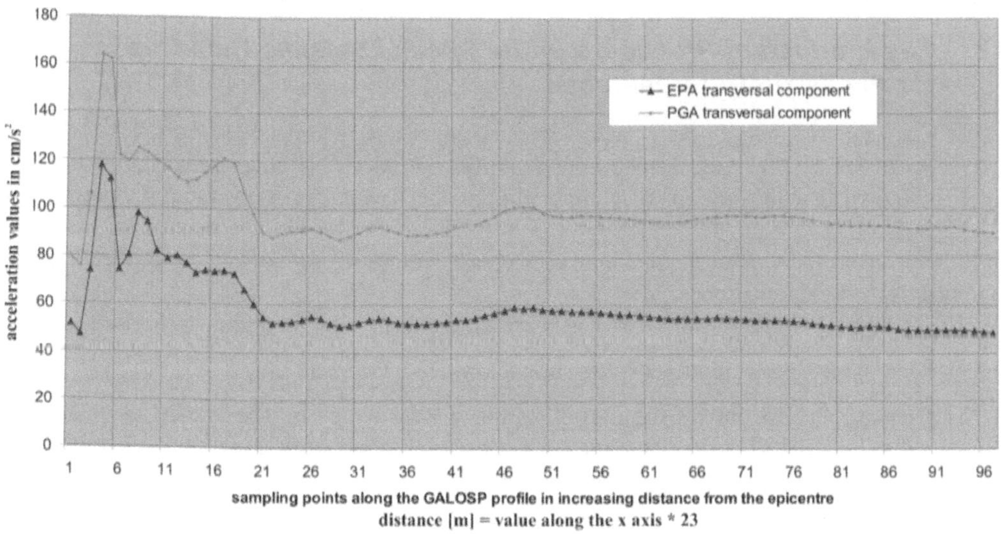

Fig. 5 Computed PGA and EPA values (transversal component) along the laterally varying profile in Debrecen, scenario No. 1 (Gálospetri epicentre).

4.3 Results, Scenario No. 2

The influences in Debrecen of an assumed earthquake occurred in the Mobile zone (epicentre was in Hosszúpályi village) were modelled by scenario No. 2. In the case of the Hosszúpályi epicentre computation (scenario No. 2), in the city centre of Debrecen, the trend of the RSR values is in good agreement with the observed decrease of the macroseismic intensities with increasing epicentral distance evaluated from the 1834 Érmellék earthquake.

Figure 6 (Figure 6 in Gribovszki and Panza, 2004) shows that the upper 200 m soil structure below Debrecen modifies the PGA values. The transversal components are six times larger than the vertical ones (and 10 times larger than the radial ones). The transversal PGA values drastically decrease from 0.38g to 0.15g, from the south-eastern part of the city to the north-western part. The maximal accelerations of the horizontal components are at the closest (southern) part of the city to Hosszúpályi village. The maximal acceleration values of the vertical component are at the centre of the city and its value is 0.05g. The maximal transversal acceleration values equal to about 8° intensity in MSK-64 scale (Bisztricsány, 1974).

Panza et al. (2008) computed for Debrecen by the Neo-Deterministic Seismic Hazard Assessment (NDSHA) 0.15–0.30 g values. These values are in good agreement with the values determined in this paper for the city centre of Debrecen by applying scenario No. 2.

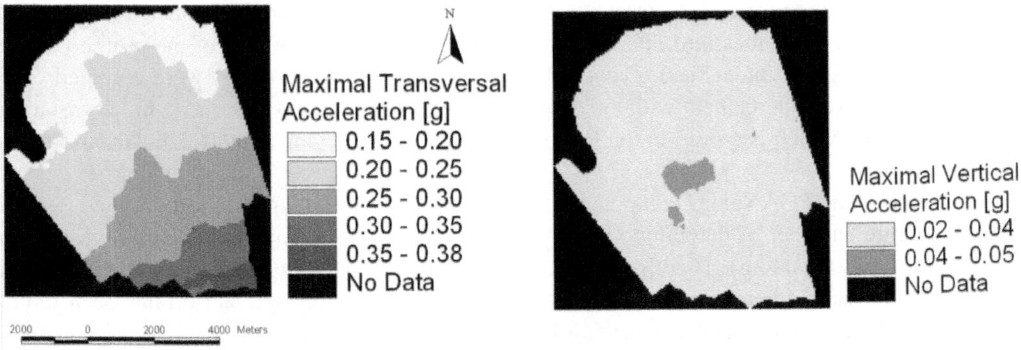

Fig. 6 Grid maps for the maximal acceleration values of the transversal and vertical components in Debrecen.

4.2 The Production Procedure of the Special Seismic Risk Map for the Territory of Debrecen

A special seismic risk map of the city of Debrecen, consistent with the 1834 Érmellék earthquake in terms of geometry (focal mechanism) but at a different place of hypocentre was produced.

It is known that the horizontal components of the waves cause the main damages of the buildings. In order to estimate the seismic risk at the city of Debrecen, the transversal component seismograms were used, since the maximal acceleration values were much higher for the transversal components than the radial ones.

The estimation of the seismic risk was based on the following, well known, formula:

$$\text{Seismic Risk} = (\text{Seismic Hazard}) \times (\text{Vulnerability}) \quad ...(1)$$

The seismic hazard part of the multiplication was provided by the synthetic seismograms computed by the DSHA hybrid technique (Fäh, 1992; Fäh et al., 1994), and the vulnerability part was estimated by using the formula:

$$T = 0.1 \times n[\text{s}] \quad ...(2)$$

where, T is the eigenperiod of a given building and n is the number of floors of the given building (Csák et al., 1981). The reason why this formula was used is because an earthquake exerts the strongest destructive effect on a building, if the building eigenfrequency coincides with the frequency of the maximal amplitude of the response spectra.

The map of the distribution of the buildings with different number of floors in Debrecen was constructed by using aerial photos and a 1: 10 000 scale topographic map (Figure 7).

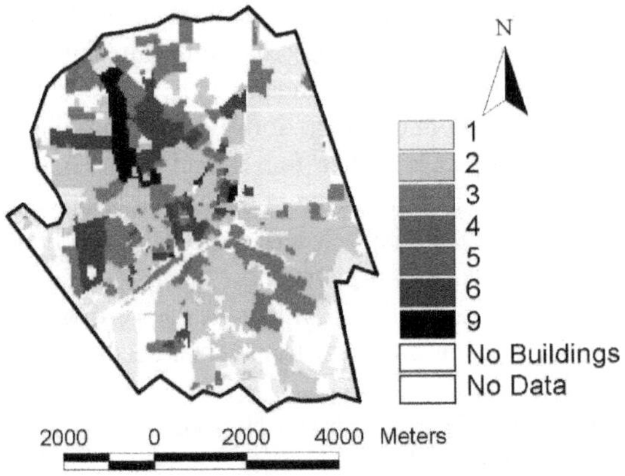

Fig. 7 Map showing the height (number of floors) of the buildings in Debrecen.

The procedure of the special seismic risk map's production summarized in one sentence is the following. If someone knows the number of floors at a given location, then by using Eq. (2), an acceleration value corresponding to the eigenperiod of the number of floors of a given building (at a given location) can be read off from the chart of spectral acceleration with 5% damping (SA 5%) computed by the hybrid technique. Details about the production procedure of the special seismic risk map can be read in Gribovszki and Panza (2004) and on Figure 12 in Gribovszki and Panza (2004) (the flow chart of the procedure of special seismic risk map s the production).

It can be seen from the seismic risk map of Debrecen city (Figure 8) that the most damageable part of the town is situated at about 1 km far south-eastward direction from the city centre. There are moderate damageable parts of the city not only on the southern part of the city, but at the city centre, and on the north too.

The estimation of the vulnerability of buildings by the eigenperiod of different storied buildings is the only one, and very simple way to characterize the vulnerability of the buildings.

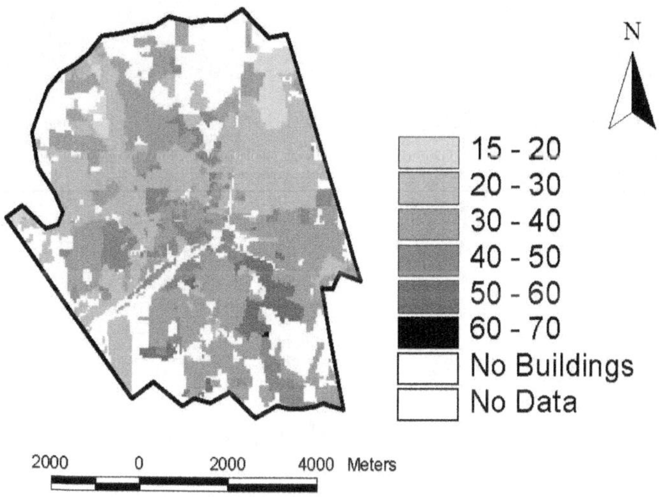

Fig. 8 Seismic risk map of Debrecen consistent with the 1834 Érmellék earthquake in terms of focal mechanism, but at a different location of the epicentre.

4.4 Results, Scenario No. 3

As the results of the computations, PGA (peak ground acceleration) grid maps of the downtown of Budapest for the three different components came into existence (Figure 9). Furthermore spectral acceleration (response spectra, SA) and RSR charts of the synthetic seismograms for the four different profiles were created as well.

The PGA grid maps (Figure 9) show that in case of all three wave components it is true that the loose, low velocity layered soil causes higher PGA values at the Pest part. The maximal PGA values manifest at the radial component at the Pest part of the downtown (100–200 cm/s^2), and these maximal PGA values are equal to 8° on MSK-64 scale (Bisztricsány, 1974). In case of the horizontal wave components the PGA results of hybrid method calculations were multiplied by the factors of 1.6 and 1.7 (compared to the 1D model PGA), where at the top of the soil, disadvantageous, loose layers (Holocene flood-plain sediments, debris covered surface) are situated. See, in more detail, in Varga et al. (2009) Final Report on pages 40–42.

Panza et al. (2008) computed for Budapest by the Neo-Deterministic Seismic Hazard Assessment (NDSHA) 0.15–0.30g values. These values are in good agreement with the highest values determined in this paper for the Pest part of the city centre in Budapest.

Szeidovitz et al. (2001) computed the designed ground acceleration values for Budapest by modal summation method taking into account the thick loose sediments at the surface as well, applying assumed scenarios of the Dunaharaszti earthquake that occurred in 1956. They found about 7° maximal intensity on the MKS-64 scale (B2 receiver: 4.9 cm/sec, radial component) for the southern Pest part of inner town of the capital, in case of the computation when the epicentre was put into Dunaharaszti village (S1 earthquake). This maximal intensity value is one degree smaller than our results. (The earthquake occurred in Dunaharaszti on 12th of January, 1956 caused 6.5° intensity in Mercalli-Cancani-Sieberg scale at Pest part of the inner town of Budapest, estimated by Szeidovitz (1986).

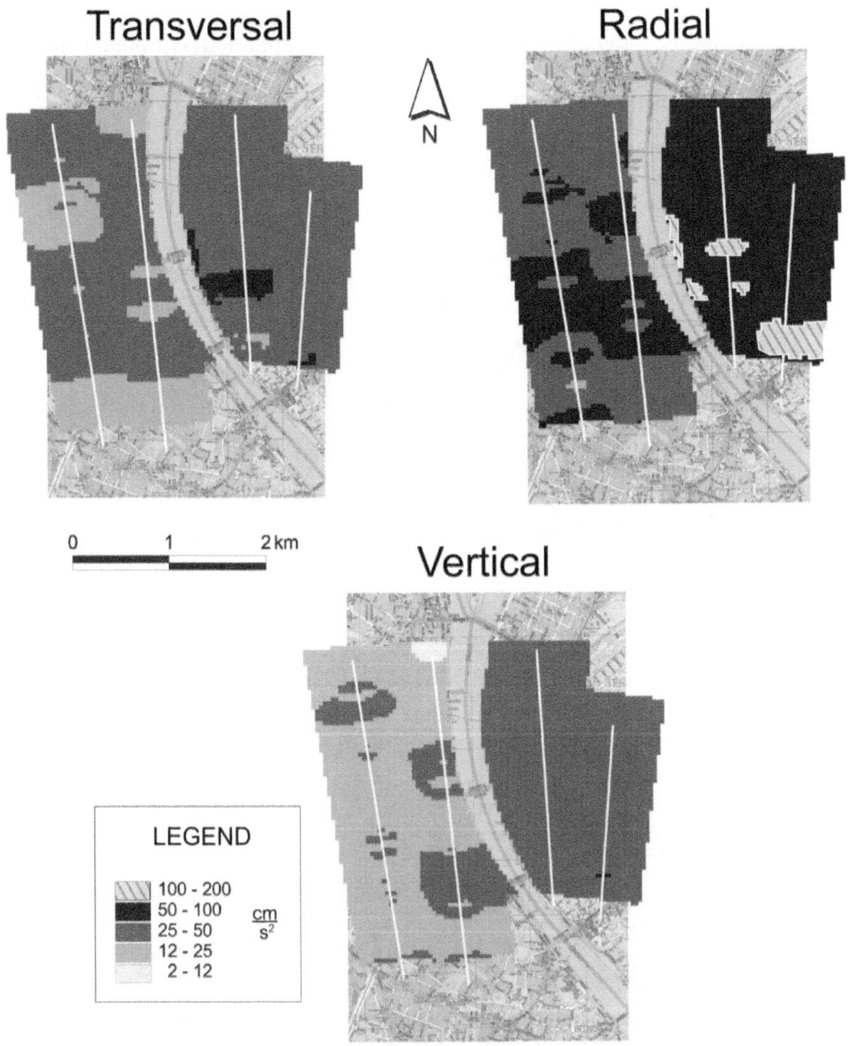

Fig. 9 Grid maps of peak ground acceleration values for the three wave components at the inner town of Budapest.

Figure 10 shows the Sa(2D) (response spectra of the synthetic accelerations taking into account the 2D laterally heterogeneous profile, as the local soil conditions) for profile 1 (Buda part in Budapest). The horizontal acceleration values differ from 10 cm/s^2 and 200 cm/s^2, and the maximal values magnify at higher than 2 Hz frequency domain. (This higher than 2 Hz resonance frequency coincides with the resonance frequencies of buildings with less than 5 or 6 floors.) At profile 1, the radial horizontal component maximal values can be found after the Gellért hill (right before 16 km), and at 16.9 and 17.7 km epicentral distance between 4–5 Hz frequency domain.

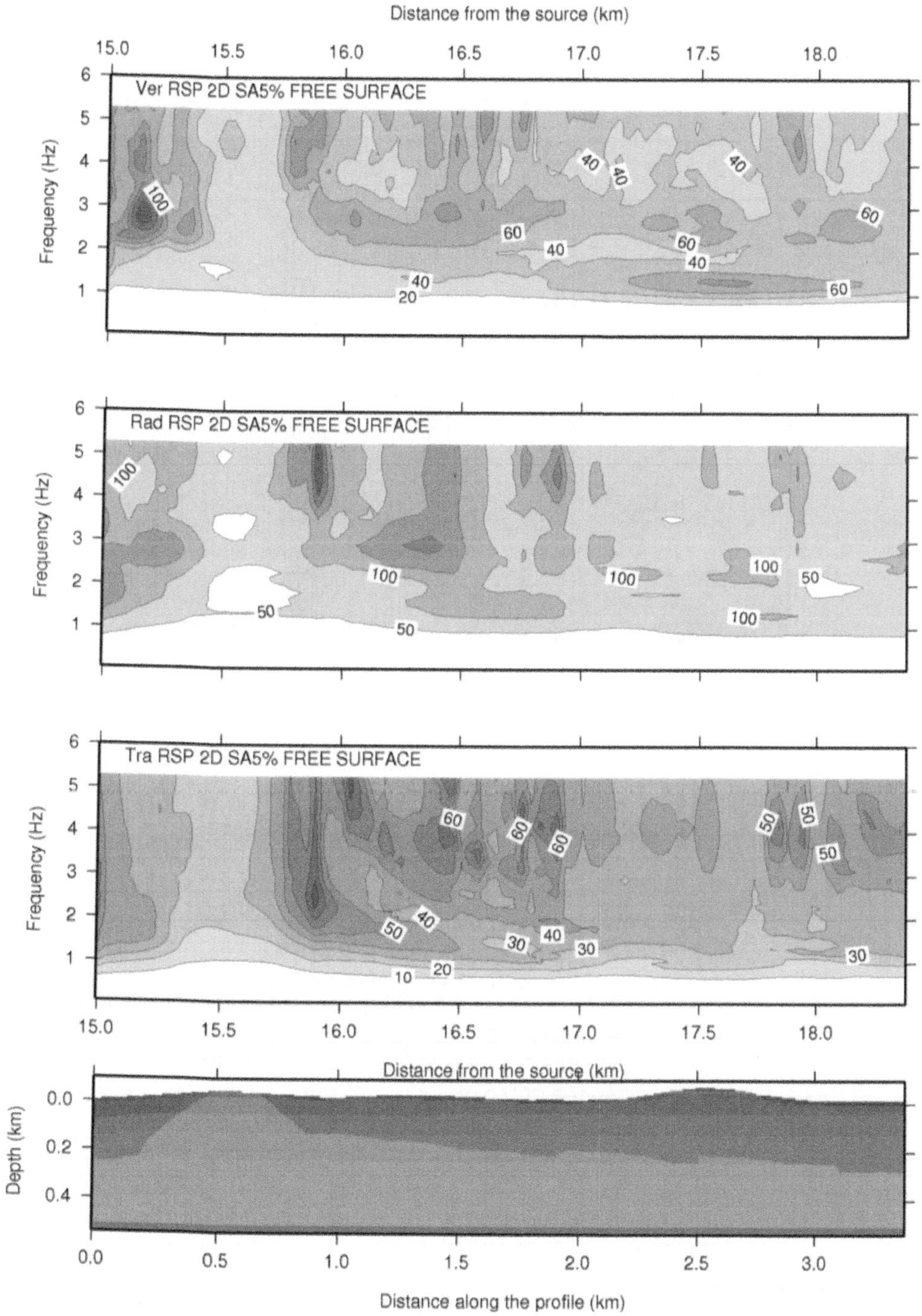

Fig. 10 Sa(2D)-Response spectra of the synthetic accelerations taking into account the 2D laterally heterogeneous profile (local soil conditions) for profile 1 (Buda part in Budapest).

For the downtown of Budapest, a special seismic risk map has been prepared (Figure 11). This special seismic risk map was created on the basis of the difference between the maximal amplitude frequencies of Sa(2D) of synthetic seismograms and the building's eigenfrequencies at every 0,1 km^2 of the downtown. The special seismic risk map shows that the buildings situated at the hilly western section of the downtown (Buda) have higher seismic risk than the ones at the flat middle and northeastern part (Pest). However, the buildings located at the southern part of Pest also have high seismic risk.

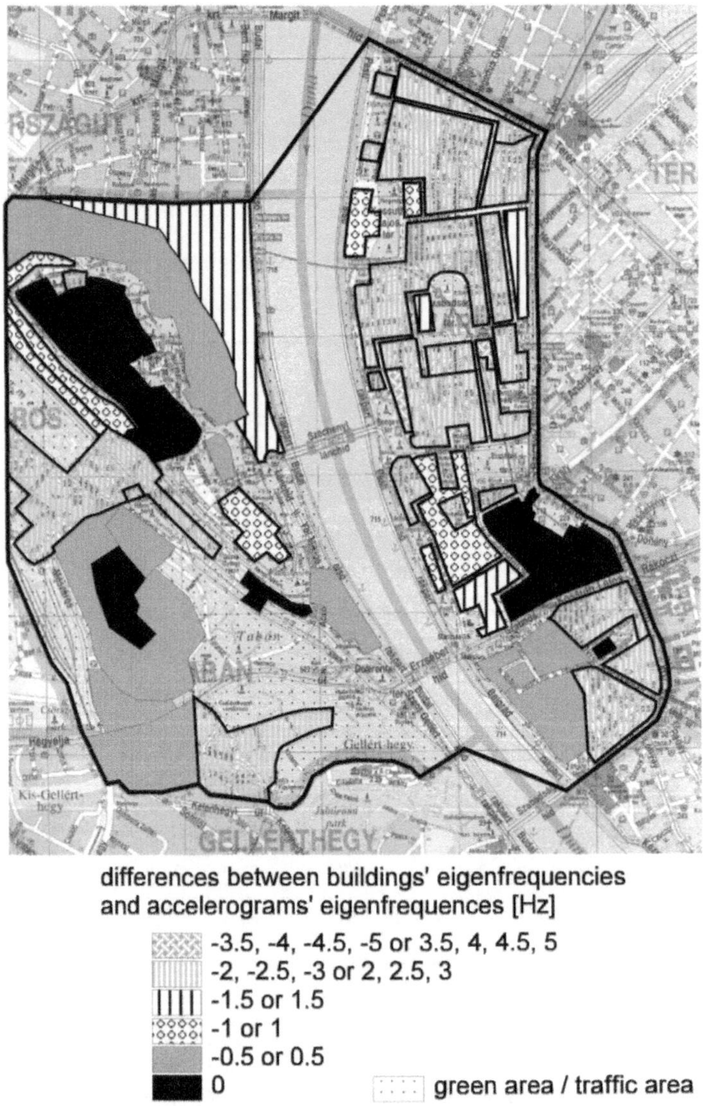

Fig. 11 Seismic risk map. Seismic Risk = frequency difference [Hz] = frequency of maximal acceleration of Response Spectra – eigenfrequency of building. The lower the difference the higher the seismic risk.

Acknowledgement

The authors express their gratitude to the staff of the Seismological Group, Department of Earth Sciences, University of Trieste for the help and professional guidance during K. Gribovszki's stay at that time.

In the late 1980s, together with Panza, we set up the Earth Science Commission of the Central European Initiative (CEI). The three working groups of the Commission (Geology, Geophysics and Geodesy) organised several technical seminars. In addition, the Commission organized five major conferences (Trieste, Sofia, Bratislava, Bucharest, and Ohrid), which mobilized specialists from across the region and played an important role in the reunification of the individual disciplines after 1989. In recognition of his activities, he was awarded the Medal of Honour by the President of the CEI in 2004.

The Hungarian authors would like to thank Prof. Panza and his wife, Rita for their friendship, and for their friendly invitations in Trieste and meetings in Hungary (Budapest and Sopron).

References

Bisztricsány, E. (1974). *Engineering Seismology*. Akadémiai Kiadó, 2015 (in Hungarian).

Bus, Z., Szeidovitz, Gy. and Vaccari, F. (2000). Synthetic Seismogram Based Deterministic Zoning for the Hungarian Part of the Pannonian Basin. *Pure Appl. Geophys., 157*: 203–219.

Csák, B., Hunyadi, F. and Vértes, Gy. (1981). *The Effects of the Earthquakes to the Buildings*. Műszaki Kiadó, Budapest (in Hungarian).

Dövényi, P., Palotai, M., Hámori, Z., Bíró, L., Tóth, T. and Surányi, G. (2008). Geological structure of the inner town of Budapest and the geological-geophysical model of the northern part of Csepel island. Final Report. *Geological Exploration and Environmental Research Ltd.*, (in Hungarian).

Fäh, D., Suhadolc, P. and Panza, G.F. (1990). Estimation of Strong Ground Motion in Laterally Heterogeneous Media: Modal Summation-Finite Differences. In: *Proceedings of the 9th European Conference of Earthquake Engineering* (pp. 100–109). Sept. 11–16. Moscow 4A.

Fäh, D., Suhadolc, P., Mueller, St. and Panza, G.F. (1994). A Hybrid Method for the Estimation of Ground Motion in Sedimentary Basins: Quantitative Modeling for Mexico City. *Bull. Seism. Soc. Am., 84*: 383–397.

Florsch, N., Fäh, D., Suhadolc, P. and Panza, G.F. (1991). Complete Synthetic Seismograms for High-frequency Multimode Love Waves. *Pure Appl. Geophys., 136*: 529–560.

Gribovszki, K. and Vaccari, F. (2004). Seismic ground motion and site effect modelling along two profiles in the city of Debrecen, Hungary. *Acta Geod. Geoph. Hung., 39*(1): 101–120.

Gribovszki, K. and Panza, G.F. (2004). Seismic microzonation with the use of GIS (Case study for Debrecen, Hungary). *Acta Geod. Geoph. Hung., 39*(2–3): 177–190.

Gribovszki, K., Schulek-Tóth, F. and Varga, P. (2010). Deterministic seismic hazard assessment of the inner town of Budapest. *Acta Geod. Geoph. Hung., 45*(3): 372–388.

Gusev, A.A. (1983). Descriptive Statistical Model of Earthquake Source Radiation and its Application to an Estimation of Short Period Strong Motion. *Geophys. J. R. Astron. Soc.*, 74: 787–808.

Győri, E., Timkó, M., Gráczer, Z. and Szanyi, Gy. (2021). Joint analysis of active and passive surface wave methods: Case studies from seismic microzonation of Budapest (pp. 1–8). In: Anon (Ed.), *Proceedings of 6th International Conference on Geotechnical and Geophysical Site Characterisation: Toward Synergy at Site Characterisation*, Budapest, Hungary. https://doi.org/10.53243/ISC2020-92.

Hegedűs, E. and Takács, E. (1998). The determination of the uniform velocity model along the deep reflection seismic profile PGT in the south-eastern part of Hungary (in Hungarian). ELGI, Budapest.

Kegyes-Brassai, O. (2015). *Earthquake Hazard Analysis and Building Vulnerability Assessment Has Been Done to Determine the Seismic Risk of Existing Buildings in An Urban Area*. PhD. Dissertation, Modelling and Development of Infrastructural Systems Multidisciplinary Doctoral School of Engineering (199 pp.) (in Hungarian).

Kegyes-Brassai, O., Wolf, Á., Szilvágyi, Zs. P. and Ray, R. (2017). Effects of local ground conditions on site response analysis results in Hungary. In: Woojin, Lee, Jong-Sub, Lee, Hyun-Ki, Kim, and Dong-Soo, Kim (Eds.) In: *Proceedings of 19th International Conference on Soil Mechanics and Geotechnical Engineering*, Seoul, South Korea: International Society for Soil Mechanics and Geotechnical Engineering (ISSMGE), 2003–2006.

Kennett, B.L.N. and Engdahl, E.R. (1991). Traveltimes for global earthquake location and phase identification. *Geophys. J. Int., 105*(2): 429–465.

Mónus, P. (1995). *Travel Time Curves and Crustal Velocity Model for the Pannonian Basin*. Technical Report. GGRI Seismological Department of HAS (in Hungarian).

Newmark and Hall. (1982). *Earthquake Spectra and Design*. Earthquake Engineering Research Institute, Oakland, California, USA.

Panza, G.F. (1985). Synthetic Seismograms: The Rayleigh Waves Modal Summation. *J. Geophys., 58*: 125–145.

Panza, G.F. and Suhadolc, P. (1987). Complete Strong Motion Synthetics. In: Bolt B. A. (Ed.), *Seismic Strong Motion Synthetics. Computational Techniques 4* (pp. 153–204). Academic Press, Orlando.

Panza, G.F., Romanelli, F. and Vaccari, F. (2000). Seismic wave propagation in laterally heterogeneous anelastic media: Theory and applications to the seismic zonation. *Advances in Geophysics, 43*: 1–95.

Panza, G.F. et al. (2002). Realistic modeling of seismic input for megacities and large urban areas. *Episodes, 25*(3): 160–184.

Panza, G.F., Kouteva, M., Vaccari, F., Peresan, A., Cioflan, C.O., Romanelli, F., Paskaleva, I., Radulian, M., Gribovszki, K., Herak, M., Zaichenco, A., Marmureanu, G., Varga, P. and Zivcic, M. (2008). Recent Achievements of the Neo-Deterministic Seismic Hazard Assessment in the CEI Region. *AIP Conference Proceedings* (*1020*: 402–409).

Szeidovitz, Gy. (1986). The Dunaharaszti earthquake, January 12, 1956. *Acta Geod. Geoph. Hung., 21*(1–2): 109–127.

Szeidovitz, Gy. (2000). Érmelléki földrengések (Earthquakes in Érmellék area). *Hungarian Geophysics, 41*(2): 78–84.

Szeidovitz, Gy., Bus, Z. and Gribovszki, K. (2001). Research for seismogenic zones in the Pannonian basin: A deterministic seismic hazard for Budapest. *Acta Geod. Geoph. Hung, 36*(4): 417–438.

Szeidovitz, Gy., Gribovszki, K. and Hajósy, A. (2002). Várható földrengések Érmellék és Nyírség területén (Expected earthquakes in the territory of Érmellék and Nyírség). *Magyar Geofizika, 434*: 161–179. (in Hungarian), ISSN: 00250120.

Tóth, L., Győri, E., Mónus, P. and Zsíros, T. (2006). Seismic Hazard in the Pannonian Region. In: Pinter, N., Grenerczy, Gy., Weber, J., Stein, S., and Medak, D. (Eds.), *The Adria Microplate: GPS Geodesy, Tectonics, and Hazards*. Springer Verlag, NATO ARW Series, *61*: 369–384.

Tóth, L., Mónus, P., Szeidovitz, Gy. and Bus, Z. (2008). *Seismic Hazard of the Inner Town of Budapest, Probabilistic Seismic Hazard Assessment.* Final Report, Archives of GGRI Seismological Department of HAS (in Hungarian).

Varga, P., Gribovszki, K., Győri, E. and Bus, Z. (2009). *Deterministic Seismic Hazard Assessment of the Inner Town of Budapest.* Final Report, GGRI Seismological Department of HAS (in Hungarian).

Zsíros, T. (1985). An Estimation of Seismic Hazard in Hungary. *Gerlands Beitr. Geophysik 94*: 111–122.

CHAPTER

14

Seismic Hazard Forecasting in Egypt: Application of Mdesign in framework of NDSHA Approach

Hany M. Hassan[1*], Mohamed ElGabry[2] and Antonella Peresan[3]

Earthquakes represent one of the most catastrophic natural hazards, with the potential to cause significant ground shaking, structural damage, loss of life, and economic disruption. In Egypt, a region of moderate seismicity, the seismic risk is influenced by both local and far-field sources, as evidenced by historical events such as the 1992 Cairo earthquake (Mw = 5.9) and the 1995 Gulf of Aqaba earthquake (Mw = 7.2). This study applies the Design Magnitude (M_{design}) approach that is calculated as $M_{design} = M_{\max} + \gamma_{EM}\, \sigma_{M}$, to the historical earthquake catalog truncated at 1900. This approach demonstrates its predictive capability for ground motion observed after 1900 in regions with well-documented historical seismicity, such as Cairo and the Nile Delta. Notably, the 1992 Cairo earthquake and its surrounding area exhibit no upward or downward triangles in the analysis, indicating that the event can be accurately Forcasted using the historical catalog alone. However, other regions, including southern Sinai, the Gulf of Suez, southeast Egypt, the southwest Western Desert, and northwest Egypt, reveal a more complex seismic hazard pattern, necessitating further investigation.

Earthquakes represent one of the most catastrophic natural hazards, with the potential to cause significant ground shaking, structural damage, loss of life, and economic disruption. In Egypt, a region of moderate seismicity, the seismic risk is influenced by both local and far-field sources (Ambraseys, 2001), as evidenced by historical events such as the 1992 Cairo earthquake (Mw 5.9) and the 1995 Gulf of Aqaba earthquake (Mw 7.2). This study applies the Design Magnitude (M_{design}) concept that is calculated as Mdesign = Mmax + γEMσM, to the historical earthquake catalog truncated in 1900. The possible integration of seismogenic nodes (Gorshkov et al., 2019; 2022) is explored as well. This approach demonstrates its predictive capability for ground motion observed after 1900 in regions with well-documented historical seismicity, such as Cairo and the Nile Delta.

Notably, the analysis shows that, following this approach, the ground shaking for the 1992 Cairo earthquake and its surroundings could be accurately enveloped based on the historical catalog alone. However, other regions, including southern Sinai, the Gulf of Suez, southeast Egypt, the southwest

[1,3] National Institute of Oceanography and Applied Geophysics, OGS, Via Treviso 55, 33100 Udine, Italy.
[1,2] National Research Institute of Astronomy and Geophysics, Helwan, Cairo 11421, Egypt.
[2,3] African Disaster Mitigation Research Center ADMiR, Cairo 11421, Egypt.
Email: elgabry@nriag.sci.eg[2]; aperesan@ogs.its[3]
* Corresponding author: helsayed@ogs.it

DOI: 10.1201/9781003684121-16

Western Desert, and northwest Egypt, reveal a more complex seismic hazard pattern, necessitating further investigation.

1. Introduction

Earthquakes are among the most catastrophic natural hazards, capable of producing significant ground shaking that leads to widespread damage to buildings and infrastructure, loss of life, economic disruption, and societal panic. Beyond the immediate effects of the fault rupture and ground motion, earthquakes can also trigger secondary phenomena such as landslides, rockfalls, liquefactions, and tsunamis. The cascading impacts of earthquakes underscore the importance of understanding and mitigating seismic risks, particularly in regions with significant exposure to such hazards.

Egypt, while generally characterized by moderate seismicity, has experienced the devastating impacts of both regional and local earthquakes throughout its long history. Regional or far-field earthquakes, such as the offshore 365 AD Crete earthquake (Mw 8.5) and the 1303 Rhodes earthquake (Mw 8.0) originated along the Hellenic Arc, are reported to cause strong impacts in Egypt due to ground shaking and associated tsunamis (Ambraseys et al., 1994; Guidoboni et al., 1994). Additionally, some local seismic sources have produced significant earthquakes, including the 1969 Shedwan Island earthquake (Mw 6.9), the 1992 Dahshur earthquake (Mw 5.9) (hereinafter Cairo earthquake), and the 1995 Gulf of Aqaba earthquake (Mw 7.2) (Hassan et al., 2017a). The occurrence of these destructive earthquakes highlights the critical need for reliable and comprehensive seismic hazard assessments to inform effective mitigation strategies and enhance earthquake preparedness in Egypt.

The seismic risk in Egypt is further amplified by the concentration of highly populated cities, industrial activities, and cultural and archaeological heritage sites along the Nile Valley and its Delta. These regions are vulnerable not only to local moderate-sized earthquakes, such as the 1992 Cairo earthquake, but also to larger-magnitude events originating from the Red Sea, the Gulfs of Suez and Aqaba, and the northern continental margin of Egypt (e.g., the 1955 Alexandria and 1998 Ras El-Hekma earthquakes) (Hassan et al., 2017b; Groshkov et al., 2019). Additionally, the eastern Mediterranean region, with its history of significant seismic activity (e.g., the 365 and 1303 earthquakes), contributes to the seismic risk profile of Egypt. Consequently, evaluating these seismic sources is essential for urban planning, infrastructure development, and the preservation of cultural heritage.

The development of reliable seismic hazard maps, based on accurate data and robust methodologies, is a cornerstone of effective seismic risk management. These maps must be able to predict the ground shaking from future earthquakes, hence they should be validated and, when necessary (e.g. if falsified), they should be updated to reflect advancements in seismic hazard assessment approaches, new data, new definitions and concepts, such as design magnitude Mdesign, and evolving understanding of regional seismicity. The use of Mdesign, known as the Panza-Rugarli Law (Rugarli et al., 2019; Wen and Wang, 2024; Zhang et al., 2025a&b), bridges the gap between seismology and engineering, offering a pragmatic and physics-based approach for seismic risk reduction. The analysis of the 2023 Al-Haouz and of the 2004 Al-Hoceima earthquakes (Hassan et al., 2025) demonstrates that the use of Mdesign definition offers a conservative and physically meaningful basis for reliable long-term earthquake ground motion forecasting and that PR law, as reported by Wen and Wang (2024), is worth further validation/falsification, in accordance with Popper's principle of falsifiability.

In this context, the Neo-Deterministic Seismic Hazard Assessment (NDSHA) approach has emerged as a particularly suitable methodology for regions like Egypt, where seismic risk is significant, but historical earthquake data may be limited (Hassan et al., 2017a; Panza et al., 2022). The NDSHA approach focuses on determining the Mdesign based on maximum credible earthquakes, enabling the generation of reliable seismic hazard maps for seismic source zones model and nodes such as the Nile Valley, Red Sea, and Gulf of Aqaba regions (ElGabry and Hassan, 2021; Gorshkov et al., 2019; 2022). These maps are invaluable for informing the design of earthquake-resistant buildings, bridges, and other critical infrastructure, thereby enhancing resilience to seismic events.

Moreover, the dynamic nature of the NDSHA approach allows for the continuous refinement of hazard maps as new earthquake data or new concepts becomes available, ensuring that seismic risk assessments remain up-to-date and relevant. This is particularly important for a country like Egypt, where the interplay of factors such as hypocenter properties, local site conditions, exposure levels, and vulnerability which may combine together and ultimately influence significantly the seismic risk profile (Hassan et al., 2017a; ElGabry and Hassan, 2021; ElGabry et al., 2021; ElGabry et al., 2022; Hassan et al., 2022, Hassan et al., 2023; Badreldin et al., 2024). For instance, past moderate to strong earthquakes in and around Egypt have demonstrated that the severity of seismic impacts is often a function of these combined factors, underscoring the need for a holistic approach to seismic hazard assessment and mitigation.

Furthermore, the development, testing and possible updating of seismic hazard maps, supported by advanced methodologies such as NDSHA, are essential for informed seismic risk mitigation strategy in Egypt and to preserve the sustainable development planning and Egypt Vision 2030. By integrating different data and validated approaches and methodologies these maps can provide a robust foundation for earthquake preparedness, urban planning, and the design of resilient infrastructure. This proactive approach is critical for safeguarding Egypt's population, economy, and cultural heritage against the potentially devastating impacts of future earthquakes.

2. Seismicity and Seismogenic Source Models for Egypt

It is well known that Egypt has a long and well-documented history preserved in a variety of sources on the effects of historical seismicity, which goes back in time about four millennia with variable levels of reliability and details. Those original documentary sources are considered as the principal source of macroseismic data for historical earthquakes. Additionally, they provided some scholars with the chance to reconsider the past seismicity of the area. Since the beginning of the 19th century, many efforts have been devoted to the establishment of reliable catalogues of historical seismicity based on the retrieval and assessment of the original sources of information (e.g., Ambraseys 1961; Maamoun et al., 1984; Ambraseys et al., 1994). The historical earthquake catalogues of Maamoun et al. (1984) and Ambraseys et al. (1994) seem to be the first comprehensive and most complete ones, in which historical earthquakes are revaluated considering, in a critical way, all available original source data.

The historical earthquake catalogues of Maamoun et al. (1984) and Ambraseys et al. (1994) give information about occurrence dates, epicenter coordinates for most earthquakes, and epicentral macroseismic intensity, (I_0), in units of the MSK scale by Medvedev–Sponheuer–Karnik (1964). Maamoun (1979) has established a regional relationship between the intensity and the magnitude, by comparing the instrumental and macroseismic data of strong earthquakes:

$$M = 0.7\ I_0 + 2.7 \log h - 3.0 \qquad ...(1)$$

Where, h is the depth and I_0 is the intensity at the epicenter (I_0).

For each seismogenic zone, Eq. 1 has been used to convert the intensity data into magnitude by fixing the focal depth that was estimated instrumentally.

The evaluation of the available information about the historical earthquakes collected from different sources by Ambraseys et al. (1994) indicated that not all the events originated within the study area and some are of dubious (seismic) origin. Moreover, Ambraseys et al. (1994) identified false reports and resolved conflicting information, particularly for events of uncertain dates and locations.

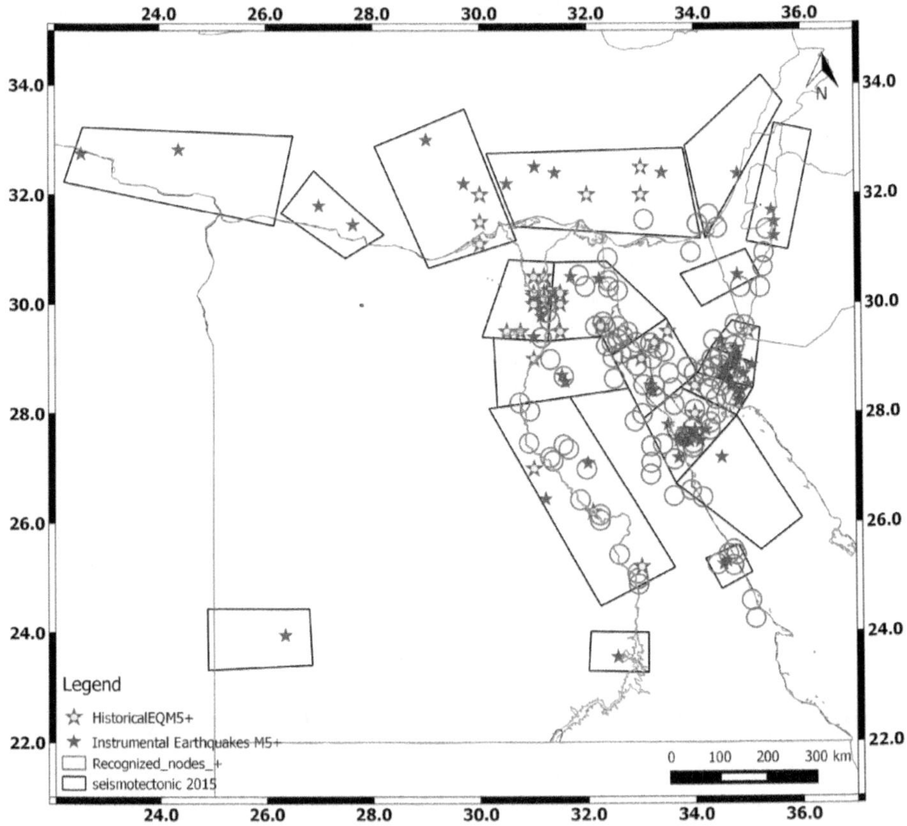

Fig. 1 Seismogenic nodes and seismo-genetic zonation for Egypt.

In 2010, a team of seismologists and historians from National Research Institute of Astronomy and Geophysics (NRIAG) has revised the historical earthquakes catalogue, and they qualitatively qualify each event in the catalogue using quality level descriptors (poor, fair, good) of the event's existence (Q_e), location (Q_l) of the maximum intensity, I_{max}, and date (Q_d) (NRIAG, 2010). More investigations need to be done to have an accurate historical earthquake catalogue, particularly for the Nile Delta and Cairo areas, since the local site effects, population distribution, and the vulnerability of buildings may result in inaccurate location and size of the earthquake. The observed damages could be due to a combination of different factors besides the earthquake ground motion: 1) proximity of the earthquake epicenter to Cairo city; 2) propagation effects; 3) site amplification; 4) high vulnerability of buildings. Recently, the earthquake catalogue for Egypt has been critically revised and updated by Hassan et al. (2017b) based on information from publications, local reports, and paleoseismological and archaeological investigations. In their work, they have excluded the historical events of poor existence and location, also they did some modifications (adding or removing events) based upon recently published paleoseismological and archaeological studies (e.g., Casciati & Borja, 2004). Accordingly, the main sources of compiled historical earthquake data used in the current work are these remarkable catalogues compiled by Maamoun et al. (1984), Ambraseys et al. (1994), NRIAG, (2010), and Hassan et al. (2017b).

Figure 1 shows that most of the historical earthquakes are located along the Nile Valley and its Delta since the population distribution can significantly blur the location of pre-instrumental earthquakes. On the other hand, it has been possible to single out a few events along the Gulf of

Suez and the continental margin of Egypt, areas of very low population density, and in barren regions such as the Eastern, Western Deserts, and the Sinai Peninsula.

The Instrumental earthquakes era in Egypt started in 1899 (~1900); nevertheless, much of the seismic activity along its remote borders was not revealed until the increase of the spatial resolution of the seismic network, deployed by NRIAG, both locally and regionally. After more than 125 years from the installation of the first seismic station, the territory of Egypt is defined as a relatively moderate seismic activity region. Also it is characterized by the occurrence of small to moderate events (intraplate events), while most, if not all, the large events take place sporadically farther east, along the northern Red Sea, Gulf of Suez entrance or Gulf of Aqaba, and to the north, the Eastern Mediterranean region (interplate events).

Despite the inherent uncertainty and spatial and temporal incompleteness that affect, generally, any historical catalog, its use for seismic hazard assessment is imperative because, very often, it represents the longest available seismological data segment and can be naturally extended further backward considering seismogenic nodes (Groshkov et al., 2019, 2022). It is worth mentioning here that a revised and updated instrumental earthquake catalog of Egypt relative to that has been compiled by Hassan et al. (2017b) then used for ground shaking computation later on in this paper. The complete catalogue that also includes non-instrumental events is used with the precaution to round to the next largest ordinal value all intermediate macroseismic intensity values, in agreement with the definition of any macroseismic scale (e.g., EMS98). The discrete integer ordinal number nature of any macroseismic scale (a sequence of natural ordinal number, i.e., a scale in which each number tells the position of something in a discrete scale of integers, such as I, II, III, IV, V, etc.) implies a straightforward estimation of the minimum error, ΔI, that may affect intensity values: one degree.

For the areas where the seismicity is not known, it is crucial to rely on information from other relevant studies. Among these studies, the formalized identification of seismogenic prone sites using Morphostructural zonation (MZ) and pattern recognition (e.g., Alekseevskaya et al., 1977; Caputo et al., 1980; Peresan et al., 2014; Gorshkov et al., 2019; Gorshkov & Soloviev, 2022; Gorshkov et al., 2022) turns out particularly useful for our purposes, since the result from MZ can be naturally incorporated in the seismic hazard computation by the NDSHA (Panza & Bela, 2019).

Seismogenic nodes capable of earthquakes with $M \geq 5.0 = \underline{\boldsymbol{M}}$, which is conventionally (e.g., D'Amico et al., 1999) taken as the lower bound magnitude of damaging earthquakes, have been identified in the north east part of Egypt using MZ and pattern recognition by Gorshkov et al. (2019) as shown in Figure 1. Morphostructural zonation (MZ) is employed to map nodes over the entire study region, which includes the Sinai Peninsula and the northern part of the Eastern Desert between the Nile River and the Red Sea, without using the existing knowledge about regional seismicity. The identification of seismogenic nodes prone to earthquakes with a magnitude of $M \geq 5.0$, as established by Gorshkov et al. (2019), has the potential to significantly enhance the accuracy of seismic hazard assessment in northeastern Egypt by expanding our knowledge of seismogenic sources. It is crucial to acknowledge that certain D nodes are situated in close proximity to critical facilities and newly developed megacities, such as the New Capital project. Therefore, it is imperative to consider these nodes when determining the seismic hazard input for these specific sites. By incorporating this valuable information into the seismic hazard computation using the NDSHA approach (Hassan et al., 2017a, b), it can contribute to the revision of seismic zoning and ultimately improve the overall effectiveness of seismic hazard studies in Egypt.

3. Application of M_{design} for Egypt in Framework of NDSHA Approach

A proper way to reduce the shortcoming due to inadequate knowledge and gaps in available seismicity and associated uncertainties is to judiciously incorporate all available reliable information obtained by different methods, newly developed techniques and strategies in the framework of a robust and reliable scenario based Seismic Hazard Analysis (SHA) like, for instance, the NDSHA (e.g., Panza et al., 2017, 2019, 2022).The standard NDSHA method is based on the computation of realistic synthetic time histories of ground motion using basic physical principles of wave's generation and propagation in inelastic vertically heterogeneous media, that represent bedrock conditions. Thus, it is possible for any area of interest to quantitatively model the hazard scenario for future earthquakes and therefore NDSHA represents a very effective quantitative tool even in regions with scarce or no historical or instrumental information available, such as North Africa and Middle East Regions. NDSHA uses scenario earthquakes characterized in terms of magnitude, distance, fault type, and source mechanism. The approach naturally supplies time histories of ground motions readily suitable for (even non-linear) analysis of normal and critical structures. This methodology has been applied to many areas world wide (e.g., Panza et al. 2001, 2012; Hassan et al., 2017b), to strategic buildings, lifelines, and cultural heritage sites, and, for the purpose of seismic microzoning, to several urban areas, in the framework of the UNESCO/IUGS/IGCP project "Realistic Modelling of Seismic Input for Megacities and Large Urban Areas" (e.g., Panza et al., 2002).

Standard NDSHA work package (https://www.xeris.it/index.html) is suited to compute the ground motion parameters at 1 and 10 Hz cut-off frequencies for a set of 1D structural models and earthquake scenarios at different spatial scales. In the framework of default NDSHA procedure (for further reading refer to Panza et al., 2012), the study region can be covered by a regular grid (usually 0.2° × 0.2°) or the ground motion computation can be carried out for specific observation sites. The key components of the hazard input for this study encompass seismogenic zones and nodes, along with geo structural models for bedrock, which are crucial for reliable seismic hazard computations (Panza et al., 2001, 2012). The parameterization of seismogenic zones involves considering their location, extent, focal mechanisms, and seismicity. A representative structural model is employed to capture the lithospheric properties of the study area. This model comprises multiple flat and elastic layers with well-defined thickness, density, P- and S-wave velocities, and attenuation, specifically tailored to reflect the bedrock conditions at the site of interest.

The design magnitude, M_{design}, is a critical parameter in NDSHA. It is defined as the maximum estimated magnitude ($M_{\max}$) plus an additional safety term ($\gamma_{EM}\sigma_M$), where γ_{EM} is a safety factor and σ_M is the standard deviation of magnitude determination (Rugarli et al., 2019). The formula for M_{design} is expressed as:

$$M_{design} = M_{\max} + \gamma_{EM}\sigma_M$$

In practical applications, γ_{EM} is often set to 2 and 3, leading to $M_{design} = M_{\max} + 0.5/0.7$ (Rugarli et al., 2019; Li et al., 2025). The primary purpose of M_{design} is to ensure that seismic hazard assessments are conservative, accounting for uncertainties in magnitude estimation and providing a safe envelope for engineering design. This approach is particularly important in regions with limited historical earthquake data, as it allows for a more robust estimation of seismic hazard.

The concept of M_{design} was developed by Panza and Rugarli (2019) as part of the NDSHA framework, which aims to address the limitations of traditional deterministic (DSHA) and probabilistic (PSHA) seismic hazard assessments. The development of M_{design} was influenced by several key milestones. First, the use of the global standard deviation of magnitude, which typically ranges between 0.2 and 0.3, provided a statistical basis for magnitude uncertainty (Båth, 1973; Bormann et al., 2007). Second, the introduction of the safety factor γ_{EM} was rooted in engineering considerations, ensuring that the design magnitude would be conservative enough to account for potential underestimation of seismic hazard (Rugarli et al., 2019). Third, the principle of uniformitarianism, as proposed by Hutton (1795), was applied to justify the use of $M_{\max}$ as a basis for M_{design}. Finally, the concept of seismogenic nodes was introduced to compensate for the lack of historical or paleo-seismological data, particularly in regions with sparse earthquake records (Gorshkov et al., 2003) and the corresponding magnitude is defined to be the lower magnitude or $M_{\min}$.

The justification for M_{design} lies in its ability to provide a physically based and conservative estimate of seismic hazard. Unlike probabilistic approaches, which rely on uncertain probability distributions, M_{design} is based on the maximum credible earthquake (MCRE), which is a property of the fault system and reflects the physical limits of seismic hazard (Rugarli et al., 2019). The safety factor γ_{EM} ensures that the design magnitude is sufficiently conservative, reducing the risk of underestimating seismic hazard. Furthermore, M_{design} is falsifiable; if an earthquake exceeds the design magnitude, the value can be adjusted, making the approach more robust and adaptable to new data (Li et al., 2025; Rugarli et al., 2019). This falsifiability is a key advantage over probabilistic methods, which cannot be definitively tested or falsified due to their reliance on probability distributions (Li et al., 2025; Panza & Bela, 2019).

The application of M_{design} in NDSHA has significant implications for seismic hazard assessment and engineering design. It is used to generate seismic hazard maps that provide key ground motion parameters, such as peak ground displacement (PGD), peak ground velocity (PGV), and design ground acceleration (DGA) or peak ground acceleration (PGA or DGA), for specific sites. These maps are essential for the seismic design of critical infrastructure, ensuring that structures can withstand the maximum credible earthquake. Additionally, the value of M_{design} can be dynamically updated if a rare, large earthquake (referred to as a "Dragon King" event) exceeds the current design magnitude, allowing for continuous improvement of seismic hazard assessments (Rugarli et al., 2019).

The NDSHA approach, with its focus on M_{design}, is particularly well-suited for regions like Egypt, where seismic risk is significant but historical earthquake data may be limited (Hassan et al., 2017a, b; Hassan et al., 2020). Egypt has several seismogenic zones and nodes, particularly in the Nile Valley, Red Sea, and Gulf of Aqaba regions, which could benefit from the application of NDSHA (Hassan et al., 2017a, b, 2020; Groshkov et al., 2019, 2022; Panza et al., 2022). By using historical earthquake data to determine M_{max} and subsequently M_{design}, it is possible to generate accurate seismic hazard maps for these regions. These maps can then be used to inform the design of buildings, bridges, and other critical infrastructure, enhancing their resilience to seismic events. Furthermore, the dynamic nature of NDSHA allows for the continuous updating of hazard maps as new earthquake data becomes available, providing a more accurate and up-to-date assessment of seismic risk in Egypt.

The primary aim of this work is to show how the use, for each cellular source, of the M_{design} affects the SHA for Egypt. For this purpose, we limit the information to the catalog truncated to 1900, assume values of $\gamma_{EM}\sigma_M = 0.5$ and 0.7 (a logical upper limit reflecting $3\sigma_M$), and we check how well NDSHA maps based on the historical record (catalog truncated to 1900) turn out adequate to forecast observed values in the subsequent hundred and more years, till 2025.

To achieve the goal we consider the M_{design} for both historical record and seismogenic nodes (Figure 1) in the NDSHA framework, then we take the ratio between a reference run and different variants computed using different inputs described in the next sections. Given two maps (X and Y) to be compared, the ratio $R_{i,j} = X_{i,j}/Y_{i,j}$ is computed and displayed at each node with i, j coordinates of the grid. If the R is ≥ 1, the value $R_{i,j}$ is plotted with an upward triangle; if R is < 1, the value $1/R_{i,j}$ is plotted instead, with a downward triangle. The color code shows the value of the ratio plotted, naturally in steps of one macroseismic intensity degree, indicated in parenthesis in the maps' legend. Only the sites where there is a variation of at least one level of macroseismic intensity (corresponding to a peak value differential of around a factor of two) are mapped.

For convenience, we divided the computations into two parts. The first part aims to investigate the relevance of considering the M_{design} only for historical earthquake records (keeping the magnitude of the nodes unchanged). The second part is focused on the application of M_{design} to the seismogenic nodes keeping the magnitude of historical earthquakes unchanged. The result from each run is then compared with the reference run computed by Hassan et al. (2017b).

4. Application of M_{design} to Historical Catalog in Egypt: Results and Discussion

The historical earthquake catalog is the longest available seismological data segment and it is an important benchmark for our experiment. The NDSHA maximum or PGA map for the study region is computed using the earthquake catalog truncated to 1900 and its variant obtained with the addition of seismogenic nodes. The comparison is made considering, as reference run, the variant 3 in Hassan et al. (2017b) obtained considering revised and updated input data: non-truncated earthquake catalog, seismotectonic zones with their representative focal mechanisms, and structural models.

The experiments carried on can be summarized as follows:

- PGA map for the reference run (i.e., Variant 3 in Hassan et al., 2017b);
- PGA map computed for the earthquake catalog for Egypt truncated to 1900 (without seismogenic nodes);
- PGA map computed using the earthquake catalog for Egypt truncated to 1900 enriched with seismogenic nodes;
- PGA map computed using the earthquake catalog truncated to 1900 after adding $M_{design} = M_{max} + 0.5 = \gamma_{EM}\sigma_M = 2\ \sigma_M$ (Båth, M., 1973; Bormann et al. 2007) to their magnitude (without including seismogenic nodes);
- PGA map computed considering seismogenic nodes and using the earthquake catalog truncated to 1900, but adding $+0.5 = \gamma_{EM}\sigma_M$ to (all) magnitudes.

4.1 Reference Run (Experiment 1.1)

The PGA map computed in Variant 3 by Hassan et al. (2017b) is adopted as a reference run (Figure 2). The earthquake sources are realistically modeled and include source rupture processes (i.e., directivity, rupture velocity, and slip distribution). In such a way we investigate the influence of adopting different models of source time function on the resulting ground motions.

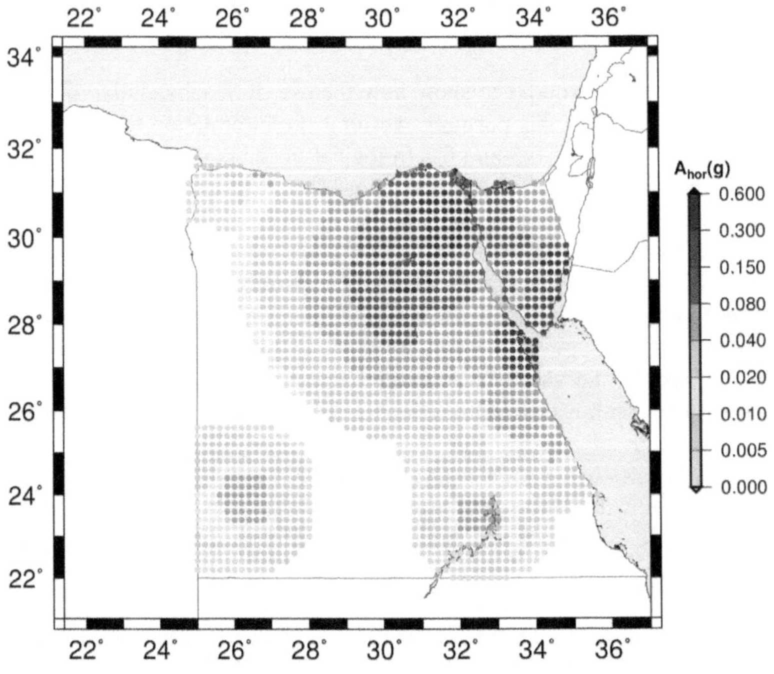

Fig. 2 Peak ground acceleration map, which is computed in Variant 3 in Hassan et al. (2017b) (Experiment 1.1).

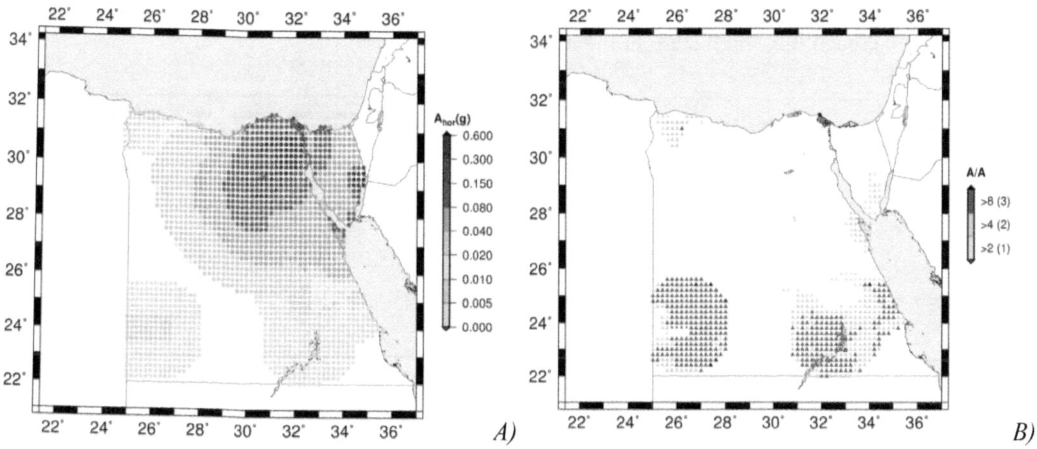

Fig. 3 (A) Peak ground acceleration map, for Experiment 1.2; (B) Map of the ratio between PGAs maps from Experiments 1.1 and 1.2, respectively = Experiment 1.1/Experiment 1.2.

4.2 PGA Map Computed Using the Earthquake Catalog Truncated to 1900 (Experiment 1.2)

In this experiment, the earthquake catalog truncated to 1900 is used to compute PGA using the same input parameters and source model used in Section 4.1 (Figure 3a). The map with the ratio between the values of the reference run (Experiment 1.1) and those of Experiment 1.2 is shown in Figure 3b. The map indicates that the use of the truncated earthquake catalog can adequately forecast the ground shaking that occurred after 1900 for the areas covered by the historical catalog. The sites of almost no differences are located along the Nile Delta and Valley and their extensions because the rivers and deltas were and still are the most favorite and suitable places for any human settlement and civilization. This can be indicated by the sites with no upward or downward triangles (i.e., the ratio between Experiments 1.1 and 1.2 is less than 2) in Figure 3b that occupy most of the northeastern part of the country (Nile Valley and Delta). It is relevant to mention that the sites where there is a change of at least one level of macroseismic intensity (corresponding to about a factor of 2 in the peak value) are mapped.

In the areas of Aswan, Western Desert, Gulfs of Suez and Aqaba the ratio is significant. The large differences observed in these areas can be naturally attributed to the lack of historical information due to the lack of population in the past.

4.3 PGA Map Computed Using the Earthquake Catalog Truncated to 1900 Enriched with Seismogenic Nodes (Experiment 1.3)

In the current experiment, the earthquake catalog truncated till 1900 (Experiment 1.3) is enriched with the addition of the seismogenic nodes which have been defined for the Northeast part of the country by Gorshkov et al. (2019) (Figure 4a).

The map of the ratio between the results of Experiments 1.1 and 1.3, shown in Figure 4b, illustrates that the inclusion of the nodes in the NE part of Egypt introduces small changes relative to the previous map (Figure 3a), because the magnitude threshold assigned to the nodes is quite small relative to the magnitude of the instrumental earthquakes occurred during the last 118 years (e.g., Aqaba 1995 earthquake $M_w = 7.3$). As shown by Gorshkov et al. (2019) the choice of this minimum threshold is based on the seismicity level in northeast Egypt. Small changes in the eastern part of the Sinai Peninsula are indicated by two triangles of downward direction which means that the nodes are controlling the hazard at those sites.

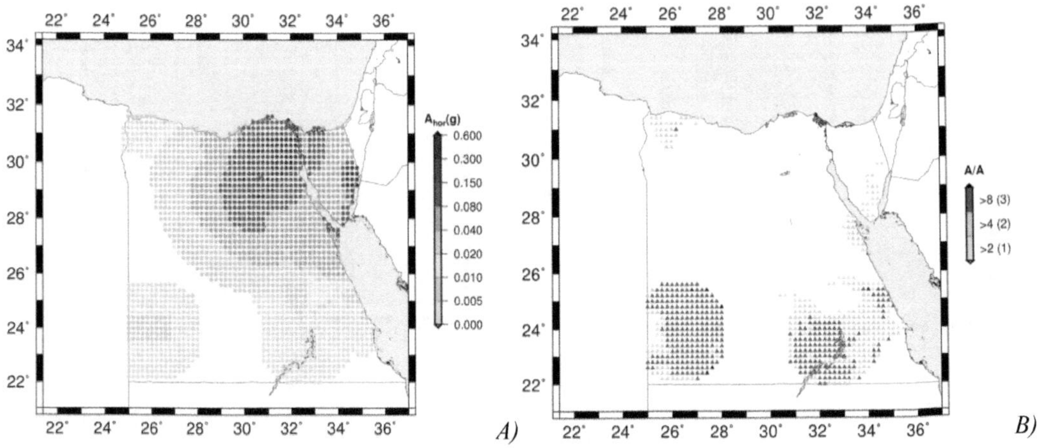

Fig. 4 (A) Peak ground acceleration map for Experiment 1.3; (B) Map of the ratio between PGAs maps from Experiments 2.1 and 2.3, respectively = Experiment 2.1/Experiment 2.3.

4.4 Earthquake Catalog Truncated to 1900 and $\gamma_{EM}\sigma_M = 0.5$ (without Seismogenic Nodes) (Experiment 1.4)

In this section, the magnitude for historical seismicity truncated to 1900 has been increased by 0.5, while the nodes are not considered (Figure 5a).

The computed ratio map between the results of Experiments 1.1 and 1.4 (Figure 5b) shows few upward and a lot of downward triangles which pinpoint that adding 0.5 to the magnitude of the historical events will result in a seismic hazard map capable of forecasting well the ground shaking occurred after 1900 for the sites where the historical data are exist.

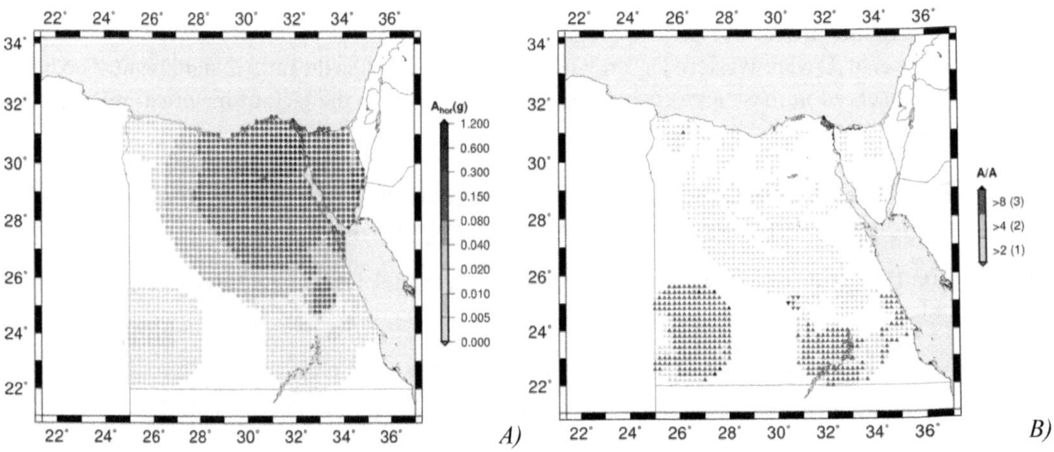

Fig. 5 (A) Peak ground acceleration map for Experiment 1.4; (B) Map of the ratio between PGAs maps from Experiments 1.1 and 1.4, respectively = Experiment 1.1/Experiment 1.4).

4.5 Earthquake Catalog Truncated to 1900, $\gamma_{EM}\sigma_M$ and with Seismogenic Nodes (Experiment 1.5)

Figure 6 a shows the PGA map developed considering both historical catalog where the magnitude has been increased by $0.5 = \gamma_{EM}\sigma_M$, and seismogenic nodes with magnitude unaltered and equal to the lower threshold defined by MZ (Gorshkov et al., 2019).

As it could be expected there is no much change in the map of the ratio between the results of Experiments 1.4 and 1.5 (Figure 6 b) due to the small magnitude of the nodes relative to the one from the instrumental catalog. Only a small change can be detected at the eastern Sinai and near the Gulf of Aqaba.

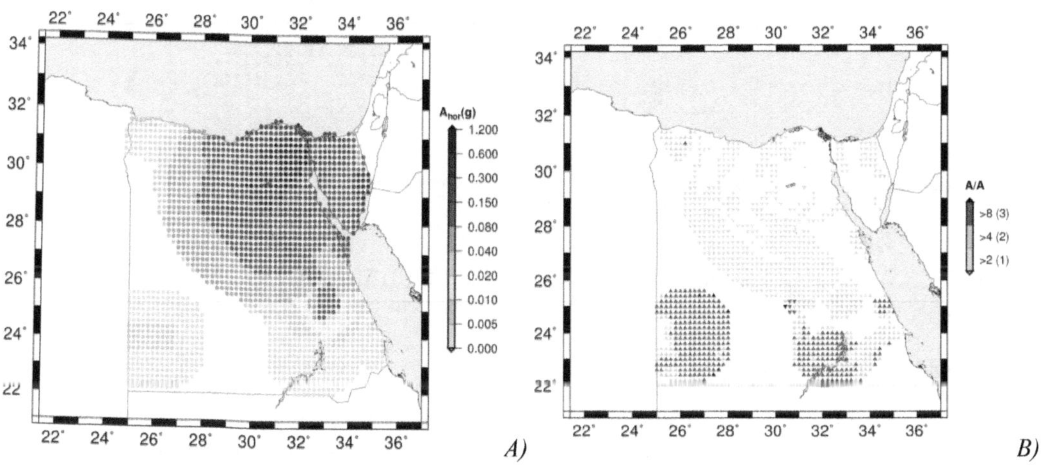

Fig. 6 (A) Peak ground acceleration map for Experiment 1.3; (B) Map of the ratio between PGAs from Experiments 1.1 and 1.5, respectively = Experiment 1.1/ Experiment 1.5.

5. Relevance to Consider M_{design} for Seismogenic Nodes in Egypt

The opportunity to test if nodes alone do enable us to obtain a trustworthy DGA map, i.e., an NDSHA map that envelopes the one derived from past earthquake history, is what drives the concept of expanding the MCRE definition to nodes. We carry out the following computations to reach the objective:

- only nodes with their magnitude, defined as the MZ minimum threshold;
- only nodes with their magnitude + 0.5, i.e., M_{design} magnitude, equal to $M_{design} = M_{max} + \gamma_{EM}\sigma_M$, with $\gamma_{EM}\sigma_M = 0.5$
- only nodes with their magnitude $\underline{\boldsymbol{M}} + 0.7$, i.e., $\gamma_{EM}\sigma_M = 0.7 = 3\sigma_M$.

The recognition of seismogenic nodes has been, so far, limited to the Northeastern of Egypt as shown by Gorshkov et al. (2019), and we obviously limit our computation to this part of the country.

5.1 Seismogenic Nodes with MZ Magnitude (Experiment 2.1).

The computed PGA map shows much less ground shaking intensity than the reference run that is indicated also by the map of the ratio between Experiments 2.1 and 3.1, which is due to the small magnitude threshold for the defined nodes, i.e., set equal to 5 (Figure 7 a, b).

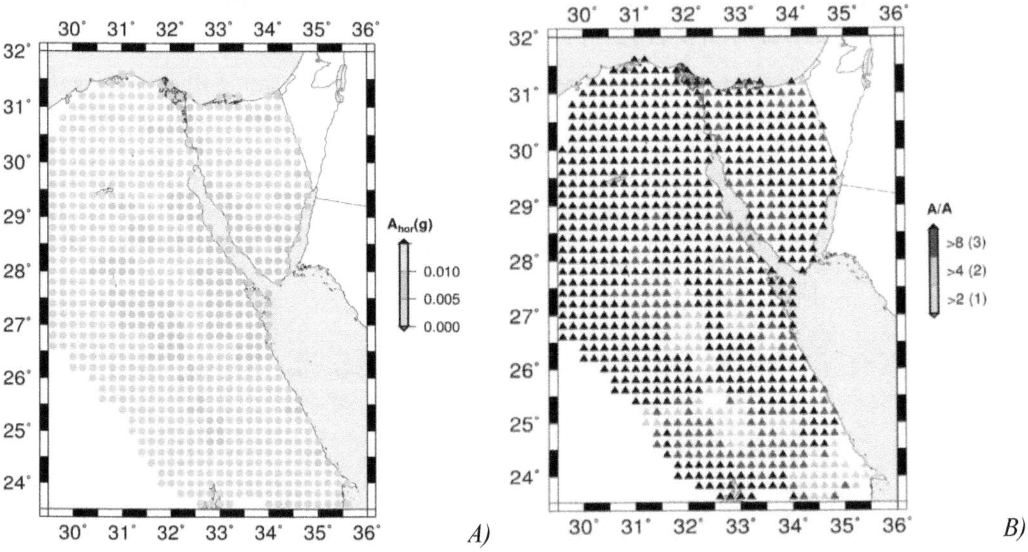

Fig. 7 (A) Peak ground acceleration map for Experiment 1.1; (B) Map of the ratio between PGAs from Experiments 1.1 and 2.1, respectively = Experiment 1.1(A)/ Experiment 2.1.

5.2 Seismogenic Nodes and $\gamma_{EM}\sigma_M = 0.5$

In this experiment, the M_{design} magnitude is defined as $\underline{\boldsymbol{M}} = 5.0$ plus $\gamma_{EM}\sigma_M = 0.5$, and the PGA map is generated just taking the seismogenic nodes into account. The map of the ratio *Experiment 1.1/Experiment 2.2* shows some locations without any markers near Gulf of Suez, Nile Valley, and Eastern Desert (sites where there is a change of at more than one level of macroseismic intensity corresponding to about a factor of 2 in the peak value are mapped), where at some other sites there are downward triangles which indicate that the nodes are the controlling scenarios. The black triangles in the west of Figure 2.2b indicate where the MZ studies have not yet been performed.

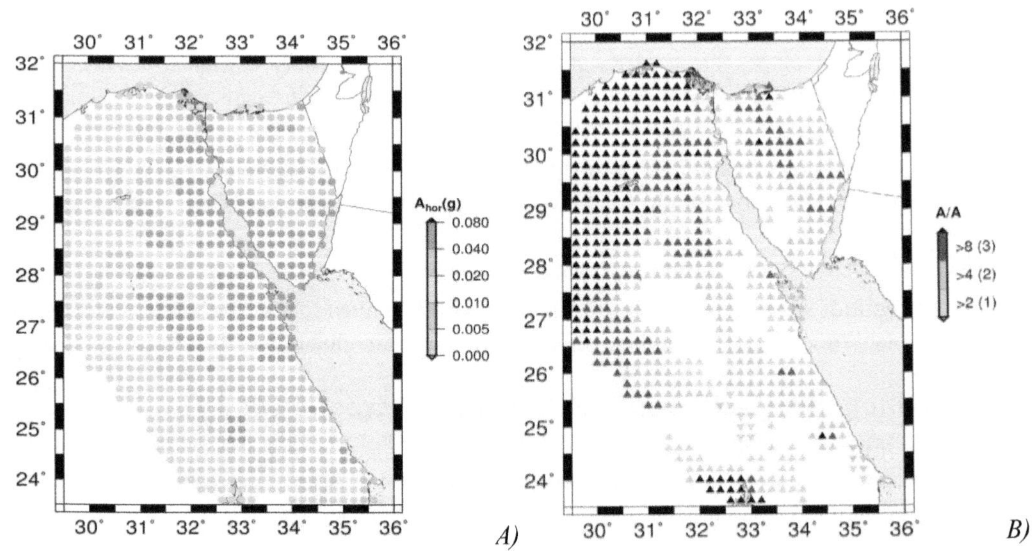

Fig. 8 (A) Peak ground acceleration map for Experiment 1.2; (B) Map of the ratio respectively Experiment 1.1/ Experiment 2.2 (1.1/2.2).

5.3 Seismogenic Nodes and $\gamma_{EM}\sigma_M = 0.7$

Quite naturally, the map of the ratio shows more sites with no triangles (upward or downward) than the map of the previous experiment (Experiment 2.2) and more downward triangles which indicates that the addition of $0.7(3\sigma_M)$ to the magnitude of the nodes, ***M*** reduces the difference between the reference experiment and the current one. It is worth mentioning that the presence of black upward triangles, which indicate a significant difference between the reference and current runs, in the north east and south of the map in Figure 9b, are due to the fact that seismogenic nodes are not defined yet for these parts of the country.

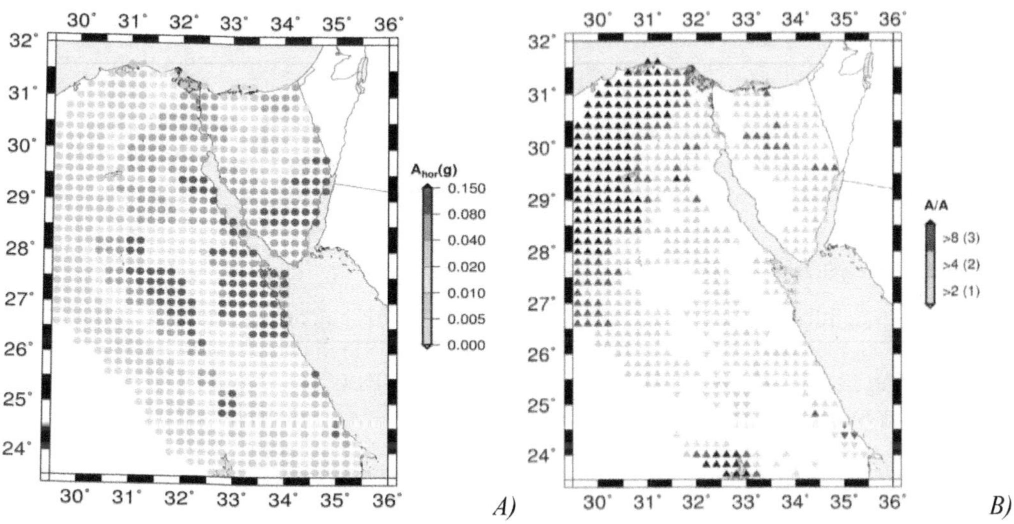

Fig. 9 (A) Peak ground acceleration map for Experiment 3.3; (B) Map of the ratio respectively Experiment 1.1/ Experiment 2.3 (1.1/2.3.).

6. Conclusions

Applying the M_{design} magnitude which is defined as $M_{design} = M_{max} + \gamma_{EM}\,\sigma_M$, to the historical catalog truncated till 1900 allows us to forecast the ground motion observed after 1900 in the areas where we have a good historical catalog (e.g., Cairo, Nile Delta) as shown in Sections 4 and 5. Also, the work here proves the feasibility of the M_{design} concept to forecast ground motion for destructive earthquakes that happened after 1900, e.g., Cairo 1992.

Sections 4 and 5 reveal an interesting pattern regarding the earthquake of 1992 and its surrounding area. No triangles, whether upward or downward, are observed in this specific region, indicating that the event can be accurately forecasted using only the historical earthquake catalog. However, a different scenario unfolds for other areas such as southern Sinai, the Gulf of Suez, southeast Egypt, southwest Western Desert, and northwest Egypt. This is evident as the majority of historical earthquakes are concentrated along the Nile Valley and its Delta, and these earthquakes tend to have higher magnitudes compared to the 1992 Cairo Earthquake.

When examining Experiment 1.3, which involved the addition of seismogenic nodes, no discernible changes are apparent. However, in Experiment 1.4, a few triangles in south Sinai and several downward triangles are observed in the north-central and eastern parts of Egypt relative to Experiments 1.2 and 1.3. This fact suggests that when 0.5 is added to the magnitude of historical events, i.e., when $\gamma_{EM}\,\sigma_M = 0.5$ is used, the resulting seismic hazard map effectively forecast, within $2\sigma_M$, the ground shaking experienced after 1900 in areas where historical data is available.

It is relevant to say that the magnitude of nodes adopted in the computations is set to be $\underline{\boldsymbol{M}} = 5$, consistent with the available information that is lacking for higher magnitude events. So even if $\gamma_{EM}\sigma_M = 0.7$ (i.e., $3\sigma_M$) is added to the node's magnitude still the magnitude of the earthquakes of reference run is still higher (e.g., Nile Delta, Gulf of Aqaba, near Cairo).

The extension of MZ study to the whole territory of Egypt and the recognition of nodes (e.g., $\underline{\boldsymbol{M}}$, > 6.0) with higher magnitude are really needed for the areas of south and western Egypt. A project that embraces teams of archeologists, architects, engineers, geologists, historians, paleoseismologists, and seismologists needs to be launched in order to enhance the historical earthquake record for Egypt by detailed and comprehensive examination of the remains of ancient sites and structures that are distributed over a wide range of time and space.

The concept of M_{design} in NDSHA represents a significant advancement in seismic hazard assessment, offering a physics-based and falsifiable approach to estimating seismic risk. Its application in regions like Egypt, where seismic risk is significant but historical data may be limited, can improve the accuracy of hazard assessments and enhance the safety of infrastructure. By incorporating M_{design} into seismic hazard computations, it is possible to reduce the risk of seismic disasters and improve the resilience of communities to future earthquakes.

Acknowledgements

The authors gratefully acknowledge Prof. Giuliano Panza for his guidance, insightful comments and fruitful discussions, which significantly contributed to the development of this work. This research study was carried out in the framework of the "KNIGHT REG FVG" project – Strengthening resilience to geo-hazards and climate challenges along the Nile corridor (Knowledge-base for NIle Geo-Hazards Tackling, 2025–2027). Moreover, the study partly benefited from funding from the RETURN project (EU: National Recovery and Resilience Plan – NRRP, Extended Partnership) and project, "Enhancing Disaster Resilience in Arab Countries through Multi-Hazards Modelling and Mapping Using Machine Learning and IoT Sensors" funded by Federation of Arab Scientific Research Councils (FASRC) and Academy of scientific research and technology (ASRT), Egypt.

References

Alekseevskaya, M., Gabrielov, A., Gel'fand, I., Gvishiani, A. and Rantsman, E. (1977). Formal morphostructural zoning of mountain territories. *Journal of Geophysics*, *43*(1), 227–233.

Ambraseys N.N., Melville C.P. and Adam R.D. (1994). *The Seismicity of Egypt, Arabia and the Red Sea: A Historical Review*. Cambridge University Press, Cambridge.

Ambraseys, N.N. (1971). Value of historical records of earthquakes. *Nature*, *232*, 375–379.

Ambraseys, N.N. (2001). Far-field effects of Eastern Mediterranean earthquakes in Lower Egypt. *Journal of Seismology*, *5*(2), 263–268.

Badawy, A., ElGabry, M. and Girgis, M. (2010). Historical seismicity of Egypt. In: *A Study for Previous Catalogs Producing Revised Weighted Catalog the Second Arab Conference for Astronomy and Geophysics*, Egypt.

Bormann, P., Liu, R.F., Ren, X., Gutdeutsch, R., Kaiser, D. and Castellaro, S. (2007). Chinese national network magnitudes, their relation to NEIC magnitudes, and recommendations for new IASPEI magnitude standards. *Bulletin of the Seismological Society of America*, *97*, 114–127.

Båth, M. (1973). *Introduction to Seismology* (395 pp.) Birkhäuser Verlag, Basel.

Caputo, M., Keilis-Borok, V., Oficerova, E., Ranzman, E., Rotwain, I. and Solovjeff, A. (1980). Pattern recognition of earthquake-prone areas in Italy. *Physics of the Earth and Planetary Interiors*, *21*(4), 305–320.

Casciati, S. and Borja, R.I. (2004). Dynamic FE analysis of South Memnon Colossus

D'Amico, V., Albarello, D. and Mantovani, E. (1999). A Physics and Chemistry of the Earth, Part A. *Solid Earth and Geodesy*, *24*, 517–521.

ElGabry, M. and Hassan, H. M. (2021). Updated seismic input for next generation of the Egyptian building code. In: *Sustainable Issues in Infrastructure Engineering: The Official 2020 Publication of the Soil-Structure Interaction Group in Egypt* (*SSIGE*) (pp. 55–79). Springer International Publishing.

Gorshkov, A.I., Hassan, H.M. and Novikova, O.V. (2019). Seismogenic nodes ($M \geq 5.0$) in Northeast Egypt and implications for seismic hazard assessment. *Pure and Applied Geophysics*, *176*, 593–610.

Gorshkov, A. and Soloviev, A. (2022). Morphostructural zoning for identifying earthquake-prone areas. In: *Earthquakes and Sustainable Infrastructure* (pp. 135–149). Elsevier.

Gorshkov, A., Hassan, H.M., Mandal, P. and Novikova, O. (2022). Identifying potential earthquake sources in continental environments. *Surveys in Geophysics*, *43*(2), 529–559.

Gorshkov, A., Kossobokov, V. and Soloviev, A. (2003). Recognition of earthquake-prone areas. In: V. I. Keilis-Borok & A. A. Soloviev (Eds.), *Nonlinear Dynamics of the Lithosphere and Earthquake Prediction* (pp. 239–310). Springer.

Hassan, H.M., Panza, G.F., Romanelli, F. and ElGabry, M.N. (2017a). Insight on seismic hazard studies for Egypt. *Engineering Geology*, *220*, 99–109.

Hassan, H.M., Romanelli, F., Panza, G.F., ElGabry, M.N. and Magrin, A. (2017b). Update and sensitivity analysis of the neo-deterministic seismic hazard assessment for Egypt. *Engineering Geology*, *218*, 77–89.

Hutton, J. (1795). Theory of the Earth. *Project Gutenberg. IV (B)*, 109–160.

Li et al. (2025). *Scientific Investigation of Continental Earthquakes and Relevant Studies*. https://doi.org/10.1007/978-981-96-2826-1_9.

Maamoun, M. (1979). Observed intensity: Epicentral distance relations in Egyptian earthquakes. *Bull. of Helwan Obs.*, No. 184.

Maamoun, M., Megahed, A. and Allam, A. (1984). Seismicity of Egypt. *HIAG Bull.*

Maamoun, M., Megahed, A. and Allam, A. (1984). Seismicity of Egypt. *HIAG Bull. IV (B)*, 109–160.

Magrin, A., Peresan, A., Kronrod, T., Vaccari, F. and Panza, G.F. (2017). Neo-deterministic seismic hazard assessment and earthquake occurrence rate. *Engineering Geology*, *229*, 95–109.

Medvedev, S.V. and Sponheuer, W. (1969, January). Scale of seismic intensity. In: *Proc. IV World Conference of the Earthquake Engineering* (pp. 143–153). Santiago, Chile, A-2.

NRIAG (2010). Historical seismicity of Egypt. In: *A Study for Previous Catalogs Producing Revised Weighted Catalog*. Internal catalog developed by the National Research Institute of Astronomy and Geophysics, Helwan, Egypt.

Panza, G. and Bela J. (2020). NDSHA: A new paradigm for reliable seismic hazard assessment. *Engineering Geology*, *275*, 105403. ENGEO Best paper Award 2023.

Panza, G.F. (2017). NDSHA: Robust and reliable seismic hazard assessment. *arXiv preprint arXiv:1709.02945*.

Panza, G.F. and Bela, J. (2019). NDSHA: A new paradigm for reliable seismic hazard assessment. *Engineering Geology*, *275*, 105403.

Panza, G.F., Alvarez, L., Aoudia, A., Ayadi, A., Benhallou, H., Benouar, D., ... and Vaccari, F. (2002). Realistic modeling of seismic input for megacities and large urban areas (the UNESCO/IUGS/IGCP project 414). *Episodes Journal of International Geoscience*, *25*(3), 160–184.

Panza, G.F., Kossobokov, V.G., Laor, E. and De Vivo, B. (Eds.). (2022). *Earthquakes and Sustainable Infrastructure: Neodeterministic (NDSHA) Approach Guarantees Prevention Rather Than Cure*. Elsevier.

Panza, G.F., La Mura, C., Peresan, A., Romanelli, F. and Vaccari, F. (2012). Seismic hazard scenarios as preventive tools for a disaster resilient society. In: *Advances in Geophysics* (*53*, 93–165). Elsevier.

Panza, G.F., Romanelli, F. and Vaccari, F. (2001). Seismic wave propagation in laterally heterogeneous anelastic media: Theory and applications to seismic zonation. In: *Advances in Geophysics* (*43*, 1–95). Elsevier.

Peresan, A., Gorshkov, A., Soloviev, A. and Panza, G.F. (2014). The contribution of pattern recognition of seismic and morphostructural data to seismic hazard assessment. *arXiv preprint arXiv:1406.2932*.

Rugarli, P., Amadio, C., Peresan, A., Fasan, M., Vaccari, F., Magrin, A., ... and Panza, G.F. (2018). Neo-Deterministic Scenario-Earthquake Accelerograms and Spectra. *Engineering Dynamics and Vibrations*, 187.

Rugarli, P., Vaccari, F. and Panza, G.F. (2019). Seismogenic nodes as a viable alternative to seismogenic zones and observed seismicity for the definition of seismic hazard at regional scale. *Vietnam Journal of Earth Sciences*, *41*, 289–304.

Said, R. (1990). The geology of Egypt: Rotterdam. Netherlands, Brookfield. including 3D soil–foundation–structure interaction. *Computers and Structures*, *82*(20).

CHAPTER

15

Research on Three-dimensional Elastoplastic Response Spectra for Seismic Isolated Structures

Aiqun Li[1,2], Ruilin Long[2], Yifeng Wu[3*], Zijian Wang[4] and Jiangdi Fu[5]

Extensive earthquake reconnaissance data and seismic observation records indicate that while horizontal isolation systems effectively mitigate structural acceleration and internal force responses under seismic excitation, critical challenges persist for three-dimensional (3D) isolation, particularly regarding excessive horizontal nonlinear displacement and uncontrolled vertical seismic responses in structures subjected to strong near-fault earthquakes. Consequently, this study develops a single-degree-of-freedom (SDOF) response spectrum analysis program, incorporating commonly used hysteretic models for characterizing both horizontal and vertical isolation behavior. This program is utilized to systematically investigate the horizontal inelastic displacement spectrum and the vertical inelastic response spectrum (encompassing acceleration and displacement ones) for 3D seismic isolated structures. The investigation specifically examines the influence of key isolation system parameters—namely, the natural period (T), ductility coefficient (μ), and stiffness ratio (α)—on the resultant inelastic response spectra. Results reveal that: (1) For near-fault pulse-like ground motions, period normalization significantly reduces the dispersion of the horizontal inelastic displacement spectrum. (2) The horizontal displacement spectrum exhibits a positive correlation with μ within the range [0, 0.6], but displays no significant correlation with μ within [0.6, 3.0], fluctuating around 4.0 cm/s^2. (3) Substantial reduction of the vertical acceleration response can be achieved by appropriately extending the structural fundamental period coupled with vertical hysteresis energy dissipation, while the induced vertical displacement remains within acceptable limits. These findings provide valuable insights for the rapid estimation of seismic responses in 3D seismic isolated structures and offer a theoretical foundation for advancing the implementation and broader application of 3D seismic isolation technology.

1. Introduction

The recurring major earthquakes worldwide have profoundly influenced the iterative evolution of structural seismic design theories, including the 1989 Loma Prieta earthquake, the 2008 Wenchuan

[1-5] School of Civil and Transportation Engineering, Beijing University of Civil Engineering and Architecture, Beijing, 100044, China.

* Corresponding author: wuyifeng@bucea.edu.cn

DOI: 10.1201/9781003684121-17

Earthquake and the 2024 Kyushu Island earthquake. A universal feature of these seismic disasters is the catastrophic damage sustained by structures in affected areas, coupled with the collapse of external transportation networks, which induces a severe "islanding effect" (J.Z. Li & Guan, 2017). This not only directly shortens the golden rescue window after earthquakes, but also triggers cascading secondary disasters, leading to enormous economic losses. In response, governments and the global engineering community have prioritized the seismic performance of structures, driving the continuous advancement of seismic design theories and technologies.

To control the seismic response of structures, base isolation technology prolongs the natural vibration period and provides additional damping by installing isolation layers at appropriate positions, thereby reducing the acceleration response and significantly decreasing the flexural-shear demands on vertical load-bearing members (Fan & Wang, 2001). Although numerous studies have demonstrated the excellent seismic performance of base-isolated structures, they may still suffer from damage, failure, or even global collapse under strong earthquakes, particularly in near-fault seismic zones. For instance, during the 1999 Duzce earthquake in Turkey, the fault passed beneath the under-construction Bolu Viaduct, causing severe damage to its seismic isolation system. This was mainly due to the low stiffness and the consequent large displacement of the isolation layer (Roussis et al., 2003). Similar cases occurred in the 2008 Wenchuan earthquake, where the lateral displacement of the main girder of Baihua Bridge reached 60 cm (Han et al., 2009), and the 2022 Luding earthquake, where post-seismic investigations revealed shear failure of some rubber bearings due to uncontrolled displacement. Given the close correlation between structural seismic damage levels and displacement, research on displacement demands has become a critical aspect for ensuring the safety of seismic isolated structures.

On the other hand, although extensive research on base-isolated structures has been conducted (Pan et al., 2024; Deng et al., 2019), most studies primarily focus on structural responses induced by horizontal components of ground motions. Even in some cases when vertical components are considered, the seismic input is often taken as 2/3 of the horizontal one (Elnashi & Papazoglou, 1997). However, historical earthquake records and disasters have shown that peak acceleration of vertical ground motion in high-intensity areas, especially in near-fault regions, can be remarkably large and even exceed those of horizontal motion (Papazoglou & Elnashi, 1997; X. Li et al., 2008; Bradley, 2011; Chen, 1985). Their spectral characteristics also differ significantly (Bozorgnia & Niazi, 1993), the vertical earthquake is prone to induce high axial forces or tensile forces in RC load-bearing members (Kim et al., 2011; Yi & Wang, 2012), thereby leading to insufficient compression-tension/shear capacity or punching failure of foundations Harrington & Liel, 2016; Tian et al., 2020). Current usual horizontal isolation devices (e.g., laminated rubber bearings, friction pendulum bearing) have vertical stiffness comparable to that of its upper and lower RC columns, failing to effectively isolate vertical ground motion (Furukawa et al., 2013). This poses potential risks to structures sensitive to vertical vibrations, such as nuclear power plants and long-span structures. Therefore, realizing the vertical base isolation effect to control structural vertical seismic responses has become a research focus and challenge in engineering seismology in recent years (Pan et al., 2024; Deng et al., 2019).

For base-isolated structures with regular geometric configurations, seismic response analysis mainly rely on numerical simulation (Zhou et al., 2022), shaking table test (Furukawa et al., 2013; Zhao et al., 2088; Liu et al., 2022) and field measurement (Sheng et al., 2022), but these involve high computational or experimental costs. Alternatively, the equivalent linear method based on elastic spectra or the elastoplastic response spectra method can be used. The former characterizes the nonlinearity of isolation devices by deformation-dependent equivalent shear stiffness and equivalent damping ratio. This method is mature but has certain theoretical errors, while the latter elastoplastic method uses a real nonlinear mechanical model to accurately predict the maximum seismic response of single-degree-of-freedom (SDOF) systems (Chopra & Goel, 2001). Current elastoplastic response spectra are generally constant-ductility or constant-strength spectra, with ductility coefficients typically less than 10 in most studies, suitable for conventional RC pier systems. However, for isolation devices such as lead-rubber bearings, the ductility coefficient often exceeds 20. Thus, for base-isolated structures that can be simplified as SDOF systems, conducting research on adaptive elastoplastic response spectra is essential for rapidly predicting seismic responses.

In summary, horizontal base isolation systems have favorable control effects on acceleration and structural internal force responses, yet the need for further in-depth research on their nonlinear displacement behavior remains. Additionally, vertical seismic actions can also lead to severe seismic disasters in structures, while current research on vertical base isolated structures is still insufficient. Thus, this study aims to conduct a systematic investigation on the horizontal elastoplastic displacement spectra and vertical elastoplastic response spectra (including acceleration and displacement) of base-isolated structures. Given the numerous researches in the field of horizontal acceleration response spectra, this paper will not focus on this topic. The expected research findings will provide new academic support for the further development and improvement of the seismic isolated system, and lay a more solid theoretical foundation for the application of base isolation technologies in engineering practice.

2. Three-Dimensional Mechanical Model of SDOF System for Base-Isolated Structures

2.1 Mechanical Model of Horizontal Seismic Isolation

Extensive investigations have been conducted on the horizontal base isolated structures, leading to the continuous emergence of various isolation devices. Typical examples include lead-rubber bearings and friction pendulum bearings. Zhang Ruijun et al. (2022) studied the effects of different loading rates on the yield characteristics and hysteretic energy dissipation of lead-rubber bearings, with the corresponding hysteretic curves shown in Figure 1. Figure 2 illustrates the theoretic bilinear hysteretic model that characterizes the horizontal nonlinear mechanical properties of such isolation devices, effectively reflecting behaviors such as yielding and energy dissipation.

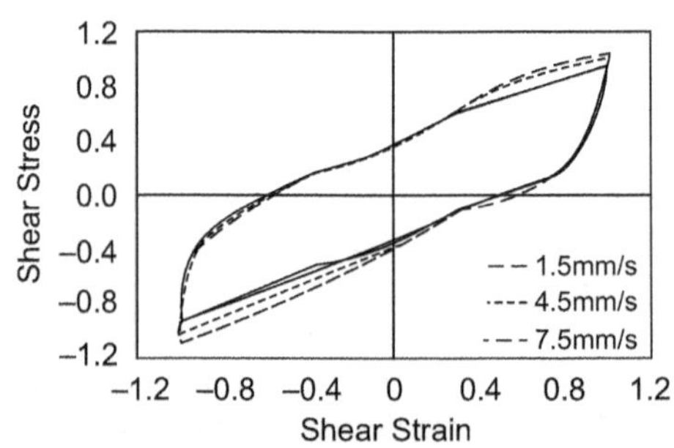

Fig. 1 Hysteretic curves of isolation bearings under varying loading rates (Zhang, R.J,2022)

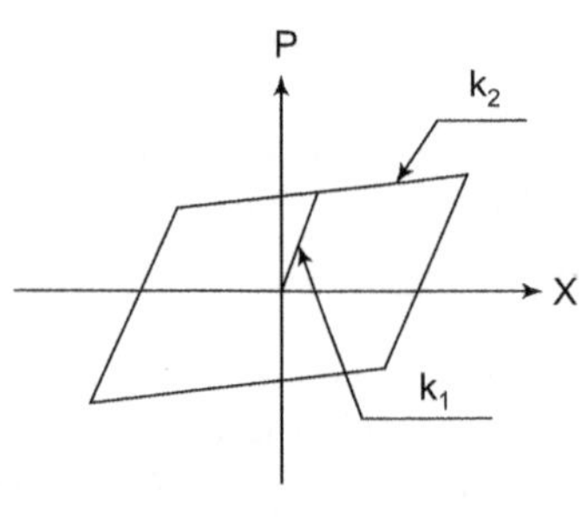

Fig. 2 Theoretic bilinear hysteretic model.

2.2 Mechanical Model of Vertical Seismic Isolation

As previously discussed, achieving vertical base isolation for structures remains a prominent research focus and challenge. Wang Wei et al. (Wang & Wang, 2022) proposed a composite disc spring isolation bearing (CDSCI) by integrating a conventional coned disc spring bearing with a viscoelastic damping layer, as illustrated in Figure 3. This device exhibits a lower vertical compression stiffness compared to conventional rubber bearings, and the friction between viscoelastic damping layers endows it with hysteretic energy dissipation capacity in the vertical direction.

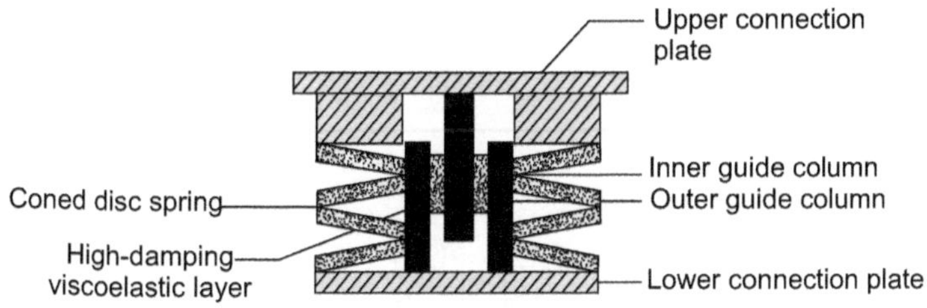

Fig. 3 Drawings of CDSCI (Wang & Wang, 2022).

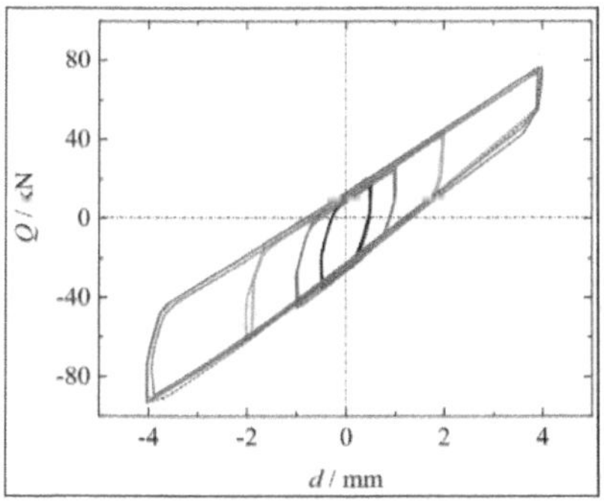

Fig. 4 Vertical hysteresis curve (Wang & Wang, 2022).

In order to incorporate the gravitational preloading effect of the superstructure, the device was pre-compressed by 10 mm, and its hysteretic curves under different vertical displacement amplitudes are shown in Figure 4. Notably, once the static equilibrium position is reached, the isolation bearing swiftly moves into the plastic stage under cyclic loading, experiencing reciprocal deformation starting from its initial static equilibrium position. Similar phenomena occur in other three-dimensional isolation devices due to gravitational preloading, as detailed in the studies by Liu Wenguang (2018) and Cao Yingri et al. (2022). To characterize the nonlinear characteristics of hysteretic curves for vertical isolation devices as described above, this study proposes an asymmetric bilinear hysteretic model (Figure 6). The derivation process of this model is illustrated in Figure 5, assuming that the force-displacement curve of the isolation device under cyclic hysteresis remains bilinear. Here, the static force transmitted from the superstructure to the isolation device is defined as G_0, with the corresponding vertical displacement as z_0. It is assumed that the yield force F_y is significantly smaller than G_0, enabling the device to oscillate around the equilibrium position (z_0, G_0). The yield displacement is denoted as z_y, and the stiffness ratio is defined as $\alpha = k_2/k_1$, where k_1 and k_2 denote the stiffness before and after yielding, respectively. Since gravitational loads have no dynamic effect, excluding the gravitational influence and shifting the origin to the equilibrium point (z_0, G_0) in Figure 5, this simplification leads to the asymmetric bilinear hysteretic model illustrated in Figure 6.

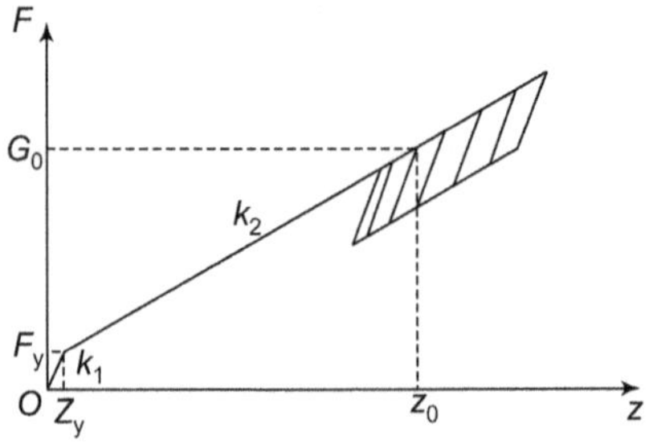

Fig. 5 The vertical hysteretic model.

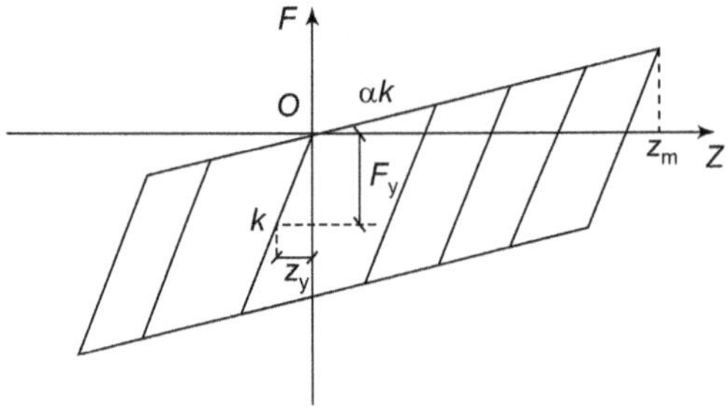

Fig. 6 The asymmetric bilinear model.

3. Selection of Near-Fault Pulse Ground Motions

Compared with far-fault ground motions, near-fault pulse ground motions generally induce larger structural internal force and displacement, this phenomenon is particularly pronounced for base-isolated structures. For this study, 100 near-fault ground motion records were selected from the Next Generation Attenuation-West 2 (NGA-West2) strong motion database released by the Pacific Earthquake Engineering Research Center (PEER) in 2013. The selected earthquakes have magnitudes (M_w) ranging from 5.8 to 7.5 and a rupture distance (R_{rup}) of less than 20 km. Partial ground motion information is listed in Table 1, with detailed magnitude and rupture distance data shown in Figure 7.

Table 1 List of part earthquake records.

RSN Number	*Earthquake Event*	*Magnitude Mw*	*Rupture Distance (km)*	*Pulse Period (s)*	*Peak Ground Velocity (cm/s)*
77	San Fernando	6.6	1.8	1.64	121.9
159	Imperial Valley-06	6.5	0.7	2.34	53.5
415	Coalinga-05	5.8	9.5	0.88	64.4
722	Superstition Hills	6.5	18.5	2.13	33.1
879	Landers	7.3	2.2	5.12	132.3
982	Northridge-01	6.7	5.4	3.16	101.5
1 045	Northridge-01	6.7	5.5	2.98	118.3
1 120	Kobe, Japan	6.9	1.5	1.55	153.2
1 161	Kocaeli, Turkey	7.5	10.9	5.99	53
1 605	Duzce, Turkey	7.1	6.6	5.94	78.9

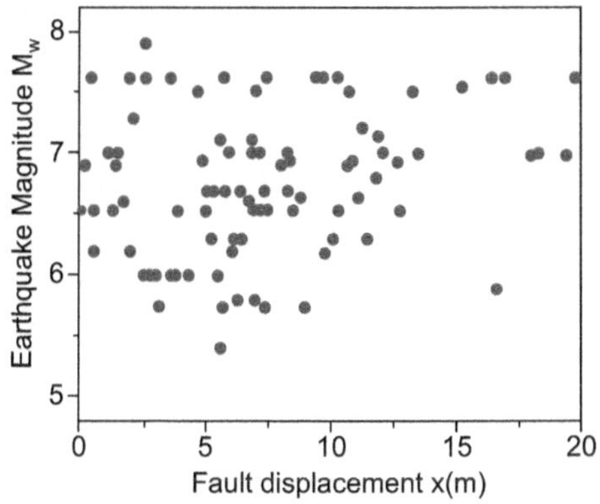

Fig. 7 Relationship between earthquake magnitude and epicentral distance.

4. Elastoplastic Response Spectrum Calculation Procedure

4.1 Equations of SDOF System

A calculation program is formulated based on the incremental form of the equation of motion for a SDOF system,

$$m\Delta\ddot{x}(t) + c\Delta\dot{x}(t) + k\Delta x(t) = -m\Delta\ddot{x}_g \qquad ...(1)$$

Where, m is the mass of the SDOF system particle, c is the damping coefficient, and k is the restoring stiffness. $\Delta\ddot{x}(t)$, $\Delta\dot{x}(t)$ and $\Delta x(t)$ denote the acceleration, velocity, and displacement increments, respectively, while $\Delta\ddot{x}_g(t)$ is the increment of ground motion acceleration.

4.2 Displacement Ductility Coefficient

Ductility reflects the capacity of a structure to undergo inelastic deformation, and hysteretic ductility is particularly crucial for seismic-resistant structures. In a hysteretic model, the displacement ductility coefficient μ is defined as the ratio of the maximum displacement to the yield displacement, expressed as:

$$\mu = \frac{x_m}{x_y} \qquad ...(2)$$

Where, x_m is the maximum displacement from elastoplastic analysis, and x_y is the yield displacement of the structure. It should be mentioned the displacement ductility coefficient μ is generally less than 10, applicable to conventional concrete columns or bridge piers, but not suitable for base-isolated devices with high ductility coefficients.

4.3 Calculation Procedure

A self-developed MATLAB program for nonlinear time-history analysis (Wu et al., 2021) was used to investigate the seismic response of SDOF systems, generating elastoplastic constant-ductility response spectra considering the effects of displacement ductility coefficient μ and stiffness ratio α. The simplified operational flowchart of the program is shown in Figure 8, and the specific steps are as follows:

(1) Input the ground motion and determine the integration step size h based on the time interval of the ground motion record.

(2) Input basic parameters, including the target displacement ductility coefficient μ_0, natural period T, integration step size h, mass m (which has no effect on the response spectra and is generally taken as 1), and stiffness ratio α.
(3) Assume the yield displacement x_y.
(4) Perform elastoplastic time-history analysis using the Newmark-β method, where the elastic damping ratio ξ is taken as 5%.
(5) Extract the maximum displacement response x_m and calculate the displacement ductility coefficient $\mu = x_m/x_y$. If μ exceeds the preset error tolerance, re-assume the yield displacement $x_y(i) = x_m/\mu_0$ and return to Step 4. When the iteration count $i > 3$, to prevent non-convergence due to numerical oscillations, use $x_y(i+1) = (x_y(i) + x_y(i-1)/2)$. If μ falls within the error range, output the results.

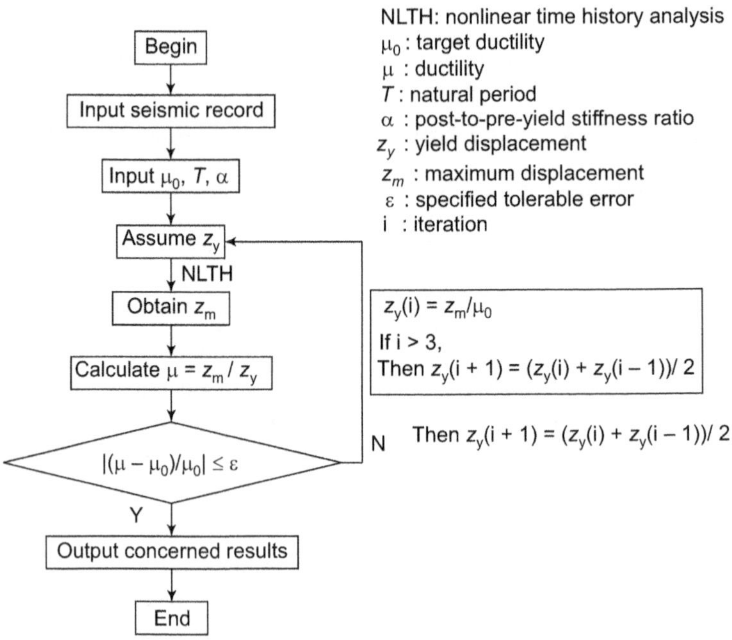

Fig. 8 The calculation procedure of the response spectra (Wu et al., 2021).

5. Factors Considered in Response Spectra

Based on the hysteretic model in Figures 1-2 and 1-6, and referring to the analysis procedure for the response spectra of SDOF systems (Figure 3-1), this study aims to investigate the effects of the natural period T, ductility coefficient μ, and stiffness ratio α of seismic isolated structures on the elastoplastic response spectra. The following assumptions and parameter settings are adopted. For horizontal base-isolation system:

- Assume the hysteretic curve of the SDOF system is bilinear with an inherent elastic damping ratio of 5%;
- The natural vibration period T of the SDOF system ranges from 0.1 T_P to 3T_P with an increment of 0.05 T_P, where T_P is the pulse period;
- For isolation devices with ductility significantly higher than that of ordinary reinforced concrete bridge piers, this study considers several relatively high ductility values, including 5, 10, 15, 20, 25, 30, 40, 50, and 70;
- The post-yield stiffness ratio α is set to 0, 0.05, 0.10, 0.15, and 0.20, respectively.

For vertical base-isolation system:

- Considering the preloading effect of gravity on isolation devices, assume an asymmetric bilinear hysteretic model;
- The ductility coefficient significantly influences the response spectra, and since the maximum displacement in the hysteretic curve of vertical isolation devices is much larger than the yield displacement, the target ductility μ_0 is designated as 1, 5, 10, 15, 20, 30, 50, 70, 100, and 300 to encompass a wide scope of ductility;
- The stiffness ratio α is 0.1, 0.2, 0.3, and 0.4 to assess its impact on the response spectra;
- Considering the restricted range of vertical ground motion period variations, the period range examined in this study spans from 0.02 s to 2 s at an interval of 0.02 s.

6. Response Spectra Analysis

6.1 Horizontal Elastoplastic Displacement Response Spectra

Before calculating the displacement spectra, this study first computed the intermediate variable, nonlinear deformation coefficient spectra C_μ, which is defined as follows,

$$C_\mu = \mu/R \quad \text{...(3)}$$

In the formula, the strength reduction factor R is defined as the ratio between the demand of the elastic structural system and that of the corresponding elastoplastic system. Currently, Veletsos, Newmark, and others (Veletsos, 1960) have developed several recommended equations for R and C_μ. The determination of the ratio C_μ is crucial, as it serves as the foundation for constructing inelastic displacement spectra. The peak displacement u_m of the inelastic SDOF system can be derived using the following expression.

$$u_m = C_\mu (T/2\pi)^2 A \quad \text{...(4)}$$

Where, A denotes the design elastic pseudo-acceleration, and T represents the natural vibration period.

Given the distinct spectral characteristics of ground motions, Mavroeidis et al. (200$) and Iervolino et al. (2012) normalized the ductility displacement coefficient C_μ using the pulse period T_P to mitigate result dispersion. Zhang et al. (2011) and Sucuoglu et al. (2019) adopted the site period T_g for normalizing the period-dependent strength reduction factor. For near-fault pulse-like ground motions, this normalization approach is also applicable to elastoplastic response spectra. Notably, period normalization preserves the statistical pattern of results while reducing scatter, enhancing the engineering utility of response spectra.

To further verify the reliability of the nonlinear displacement coefficient spectra calculation method constructed in this paper, Figure 9 compares the normalized results calculated using 100 seismic records in this study with the deformation coefficient spectrum given by Iervolino et al. (2012). Considering the limited range of μ studied by Iervolino et al. (2012), only the comparison results for $\mu = 2$, 4, and 6 are presented.

As is illustrated in Figure 9, the variations of C_μ exhibit good agreement across most of the period range, with only slightly larger discrepancies near $T/T_P = 1$, possibly arising from different selections of pulse-like records.

The simplified nonlinear deformation coefficient spectrum is illustrated in Figure 10. Within the 0-0.3 normalized period interval, the influences of μ and α are carefully investigated. Therefore, Figure 10(a) consists of five parts, each representing different α ratios, and each part's results for distinct μ values are illustrated by the curves of varying colors. Figure 10(b) shows the change of the normalized period from 0.3 to 3. When the period is less than 0.6, only the influence of ductility is considered. When the normalized period is greater than 0.6, both influencing factors are ignored, and only one curve is plotted in the figure.

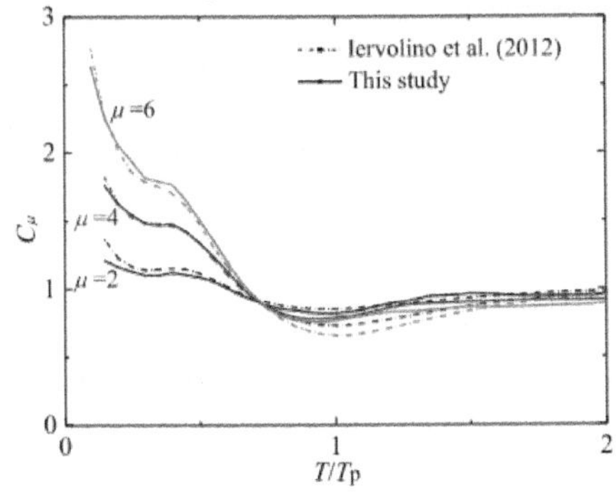

Fig. 9: Comparison of the C_μ between different studies.

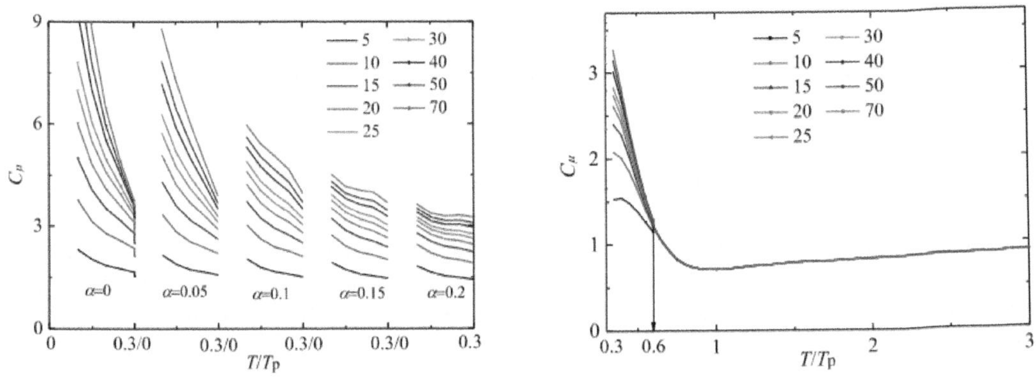

Fig. 10 C_μ with Normalized Periods (a) 0–0.3, (b) 0.3–3

According to Equation 4, prior to constructing the inelastic displacement spectra, the period-normalized acceleration response spectrum S_a is also required to be established. Given the extensive existing research on this topic and the availability of acceleration response spectra, this paper will not elaborate further. Taking the PGA as 2m/s² as an example, after obtaining C_μ and S_a as functions of T/T_p, the normalized displacement spectra with the horizontal axis T/T_p and the vertical axis S_d/T_p^2 can be derived using Equation 4. As observed in Figure 11, S_d/T_p^2 as a function of T/T_p exhibits a positive correlation with both μ and T/T_p within the range of 0 to 0.6. Conversely, when T/T_p exceeds 0.6, S_d/T_p^2 demonstrates no distinct relationship with the normalized period, instead fluctuating around 4.0 cm/s².

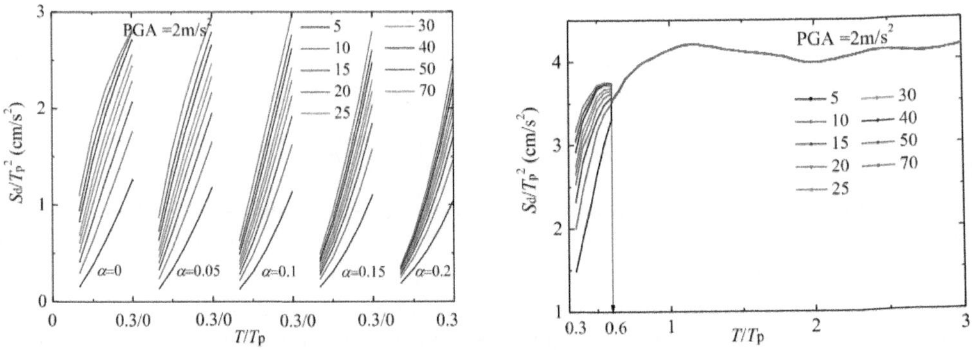

(a) (b)

Fig. 11 Inelastic displacement spectra with the normalized period of (a) 0–0.3, (b) 0.3–3.

6.2 Vertical Isolation Elastoplastic Response Spectrum

To calculate the response spectra of the asymmetric bilinear hysteretic model, a self-programmed code is developed, and its accuracy must be verified. Since common software does not include this asymmetric model, a bilinear spring is used to simulate the ideal model in Figure 1-5, and an SDOF analysis model in ANSYS is established (Figure 12). Taking the RSN159 ground motion (Imperial Valley-06, 1979, Agrarias station) as an example, the time-history of displacement with a natural period $T = 1.0$s, $\mu_0 = 5$, and $\alpha = 0.2$ is computed using both ANSYS and the self-programmed MATLAB code (Figure 13). The results show perfect agreement in displacement, validating the numerical precision of the self-programmed code. Furthermore, comparisons of maximum displacement responses under different ductility coefficients (Figure 14) confirm consistency between the two methods, with the self-programmed code demonstrating significantly higher efficiency (much shorter computation time for Figure 14 than ANSYS). These analysis results fully verify the reliability and efficiency of the self-developed elastoplastic response spectra code.

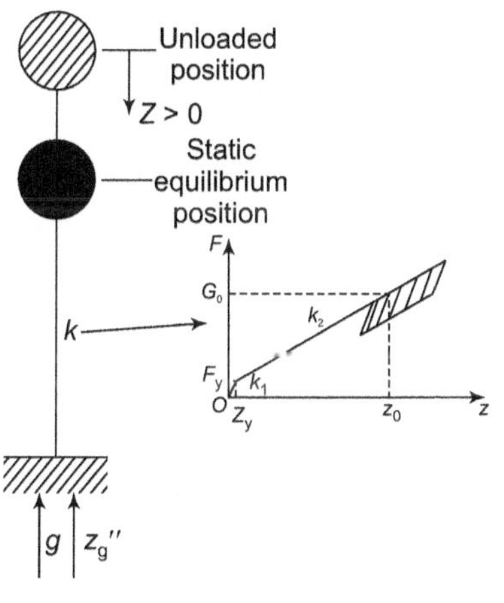

Fig. 12 The verification SDOF model.

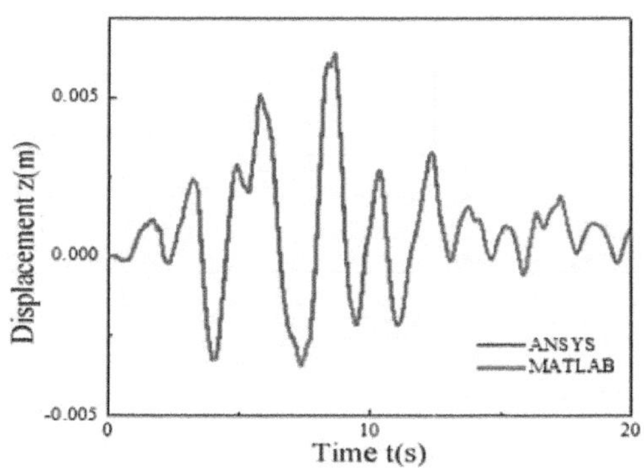

Fig. 13 Validation of Elastoplastic Displacement Time-History.

In this study, the peak ground acceleration (PGA) for all vertical ground motions is set to 1 m/s^2. Following the program configuration and parameter analysis described above, the study also involves 100 seismic records, 10 target ductility levels, four stiffness ratios, and 100 period points, resulting in a total of $100 \times 100 \times (4 \times 9 + 1) = 370000$ analysis cases. The self-developed program is employed to compute the constant-ductility response spectra for both acceleration and displacement.

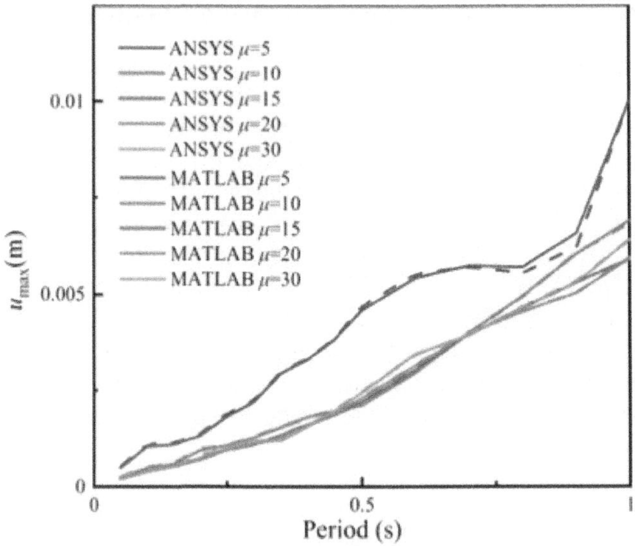

Fig. 14 Comparison of the displacement spectra.

(1) Analysis of acceleration spectra

For different ductility coefficients μ and post-yield stiffness ratios α, the vertical acceleration response spectra within the period range of 0–2s are presented in Figure 16. The curve for $\mu = 1$ shows that in the period range of 0–0.1s, the vertical elastic acceleration response rapidly rises from 1 m/s^2 to 2.53 m/s^2 and then gradually declines. Even when the vertical natural vibration period is extended to 0.5s, the spectrum value at this point is 1.01 m/s^2, still exceeding the input peak ground acceleration of 1 m/s^2. Under such circumstances, the vertical stiffness of the isolation device is inherently low, potentially causing uncontrolled deformation of the bearing under self-weight. In Figure 15(a), dashed lines illustrate the elastic acceleration spectra for damping ratios of 10%, 15%, and 20%. Results reveal that spectrum values diminish with increasing damping ratios, though the reduction remains subtle. Hence, vertical isolation cannot be adequately achieved by merely extending the natural vibration period or enhancing the viscous damping ratio.

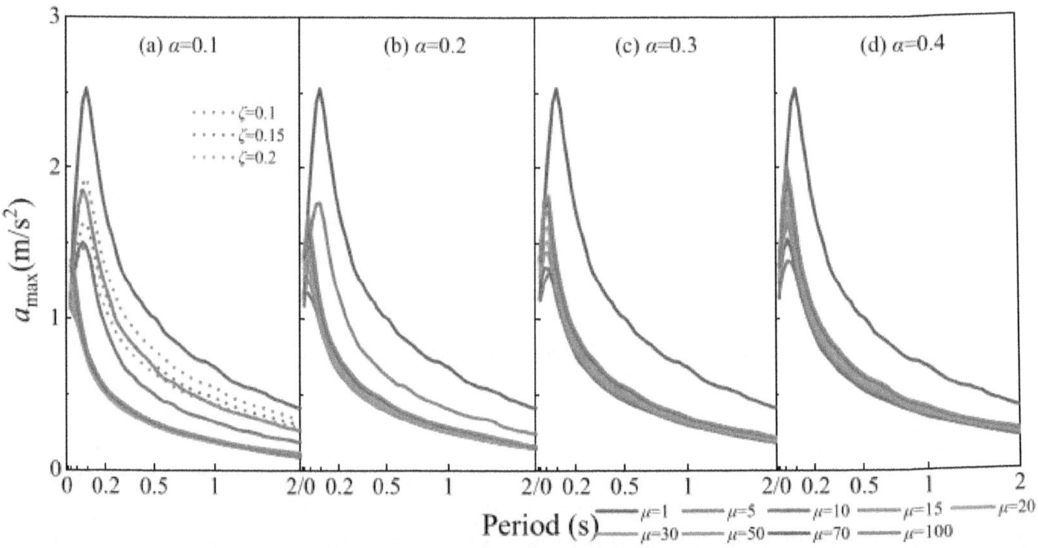

Fig. 15 The vertical inelastic acceleration spectra.

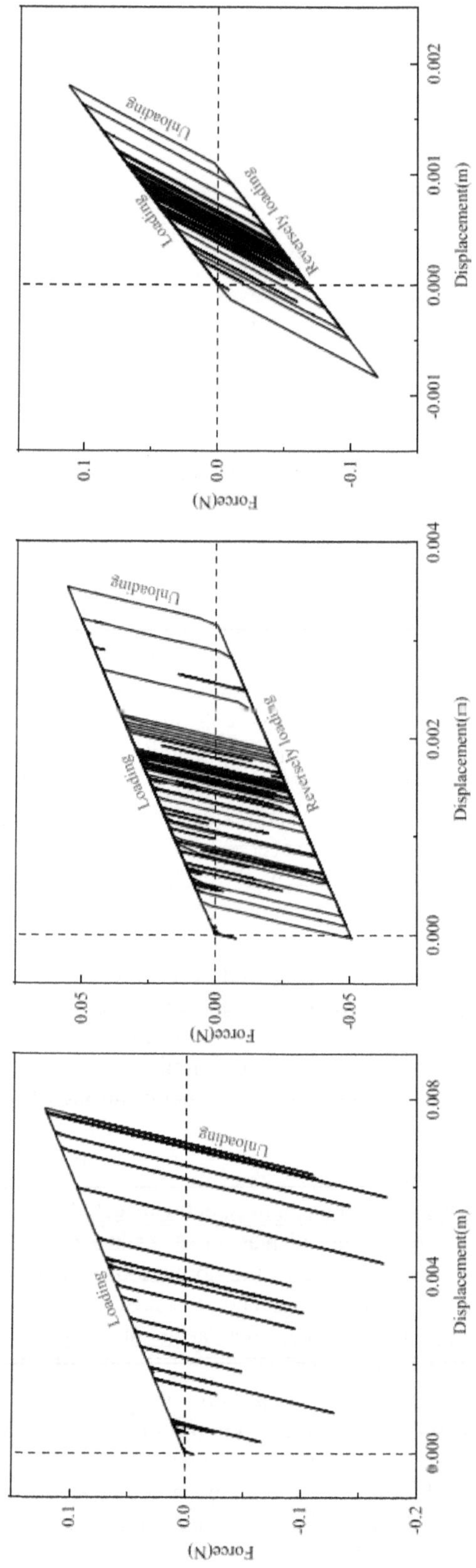

(a) case 1 (μ=5, α=0.1, T=0.5s)

(b) case 2 (μ=20, α=0.1, T=0.5s)

(c) case 3 (μ=20, α=0.4, T=0.5s)

Fig. 16 The hysteresis curves of three typical cases.

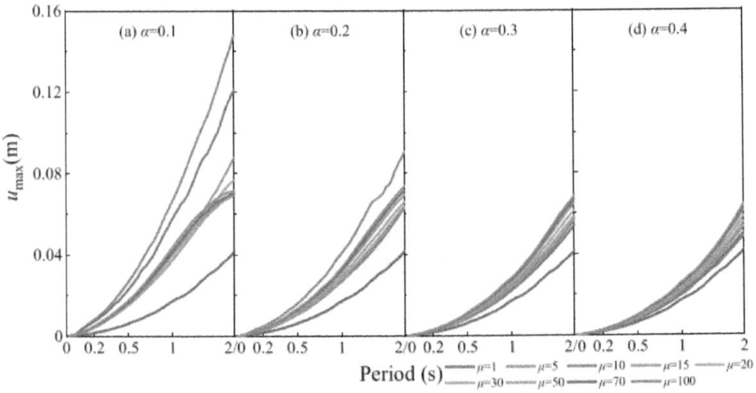

Fig. 17 The vertical inelastic displacement spectra.

For the constant-ductility elastoplastic acceleration spectra, as the ductility increases, its value drops sharply over the entire period range, and the period corresponding to the maximum value gradually shifts leftward and decreases. Within Figure 15(a), spectral values at $\mu = 5$ and $\mu = 10$ are markedly higher than all values for $\mu \geq 15$, with the impact of ductility on the acceleration spectrum weakening as μ increases further. Figure 15(b) exhibits a similar trend but with a nuance: only the $\mu = 5$ spectral value exceeds those at other ductility levels. In Figures 15(c) and 15(d), as μ increases, the spectral values also increase, and the inelastic acceleration spectrum is positively correlated with the ductility coefficient μ. Notably, such a trend is absent in horizontal response spectra. The following hysteresis curves under three typical working conditions explains the above phenomenon, with specific working condition parameters marked in Figure 16. The ground motion input is calculated using the record from RSN159 (Imperial Valley-06, 1979, Agrarias station). The primary distinction among the three sets of hysteresis curves is that in Figure 15(a), only loading and unloading phases are observed, with no reverse loading states. The curves fail to produce reciprocating hysteresis loops, unable to effectively dissipate seismic energy, leading to larger displacement responses. Therefore, when α is small, an isolation system with $\mu \geq 15$ is preferred. At this time, the curves include not only unloading states but also reverse loading states, and the device can dissipate energy through reciprocating hysteresis loops to reduce seismic responses, as shown in Figures 16(b) and 16(c). Additionally, $\mu \geq 15$ means $z_y \leq z_m/15$ and $F_y \leq k_1 z_m/15$, which limits the magnitude of the yield force F_y. That is to say, in Figure 1-5, it is necessary to assume that the yield force F_y is significantly smaller than G_0, aiming to ensure that the isolation device oscillates near the equilibrium position and dissipates energy through hysteresis to significantly reduce seismic responses.

(2) Analysis of displacement spectra

Figure 17 presents the elastoplastic displacement spectra affected by T, a α, and μ. With the increase of ductility, in Figure 17(a), a larger displacement response is observed when $\mu < 15$. As elaborated in the previous section, this is due to the absence of hysteresis loops, which hinders the dissipation of vertical ground motion energy. The existence of reverse unloading reduces the maximum displacement response, making the peak displacement in Figure 17(a) notably larger than those in Figures 17(b) and (c). It is evident that in Figures 17(c) and (d), the inelastic displacement spectra exhibit a positive correlation with μ, albeit with a negligible effect. Moreover, the elastic displacement spectrum has the smallest response among all cases. The displacement spectrum increases rapidly as T rises. When α grows from 0.1 to 0.4, the inelastic displacement response gradually decreases. Taking $T = 0.1$s and $\alpha = 0.1$ as an example, the inelastic displacement response is relatively small in the short period, measuring only 0.0017 m.

7. Summary and Conclusions

This paper developed a corresponding elastoplastic response spectra analysis program using MATLAB. The elastoplastic response spectrum values of three-dimensional isolation structures under various working conditions were calculated for 100 seismic records. The main conclusions are as follows:

- For constant-ductility elastoplastic displacement response, spectra of horizontal seismic isolation structures, normalizing the period T using the pulse period T_p can effectively reduce the dispersion of displacement spectra. Taking the peak ground acceleration (PGA) as 2 m/s² as an example, S_d/T_p^2 as a function of T/T_p exhibits a positive correlation with both μ and T/T_p within the range of 0 to 0.6. However, when T/T_p exceeds 0.6, S_d/T_p^2 demonstrates no distinct relationship with the normalized period, instead fluctuating around 4.0 cm/s².
- For constant-ductility elastoplastic response, spectra of vertical seismic isolation structures, merely extending the natural vibration period or increasing the viscous damping ratio has an insignificant effect on reducing the vertical acceleration response; the elastic displacement spectrum has the smallest response compared to other inelastic ones, and its spectral value increases rapidly with the increase of the period; it is positively correlated with ductility and has smaller spectral values during brief periods; when the post-yield stiffness ratio α is small, the displacement response with low ductility coefficients is large, so it is recommended that the ductility coefficient be greater than 15 to reduce the displacement response and effectively control the acceleration response, it is recommended that α be taken as 0.1 to 0.2.

References

Bozorgnia Y, and Niazi M. (1993). Distance scaling of vertical and horizontal response spectra of the Loma Prieta earthquake. *Earthquake Engineering and Structural Dynamics*, *22*(8): 695–707.

Bradley, B.A. and Cubrinovski, M. (2011). Near-source strong ground motions observed in the 22 February 2011 Christchurch earthquake. *Seismological Research Letters*, *82*(6): 853–865.

Cao, Y,R., Pan, P., Sun, J.B., et al. (2022). Mechanical properties and isolation effectiveness of 3D disc spring-single friction pendulum isolation (vibration isolation) system. *Journal of Building Structures*, *43*(7): 44–53.

Chen, N.Y. (1985). Engineering case studies on building damage caused by vertical seismic forces. *Engineering Mechanics*, *2*(4): 130–133.

Chopra, A.K. and Goel, R.K. (2001). Direct displacement-based design: use of inelastic vs. elastic design spectra. *Earthquake Spectra*, *17*(1): 47–64.

Deng, C.G., Zeng, M. and Xu, Z.G. (2019). State-of-the-art review on vertical seismic isolation technology in structural engineering. *Journal of Architecture and Civil Engineering*, *36*(4).

Elnashai, A.S. and Papazoglou, A.J. (1997). Procedure and spectra for analysis of RC structures subjected to strong vertical earthquake loads. *Journal of Earthquake Engineering*, *1*(01): 121–155.

Fan, L.C. and Wang, Z.Q. (2001). *Qiao liang jian ge zhen she ji* [Bridge seismic isolation design]. Renmin Jiaotong Chubanshe.

Furukawa, S., Sato, E., Shi, Y., et al. (2013). Full-scale shaking table test of a base-isolated medical facility subjected to vertical motions. *Earthquake Engineering and Structural Dynamics*, *42*(13): 1931–1949.

Han, Q., Du, X., Liu, J., et al. (2009). Seismic damage of highway bridges during the 2008 Wenchuan earthquake. *Earthquake Engineering and Engineering Vibration*, *8*(2): 263–273.

Harrington, C.C. and Liel, A.B. (2016). Collapse assessment of moment frame buildings, considering vertical ground shaking. *Earthquake Engineering and Structural Dynamics*, *45*(15): 2475–2493.

Iervolino. I., Chioccarelli, E. and Baltzopoulo, G. (2012). Inelastic displacement ratio of near-source pulse-like ground motions. *Earthquake Engineering and Structural Dynamics*, *41*(15): 2351–2357.

Kim, S.J., Holub, C.J. and Elnashai, A.S. (2011). Experimental investigation of the behavior of RC bridge piers subjected to horizontal and vertical earthquake motion. *Engineering Structures*, *33*(7): 2221–2235.

Li, J.Z. and Guan Z.G. (2017). Development of bridge seismic design theory: From seismic resistance and reduction design to post-earthquake recoverable design. *China Journal of Highway and Transport*, *30*(12): 1–9, 59.

Li, X., Zhou, Z, Yu H, et al. (2008). Strong motion observations and recordings from the great Wenchuan Earthquake. *Earthquake Engineering and Engineering Vibration*, *7*: 235–246.

Liu, W., Xu, H., He, W., et al. (2018). Static test and seismic dynamic response of an innovative 3D seismic isolation system. *Journal of Structural Engineering, 144*(12): 04018212.

Liu, W.G., Li, J.L., Xu, H., et al. (2002). Rocking dynamic model of 3D isolation structure and verification via shaking table test. *Journal of Vibration Engineering, 35*(5): 1200–1210.

Mavroeidis, G.P., Dong, G. and Papageorgiou, A.S. (2004). Near-fault ground motions, and the response of elastic and inelastic single-degree-of-freedom (SDOF) systems. *Earthquake Engineering and Structural Dynamics, 33*(9): 1023–1049.

Pan, P., Zeng, Y., Cao, Y.R., et al. (2024). State-of-the-art review on seismic isolation technology for building structures. *Engineering Mechanics, 41*(5): 39–54.

Papazoglou, A.J. and Elnashai, A.S. (1996). Analytical and field evidence of the damaging effect of vertical earthquake ground motion. *Earthquake Engineering and Structural Dynamics, 25*(10): 1109–1137.

Roussis, P.C., Constantinou, M.C., Erdik, M., et al. (2003). Assessment of performance of seismic isolation system of Bolu Viaduct. *Journal of Bridge Engineering, 8*(4): 182–190.

Sheng, T., Liu, G., Bian, X., et al. (2022). Development of a three-directional vibration isolator for buildings subject to metro-and earthquake-induced vibrations. *Engineering Structures, 252*: 113576.

Sucuogluu, H., Diclelil, M. and Nurtug, A. (1994). An analytical assessment of elastic and inelastic response spectra. *Canadian Journal of Civil Engineering, 21*(3): 386–395.

Tian, Ying, Xiang Liu and Sara George. (2020). Effects of vertical ground motion on seismic performance of reinforced concrete flat-plate buildings. *Journal of Structural Engineering, 146*(12) : 04020258.

Veletsos, A.S. (1960). Effect of inelastic behavior on the response of simple systems to earthquake motions. In: *Proc. of 2nd WCEE*, Tokyo and Kyoto, 1960: 895–912.

Wang, W., Wang, X. and Li A. (2022). Coned disc spring compound vertical isolation: testing and modelling. *Journal of Earthquake Engineering, 26*(9): 4877–4909.

Wu, Y.F., Zhang, G.D., Li, A.Q., et al. (2021). Constant-ductility inelastic displacement spectra for generalized flag-shaped models under far-field ground motions. *Journal of Southeast University* (English Edition), *51*(5): 1–9.

Yi, W.J. and Wang, W.Y. (2012). Influence of vertical ground motion in near-fault regions on seismic performance of RC frame structures. *China Civil Engineering Journal, 45*(10): 81–88.

Zhang, H.M. and Yi, W.J. (2011). Strength reduction factor spectra for SDOF structures under near-fault earthquakes. *China Civil Engineering Journal, 44*(10): 53–59.

Zhang, R.J. (2022). *Temperature Dependence of Isolation Bearings and Research on Novel Fully Prefabricated Frame Structural System* [D]. Nanjing: Southeast University.

Zhao, Y.M., Su, J.Y., Zhou, X.Y., et al. (2008). Shaking table test and numerical simulation of disc spring vertical isolation system. *Journal of Building Structures, 29*(6): 99–106.

Zhou, Y., Chen, P. and Mosqueda, G. (2022). Numerical studies of three-dimensional isolated structures with vertical quasi-zero stiffness property. *Journal of Earthquake Engineering, 26*(7): 3601–3622.

CHAPTER

16

Dynamic Response and Shear Characteristics of Rubber Fiber Reinforced Calcareous Sand

Jiajun Shu[1], Wenbo Wang[2], Tao Li[3*] and Bo Liu[4*]

Calcareous sand, a typical foundation soil in marine and island engineering, is characterized by strong particle breakage and high compressibility, making it vulnerable to cumulative deformation and liquefaction under cyclic loading, such as earthquakes and wave action. To address the limitations of conventional reinforcement methods in island environments, this study systematically investigated the effects of rubber fiber length (5–20 mm), rubber fiber content (0.3–1.2%), and the cyclic stress ratio (0.2–0.3) on the dynamic shear behavior of calcareous sand through a series of dynamic triaxial tests. The results indicate that the deformation process of calcareous sand under cyclic loading can be divided into three stages: vibratory compaction, shear deformation, and failure. Rubber fibers significantly enhance the liquefaction resistance of calcareous sand by filling pores and increasing interparticle interlocking. Under optimal conditions—a fiber length of 15 mm and a content of 0.9%—the liquefaction resistance of the sample nearly doubled compared with that of unreinforced calcareous sand. However, excessively long fibers or excessively high fiber content may lead to uneven distribution and deteriorated performance. The evolution of the dynamic pore water pressure demonstrates that rubber fibers effectively delay pore pressure buildup, reducing the rate of increase in pore pressure by more than 40% under optimal conditions. Analysis of the dynamic shear modulus further revealed that the inclusion of fibers increased the initial stiffness of the sand by approximately 20%, although the modulus degradation rate accelerated under high cyclic stress ratios.

1. Introduction

Calcareous sand is a distinctive type of foundation material predominantly distributed in coral reef islands and continental shelf platforms in tropical and subtropical marine environments. It is commonly found in regions such as South China Sea islands and the Great Barrier Reef in Australia. The particles of calcareous sand are primarily composed of calcium carbonate, which originates from the skeletal remains of marine organisms, including corals, shells, and algae. This material is characterized by irregular particle shapes, high susceptibility to fragmentation, weak interparticle cementation, and significant compressibility (Shahnazari & Rezvani, 2013; Wang et al., 2024).

[1-4] School of Mechanics and Civil Engineering, China University of Mining and Technology-Beijing, Beijing 100083, China.

* Corresponding author: lit@cumtb.edu.cn; liub@cumtb.edu.cn

DOI: 10.1201/9781003684121-18

Owing to these characteristics, calcareous sand is frequently employed in island infrastructure projects as a natural foundation material or as a filler for artificial islands, seawalls, harbors, and offshore platforms. However, island regions—often located within tectonically active zones or along seismic wave propagation paths—face an elevated risk of intense ground shaking. Consequently, the engineering behavior of calcareous sand foundations under strong seismic events and prolonged aftershock sequences has become a critical concern. Seismic loading, which is characterized by high frequency and intensity, profoundly impacts the structural integrity of calcareous sand foundations. Specifically, earthquake-induced dynamic loads can lead to a rapid increase in excess pore water pressure, triggering liquefaction while simultaneously promoting frequent interparticle collisions. These interactions accelerate particle breakage, thereby disrupting the original granular fabric, reducing the effective stress, and diminishing the strength reserve of the soil matrix (Li et al., 2024; Tang et al., 2018; Guo et al., 2017). Compared with siliceous sands, calcareous sands exhibit a more pronounced tendency toward cumulative plastic deformation and a more rapid decrease in shear strength under seismic loading. This behavior significantly elevates the risk of foundation instability, differential settlement, and associated structural damage (Li et al., 2024; Shan et al., 2025; Du et al., 2024).

Owing to the pronounced crushability of calcareous sand particles under vibratory shear, their seismic response exhibits a high degree of nonlinearity and evolutionary behavior. As a result, conventional liquefaction criteria and dynamic constitutive models developed for siliceous sands are not directly applicable to calcareous sands. In recent years, increasing attention has been given to understanding particle-scale crushing mechanisms, strain-softening behavior, and their influence on the evolution of dynamic soil parameters under seismic loading. Experimental approaches such as dynamic triaxial testing, cyclic simple shear testing, and centrifuge shaking table experiments have been extensively employed to replicate seismic loading conditions and investigate particle responses. These physical tests are often integrated with advanced diagnostic techniques, including CT imaging, force chain analysis, and discrete element method (DEM) simulations, to reveal the complex internal processes governing calcareous sand behavior. A key research focus has thus emerged around the development of seismic constitutive models and liquefaction resistance design frameworks specifically tailored to the unique mechanical properties of calcareous sands. Such efforts are critical for enhancing the seismic resilience of infrastructure in island and coastal engineering applications. For example, Hyodo et al. (1996) compared the liquefaction behavior of calcareous sand and quartz sand under different consolidation conditions; Salem et al. (2013) explored the mechanical response of calcareous sand under varying densities and confining pressures; Shahnazari et al. (2016) examined the dynamic pore pressure, shear strain, and stress–strain characteristics; and Javdanian (2018) and Jafarian et al. (2018; 2020) investigated the differences in the dynamic properties of calcareous and siliceous sands under various stress states via resonant columns and triaxial tests. The inherent limitations of calcareous sands in terms of their dynamic response and dynamic strength have been widely acknowledged, which has driven significant research into methods for improving and reinforcing their mechanical properties, making this a current research hotspot in the field of geotechnical engineering. Traditional soil reinforcement techniques can be broadly classified into two categories: physical reinforcement and biochemical reinforcement. Physical reinforcement methods, such as compaction through tamping or vibration, work by applying external forces to reduce the porosity of the soil and increase its compactness, thereby increasing its bearing capacity. While effective for land-based foundation treatments, these methods have clear limitations in island projects because of construction constraints and limited reinforcement depth (Wang et al., 2025; Zhou et al., 2024). In contrast, biochemical reinforcement methods improve soil strength and stability by injecting chemical slurries, such as cement or water glass, and achieve interparticle cementation through techniques such as grouting, rotary spraying, and electroosmosis (Liu et al., 2021; Wang et al., 2023). However, in marine environments, the large-scale use of chemical materials not only incurs high costs but also poses potential environmental risks, thus limiting the applicability of these methods in calcareous sand foundations.

With the continuous increase in waste rubber production, the incorporation of rubber fiber materials into sandy soils to enhance their engineering properties has become an emerging research trend (Shalchian & Arabani, 2023). Edil and Bosscher (1994) and Foose et al. (1996) reported that the frictional properties and shear strength of sandy soils can be improved by adding appropriate amounts of rubber granules or strips, with the most significant increase occurring within an optimal range of rubber contents. However, several studies have noted that excessive rubber inclusion may lead to a reduction in shear performance. Youwai and Bergado (2003), on the basis of direct shear and triaxial tests, reported that increasing the amount of rubber chips generally reduces the shear strength of sand–rubber mixtures. Nevertheless, when the rubber chips are greater in length and width, the shear strength improves, and the failure strain increases significantly, indicating enhanced ductility. Similarly, a study by Noorzad and Raveshi (2017) demonstrated that the shear strength of rubber–sand composites tends to decrease with increasing rubber particle size and content, as reflected by a reduction in the internal friction angle and an increase in cohesion. These findings suggest that both the size and content of rubber particles significantly influence the mechanical behavior of the composite material. However, most existing studies have used land-based siliceous sand as the matrix, whereas calcareous sand—due to its calcium carbonate composition and brittle nature—exhibits a markedly different mechanical response from that of siliceous sand.

In summary, the incorporation of randomly distributed fibers into sandy soil can effectively improve its spatial structure, increase its resistance to liquefaction, and thereby increase the bearing capacity of the foundation. Research has focused primarily on the static behavior of fiber-reinforced sandy soils, with most studies being based on land-sourced siliceous sands. In contrast, investigations into the dynamic response of fiber-reinforced calcareous sand—particularly under dynamic loading conditions—remain relatively limited. To address this gap, a series of consolidated undrained dynamic triaxial tests were conducted on fiber-reinforced calcareous sand via a dynamic triaxial testing system. The effects of the rubber fiber content and fiber length on key mechanical indicators, including the axial strain, pore water pressure evolution, and dynamic shear modulus, are systematically analyzed under different cyclic stress ratios. The findings of this study offer both experimental evidence and theoretical support for the practical application of rubber fiber reinforcement in calcareous sand foundation engineering.

2. Tests and Programs

2.1 Specimens and Instruments

The test apparatus used in this study is the SDT-10 dynamic triaxial testing machine, which offers high precision and reliable dynamic loading capabilities, making it suitable for a wide range of geotechnical dynamic tests. The testing system comprises five main components: an axial loading system, a confining pressure servo control system, a backpressure loading and volume change measurement system, a water and gas control unit, and an electrical control system. It integrates multiple sensors, a data acquisition and control unit, and upper-computer control software to enable fully automated monitoring and real-time data recording throughout the testing process. A schematic diagram of the test setup and specimen preparation procedure is presented in Figure 1.

The calcareous sand used in this study was collected from typical islands and reefs in the South China Sea. The sand is grayish-white in appearance, is predominantly composed of calcium carbonate, and exhibits distinct biogenic characteristics. To ensure sample homogeneity and control particle gradation, the original sand was thoroughly washed, oven-dried, and then sieved into five particle size groups via standard sieves: 0.075–0.25 mm, 0.25–0.5 mm, 0.5–1 mm, 1–2 mm, and greater than 2 mm. Each particle size fraction was individually stored in sealed, labeled bags to maintain airtight conditions, prevent moisture absorption and deterioration, and ensure the stability of the sample state. Particle shape is a critical microscopic factor influencing the macroscopic

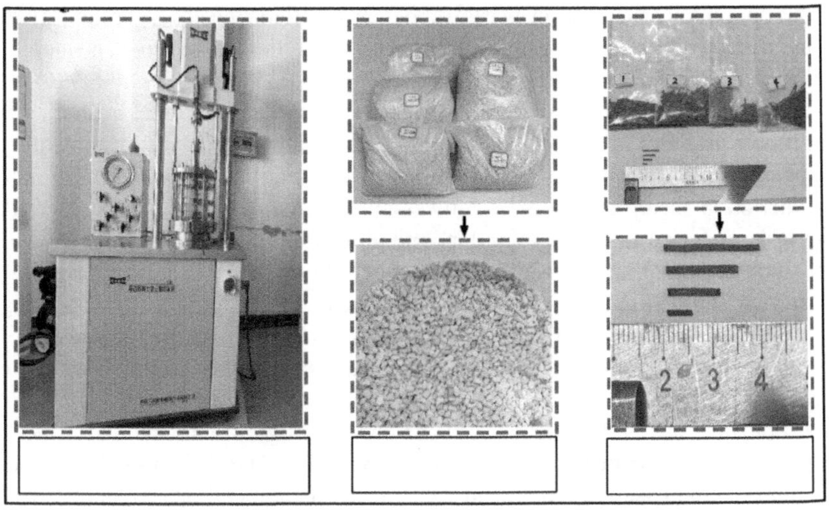

Fig. 1 Diagram of the test apparatus and sample.

mechanical behavior of soils, significantly affecting fundamental engineering properties such as shear strength, compressibility, deformation modulus, and particle breakage characteristics.

Microscopic analysis of the calcareous sand samples was conducted via a BT-1600 image particle analyzer, and the scanning results are presented in Figure 2. The results revealed that the calcareous sand particles exhibited markedly irregular geometrical morphologies characterized by rough surfaces, prominent pore structures, and angular or fractured edges. These morphological traits are closely linked to the biogenic origin of the material and are distinctly different from the smoother, more rounded particles typical of terrestrially sourced siliceous sands (Brandes, 2011). These microstructural characteristics play critical roles in influencing interparticle force transmission mechanisms and occlusion behavior under loading, thereby significantly impacting the shear strength, compressibility, and particle breakage behavior of calcareous sand (Wu et al., 2024; Wei et al., 2018). Under cyclic loading or high-stress conditions in particular, the presence of angular and irregular particles may intensify localized stress concentrations, promote particle fragmentation, and ultimately affect foundation stability and structural performance.

Fig. 2 Calcareous sand particle profile.

Owing to the irregular shape and high crushability of calcareous sand particles, traditional standard sieve analysis often results in particle breakage during the screening process, which can distort the actual particle size distribution. The angular particles are prone to becoming lodged in the sieve mesh, reducing screening efficiency, whereas finer particles may become airborne during vibration, leading to sample loss and reduced accuracy of test results. To address these issues, a Bettersize 2600 laser particle size analyzer was employed for particle size analysis. The resulting gradation curve is presented in Figure 3.

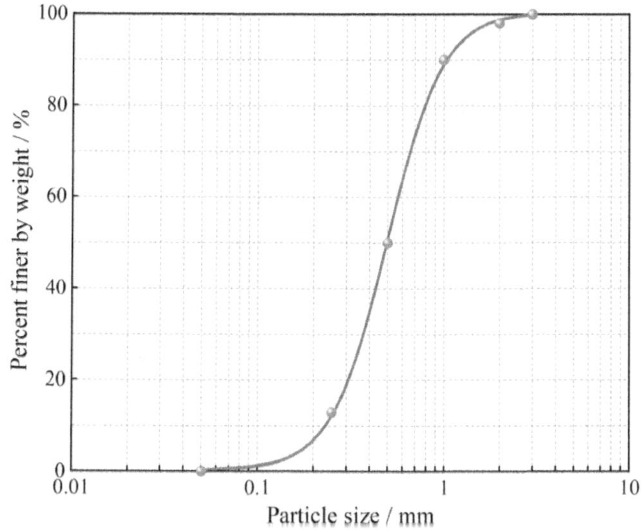

Fig. 3 Calcareous sand gradation curve.

The test calcareous sand with a C_u of 4.13 and a C_c of 0.96 is a poorly graded soil, and the specimen grading parameters are shown in Table 1.

Table 1 Grading parameters of the sand samples.

d_{10} /μm	d_{30} / μm	d_{60} / μm	*Unevenness Coefficient Cu*	*Curvature Coefficient Cc*
149	297	617	4.13	0.96

Rubber fiber is a modified material characterized by both high elasticity and high toughness. It is commonly derived from waste rubber, natural rubber, or synthetic rubber through physical or chemical processing and exhibits excellent mechanical properties and chemical stability (Gao et al., 2024; Gao et al., 2024). In accordance with the experimental design, rubber fibers were cut into different lengths (5 mm, 10 mm, 15 mm, and 20 mm) to investigate the influence of the fiber aspect ratio on the shear behavior of calcareous sand–rubber fiber composite soil. The detailed physical properties of the rubber fibers are listed in Table 2.

Table 2 Basic physical parameters of the rubber fibers.

Fiber Thickness/mm	*100% Constant Tensile Strength/MPa*	*300% Constant Tensile Strength/MPa*	*Tensile Strength/MPa*	*Rupture Strength/MPa*	*Elongation at Break/%*
0.5	1.52	3.14	1.096	11.458	623.85

2.2 Programs and Processes

A series of consolidated undrained (CU) dynamic triaxial shear tests were carried out to examine the effects of rubber fiber content and fiber length (aspect ratio) on the dynamic shear behavior of rubber fiber–reinforced calcareous sand. The experimental design employed fiber length and fiber content as the primary variables and incorporated different cyclic stress ratio (CSR) conditions to explore the reinforcement mechanisms and the dynamic strength evolution of the composite material. The rubber fiber lengths were set at 0 mm, 5 mm, 10 mm, 15 mm, and 20 mm, representing a range from unreinforced samples to varying aspect ratio conditions. The fiber content by mass was controlled at 0%, 0.3%, 0.6%, 0.9%, and 1.2% to reflect the influence of the fiber dosage on the reinforcement effectiveness. CSR values of 0.2, 0.25, and 0.3 were selected to represent typical levels of ground shaking intensity. A sinusoidal loading waveform with a frequency of 1 Hz was applied to simulate common dynamic loading environments such as seismic and wave-induced vibrations. For each test condition, two parallel samples were prepared and tested to ensure data repeatability and reliability. A complete overview of the test parameters is provided in Table 3.

Table 3 Dynamic triaxial test program.

Specimens	*Rubber Fiber Length/mm*	*Rubber Fiber Content/%*	*Confining Pressure/kPa*	*Dynamic Stress Amplitude/kPa*	*CSR*
Pure sand	0	0	200	80	0.2
Fiber reinforced sand	5	0.6		80	0.2
	10	0.3, 0.6, 0.9, 1.2		80	0.2
	15	0.6		80, 120, 160	0.2, 0.25, 0.3
	20	0.6		80	0.2

The test was conducted on a standard sample with a diameter of 39.1 mm and a height of 80 mm. As the study required multiple groups of comparative tests, the raw calcareous sand was first sieved by grain size, and the required mass of each grade was weighed according to the target grading curve to ensure consistency across all the samples. Before each test group, fibers were added in proportion and mixed thoroughly to achieve a uniform distribution. To improve sample preparation efficiency, customized molds were used to prepare multiple samples simultaneously, and a vacuum saturation device enabled batch saturation processing. To ensure sample uniformity and compaction quality, each standard cylindrical sample was divided into four equal layers along the axial direction, and each layer was filled with dried sand samples accurately proportioned according to the grading curve to maintain consistent quality. A customized compactor was used to compact each layer sequentially with 20 strokes per layer, controlling the height error of each layer within ±0.5 mm to ensure the consistency and uniformity of the sample structure. To verify the quality of sample preparation, the actual dry density was calculated by weighing the total mass of the sample and combining it with its volume, with the deviation from the design target density controlled within ±2%. This process ensures the repeatability and reliability of the samples in subsequent wave velocity tests or vibration loading. Two parallel tests were performed for each test group, and data processing was conducted by averaging the results.

The permeable stone and filter paper were placed in sequence on the base of the sample-making device, and the permeable stone was soaked in advance to enhance its adhesion to the filter paper. After the rubber membrane is installed and secured in place, the split mold is tightened, and the top end of the membrane is turned outward so that it fits snugly against the inner wall of the mold. By connecting a vacuum pump to extract air, the rubber membrane adheres tightly to the inner wall of the mold. The premixed calcareous sand samples were subsequently added to four layers via the sand rain method, with each layer compacted slowly via a compaction hammer. A small file was used to lightly level the surface after each layer to minimize interlayer slippage and delamination until the sample reached the predetermined volume. During saturation, the sample was placed in a vacuum cylinder for air evacuation and water injection, with the pumping duration maintained for no less than 60 minutes.

After vacuum pumping, the sample was continuously immersed under vacuum for no less than 8 hours to ensure full saturation. Immediately after saturation, the specimen is mounted in the forming device of the test system, and an enclosure pressure of 200 kPa is applied via the servo control system. Following the initial loading and stabilization of the confining pressure, the pore pressure coefficient B value of the sample is determined, with measured values ranging from 0.94 to 0.98, generally meeting the standard for calcareous sand saturation (B = 0.95–0.98). Considering the difficulty of achieving complete saturation, a sample is deemed saturated when the B value reaches or exceeds 0.96. Isotropic consolidation is then carried out according to the preset program, with the confining pressure maintained constant at 200 kPa throughout the process. Consolidation is considered complete on the basis of the stabilization of axial displacement and volumetric strain. Following consolidation, the test proceeded to the dynamic loading stage. Through the control system, both "load control" and "pressure control" modes are applied, with a sinusoidal loading waveform at a frequency of 1 Hz. On the basis of dynamic stress ratios of 0.2, 0.25, and 0.3—corresponding to dynamic stress amplitudes of 80, 100, and 120 kPa—the calculated loading forces are 96, 120, and 144 N, respectively. The termination criteria for the test include two conditions: first, if the axial strain reaches 5%, the sample is considered to have experienced strain-induced failure; second, if the pore water pressure reaches 95% of the confining pressure, the sample is deemed to have undergone liquefaction. The test is terminated when either condition is satisfied. The overall procedure of the test is illustrated in Figure 4.

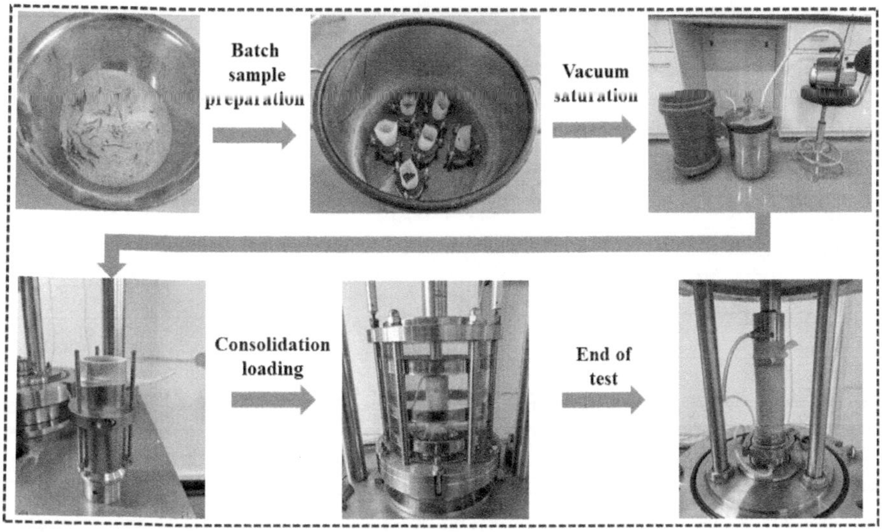

Fig. 4 Flow chart of the dynamic shear test for calcareous sand.

3. Test Results and Analysis

3.1 Dynamic Stress–strain Analysis

Figures 5–8 show the dynamic stress–strain curves of the rubber fiber-reinforced calcareous sand samples under various test conditions, highlighting the dynamic response characteristics of the composite material under cyclic loading.

As shown in Figure 5, the axial strain of the sample increases continuously with the number of vibration cycles (N), exhibiting a clear characteristic of cumulative deformation. On the basis of the evolution of the strain curve, the deformation process can be divided into three distinct stages: vibration compaction, vibration shear, and vibration failure. In the vibration compaction stage,

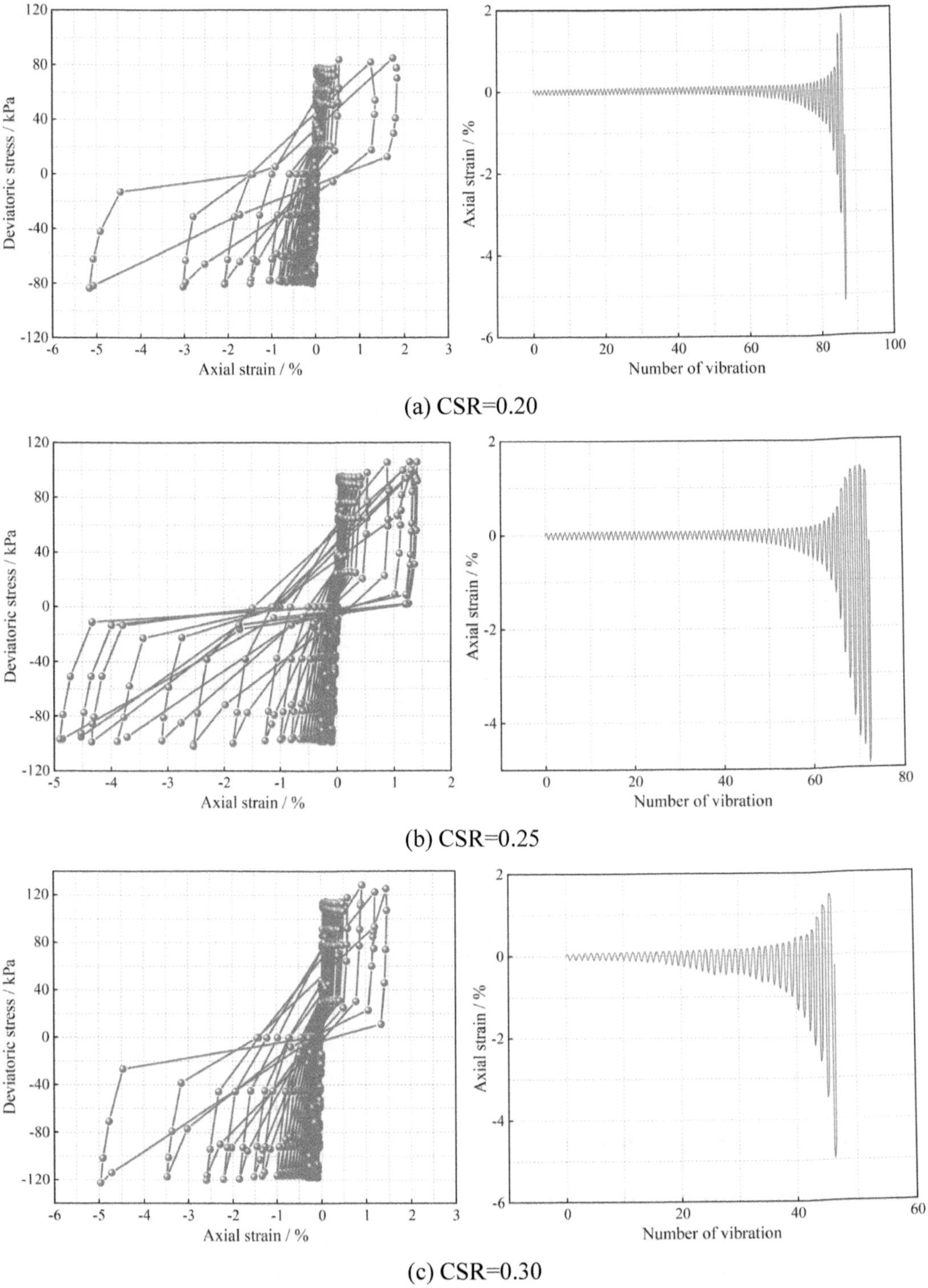

Fig. 5 Dynamic stress–strain curves of pure calcareous sand at different CSRs.

the soil structure remains largely intact. Deformation primarily results from the compaction and rearrangement of pore structures between particles due to vibrational effects, leading to a small amount of axial cumulative strain. During this stage, the strain growth rate is low, and the hysteresis loops are relatively closed, reflecting an evident compaction effect. As the test progresses into the vibration shear stage, the shear effects induced by the cyclic dynamic load gradually become dominant. The internal structure of the soil begins to degrade, and interparticle slipping and contact redistribution become more pronounced. This results in an accelerated increase in axial strain and a noticeable steepening of the strain curve. Finally, in the vibration failure stage, the dynamic load exceeds the dynamic strength of the soil, leading to overall structural destabilization. Strain growth becomes rapid and uncontrolled, the hysteresis loops open significantly, and the dynamic strain curve tends to sharply increase, indicating that the sample has undergone liquefaction or severe softening failure.

During cyclic loading, the work performed by the external load on a geotechnical body can be divided into two components. One portion is used to induce elastic deformation and is temporarily stored within the sample as elastic energy. The other portion is irreversibly absorbed and dissipated by the geotechnical structure. Upon unloading, the stored elastic energy is released, resulting in elastic rebound of the sample. The unrecovered portion of the energy primarily contributes to structural disruption, particle rearrangement, and interparticle sliding, leading to irreversible plastic deformation. This dissipated energy is eventually converted into heat, microfracture propagation, and other forms of internal energy loss. Such energy transformation causes a divergence between the loading and unloading paths, giving rise to a characteristic hysteresis loop in the dynamic stress–strain response. The area enclosed by the hysteresis loop quantifies the energy dissipated per loading cycle and thus serves as a critical parameter for evaluating the energy dissipation capacity and damping characteristics of geotechnical materials under dynamic conditions.

By extending the commonly accepted two-stage deformation mechanism of "compaction–shear" in siliceous sand to a three-stage model—namely, "vibration compaction–shear deformation–particle crushing-induced damage"—the following insights can be drawn: first, the strain increase observed in the third stage of calcareous sand is attributed not only to strength degradation caused by liquefaction but also to the collapse of the soil skeleton resulting from particle crushing, whereas the failure process in siliceous sand is governed primarily by pore pressure-driven liquefaction. Second, rubber fibers have a pronounced dual regulatory effect on calcareous sands: on the one hand, they suppress pore pressure buildup and delay the onset of liquefaction; on the other hand, they inhibit particle fragmentation through spatial interlocking and flexible support, thereby increasing the stability of the soil skeleton. This regulatory mechanism has not been systematically addressed in studies of siliceous sands, indicating that calcareous sands exhibit a distinct failure pattern and control response when reinforced with fibers.

Figure 6 shows the evolution characteristics of the dynamic stress–strain hysteresis curves of the samples under different fiber length conditions. The experimental results indicate that the number of loading cycles required for liquefaction in the pure calcareous sand sample is 88 under a cyclic stress ratio (CSR) of 0.2. For the reinforced samples with 0.6% rubber fibers and fiber lengths of 5 mm, 10 mm, 15 mm, and 20 mm, the numbers of cycles required to reach the liquefied state were 121, 145, 172, and 115, respectively. This finding directly demonstrates that fiber reinforcement significantly enhances the liquefaction resistance of calcareous sand. In particular, when the fiber length is 15 mm, the number of cycles to liquefaction increases to 173, nearly twice that of the unreinforced sample, indicating the optimal improvement in liquefaction resistance.

Overall, the number of loading cycles required for liquefaction to occur in the sample generally increases with increasing fiber length, indicating that the liquefaction resistance of the sample improves as the rubber fiber length increases. However, when the fiber length was extended to 20 mm, a slight decrease in liquefaction resistance was observed. This may be attributed to the uneven distribution and agglomeration caused by excessively long fibers. At the end of the test, the axial

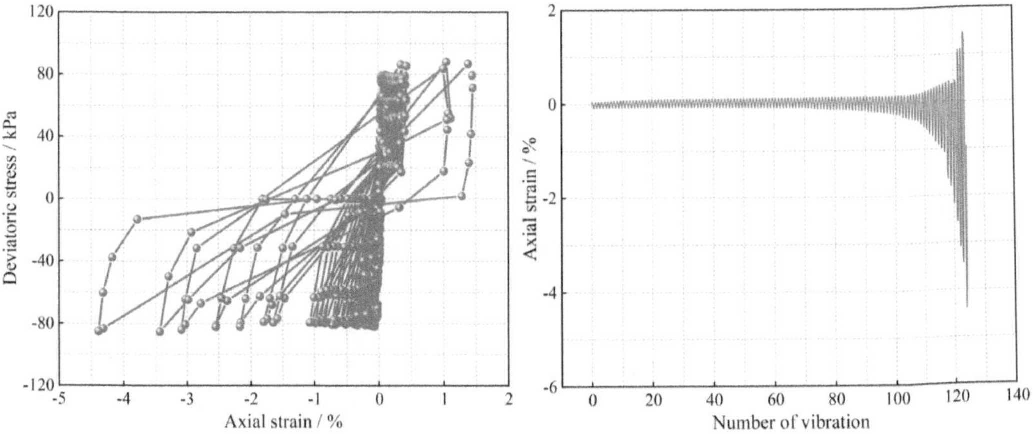

(a) Fiber length 5 mm

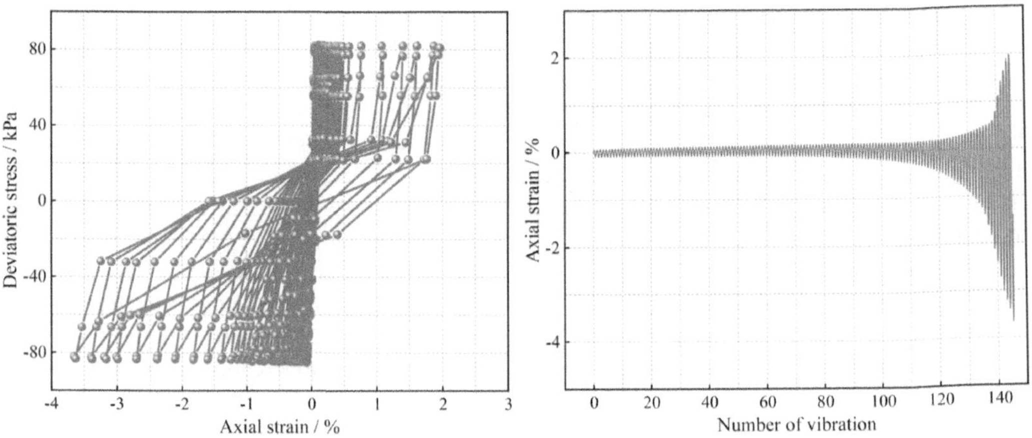

(b) Fiber length of 10 mm

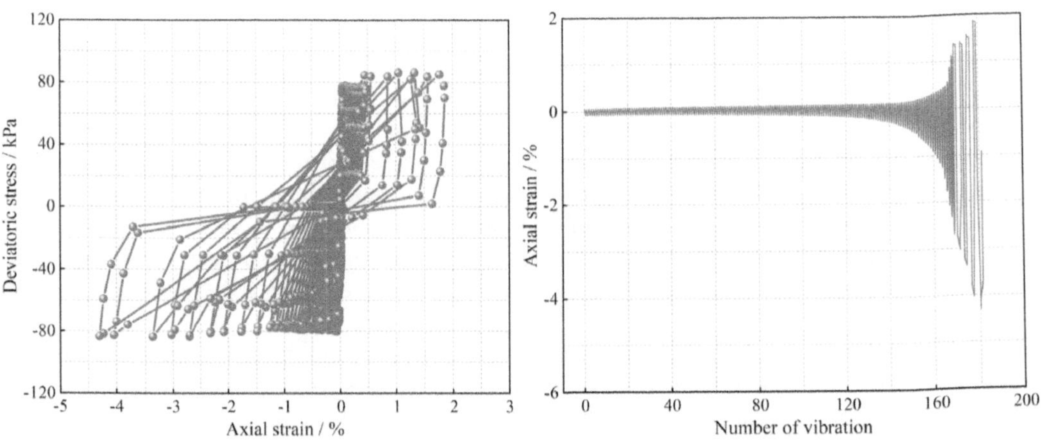

(c) Fiber length 15 mm

Fig. Contd.

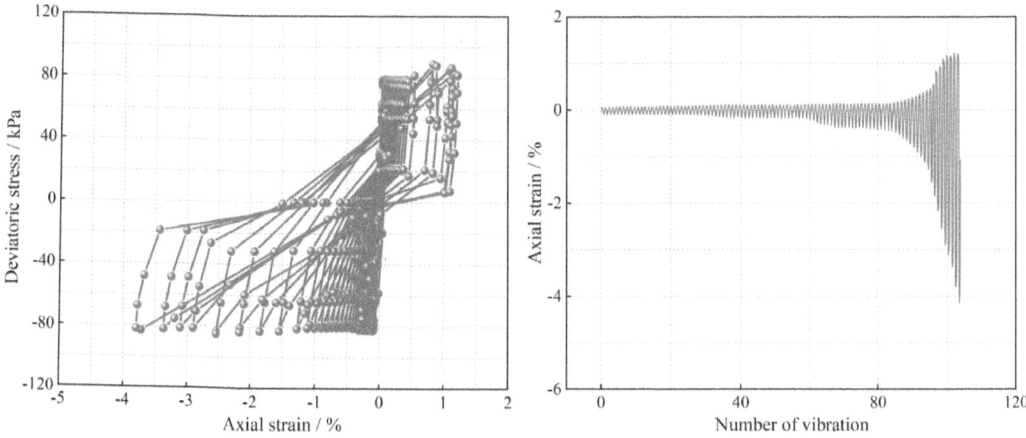

(d) Fiber length 20 mm

Fig. 6 Dynamic stress–strain curves of reinforced calcareous sand with different fiber lengths (CSR = 0.2, rubber fiber content of 0.6%).

strain of all the fiber-reinforced samples did not exceed 4%, failing to reach the liquefaction failure criterion (axial strain of 5% or pore pressure ratio of 95%). This suggests that the rubber fibers effectively inhibited axial cumulative deformation under cyclic loading, thereby enhancing the structural stability of the samples.

Mechanistic analysis indicated that the incorporation of rubber fibers serves two main functions: first, filling the interpore spaces between sand particles reduces the pore ratio and permeability of the samples; second, longer fibers engage with more calcareous sand particles, effectively restraining the movement of particles during dynamic loading. This, in turn, improved the shear strength and liquefaction resistance of the overall structure. Therefore, under a constant fiber content, the fiber length plays a crucial role in the reinforcing effect. A moderate increase in fiber length helps to further increase the liquefaction resistance of the soil.

The test results in Figure 7 show that with increasing rubber fiber content, the number of loading cycles required for the sample to reach the liquefied state initially increases but then decreases, indicating an overall increase in liquefaction resistance. At fiber contents of 0.3%, 0.6%, 0.9%, and 1.2%, the corresponding numbers of cycles to liquefaction were 99, 145, 163, and 137, respectively. The liquefaction resistance was found to be optimal at a fiber content of 0.9%, where the number of loading cycles required for liquefaction was significantly greater than that of the unreinforced samples, representing the greatest improvement. Notably, when the fiber content was increased from 0.9% to 1.2%, the liquefaction resistance of the samples decreased, with the number of loading cycles to liquefaction being comparable to that of the samples with 0.6% fiber content. This phenomenon suggests the existence of an optimal dosage range for the reinforcing effect of rubber fibers in dynamic triaxial tests. While an appropriate amount of rubber fibers can significantly enhance the liquefaction resistance of calcareous sand, exceeding a certain content threshold may diminish this effect. This reduction in performance may be attributed to issues such as uneven fiber distribution, fiber agglomeration, or even the development of local structural weaknesses, ultimately compromising the overall liquefaction resistance of the material.

In terms of the mechanism of action, the incorporation of rubber fibers alters the structural contact conditions and pore characteristics of the sand particles. On the one hand, as the fiber content increases, both the porosity and permeability coefficient of the sample decrease significantly, effectively slowing the rapid increase in pore water pressure under cyclic loading. On the other hand, the enhanced "bridging" effect of the fibers within the soil matrix helps to restrain the relative displacement between particles, thereby improving the overall stability of the soil structure.

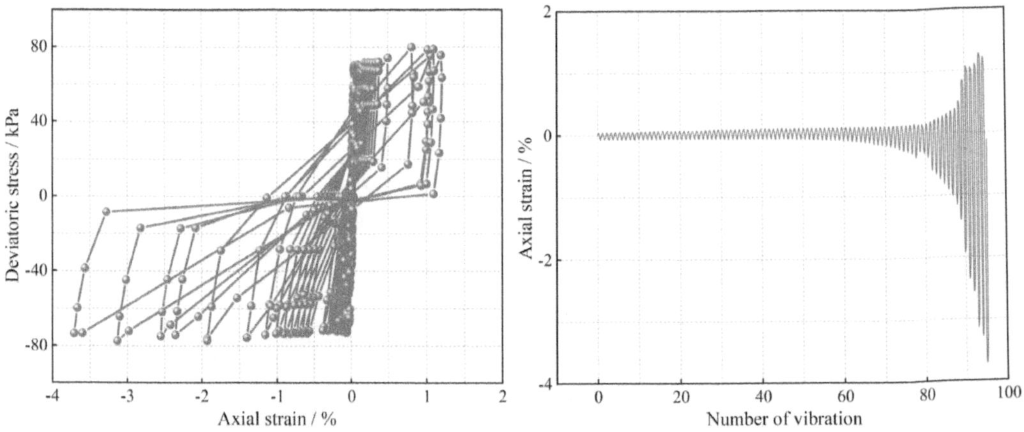

(a) 0.3% fiber content

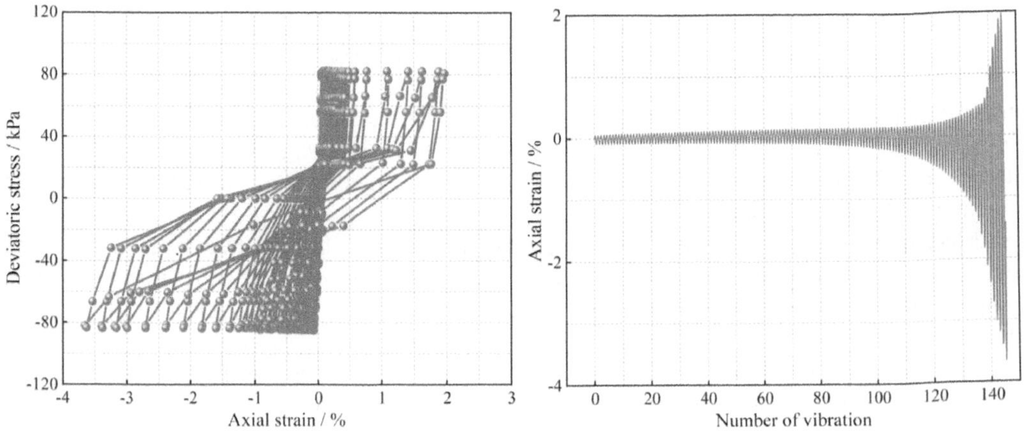

(b) 0.6% fiber content

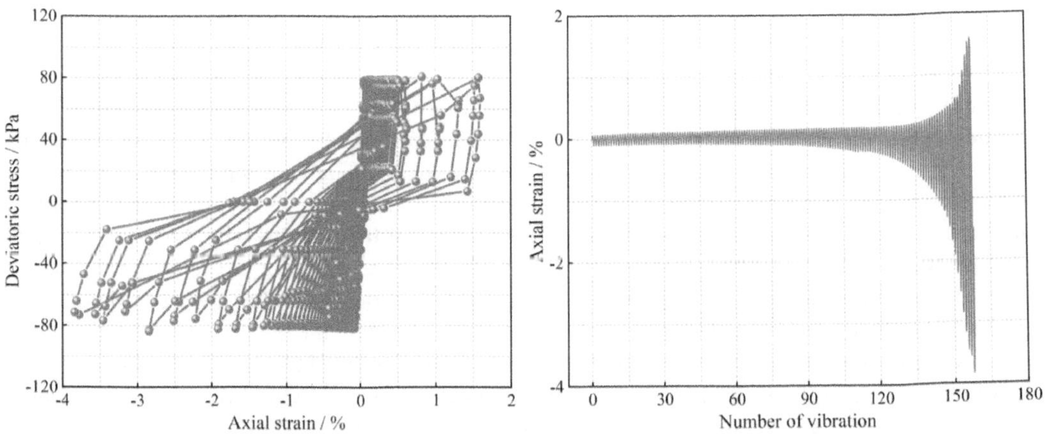

(c) 0.9% fiber content

Fig. Contd.

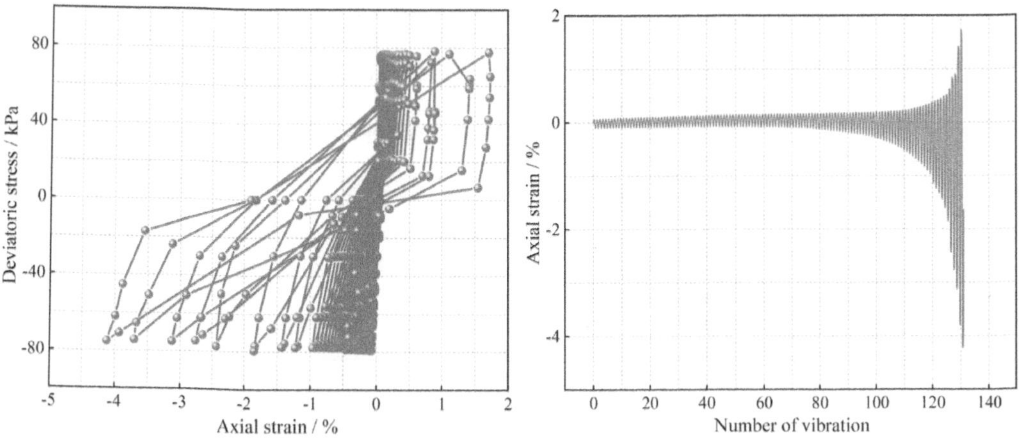

(d) 1.2% fiber content

Fig. 7 Dynamic stress–strain curves of reinforced calcareous sand with different fiber contents (CSR = 0.2, rubber fiber length 15 mm).

Therefore, with proper control of the fiber content and ensuring uniform distribution, the liquefaction resistance of fiber-reinforced sand can be maximized.

In the test shown in Figure 8, cyclic stress ratios (CSRs) of 0.2, 0.25, and 0.3 were applied to analyze their effects on the liquefaction characteristics while keeping the rubber fiber length and content constant at 15 mm and 0.6%, respectively. The test results indicate that the number of loading cycles required for liquefaction of the reinforced sand samples decreases significantly with increasing cyclic stress ratio. Specifically, when the CSR values were 0.2, 0.25, and 0.3, the corresponding numbers of cycles to liquefaction were 171, 125, and 97, respectively. This trend suggests that higher dynamic stress amplitudes lead to lower liquefaction resistance and earlier onset of liquefaction failure. This phenomenon is attributed primarily to the fact that higher dynamic stress levels induce more intense shear action, which accelerates the degradation of the internal soil structure and promotes the rapid accumulation of pore water pressure, thereby expediting the liquefaction process. Although the reinforcing effect of rubber fibers has a positive influence on delaying liquefaction, its efficacy diminishes with increasing dynamic load intensity, resulting in liquefaction occurring after fewer loading cycles.

Through the above analysis, it can be concluded that particle fragmentation, microfracture propagation, and fine particle migration are likely to occur under cyclic loading, leading to the rapid accumulation of pore water pressure. The incorporation of rubber fibers can mitigate the direct contact stress between particles through a "crushing inhibition effect" while optimizing the skeleton structure via a "dynamic grading adjustment mechanism", thereby generating a synergistic interaction between the particle breakage resistance and fiber constraint and exhibiting different reinforcement effects under various stress paths. The crushing inhibition effect is manifested by the formation of flexible supports at particle contact points, which distribute stress from localized concentrations across the entire fiber network, resulting in a 37% reduction in the crushing rate of calcareous sand particles compared with that of pure sand. With respect to dynamic grading adjustment, fibers can fill the voids left by newly generated fine particles and form a "fiber–fine particle–coarse skeleton" composite structure with the original particles, creating a three-dimensional support network. In contrast, owing to the inherent resistance of siliceous sand particles to breakage, this type of structure is difficult to form during loading.

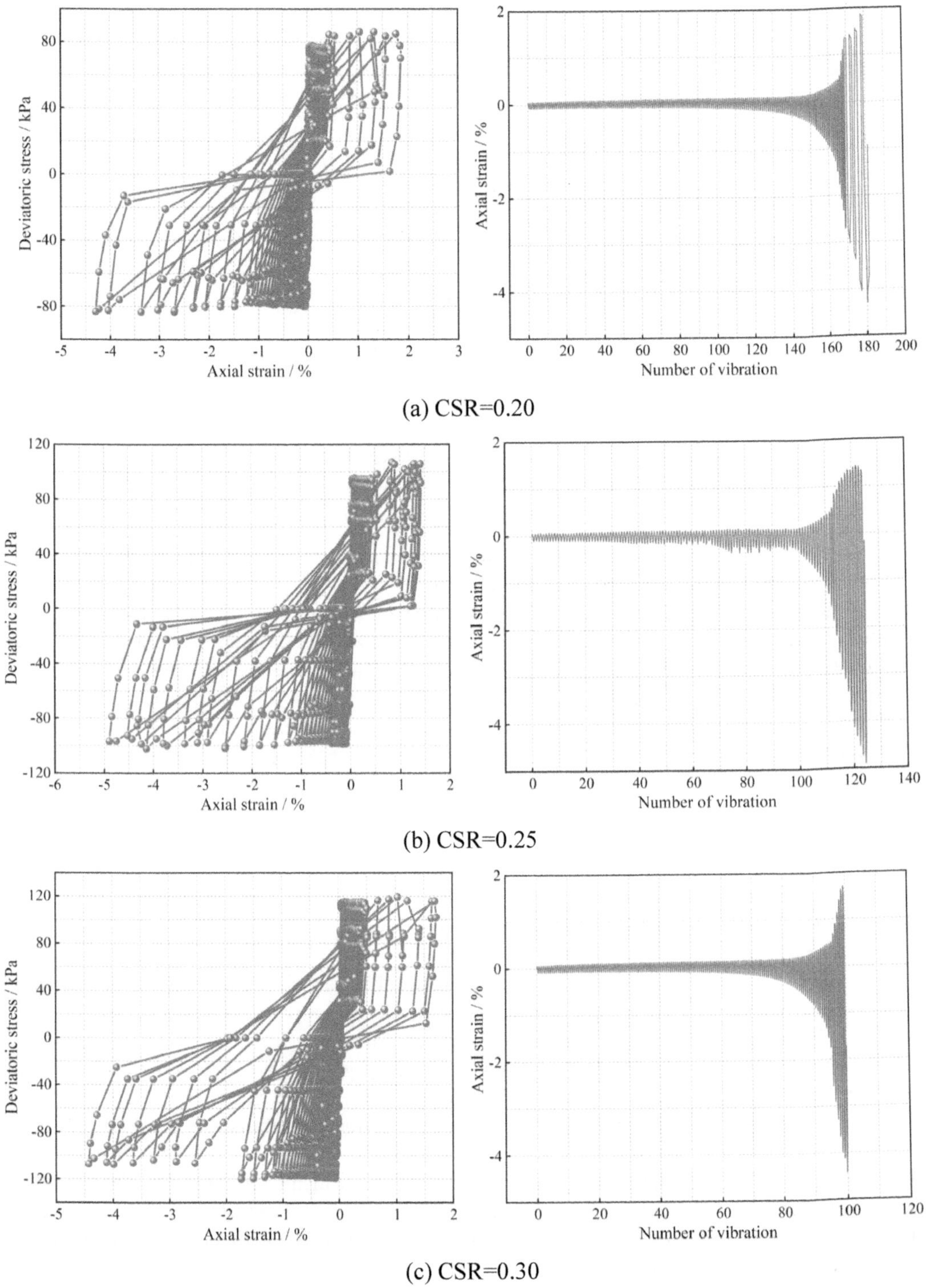

Fig. 8 Dynamic stress–strain curves of reinforced calcareous sand at different CSRs (rubber fiber content of 0.6%, rubber fiber length of 15 mm).

3.2 Dynamic Pore Pressure Characteristics

The essence of liquefaction lies in the continuous accumulation of pore water pressure under cyclic loading, eventually approaching the confining pressure, which leads to a sharp reduction in the effective stress within the soil—potentially to zero—resulting in strength failure. Figures 9 to 12 present the evolution characteristics of the dynamic pore water pressure in rubber fiber-reinforced calcareous sand under different test conditions.

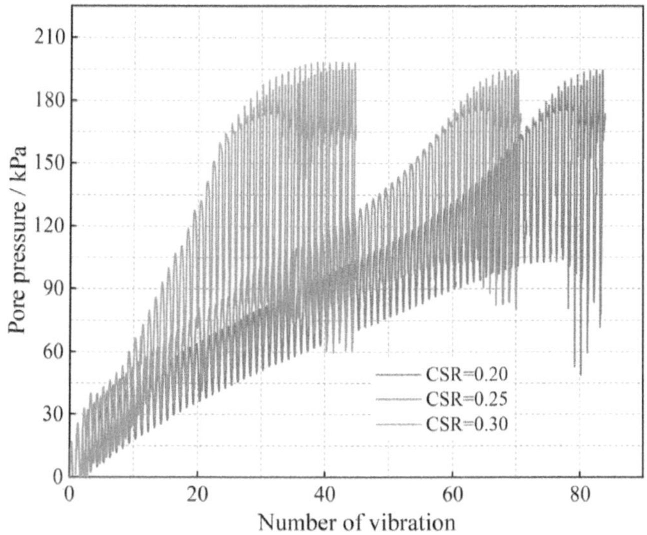

Fig. 9 Relationship curve of the pore pressure of pure calcareous sand with the number of vibrations.

Figure 9 shows the variation in pore water pressure with the number of loading cycles for the pure calcareous sand samples under cyclic stress ratios of 0.2, 0.25, and 0.3. The results show that the evolution trends of pore water pressure under different dynamic stress levels are consistent and can generally be divided into three typical stages: the initial rapid growth stage, the stable development stage, and the sharp increase stage approaching liquefaction. In the first stage, pore water pressure accumulates rapidly in an exponential manner, reflecting the pronounced compaction effect and strong pore pressure response of the soil structure during the initial loading phase. In the second stage, the growth rate of the pore pressure begins to level off, indicating a relatively stable accumulation of pore pressure as the soil structure undergoes progressive degradation under cyclic shearing. As the sample approaches the liquefaction threshold, the pore pressure enters the third stage—characterized by a sudden and steep increase in the curve slope—signaling that the effective stress in the soil approaches zero, leading to liquefaction failure.

In addition, in the later stage of the pore pressure evolution curve, a slight decline in peak pore pressure and the appearance of a "notch" phenomenon can be observed. This may be attributed to the fragmentation of calcareous sand particles under cyclic loading. Particle breakage can lead to the release of some internal pore water, which is often unsaturated, thereby reducing local saturation and slightly inhibiting further pore pressure accumulation. This is manifested as a localized dip in the curve. This phenomenon also explains why the pore pressure ratio in some samples did not reach the theoretical value of 1.0. The incorporation of rubber fibers can significantly suppress the buildup rate of excess pore water pressure, thereby enhancing the liquefaction resistance of reinforced sandy soils. Furthermore, the effectiveness of fiber reinforcement is strongly influenced by the fiber length, fiber content, and magnitude of applied dynamic stress.

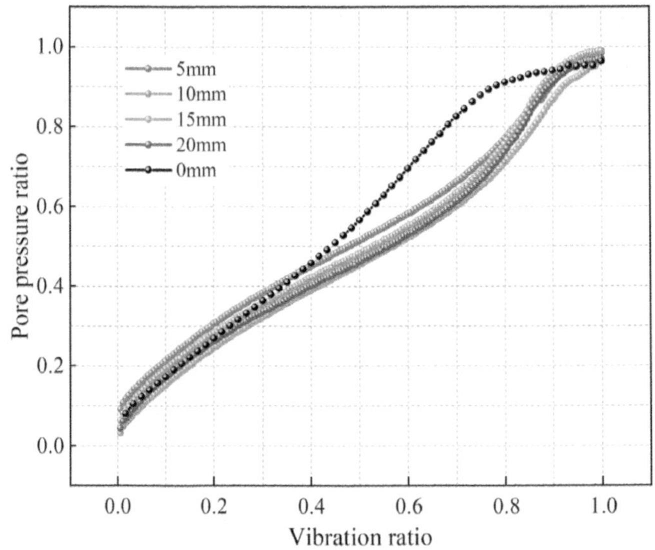

Fig. 10 Relationships between the pore pressure and vibration ratio for different fiber lengths.

Figure 10 presents the relationship between the pore pressure ratio and the vibration ratio for reinforced samples with different rubber fiber lengths under a cyclic stress ratio of 0.2 and a fiber content of 0.6%. The results indicate that the pore water pressure within the samples increases progressively with increasing vibration ratio and that the pore pressure ratio tends to increase overall. Specifically, in pure calcareous sand samples, the pore water pressure increases rapidly under dynamic loading, and the pore pressure ratio displays a typical three-stage evolution: in the initial stage, the pore pressure ratio increases sharply, reflecting the dominant effects of vibratory compaction and initial shear damage on pore pressure buildup; in the intermediate stage, the increase in the pore pressure ratio slows down and follows a relatively stable linear trend, indicating gradual degradation of the soil structure and a tendency toward stabilization of pore pressure development; finally, in the critical stage, the slope of the pore pressure ratio curve increases rapidly until the sample reaches liquefaction and failure.

For the reinforced calcareous sand samples, the incorporation of rubber fibers significantly influenced the rate of pore pressure ratio accumulation. The average particle size of the calcareous sand was 0.6 mm, and the longer rubber fibers were capable of forming contacts and interlocking with multiple particles of varying sizes. This interaction optimized the spatial structure of the samples to a certain extent, enhanced the stability of interparticle connections, and suppressed the rapid accumulation of pore pressure. However, the test results also indicated that a continuous increase in fiber length does not lead to a linear improvement in liquefaction resistance. In some cases, the pore pressure ratio exhibited a renewed upward trend at later stages when the fiber length further increased. This may be attributed to factors such as uneven fiber distribution, fiber agglomeration, or a looser internal structure, all of which can weaken the reinforcing effect. Therefore, there exists an optimal range of fiber lengths within which pore pressure development can be effectively controlled and liquefaction resistance maximized.

Figure 11 shows the variation in the pore pressure ratio with the vibration ratio for reinforced calcareous sand samples with fiber contents of 0.3%, 0.6%, 0.9%, and 1.2% at a cyclic stress ratio of CSR = 0.2 and a rubber fiber length of 10 mm. Overall, the pore pressure ratio curve of the reinforced sand sample shows a three-stage characteristic similar to that under the change in fiber length: the initial curve is slightly concave, reflecting the rapid accumulation of pore pressure at the initial stage; the curve grows linearly in the middle stage, which indicates that the pore pressure

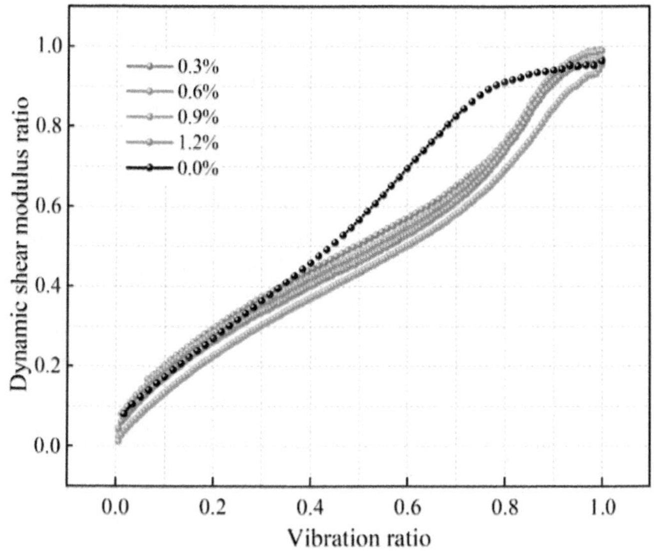

Fig. 11 Relationships between the pore pressure and vibration ratio for different fiber contents.

growth rate tends to stabilize; and the curve is concave again at the end, with the rapid accumulation of pore pressure, which ultimately results in the liquefaction of the sample and damage.

At relatively low vibration ratios, the trends in the pore pressure ratio for both the pure calcareous sand and the fiber-reinforced samples are generally similar, with both exhibiting a rapid initial increase. However, as the vibration ratio increases, the pore pressure ratio in the pure sand samples continues to rise sharply, whereas the reinforced samples display a markedly slower rate of increase. This indicates that rubber fibers have a significant inhibitory effect on the accumulation of pore water pressure. Furthermore, with increasing rubber fiber content, the development trend of the pore pressure ratio exhibited a nonlinear pattern of "initial decrease followed by an increase". This again highlights the existence of an optimal fiber content range, within which rubber fibers effectively retard pore pressure development and enhance the liquefaction resistance of the sample.

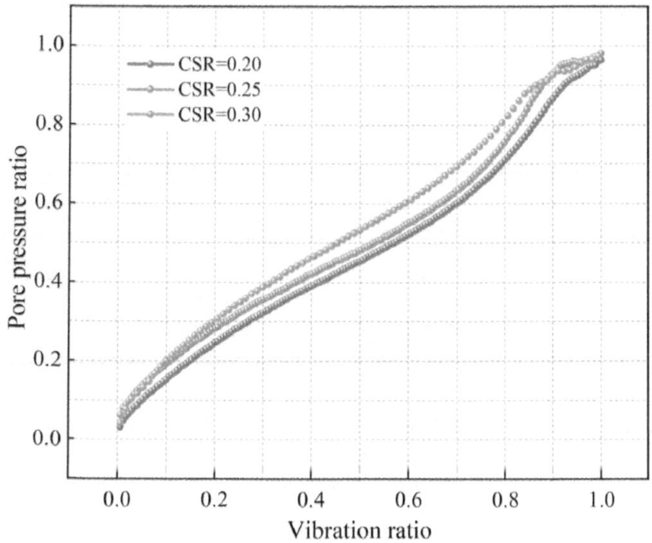

Fig. 12 Relationships between the pore pressure and vibration ratio under different CSRs.

When the fiber content exceeds this optimal value, the reinforcing effect may diminish due to issues such as uneven fiber distribution or agglomeration, ultimately leading to a higher pore pressure ratio.

Figure 12 shows the curves of the pore pressure ratio versus the vibration ratio for samples with different cyclic stress ratios (CSRs) at a fiber length of 15 mm and a fiber content of 0.6%. The results show that the pore pressure ratio of fiber-reinforced calcareous sand samples under lower stress ratios follows a typical three-stage pattern: in the initial stage, the pore pressure increases rapidly, followed by a stable stage where the growth rate decreases, reflecting the progressive destruction of the soil structure. Finally, as the sample approaches the critical state of liquefaction, the pore pressure ratio accelerates, culminating in liquefaction failure. In contrast, under higher stress ratios, the pore pressure ratio increases more linearly, with an overall growth rate significantly higher than that observed under lower stress ratios. This trend suggests that under greater cyclic loading, even with the addition of rubber fibers, the excess pore water pressure in the reinforced sandy soils is prone to rapid accumulation, causing the sample to experience liquefaction sooner. This highlights the limited effectiveness of fiber reinforcement at higher stress levels and its reduced capacity to control pore water pressure.

On the basis of an in-depth study of the dynamic pore pressure development behavior in pure sand and chalk-containing sandy soils, Seed initially proposed a representative inverse hyperbolic model for dynamic pore pressure development, as shown in Equation (1). Booker and Baziar further simplified the model and developed their respective empirical formulations to enhance its applicability in the dynamic response analysis of various soil types (Li et al., 2023), as presented in Equations (2) and (3) below.

$$u/\sigma_c' = 1/2 + 1/\pi\, arc\left[2(N/N_f)^{\frac{1}{\theta}} - 1\right] \qquad ...(1)$$

$$u/\sigma_c' = 2/\pi\, arc(N/N_f)^{\frac{1}{2\theta}} \qquad ...(2)$$

$$u/\sigma_c' = 2/\pi\, arc(N/N_f)^{\frac{1}{2\theta}} + \beta\sqrt{\left[1-\left(2\cdot(N/N_f)-1\right)^2\right]} \qquad ...(3)$$

where, u is the dynamic pore pressure, σ_c' is the initial effective circumferential pressure, N is the number of cyclic vibrations, N_f is the number of cyclic destructive vibrations, θ and β are model parameters.

Notably, the above three dynamic pore pressure development models were originally established primarily for sandy soil systems composed of pure sand or with a minor silt content, and their applicability after the incorporation of rubber fibers remains somewhat controversial. To address this, on the basis of the results of dynamic triaxial tests on rubber fiber-reinforced calcareous sand under varying dynamic stress levels, the predictive capabilities of three typical models—Seed, Booker, and Baziar—were systematically evaluated using a dynamic pore pressure ratio of 0.9 as the liquefaction criterion, with the results presented in Figure 13. The findings indicate a clear divergence in model prediction accuracy with increasing dynamic stress ratios: at lower stress levels, all three models exhibit relatively consistent predictions; however, under medium to high stress ratio conditions, the Seed and Booker models remain in close agreement, whereas the Baziar model shows significant deviation from the other two. This discrepancy may be attributed to the nonlinear evolution term of the Baziar model, which does not consider the fiber-induced pore pressure hysteresis effect, highlighting the limitations of traditional models in capturing the dynamic behavior of fiber-reinforced soils.

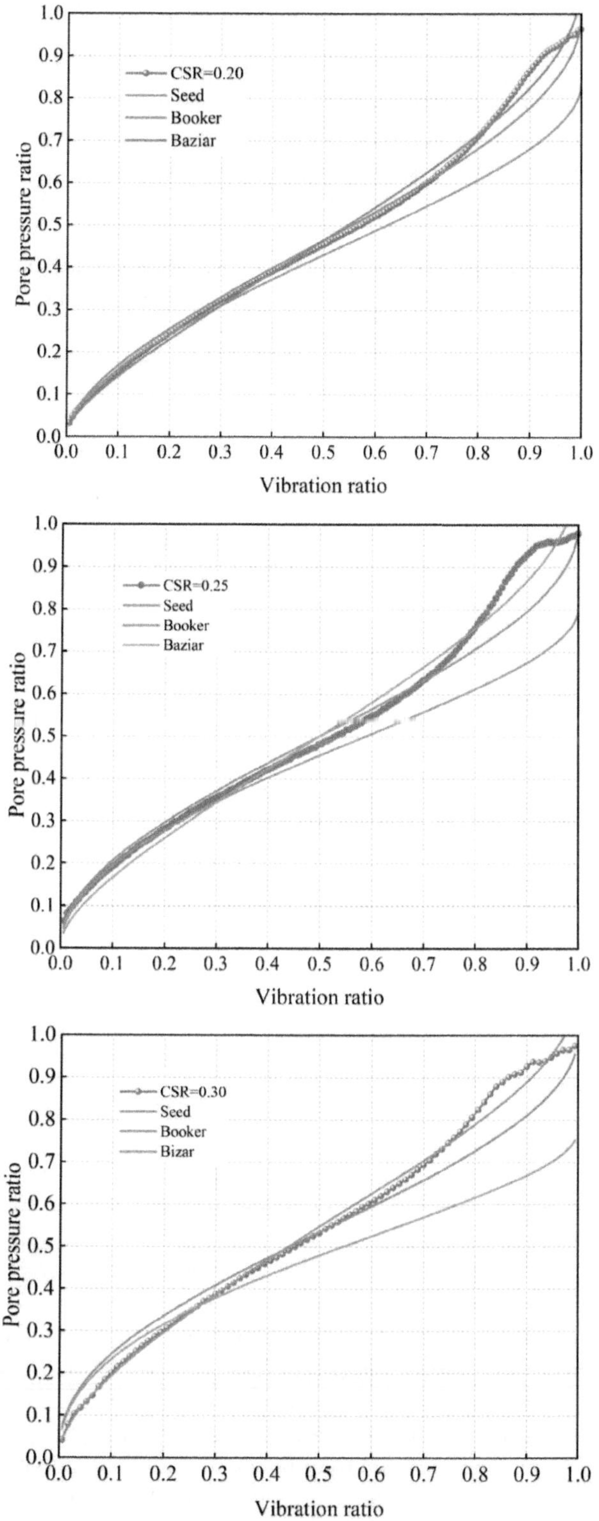

Fig. 13 Prediction results of the pore pressure of reinforced sand under different models.

In summary, calcareous sands exhibit a pronounced pore water pressure response in the early stages of vibratory loading because of their unique biogenic pore structure. The incorporation of rubber fibers influences the evolution of pore pressure primarily through two mechanisms: the permeation path regulatory mechanism and the saturation dynamic equilibrium mechanism. In reinforced calcareous sand, rubber fibers exhibit a nonlinear contact force transfer mechanism and a flexible energy dissipation effect. Their high elasticity enables substantial reversible deformation under cyclic loading, with dynamic energy dissipation achieved by enlarging the hysteresis loop area. Compared with the pure sand samples, the samples containing 0.9% rubber fibers demonstrated significantly greater energy dissipation, which was attributed mainly to the elastic deformation of the fibers. In contrast, the rigid fibers commonly used in siliceous sands lack notable elastic deformation capacity and rely primarily on frictional slip mechanisms for energy dissipation.

3.3 Dynamic Modulus Characteristics

As a key index for characterizing the dynamic properties of sandy soils, the dynamic modulus not only serves to assess the deformation resistance of reinforced sandy soils but also provides essential data for site seismic performance analysis and simulation of the dynamic response of soil layers. This parameter directly reflects the material's stiffness characteristics: the higher its value is, the smaller the material's deformation under external loads. The results of the analysis are shown in Figures 14–17.

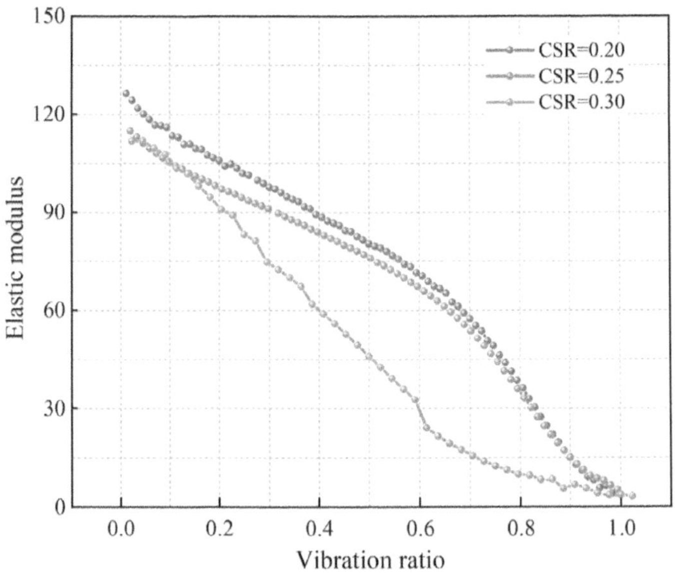

Fig. 14 Dynamic elastic modulus and vibration ratio curves.

Figure 14 shows the variation in the dynamic modulus of elasticity with the number of vibration cycles for pure calcareous sand samples under different cyclic stress ratios. As shown in the figure, the dynamic modulus of elasticity gradually decreases with increasing number of cycles. This phenomenon is attributed primarily to the continuous densification of the samples under cyclic loading. Specifically, during repeated loading, residual strain accumulates, and the sand body undergoes continuous compression, leading to reduced porosity and a more compact structure, which causes changes in stiffness. The decay rate of the dynamic modulus differs significantly under varying stress amplitudes. The test results show that at small strain levels, the dynamic modulus decreases only slightly, indicating some retention of stiffness. However, once the strain level exceeds a certain critical threshold, the material stiffness degrades rapidly, and the modulus of elasticity decreases significantly. This reflects the increased susceptibility of the structure to damage under high stress perturbations. A comparison of the response of the fiber-reinforced sand and pure sand samples further highlights the influence of fiber reinforcement on the stiffness decay process.

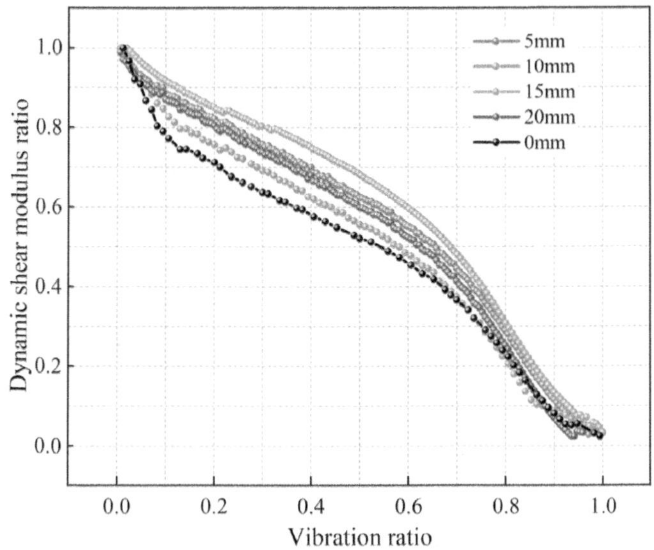

Fig. 15 Shear modulus versus vibration ratio curves for reinforced calcareous sand with different fiber lengths.

The curves in Figure 15 exhibit distinct stage characteristics. At lower vibration ratios, the dynamic shear modulus decreases slowly and linearly, indicating that the samples maintain strong stiffness retention in the initial cycling stage. However, once the vibration ratio exceeds a certain threshold, the modulus decreases at a significantly faster rate, and the curve transitions to a steep decline until the sample experiences liquefaction damage. This behavior suggests that the stiffness of rubber fiber-reinforced sand degrades more slowly under cyclic loading, and its mechanical properties remain relatively stable. When combined with the previously discussed hysteresis loop evolution, it is evident that during the stiffness stabilization stage, the shape of the hysteresis curves of each loading cycle changes little, demonstrating greater energy dissipation stability. In the later stages, as the pore water pressure rapidly increases, the hysteresis curves significantly tilt, and the stiffness of the sample substantially decreases. The internal structure loosens, and the effective stress decreases sharply. Compared with the unreinforced pure calcareous sand samples, the fiber-reinforced sand samples exhibited greater overall shear stiffness, indicating that the rubber fibers enhanced the resistance of the samples to shear deformation to some extent. Furthermore, the overall trends of the curves for the samples with different fiber lengths are quite similar, with only slight variations between them. This suggests that the effect of the fiber length on the evolution of the shear modulus is relatively limited within the range of this test.

Figure 16 shows the variation in the dynamic shear modulus ratio with respect to the vibration ratio for the samples with different rubber fiber contents. Generally, the evolution trend of the curves for each sample mirrors the pattern observed under varying fiber lengths in Figure 15, exhibiting characteristics of a slow decrease in the shear modulus in the early stage followed by accelerated decay in the later stage. As the fiber content increases, the initial shear modulus of the sample first increases but then decreases, indicating that within a certain range, the fiber content effectively enhances the shear stiffness of the reinforced sand, thereby improving its shear resistance. However, when the fiber content exceeds a certain threshold, the excess fibers may cause disorder in the

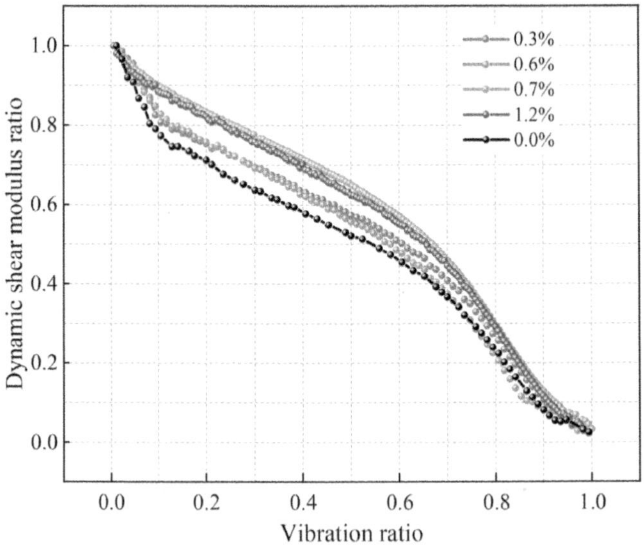

Fig. 16 Shear modulus versus vibration ratio for reinforced calcareous sand with different fiber contents.

interparticle connections or result in fiber agglomeration, which weakens the overall stiffness of the sample. Notably, the spacing between the shear modulus ratio–vibration ratio curves under different fiber content conditions is relatively small, suggesting that the effect of the fiber content on the shear modulus decay process is limited. The modulus degradation rate generally remains consistent under the present test conditions. This further confirms that there is an optimal range of fiber contents in which rubber fibers enhance the shear performance of calcareous sand.

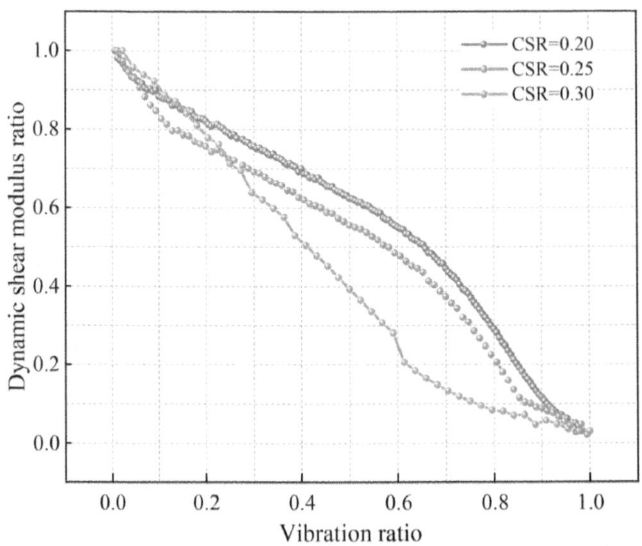

Fig. 17 Shear modulus versus vibration ratio for reinforced calcareous sand with different CSRs.

Figure 17 shows the variation rule of the dynamic shear modulus ratio of the samples with the vibration ratio under different cyclic stress ratios. The results show that the shear modulus ratio of the samples under lower stress ratio conditions decreases gradually with increasing number of loading cycles, tends to decrease overall, and finally tends to stabilize, remaining at approximately 50% of the initial modulus. When the vibration ratio exceeds the critical threshold of 0.8, the modulus ratio significantly accelerates, indicating that the sample is close to the liquefied state and that the structural stiffness degrades rapidly. In contrast, under the higher stress ratio condition, the shear modulus of the sample decreased drastically at the beginning of loading, showing an obvious characteristic of a sudden decrease in stiffness, which indicates that the larger stress amplitude caused more intense disturbance to the soil structure and accelerated the process of shear stiffness decay. This implies that the stress ratio is an important parameter influencing the evolutionary characteristics of the shear modulus of reinforced sand and that higher stress levels significantly weaken the ability of a material to resist shear deformation, leading to earlier and more drastic decay of the shear modulus and thus accelerating the onset of liquefaction damage.

In summary, the application of rubber fibers in marine environments has strong engineering feasibility, particularly in terms of corrosion resistance and long-term service performance. The experimental results show that the mechanical properties of rubber fibers remain stable after prolonged immersion in a high-salinity environment, with a tensile strength retention rate approaching 90%, indicating the material's capacity to maintain its reinforcing effect throughout the service life of a typical marine engineering project. Additionally, biodegradation tests revealed that the fibers experienced a very low rate of microbial erosion in island environments—significantly lower than that of conventional sand and soil materials—demonstrating good biocompatibility and structural stability within the marine ecosystem.

In terms of resource utilization, the use of rubber fibers not only facilitates the large-scale disposal of used tires and the efficient reuse of solid waste but also offers significant energy-saving and carbon-reducing benefits. Engineering estimates indicate that incorporating rubber fibers into large-volume blow-fill geotechnical applications can convert substantial amounts of waste rubber into valuable resources, with a high utilization rate, thereby effectively mitigating the environmental burden caused by tire accumulation. Compared with traditional reinforcement materials, rubber fibers require significantly less energy per unit during manufacturing, substantially reducing carbon emissions while achieving comparable reinforcement effects, thus demonstrating strong environmental and economic advantages.

With respect to environmental risk control, experimental analyses have shown that the heavy metal content of rubber fibers is low and that the material exhibits excellent chemical stability in aqueous environments, with no significant release of harmful elements detected. Additionally, under simulated oceanic dynamics, the microplastic release rate is extremely low—substantially lower than the levels of plastic deposition commonly found in natural waters—and such particles can be effectively adsorbed by natural calcareous sands, thereby reducing the risk of their transport and accumulation within the ecosystem. Overall, rubber fibers demonstrate strong engineering adaptability and environmental compatibility in complex marine environments, indicating broad prospects for large-scale applications.

4. Conclusions

(1) Under isobaric consolidation conditions, the deformation process of a sample under cyclic loading can be divided into three distinct stages: the vibratory compaction stage, the vibratory shear stage, and the vibratory damage stage. During the initial stage of loading, the sample exhibited strong deformation resistance, with small axial strain amplitudes. The hysteresis curves corresponding to different loading cycles nearly overlap, indicating that the material is in the linear elasticity or initial nonlinear stage. As the loading progresses, the cumulative plastic strain gradually increases, and the area of the hysteresis loop expands significantly, reflecting a heightened energy dissipation and stiffness degradation. At this point, the soil structure enters a stage of nonlinear development.

(2) The incorporation of rubber fibers significantly enhanced the liquefaction resistance of calcareous sand. The experimental results demonstrated that with increasing fiber content and fiber length, the liquefaction vibration cycles of the samples increased, thereby improving their liquefaction resistance. However, when the fiber content or length surpassed a certain threshold, the number of liquefaction vibration cycles decreased, indicating the presence of an optimal parameter range for maximizing the anti-liquefaction effect. Additionally, the cyclic stress ratio had a significant influence on the liquefaction behavior of the fiber-reinforced sand. As the stress ratio increased, the number of liquefaction vibration cycles decreased sharply, suggesting that higher stress levels diminished the reinforcing effect of the fibers.

(3) The development of pore water pressure typically follows three stages. In the initial stage, the pore pressure increases gradually and monotonically with an increasing number of loading cycles, with small fluctuations. In the middle and later stages, the pore pressure enters an accelerated growth phase, accompanied by a significant decrease in the sample stiffness, indicating a pronounced softening trend. For the pure calcareous sand samples, the pore pressure increased rapidly during this phase, approaching liquefaction quickly. In contrast, the incorporation of rubber fibers effectively slowed the accumulation of pore pressure, significantly reducing the overall growth rate of the pore pressure ratio–vibration ratio curve for the reinforced sand. The pore pressure development characteristics were influenced by both the fiber content and length, following a "first suppressed, then increased" trend. Under high cyclic stress ratios, the pore pressure growth rate accelerated considerably, increasing the likelihood of liquefaction damage.

(4) The dynamic shear modulus ratio versus vibration ratio curves for all the sample groups display similar evolutionary patterns. At lower vibration ratios, the modulus ratio decreases slowly, indicating that the material stiffness remains relatively stable. However, once the vibration ratio exceeds a critical threshold, the shear modulus rapidly decays, eventually approaching zero, indicating the onset of liquefaction damage. Compared with pure calcareous sands, fiber-reinforced calcareous sands have higher shear modulus values and better stiffness retention. However, the modulus remains relatively unchanged across a range of fiber lengths and content variations. The evolution of the dynamic shear modulus is more sensitive to the cyclic stress ratio, with the decay rate accelerating significantly as the stress ratio increases.

Credit authorship contribution statement

- **Jiajun Shu:** Writing-original draft, Investigation.
- **Wenbo Wang:** Writing–original draft, Data curation, Visualization.
- **Tao Li:** Writing–review & editing, Data curation, Funding acquisition.
- **Bo Liu:** Writing-review & editing, Methodology, Funding acquisition.

Declaration of Competing Interest

The authors declare that they have no known competing financial interests or personal relationships that could have appeared to influence the work reported in this paper.

Data availability

Data will be made available on request.

Acknowledgements

This research acknowledges the financial support provided by the National Natural Science Foundation of China (Grant NO. 51508556, 42172319), Major Science and Technology Project of Xinjiang Uygur Autonomous Region (Grant NO: 2024A01003).

References

Brandes, H.G. (2011). Simple shear behavior of calcareous and quartz sands. *Geotechnical and Geological Engineering, 29*(1), 113–126. https://doi.org/10.1007/s10706-010-9362-x.

Du, W., Gao, J., Zhang, C., Ma, Q., Zhang, J. and Cheng, A. (2024). A numerical study on the rock breakage mechanism with TBM disk cutter in calcareous cemented sand pebble stratum using a coupled FEM–DEM method. *Rock Mechanics and Rock Engineering, 58*(2), 2319–2347. https://doi.org/10.1007/s00603-023-03537-7.

Edil, T.B. and Bosscher, P.J. (1994). Engineering properties of tire chips and soil mixtures. *Geotechnical Testing Journal, 17*(4), 453–464. https://doi.org/10.1520/GTJ10318J.

Foose, G.J., Benson, C.H. and Bosscher, P.J. (1996). Sand reinforced with shredded waste tires. *Journal of Geotechnical and Geoenvironmental Engineering, 122*(9), 760–767. https://doi.org/10.1061/(ASCE)0733-9410(1996)122:9(760).

Gao, J., Pan, L., Bian, X., Wang, J. and Li, Y. (2024). Study on dynamic shear characteristics of calcareous sand reinforced with rubber and geogrid. *Bulletin of Engineering Geology and the Environment, 83*(11), 445. https://doi.org/10.1007/s10064-024-03923-4.

Gao, J., Zhu, G., Wang, J., Yang, Y. and Li, Y. (2024). Study on the dynamic characteristics of geogrids combined with rubber particles reinforced with calcareous sand. *Acta Geotechnica, 19*(10), 7031–7047. https://doi.org/10.1007/s11440-023-02127-2.

Guo, L., Zhan, W., Zhang, F., Zhang, J., Yao, Y. and Cheng, G. (2017). The influence of earthquakes on Zhubi Reef in the Nansha Islands of the South China Sea. *Acta Oceanologica Sinica, 36*(3), 99–108. https://doi.org/10.1007/s13131-017-1045-2.

Hyodo, M., Aramaki, N., Itoh, M. and Hyde, A.F.L. (1996). Cyclic strength and deformation of crushable carbonate sand. *Soil Dynamics and Earthquake Engineering, 15*(5), 331–336. https://doi.org/10.1016/S0267-7261(96)00027-6.

Jafarian, Y. and Javdanian, H. (2020). Dynamic properties of calcareous sand from the Persian Gulf in comparison with siliceous sands database. *International Journal of Civil Engineering, 18*(2), 245–249. https://doi.org/10.1007/s40999-020-00430-4.

Jafarian, Y., Javdanian, H. and Haddad, A. (2018). Dynamic properties of calcareous and siliceous sands under isotropic and anisotropic stress conditions. *Soils and Foundations, 58*(1), 172–184. https://doi.org/10.1016/j.sandf.2017.11.005.

Javdanian, H. and Jafarian, Y. (2018). Dynamic shear stiffness and damping ratio of marine calcareous and siliceous sands. *Geo-Marine Letters, 38*(4), 315–322. https://doi.org/10.1007/s00367-018-0544-7.

Li, T., Li, Y., Shu, J., Zhao, H. and Liu, B. (2024). Effect of rubber particles on mechanical properties of calcareous sand under large displacement shear. *Marine Georesources and Geotechnology*, 1–13. https://doi.org/10.1080/1064119X.2024.2317826.

Li, T., Tang, X., Zeng, L. and Yang, G. (2023). Dynamic pore pressure characteristics of saturated sand–clay and sand–silt–clay mixtures. *Chinese Journal of Geotechnical Engineering, 45*(S2), 44–50. https://doi.org/10.11779/CJGE2023S20044.

Li, T., Shu, J., Wu, B., Huang, J. and Deng, Z. (2024). Fracture failure characteristics of shallow coral reef under wave dynamic loading. *Marine Georesources and Geotechnology*, 1–11. https://doi.org/10.1080/1064119X.2024.2310009.

Liu, S., Dong, B., Jin, Y. and Cai, Y. (2021). Evaluation of biostimulation efficacy on the reinforcement of calcareous sand. *Journal of Testing and Evaluation, 49*(6), 4181–4200. https://doi.org/10.1520/JTE20200385.

Noorzad, R. and Raveshi, M. (2017). Mechanical behavior of waste tire crumbs–sand mixtures determined by triaxial tests. *Geotechnical and Geological Engineering, 35*(4), 1793–1802. https://doi.org/10.1007/s10706-017-0207-5.

Salem, M., Elmamlouk, H. and Shehab, A. (2013). Static and cyclic behavior of North Coast calcareous sand in Egypt. *Soil Dynamics and Earthquake Engineering, 55*, 83–91. https://doi.org/10.1016/j.soildyn.2013.09.006.

Shahnazari, H., Jafarian, Y., Tutunchian, M.A. and Rezvani, R. (2016). Undrained cyclic and monotonic behavior of Hormuz calcareous sand using hollow cylinder simple shear tests. *International Journal of Civil Engineering, 14*(4), 209–219. https://doi.org/10.1007/s40999-016-0028-x.

Shahnazari, H. and Rezvani, R. (2013). Effective parameters for the particle breakage of calcareous sands: An experimental study. *Engineering Geology, 159*, 98–105. https://doi.org/10.1016/j.enggeo.2013.03.008.

Shalchian, M.M. and Arabani, M. (2023). Application of plant-derived fibers in soil reinforcement on experimental, numerical, and case study scales: A review. *Bulletin of Engineering Geology and the Environment, 82*(1), 1–28. https://doi.org/10.1007/s10064-022-02793-y.

Shan, Y., Li, J., Zhou, H., Li, Y., Yuan, J. and Cui, J. (2025). Liquefaction resistance of MICP treated silica and calcareous sand with carbon fiber. *Bulletin of Engineering Geology and the Environment, 84*(3), 159. https://doi.org/10.1007/s10064-024-03891-9.

Tang, Q., Zhang, J., Feng, Y., Li, J., Yao, Y. and Zhan, W. (2018). Numerical simulation for shallow strata stability of coral reef in the southwest of Yongshu Reef (South China Sea). *Journal of Ocean University of China, 17*(4), 763–772. https://doi.org/10.1007/s11802-018-3678-3.

Youwai, S. and Bergado, D.T. (2003). Strength and deformation characteristics of shredded rubber tire–sand mixtures. *Canadian Geotechnical Journal, 40*(2), 254–264. https://doi.org/10.1139/t02-096.

Wang, R., Li, R., Wang, H. and Luo, H. (2025). Reinforcement mechanism and effect of self-compacting expansive piles on calcareous sand foundations. *Ocean Engineering, 328*, 121051. https://doi.org/10.1016/j.oceaneng.2023.121051.

Wang, D.L., Tang, C.S., Pan, X.H., Liu, B. and Shi, B. (2023). Coupling effect of fiber reinforcement and MICP stabilization on the tensile behavior of calcareous sand. *Engineering Geology, 317*, 107090. https://doi.org/10.1016/j.enggeo.2023.107090.

Wang, X., Cui, J. and Wang, W. (2024). Dilatancy behaviors of calcareous sand considering particle breakage. *Bulletin of Engineering Geology and the Environment, 83*(4), 146. https://doi.org/10.1007/s10064-023-03697-w.

Wei, H., Zhao, T., Meng, Q., Wang, X. and He, J. (2018). Experimental evaluation of the shear behavior of fiber-reinforced calcareous sands. *International Journal of Geomechanics, 18*(12), 04018160. https://doi.org/10.1061/(ASCE)GM.1943-5622.0001333.

Wu, Q., Ding, X. and Zhang, Y. (2024). Experimental and numerical investigation on the dynamic behavior of slopes in calcareous sand during earthquakes. *Acta Geotechnica, 19*(6), 3423–3446. https://doi.org/10.1007/s11440-023-01945-6.

Zhou, L., Chen, J.F., Zhu, Y. and Yao, T. (2024). Liquefaction and post-liquefaction behaviors of unreinforced and geogrid reinforced calcareous sand. *Geotextiles and Geomembranes, 52*(3), 286–303. https://doi.org/10.1016/j.geotexmem.2024.01.008.

CHAPTER

17

Energy-Based Engineering Analysis Applied to Physics-based Seismic Time Histories

A Case Study

Fabrizio Mollaioli[1*] and Fabio Romanelli[2]

This work investigates the application of energy-based engineering analysis to physics-based seismic time histories, addressing the complexities of seismic risk assessment and performance-based earthquake engineering (PBEE). The study emphasizes the significance of quantifying the variability of structural responses to ground motion inputs, particularly focusing on understanding the ground motion characteristics associated with seismic events. Two main methodologies for defining seismic ground motion—engineering and seismological—are explored, highlighting the challenges posed by the limited number of recorded motions available for response-history analysis. The research advocates for the generation of synthetic signals, validated against observed intensities, using Neo-Deterministic Seismic Hazard Analysis (NDSHA) to create realistic scenario-based synthetic time series. This study introduces an energy-based approach to characterize seismic motion, positing that energy demand metrics offer reliable indicators for seismic hazard estimation and structural design. The findings illustrate the correlation between energy-based parameters and structural damage potential, as well as the effectiveness of energy analysis in evaluating seismic performance across diverse structural systems.

1. Introduction

One of the main challenges in seismic risk assessment and performance-based earthquake engineering (PBEE) is quantifying the variability of seismic responses of structures subjected to ground motion inputs (Moehle & Deierlein, 2004; O'Reilly & Calvi, 2019). The two main sources of input uncertainty that control structural responses to seismic excitations are the variability of ground motion characteristics and structural parameters. Systematic assessment of uncertainty across various scenarios is made possible by evaluating several seismic response indicators.

[1] Department of Structural and Geotechnical Engineering, Sapienza University of Rome, Rome, Italy. https://orcid.org/0000-0002-1828-1459.

[2] Department of Mathematics, Informatics and Geosciences, University of Trieste, Trieste, Italy. https://orcid.org/0000-0002-7478-8457.

Email: romanel@units.it[2]

* Corresponding author: fabrizio.mollaioli@uniroma1.it

DOI: 10.1201/9781003684121-19

In this context, determining the ground motion features linked to upcoming earthquakes, both locally and regionally, is a common seismic hazard concern. A description of the ground shaking intensity caused by an earthquake of a specific distance and magnitude (referred to as the "ground shaking scenario") or probabilistic maps of pertinent parameters characterizing the ground motion can be used as the input for the ensuing seismic risk analysis.

Two primary methods can be used to more accurately define the seismic ground motion caused by an earthquake of a specific magnitude and source-to-site distance. The first (referred to as the engineering technique) is based on the grouping of accelerograms with similar source, path, and site effects and the analysis of the strong motion datasets that are currently available and gathered by seismic networks. Estimating realistic source-to-site transfer functions is a key component of this methodology, usually performed with ground motion prediction (or attenuation) equations.

The second strategy (known as the seismological method) is founded on modeling techniques that aim to accurately replicate the ground motion connected to the specified earthquake scenario. These approaches have been developed based on knowledge of the seismic source process and the propagation of seismic waves.

Although the number of severe motions recorded has significantly increased in recent decades, particularly in North America, Japan, and Taiwan, the deployment of a dense seismic network in each earthquake-prone location is lacking. Therefore, the number of ground motions necessary to be selected to enable response-history analysis is usually not sufficient, forcing us to use scale factors that allow to vary only the amplitude of the ground motion without modifying its duration and frequency content.

Because of their close relationship to actual seismic events, recorded motions are frequently considered to be realistic depictions of earthquake processes and are therefore preferred in design contexts. However, they may not cover the entire range of possible earthquake features, and their availability is sometimes limited, particularly for uncommon events or certain seismic scenarios (Panza et al., 2001, 2003). On the other hand, a wide range of seismic scenarios can be simulated with regulated spectral properties thanks to the high degree of freedom provided by synthetic motions.

Therefore, it seems clear that such situations require the production of synthetic signals that are validated against observed intensities that take into account the path, the source characteristics, and the local geological and geotechnical conditions in order to be used as seismic input in a subsequent engineering analysis.

In this context, on the basis of Neo-Deterministic Seismic Hazard Analysis (NDSHA) it is possible to construct scenario-based realistic and duly validated synthetic time series, accounting for source, propagation, and site effects. In this way, standard convolutive methods are not required because NDSHA heavily relies on a lot more data than typical seismic hazard investigations. The NDSHA procedure provides strong ground motion parameters based on the seismic wave propagation modeling at different scales, regional, national, and urban—accounting for a wide set of possible seismic sources and for the available information about geophysical structural models (e.g., Panza et al., 2012; Rugarli et al., 2019; Panza et al., 2021; Ricci et al., 2025). The scenario-based approach is based on physical modeling techniques that supplement observable data and can be put through a formal validation process.

The other fundamental aspect that needs to be taken into account is to find suitable criteria that accurately describe the potential for earthquake destruction, necessary for characterizing seismic motion in earthquake-prone areas. It is therefore necessary to identify the ground motion parameters that characterize the severity and the damage potential of the earthquake ground motion in order to quantify the ground motion expected at a specific site, which would drive the structure to its critical response. In this situation, energy-based metrics are a suitable way to draw attention to the potentially harmful nature of these signals (Mollaioli et al., 2019; Benavent et al., 2021).

Beyond the possibility of designing earthquake-resistant structures by balancing energy demands and supplies, energy-based methodologies enable the proper characterization of the various time histories (i.e., impulsive, periodic, with long-duration pulses, etc.) that may correspond to an earthquake ground shaking while simultaneously taking into account a structure's dynamic response. Energy also accounts for duration effects because it is a cumulative measure of ground shaking. Moreover, the stability of the overall energy input and the fact that an energy-based approach is independent of other factors that can be disregarded for design reasons, such as strength, damping level, hysteretic behavior, etc., are its important advantages. Thus, energy demand parameters can be regarded as trustworthy instruments for seismic hazard estimation, earthquake scenario selection, and earthquake design for a variety of engineering assessments.

2. Seismic Energy-based Parameters

The energy balance for a generic structural system subjected to a ground motion can be expressed in terms of relative (E_{Ir}) or absolute (E_{Ia}) input energy, given in Eq. (1a–b):

$$E_{Ir} = E_{kr} + E_{\xi} + E_s + E_H \quad ...(1a)$$

$$E_{Ia} = E_{ka} + E_{\xi} + E_s + E_H \quad ...(1b)$$

where, E_{kr} and E_{ka} are the relative and absolute kinetic energies, respectively, E_{ξ} is the damping energy, E_s is the elastic strain energy and E_H is the hysteretic energy.

Energy-based design (EBD) is based on the fact that the total input energy (E_{Ir} or E_{Ia}) depends almost completely on mass and period of the structure, being approximately the same for a multi-degree-of-freedom system (MDOF) of total mass and fundamental period as in a single-degree-of-freedom system (SDOF) of the same mass and period. In other words, the total input energy is barely affected by the viscous damping ratio ξ, the strength, the post-yield stiffness, the hysteretic behavior, the degree of inelastic action, the number of degrees of freedom of the system. This stability of the total input energy was shown by several authors (Housner, 1956; Akiyama, 1985; Uang & Bertero, 1990; Fajfar & Vidic 1994; Sucuoglu & Nurtug 1995; Decanini & Mollaioli, 1998, 2001; Donaire-Avila et al., 2017; Mollaioli et al., 2019; Cheng et al., 2014, 2020, 2021; Benavent-Climent et al., 2021).

To describe the magnitude, frequency content, and duration of ground motions, numerous parameters have been proposed. Peak ground acceleration (PGA) and peak ground velocity (PGV) are the most widely used metrics for quantifying the amplitude of ground motions, and they are derived directly from the acceleration time histories. First, nevertheless, it was demonstrated that there is no strong relationship between PGA and damage from earthquake ground motion. Since the majority of the impulse is absorbed by the structure's inertia with minimal deformation, high PGA may actually be linked to high frequency pulses that do not cause much damage. Second, velocity is another quantity that is helpful for characterizing the amplitude of ground motion, despite being less sensitive to the higher-frequency components of the motion.

A parameter that takes into account duration and amplitude is the Arias Intensity, I_A, defined by the Eq. (2) (Arias, 1970):

$$I_A = \frac{\pi}{2g} \int_{t_i}^{t_f} a(t)^2 \, dt \quad ...(2)$$

where, g is the acceleration due to gravity, t_i is the beginning of the record, t_f is the total duration of the record and $a(t)$ represents the acceleration time history of the input ground motion. Although it does not reflect the frequency content and has a tendency to overstate the intensity of long-duration motions with high amplitude and a wide range of frequency content, it has often been used to assess damage potential.

Another interesting measure is the Specific Energy Density (SED) (Mollaioli et al., 2019), which is defined Eq. (3) as the integral of the square velocity time history, $v(t)$, in the interval $t_i - t_f$:

$$\text{SED} = \int_{t_i}^{t_f} v(t)^2 \, dt \qquad ...(3)$$

Even though this parameter does not include the contribution of the frequency content, it gives a better insight into the characterization of the most damaging pulses of the considered records. This aspect will be more detailed later.

However, specifically, when inelastic behaviour is present, a rational methodology for evaluating the destructiveness potential of earthquakes should simultaneously consider the relevant properties of the ground motion (i.e., amplitude, frequency content, duration, etc.) and the earthquake-resisting structure. Due to its direct correlation with the number and amplitudes of oscillator response cycles, Input Energy implicitly captures the effects of inelasticity and ground motion duration that are missed by the more traditional spectral parameters. Generally, it is also possible to assert that even though energy-based parameters can be considered well correlated to structural damage, all the velocity-based parameters give a quite good picture of the damage potential of earthquake ground motion.

As already underlined, duration was also considered by the Arias Intensity parameter. However, as it is possible to observe in Figure 1, where the normalized time histories of I_A and the SED are compared with those of acceleration, velocity, and displacement, a slow increase of the Arias Intensity was observed with two larger increases in correspondence of the maximum acceleration spikes. For the Energy Flux (SED parameter), which is a measure of energy demand, a different pattern is shown. The late pulse that corresponds to both maximum velocity and displacement in this instance is linked to a notable jump in its time history. The correspondence with the maximum implies the possibility of significant damage.

Given the fact that the energy-based approach's key components represent a clear description of the damage potential of earthquake ground motion, it's also important to look at the maximum and relative displacements, since they highlight the damage in non-structural elements, govern second order effects, and are indicators of damage (along with hysteretic energy). Finding reliable correlations between energy and displacement demands and supplies is a key concern for both the assessment of seismic vulnerability and the establishment of more consistent seismic code provisions.

While the global and local deformations of a structure are directly linked to its seismic performance, the energy balance formulation seems to be a much more effective concept because it allows for a rational assessment of the energy absorption and dissipation mechanisms that can be successfully implemented to balance the energy imparted to the structure. Given that the two elements may be regularly combined in a performance-based seismic design process, knowing how input and dissipated energy related to displacement needs becomes an essential precondition.

As shown by Mollaioli et al. (2011), for SDOF and MDOF systems, there exists a clear relationship between the global and local displacement demand (i.e., maximum displacement, $\ddot{a}_{\max}$ and Maximum Inter-story Drift, $IDI_{\max}$) and the input (relative, E_{Ir} or absolute, E_{Ia}) and hysteretic (E_H) energies.

In general, the signal energy content has a greater impact on the spectral shape of the inter-story drift index. This situation arises from comparing the energy and drift spectra (Mollaioli et al., 2011), which show agreement in the overall trend as well as spectral coincidence of the periods when the largest values are reached. The displacement spectral forms, on the other hand, do not exhibit comparable correspondences. This should indicate that while a large local deformation demand is likely to happen, a strong energy demand imposed on a structural system does not always result in a strong top displacement. The trend to assume the same shape of the input energy spectra is proven by the frequent correspondence of periods between energy and drift peaks and valleys. The maximum drift is greatly influenced by both the amount of energy input and the way the energy is imparted to the structure. A greater rate of input in the energy demand time history strongly affects the drift demand (Angelucci et al., 2023, 2024).

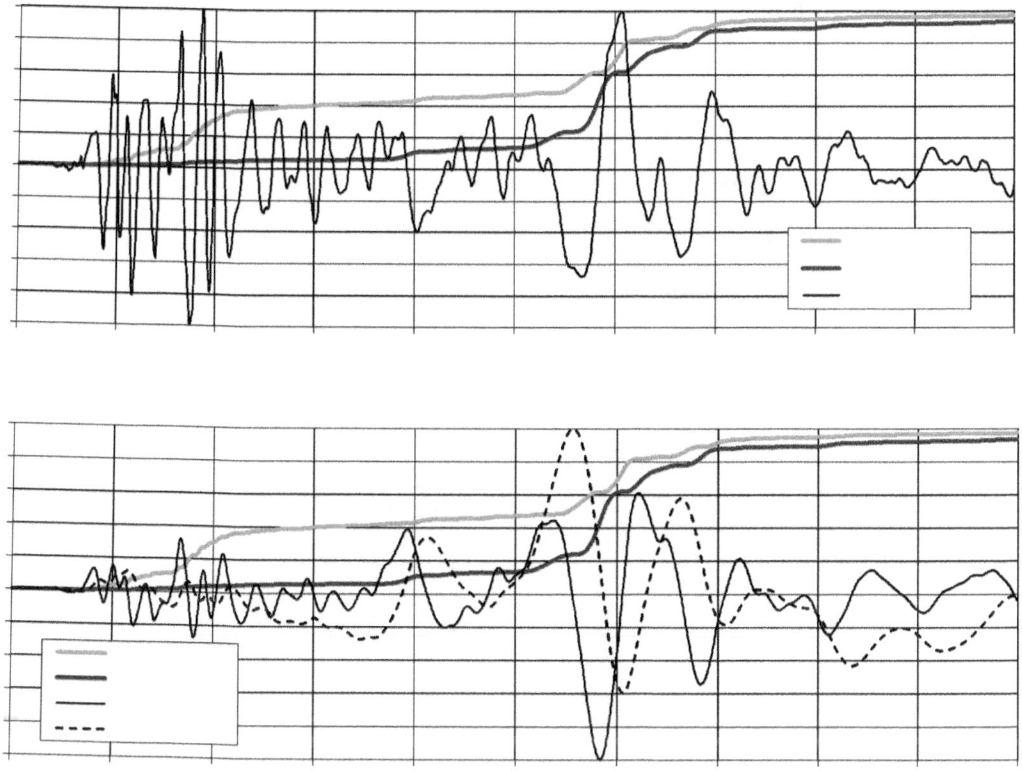

Fig. 1 Normalized time-histories of Arias Intensity (I_A), Specific Energy Density (SED) with acceleration (top), velocity and displacement (bottom) time histories corresponding to the Gilroy #6 record (1984 Morgan Hill Earthquake, $M_w = 6.2$).

3. Selection of Ground Motion Time Histories

Ricci et al. (2025) performed a novel Local Seismic Response study procedure investigating the seismic wave directivity effects on the seismic wavetrains and their local changes (e.g., amplification/reduction), which propagate from the seismic source up to the surface of a specific site. They considered a set of earthquake scenarios, with magnitudes ranging from 6.0 to 6.9, using the information available (i.e., focal and geometrical properties) in the Italian Database of Individual Seismogenic Sources (DISS Working Group, 2021); the scenarios were used with a 2D section representative of the urban area of Chieti city center.

To model the seismic wavefields radiated from the extended sources, the simulation of the kinematic source rupture process was performed using the PULSYN algorithm, based on the theory presented in Gusev (2011). PULSYN generates many stochastic realizations of the rupture model (slip distribution and rupture velocity) for each source-to-site path. It produces a set of spectra (amplitude and phase) of the equivalent source time functions (accounting for both the rupture process and the related directivity effects) ready to be convolved with the Green's Functions computed by the hybrid method at the selected sites.

In the specific case of Chieti sites, 100 realizations of unilateral rupture and 100 of bilateral rupture were considered for each of the considered sources and the multi-scenario results were summarized performing the MCSI procedure (e.g., Fasan et al. (2016); Rugarli et al. (2019)). Ricci et al. (2025) used the 6 April 2009 L'Aquila earthquake recordings at the Chieti seismic station (CHT) to benchmark the MCSI-NDSHA method for seismic input prediction. In fact, the blind ground-shaking scenario prediction (i.e., relying solely on the fault's location and geometry), successfully compared the recorded and synthetic accelerograms considering the resultant horizontal acceleration response spectrum.

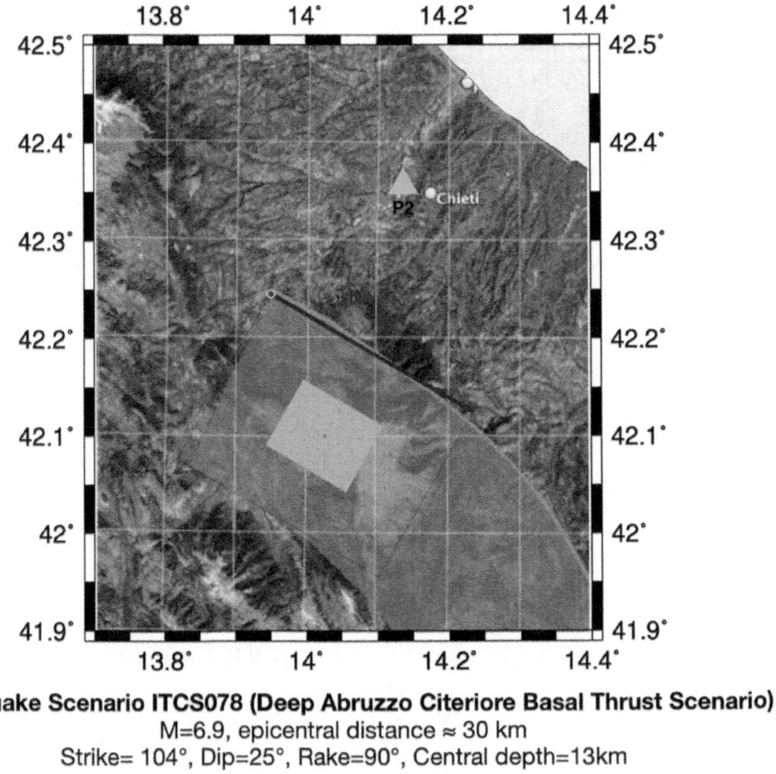

Fig. 2 Earthquake scenario and source site configuration.

In this work, we analyze the acceleration time histories generated by source S1 (see Figure 1 and Table 3 of Ricci et al. (2025)), with a magnitude $M_w = 6.9$, simulated at site P2 (see Figure 4 and Table 5 of Ricci et al. (2025)), at a horizontal distance from the center of the fault of about 30 km (see Figure 2), since they represent the source-site configuration providing the largest seismic input in terms of spectral accelerations (see Figure 11 of Ricci et al. (2025)). Specifically, we have considered a set of 100 kinematic rupture realizations for a bilateral rupture in forward and neutral directivity condition to investigate what is the effect of near-source condition, keeping all the other parameters (e.g., magnitude, distance, focal mechanism, and site conditions) fixed. The response spectra obtained from the acceleration time histories (maximum horizontal component) for the 100 realizations with bilateral rupture in forward (B0) and neutral directivity (B90) are shown, respectively in Figure 3.

For both directivity conditions, we have then conditionally selected (T = 0.5 s) for energy-based analyses three sets of 11 accelerograms, at the 50th, 64th, and 95th percentiles. We show in Figure 4 the sets of accelerograms corresponding to the 95th percentile for B0 and B90 conditions (a and b, respectively). In Figure 5 are shown, as a pictorial example, the first four corresponding rupture realizations for B0 and B90 (a and b, respectively).

A second group made of real ground motion records was selected in order to compare their responses with that of the considered synthetic records. The choice was made from a general database of worldwide records. Two bins have been considered: source-to-site distance between 20–40 km; magnitude Mw between 6.7–7.3. It is important to recall that the synthetic ground

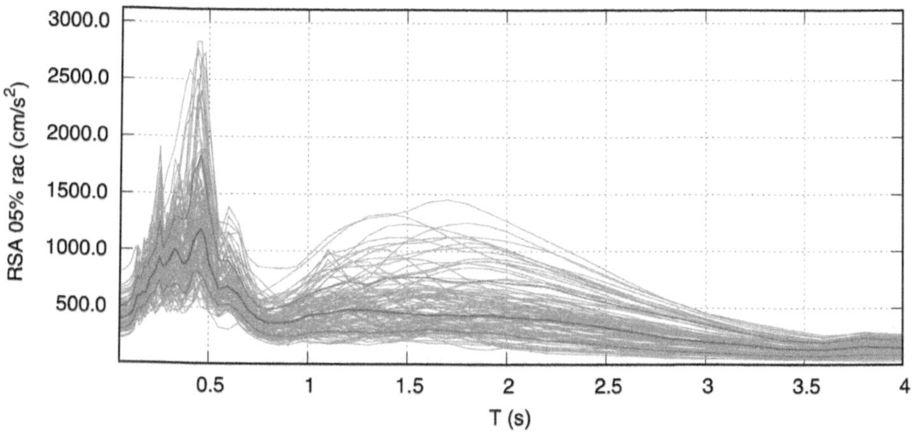

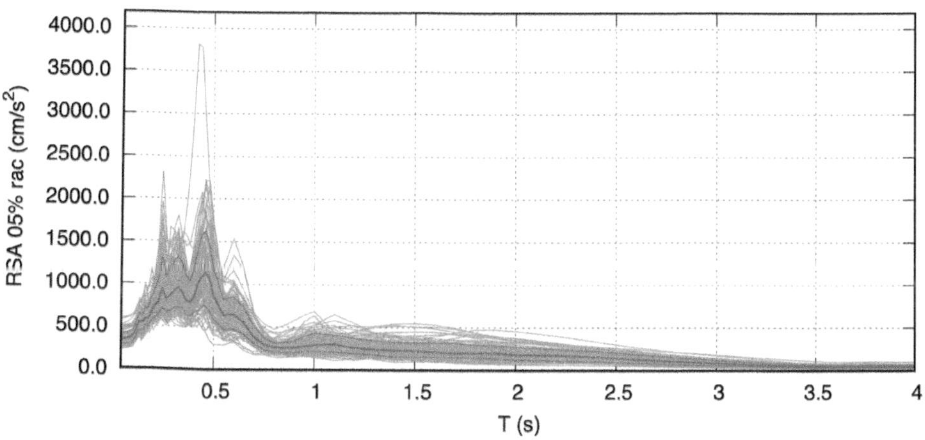

Fig. 3 Acceleration response spectra (RSA, 5% damping) obtained from the acceleration time histories (maximum horizontal component (rac)) for the 100 realizations with bilateral rupture in a) forward (B0) and b) neutral directivity (B90) are shown. The thick solid line shows the median (50th percentile), while the thin solid lines indicate the standard deviation (64th and 36th percentiles).

motions are derived for a distance of 30 km and a magnitude of 6.9. PGA values range from 40 to 568 cm/s^2, while PGV ones are between 6–49 cm/s. The earthquakes considered are indicated in Table 1.

Table 1 Selected recorded earthquake events.

Earthquake Name	*YEAR*	*MODY*	*MW*
Irpinia, Italy	1980	1123	6.9
Spitak, Armenia	1988	1207	6.8
Loma Prieta, USA	1989	1018	6.9
Cape Mendocino, USA	1992	0425	7.0
Landers, USA	1992	0628	7.3
Northridge, USA	1994	0117	6.7
Kobe, Japan	1995	0116	6.9
Hector Mine, USA	1999	1016	7.1
Duzce, Turkey	1999	1112	7.1

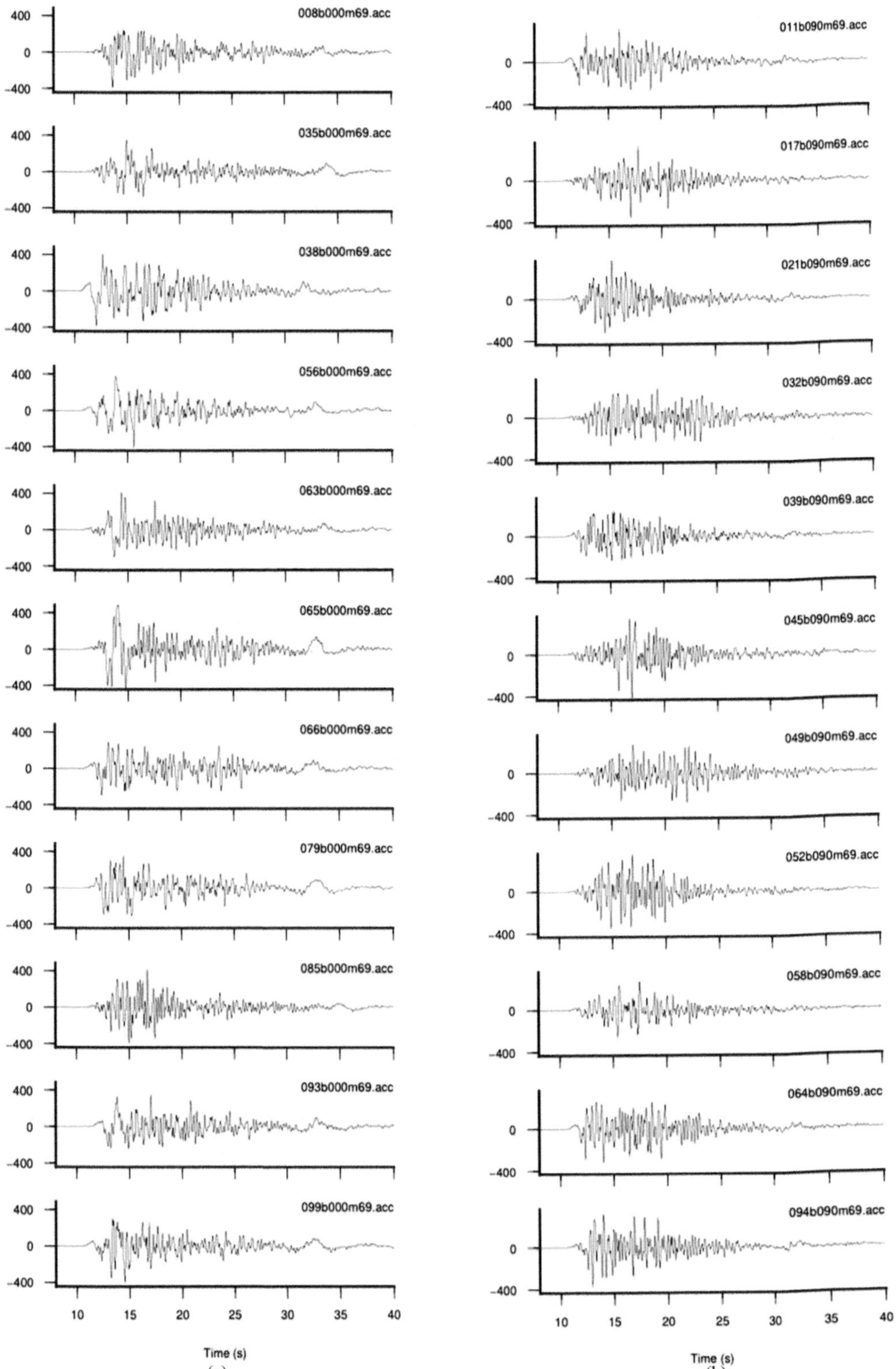

(a) (b)

Fig. 4 The two sets of 11 accelerograms correspond to the 95th percentile for B0 (a) and B90 (b) directivity conditions. For each signal, the first three digits of the label name refers to the number of rupture realization.

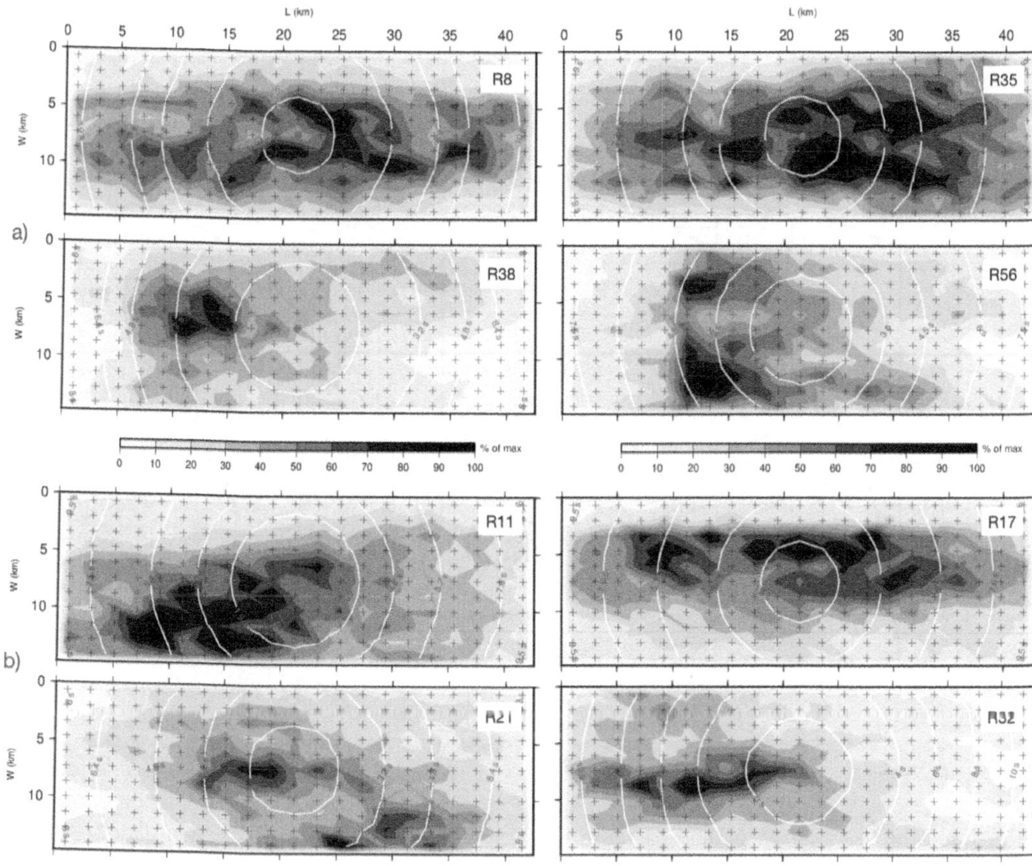

Fig. 5 The rupture realizations responsible for the first four accelerograms of the B0 (top) and B90 (bottom) accelerograms (corresponding to the 95th percentile) shown in Figure 4 (a and b). For each labeled rupture, the slip distribution over the fault surface is shown (scaled shading indicates the percentage relative to maximum slip). Crosses indicate the location of the sub-sources, a filled circle marks the nucleation point, and isochrone contours show the rupture process.

4. Definition of Structural Systems

In order to obtain a spectral representation of the most important seismic energy and deformation demand parameters in multi-story frames, in this work we have conducted a number of analyses on various structures subjected to synthetic and recorded ground motions of various characteristics. This was done by selecting an appropriate simplified shear-type model (ESTM) that could provide a reasonably accurate estimate of the seismic response of MDOF systems at both the global and local level.

The shear deformation of an n-story frame is depicted in Figure 6a. The values $(EI_b)_i$ and $(EI_c)_i$ for the *i-th* story, respectively, indicate the flexural stiffness of the vertical and transverse components (beams and columns). Figure 6b shows the lumped MDOF system, where u_i is the *i-th* story lateral displacement with respect to the ground, and m_i, k_i, and c_i are the *i-th* story mass, stiffness, and damping, respectively. The most important local displacement and energy response parameters can be directly assessed using the above-described methodology: the inter-story drift ratio (or index), or IDI, which is the inter-story drift normalized by the story height, the story maximum displacement u_i, and the dissipated story energy E_{Hi}. The inelastic behavior of joints and beams can only be indirectly and roughly described by such a response parameter, despite the fact that it offers useful information on the degree of damage at the terminal portions of each story's columns.

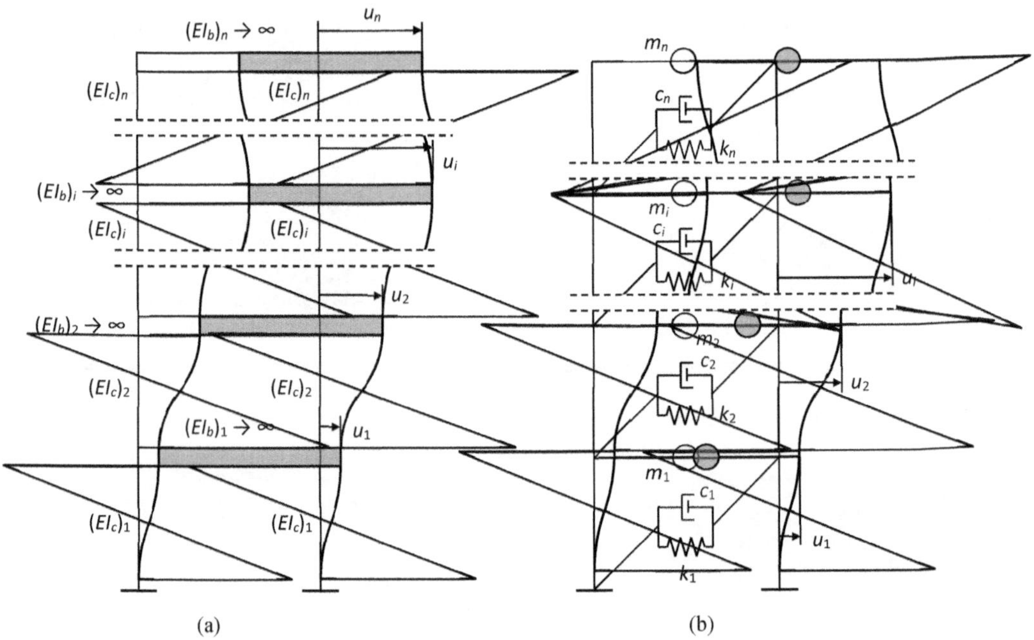

Fig. 6 Shear-type model for *n*-story frame structures: (a) shear deformation; (b) lumped MDOF system.

Since the assumption of a shear deformation in the description of the lateral deflection of the frame structure implies that the translational degrees of freedom only need to be handled by the integration procedure, using a shear-type model agrees to significantly simplify the numerical solution of the equations of motion. However, because the constructive details are not directly required, even though few data are needed to describe the mechanical properties of the systems, the model's significant simplification enables the performance of broad parametric analyses on the inelastic seismic response of various multi-story frame structures and the acquisition of a spectral representation of the most important seismic demand parameters.

Ten multi-story structural systems have been modeled using two-dimensional two-bay generic frames with varying numbers of stories (N = 2, 4, 6, 8, 10, 12, 14, 16, 20, 24), constant story height (h = 3.2 m), and beam span (w = 5.0 m) in order to capture the influence of a few key structural parameters on the seismic response of a wide range of current structural typologies.

The MDOF systems were subjected to an inverted-triangular distribution of equivalent seismic forces in order to assess the stories' yielding strengths. Additionally, a concentrated force acting at the top of the structures with fundamental periods longer than 0.7 seconds was used. The same hysteresis models with degrading stiffness have been used to describe the inelastic behaviour at each story. The analyses have been performed for a base shear seismic coefficient C_y of the bare frame equal to 0.08 that corresponds to only gravity loaded frames. Rayleigh damping was used to obtain a 5% of critical damping in the first and in the second modes.

5. Results of the Analyses

The dynamic response of the considered simplified models is evaluated for both sets of time histories, i.e., synthetic and real ground motions. In Figure 7, the $IDI_{\max}$ spectra for the Mean and

Mean + one Standard Deviation (SD) for the three subsets (50th, 64th, and 95th percentiles) of 11 synthetic time histories B0, are shown (B01150, B01164, B01195, respectively). It is possible to observe that the drift demand for the Mean + SD of the two groups B01184 and B01195 are very close, and notably higher than that of the B01150 group. It is quite clear that for a noteworthy number of the structural models (i.e., from 4-story to 12-story buildings) considered, the IDI_{max} values are higher than 2%, that corresponds to a significant damage.

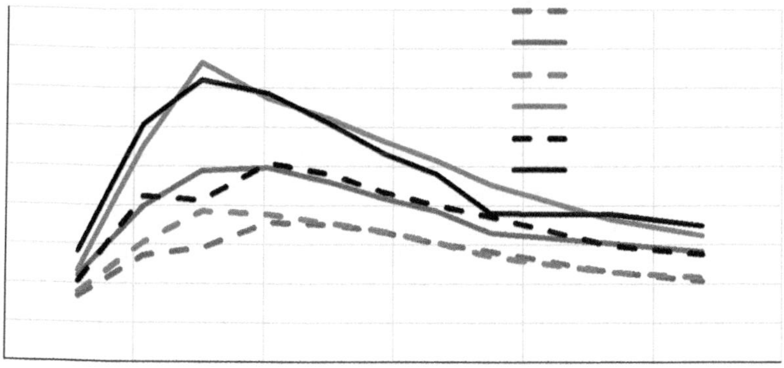

Fig. 7 Maximum Inter-story Drift spectra. Mean and mean + SD for the three subsets of synthetic signals B01150, B01184, and B01195.

It was demonstrated (Mollaioli et al., 2011) that from the comparison of energy and drift spectra a clear agreement in correspondence of the periods of the maximum values is attained. The displacement spectral shapes, on the other hand, do not exhibit a comparable correspondence. This should indicate that while a large local deformation demand is likely to happen, a strong energy demand placed on a structural system does not always result in a strong top displacement.

A satisfactory correlation between the maximum inter-story drift, IDI_{max}, and the total energy dissipated by the structures, $E_{H,max}$, is highlighted in Figure 8. This correlation is explainable by the fact that the shapes of the IDI_{max} spectra are more influenced by the energy demand, or in this case by $E_{H,max}$, than classical acceleration and displacement spectra.

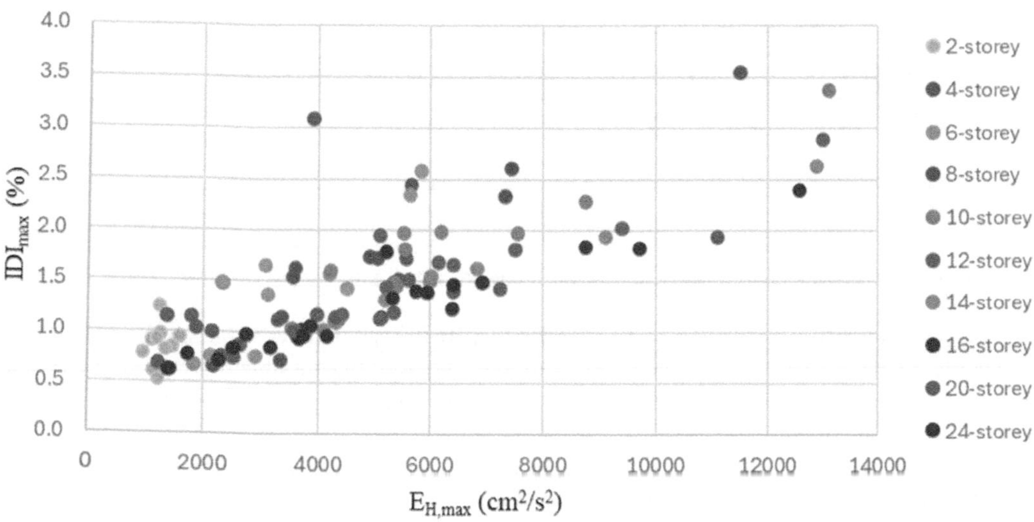

Fig. 8 IDI_{max} vs. $E_{H,max}$ for B01150 signals.

It was found that similar values of energy and drift are achieved at close period values, i.e. a strong local displacement demand occurs when a strong local energy demand is imposed. The most difficult problem and the largest information gap is determining the energy capacity of structural elements under seismic loadings. Characterizing energy dissipation capacity is challenging due to a number of factors, including the random nature of ground motions and their reliance on the hysteretic loading path. Without this knowledge, once the energy demand is known, its correlation with the drift demand permits to evaluate the maximum local deformation demand so as to assess the level of damage.

By focusing on dissipated hysteretic energy and inter-story drift, the second objective is to evaluate the extent to which the simulated ground motions can reliably replicate the structural response to real records when used in Non-Linear Time History Analyses (NLTHAs). In other words, the second validation is based on the comparison of the results of nonlinear dynamic analyses of the considered structural systems using the selected physically simulated accelerograms and the subset of real accelerograms.

In Figures 9, 10, and 11, it is possible to observe a significant correlation between dissipated energy and drift demand for 4-story, 6-story, and 8-story frames. This seems more evident for the 6-story frame where the drift demand is larger than the other two cases. Looking at the 4-story case the dispersion seems higher than the 6-story and 8-story models. It is also possible to note a reduced dissipated energy demand for the real records. It is important to note that the database of real accelerograms contains a large variability of situations, because the data cannot be grouped around values of magnitude and distance close to that of the physics-based time histories due to the lack of sufficient data.

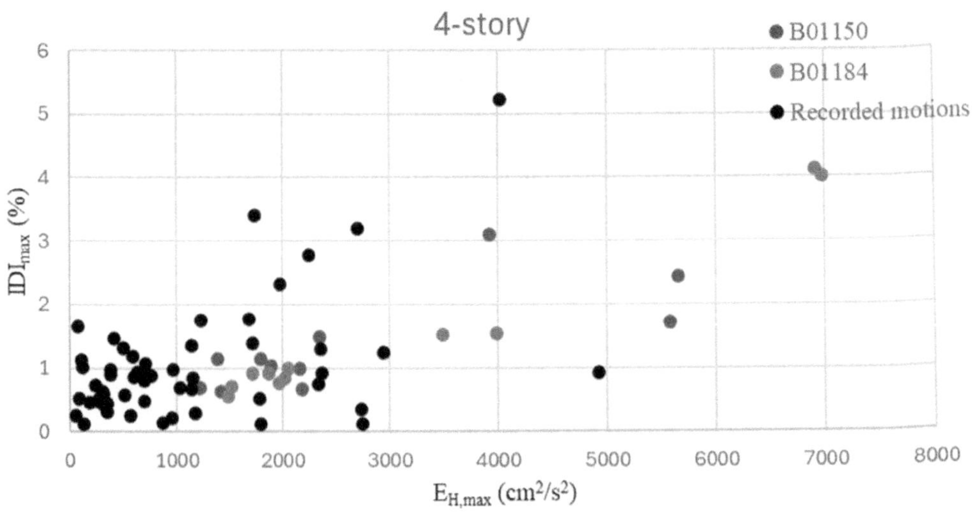

Fig. 9 Comparison of IDI_{max} and $E_{H,max}$ values for recorded and synthetic records and a 4-story frame.

In the case of the 6-story frame, the correlation is better defined even though the real records and the synthetic ground motions present two different slopes of the possible regression line (Figure 10). It means that in the first case for a selected energy demand, it is possible to obtain higher drifts than in the second case.

For the 8-story frame the dispersion is lower for the physics-based ground motions (Figure 11). Similar results have been detected for the other frames with some differences mainly for the taller ones where in some cases the response is elastic.

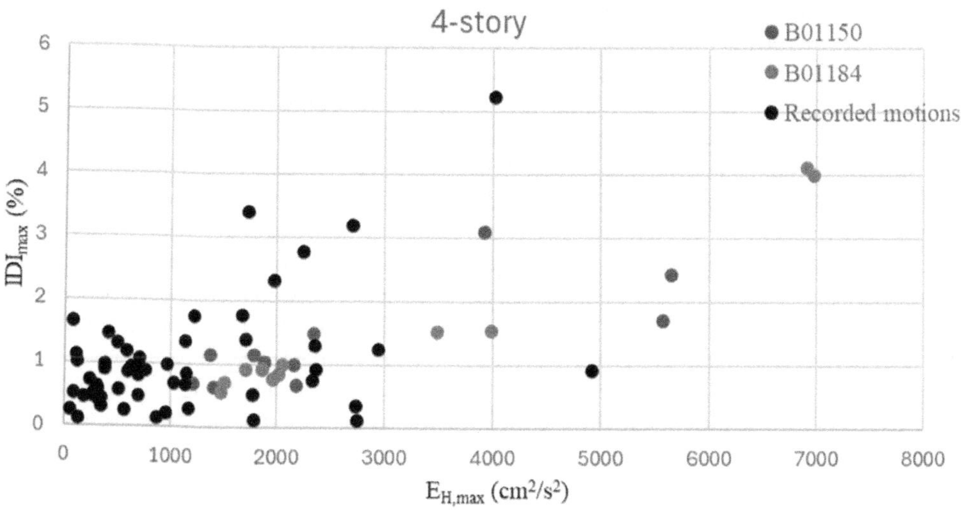

Fig. 10 Comparison of IDI_{max} and $E_{H,max}$ values for recorded and synthetic records and an 6-story frame.

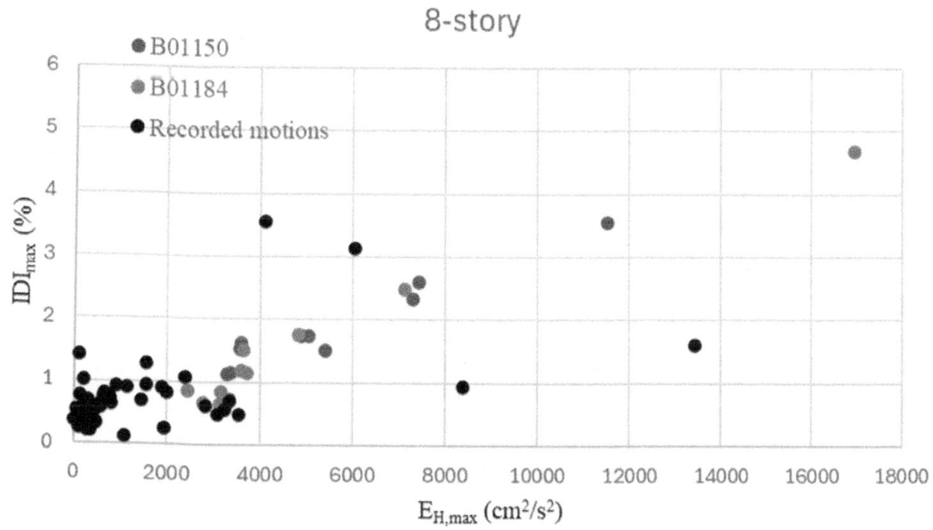

Fig. 11 Comparison of IDI_{max} and $E_{H,max}$ values for recorded and synthetic records and a 8-story frame.

6. Conclusions

The energy-based parameters offer valuable insights into seismic energy and deformation demands of structural systems. The correlation between IDI_{max} and $E_{H,max}$ demonstrates the relationship between energy demand and displacement demand, crucial for assessing damage. The correlation between energy and drift demand highlights the potential of simulated ground motions in NLTHAs. These findings contribute to our understanding of seismic wavefields impact on structural systems, emphasizing the importance of directivity effects and energy-demand parameters in seismic engineering. Comparing real and synthetic ground motions shows promising correlations, highlighting the potential for physics-based time histories to replicate structural response in non linear dynamic analyses. These findings emphasize the importance of energy-based parameters in evaluating seismic vulnerability and developing robust seismic code provisions. In fact, a significant amount of research is required to incorporate energy-based methodologies throughout current seismic design codes. Until now, the energy-based approach has been implemented only in the Japanese seismic code (2009) and it is going to be implemented in the Second Generation of Eurocode 8 for the seismic design of new buildings and for the retrofit of structures with displacement-dependent dampers. Therefore, widespread research on performance energy-based design methodologies is needed, as it lacks a common thread for defining key features for: (a) the characterization of energy-based hazard analysis, (b) the definition of specific energy-based engineering demand parameters, and (c) the establishment of energy-based damage measures and energy-based measures of economic loss. All of these features are necessary to take into account explicitly the energy-based formulation of performance criteria. In other words, there is a need to evaluate the energy input into a structure for different levels of earthquake ground motion. In such a way, one can determine how the input energy is going to be distributed throughout the structure, in order to control the dissipation demand at each level of the structure, and get knowledge of the energy absorption capacity of its structural elements.

Acknowledgements

Figures 2–5 have been created using the XeRiS Web-App (see Vaccari and Magrin, 2022). The first author would like to thank Prof. Giuliano Panza for having introduced him to the field of seismology, as a simple engineer, and for the consideration and friendship shown over the years. The second author would like to thank Prof. Giuliano Panza for having introduced him to the field of earthquake engineering, as a simple seismologist, and in particular (a long time ago) with the group of Prof. Luis Decanini.

References

Akiyama H. (1980). *Earthquake-resistant Limit-state Design for Buildings.* University of Tokyo Press (1980). English version 1985.

Angelucci G., Mollaioli F. and Quaranta G. (2023). Correlation between energy and displacement demands for infilled reinforced concrete frames. *Frontiers in Built Environment, 9*: 1198478. doi: 10.3389/fbuil.2023.1198478.

Angelucci G., Mollaioli F., Quaranta G. and Kunnath S.K. (2024). Interpretable machine learning models for displacement demand prediction in reinforced concrete buildings under pulse-like earthquakes. *Journal of Building Engineering, 95* (2024): 110124, https://doi.org/10.1016/j.jobe.2024.110124.

Arias A. (1970). A measure of earthquake intensity, In: *Seismic Design for Nuclear Power Plants* (pp. 438–483). IAEA.

Benavent-Climent A., Donaire-Ávila J. and Mollaioli F. (2021). Key Points and Pending Issues in the Energy-Based Seismic Design Approach. In: *Proceedings of the International Workshop on Energy-Based Seismic Engineering, IWEBSE 2021* (pp. 151–168.) Amadeo Benavent-Climent and Fabrizio Mollaioli (Eds.). May 24–26, 2021.

Cheng Y., Lucchini A. and Mollaioli F. (2014). Proposal of new ground-motion prediction equations for elastic input energy spectra., *Earthquakes and Structures, 7*(4): 485–510 (DOI: http://dx.doi.org/10.12989/eas.2014.7.4.485).

Cheng Y., Lucchini A. and Mollaioli F. (2020). Ground-Motion Prediction Equations for Constant-Strength and Constant-Ductility Input Energy Spectra. *Bulletin of Earthquake Engineering, 18*: 37–55. https://doi.org/10.1007/s10518-019-00725-x.

Cheng Y., Mollaioli F. and Donaire-Ávila J. (2021, August). Characterization of Dissipated Energy Demand. *Soil Dynamics and Earthquake Engineering, 147* (https://doi.org/10.1016/j.soildyn.2021.106725).

DISS Working Group. (2021). *Database of Individual Seismogenic Sources (DISS), Version 3.3.0: A Compilation of Potential Sources for Earthquakes Larger than M 5.5 in Italy and Surrounding Areas.* Istituto Nazionale di Geofisica e Vulcanologia (INGV). https://doi.org/10.13127/diss3.3.0.

Donaire-Ávila J., Benavent-Climent A., Lucchini A. and Mollaioli F. (2017). Energy-based seismic design methodology: A preliminary approach. In: *16th World Conference on Earthquake Engineering, 16WCEE 2017*, Santiago Chile, 9–11 January, Paper N° 2106.

Fajfar P, and Vidic T. (1994). Consistent inelastic design spectra: Hysteretic and input energy. *Earthq. Eng. Struct. Dynamics, 23*(5): 523–537.

Fasan M., Magrin A., Amadio C., Romanelli F., Vaccari F. and Panza, G.F. (2016). A seismological and engineering perspective on the 2016 Central Italy earthquakes. Int. J. Earthq. Impact Eng. 11 (4), 395–420.

Gusev A.A. (2011). Broadband Kinematic Stochastic Simulation of an Earthquake Source: A Refined Procedure for Application in Seismic Hazard Studies. *Pure and Applied Geophysics, 168*(1–2): 155–200. https://doi.org/10.1007/s00024-010-0156-3.

Housner, G.W. (1956). Limit design of structures to resist earthquakes. In: *Proceedings of the 1st World Conference on Earthquake Engineering* (pp. 1–20). Berkeley California (1956).

Moehle J. and Deierlein G.G. (2004). Deierlein: A framework methodology for performance-based earthquake engineering. In: *13th World Conference on Earthquake Engineering* (*679*: 12). WCEE. Vancouver.

Mollaioli F., Mura A. and Decanini L. (2007). Assessment of the deformation demand in multi-story frames. *Journal of Seismology and Earthquake Engineering*, Winter, *8* (4): 195–211.

Mollaioli F., Bruno S., Decanini L. and Saragoni R. (2011, January). Correlations between energy and displacement demands for performance-based seismic engineering. *Pure and Applied Geophysics*, *168*(1–2): 237–259. DOI 10.1007/s00024-010-0118-9.

Mollaioli F., Donaire-Avila J., Lucchini A. and Benavent-Climent A. (2019). On the importance of energy-based parameters. In: *COMPDYN 2019 7th ECCOMAS Thematic Conference on Computational Methods in Structural Dynamics and Earthquake Engineering* M. Papadrakakis, and M. Fragiadakis (Eds.), Crete, Greece, 24–26 June.

Panza G.F., Romanelli F., Vaccari F., Decanini L. and Mollaioli F. (2001). Contribution of the Deterministic Approach to the Characterization of Seismic Input. In: *Proceedings of the OECD_NEA Workshop on the Engineering Characterization of Seismic Input* (pp. 655–715). Brookhaven National Laboratory, Upton; New York, USA.

Panza G.F., Romanelli F., Vaccari F., Decanini L. and Mollaioli F. (2003). Seismic ground motion modelling and damage earthquake scenarios: A bridge between seismologist and seismic engineers. In: *OECD/NEA Workshop on The Relation between Seismological Data and Seismic Engineering* (pp. 241–266). 16–18 October 2002, Istanbul Turkey, NEA/CSNI/R(2003)18.

Panza, G.F., La Mura, C., Peresan, A., Romanelli, F. and Vaccari, F. (2012). Seismic hazard scenarios as preventive tools for a disaster resilient society. *Adv. Geophys., 53*: 93–165.

Panza, G.F., La Mura, C., Romanelli, F. and Vaccari, F. (2021). Earthquakes, Strong-Ground Motion. In: Gupta, H.K. (Eds.), *Encyclopedia of Solid Earth Geophysics. Encyclopedia of Earth Sciences Series*. Springer, Cham. https://doi.org/10.1007/978-3-030-10475-7_141-2.

O'Reilly G.J. and G.M. Calvi. (2019). Conceptual seismic design in performance-based earthquake engineering. *Earthquake Engineering and Structural Dynamics*, *48*(4): 389–411.

Ricci, A., Romanelli, F., Vaccari, F., Boncio, P., Venisti, N., Faraone, C., Vessia, G. and Panza, G.F. (2025). Comparison between the Neo-deterministic Seismic Hazard and FEM approach to assessing 2D local seismic response at Chieti's city site (Abruzzo, Italy). *Engineering Geology, 347*.

Rugarli, P., Amadio, C., Peresan, A., Fasan, M., Vaccari, F., Magrin, A., Romanelli, F. and Panza, G.F. (2019). Neo-deterministic scenario-earthquake accelerograms and spectra: an NDSHA approach to seismic analysis. In: Jia, J., and Paik, J.K. (Eds.), Chapter 6, *Engineering Dynamics and Vibrations: Recent Developments* (pp. 187–241). CRC Press, Boca Raton, Florida, USA, ISBN 978-1-4987-1926-1. https://doi.org/10.1201/9781315119908-6.

Sucuoglu H. and Nurtug A. (1995). Earthquake ground motion characteristics and seismic energy dissipation. *Earthq. Eng. Struct. Dynamics, 24*(9): 1195–1213.

Uang C.M. and Bertero V. (1990). Evaluation of seismic energy in structures. *Earthquake Engineering and Structural Dynamics, 19*: 77–90.

Vaccari, F. and Magrin, A. (2022). A user-friendly approach to NDSHA computation. In: Panza, G.F., Kossobokov, V.G., Laor, E., and De Vivo, B. (Eds.), *Earthquakes and Sustainable Infrastructure* (1st Edition): *Neodeterministic (NDSHA) Approach Guarantees Prevention Rather than Cure* (pp. 215–237). Elsevier.

CHAPTER

18

The Effect of The Eastern Mediterranean Offshore Seismogenic Zones on the Seismic Hazard Assessment of The Lebanese Restraining Bend

Tony S. Nemer[1*], Franco Vaccari[2] and Mustapha Meghraoui[3]

1. Introduction

Nemer et al. (2023) performed the first neo-deterministic seismic hazard assessment (NDSHA) for the Lebanese Restraining Bend (LRB) along the Dead Sea Transform Fault. They used the known active tectonic structures onshore and the proposed structures offshore in order to designate the seismogenic zones required for the modeling. However, recent seismic interpretation studies in the Levant Basin offshore Lebanon have provided valuable insights into the earthquake geology and tectonics of the eastern Mediterranean region, and these studies have highlighted new potential sources of the 9 July 551 (M 7.2) tsunamigenic earthquake (Faysal et al., 2023; Sarieddine et al., 2025). As these potential offshore sources and their associated seismogenic zones are different from what was used in Nemer et al. (2023), we updated the input data used in the 2023 modeling, and we performed a new NDSHA of the same region. The modified input data replaced the previously proposed offshore seismogenic zone with two new offshore zones (Figure 1). Our results indicate that the change in the offshore seismogenic zones has little influence on the results of Nemer et al. (2023), showing only minor changes in the modeled ground displacement, velocity, and acceleration. In addition, we performed another modeling considering M_{design} (Rugarli et al., 2019; Rugarli, 2022; Zhang et al., 2025) instead of the smoothed M_{max} (the maximum magnitude available in the seismicity catalogue), whereby $M_{design} = M_{\max} + 0.7$ with no smoothing applied to the magnitudes (Zhang et al., 2025).

1 Department of Geology, American University of Beirut, Riad El-Solh, Beirut 1107 2020, Lebanon.
2 Dipartimento di Matematica, Informatica e Geoscienze, Università di Trieste, via Weiss, 4, 34128 Trieste, Italy.
3 Institut de Physique du Globe, EOST - Université de Strasbourg 5, rue René Descartes 67084 Strasbourg cedex, France. Email: vaccari@units.it[2]
* Corresponding author: tn01@aub.edu.lb

DOI: 10.1201/9781003684121-20

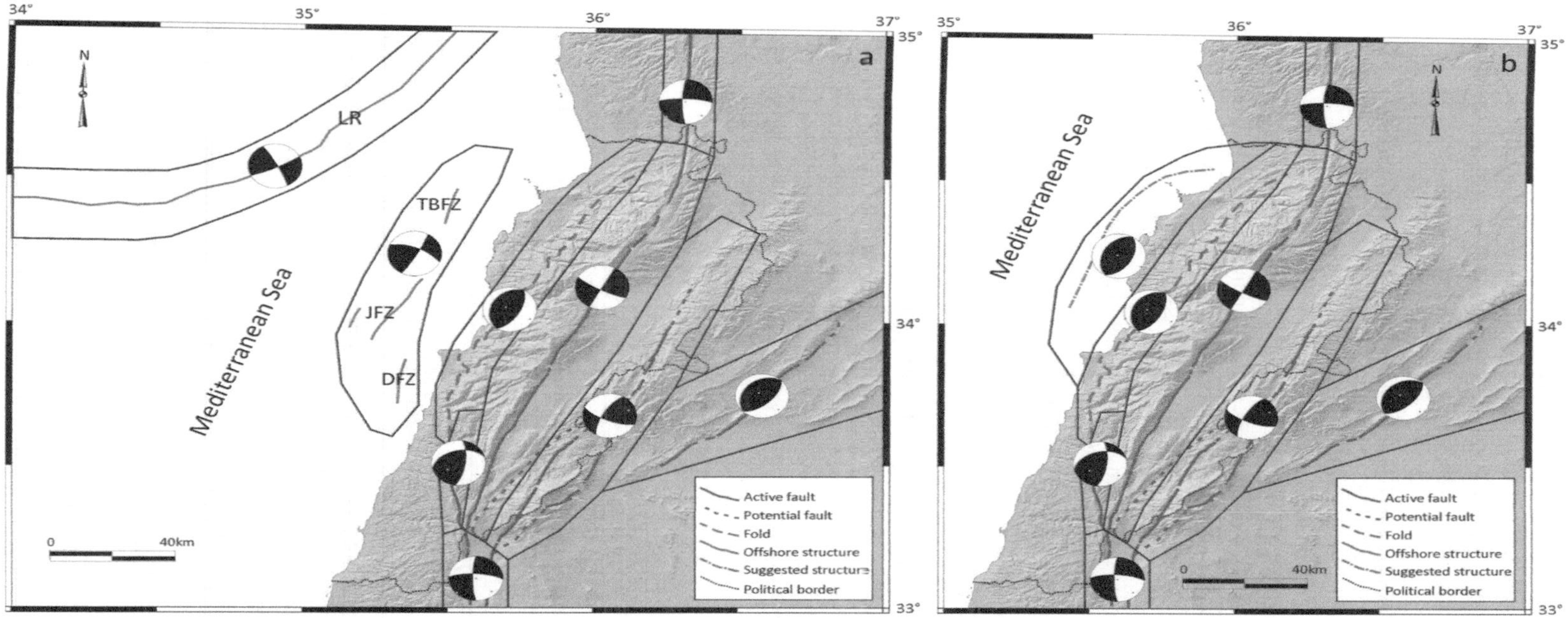

Fig. 1 The study area with: main faults, folds, offshore structures, and seismogenic zones each associated with its focal mechanism solution that was used in the modeling. Distinct line styles and patterns are used to differentiate structural elements and zones. (a) Seismogenic zones used in the current study. DFZ Damour fault zone, JFZ Jounieh fault zone, LR Latakia Ridge, TBFZ Tripoli-Batroun fault zone. (b) Seismogenic zones used in Nemer et al. (2023); modified from Nemer et al. (2023).

2. Input Data

As in Nemer et al. (2023), the input data of the new NDSHA for the LRB consist of four principal sets: (1) the seismogenic zones, (2) the focal mechanism solutions, (3) the seismicity catalogue, and (4) the velocity structural models.

2.1 Seismogenic Zones and Focal Mechanism Solutions

Two new seismogenic zones have been introduced offshore the LRB: (a) one zone that is close to the shoreline of Lebanon and includes three fault zones (Tripoli-Batroun, Jounieh, and Damour) based on the 3D seismic study of Faysal et al. (2023); (b) another seismogenic zone that includes the Latakia Ridge as a major thick-skinned tectonic structure between Lebanon and Cyprus (Faysal et al., 2023; Sarieddine et al., 2025; Figure 1a). Each new seismogenic zone was assigned one focal mechanism representative of its tectonic setting, as required by the NDSHA for the generation of the synthetic seismograms. The other seismogenic zones that are located onshore have not changed, and they were assigned the same focal mechanisms that were used in Nemer et al. (2023) (Figure 1).

2.2 Seismicity Catalogue and Velocity Structural Models

The seismicity catalogue used in Nemer et al. (2023) was used here, and we added to it the offshore events that were recorded by the European-Mediterranean Seismological Center (EMSC) between 2019–2024. In addition, as the source of the 9 July 551 (M 7.2) earthquake is still debatable (e.g., Salomon et al., 2024), the source of the 551 earthquake was presumably located in the two new offshore seismogenic zones in order to model its effect regardless where it actually took place. This approach is not considered excessive in the applied NDSHA as, by definition, the modeling simulates the sources of the strongest earthquakes within each seismogenic zone, whether they are factual from instrumental seismicity or suggested from historical events. Concerning the velocity structural models, those used in Nemer et al. (2023) were used here with no modification.

3. Computations

We followed the same computations that were used in Nemer et al. (2023), and we performed two runs: the *2023_new-seis-zones* that uses the new seismogenic zones with the 2023 model's input, and the M_{design}*_new-seis-zones* that uses the new seismogenic zones with the M_{design} applied to the seismicity catalogue.

3.1 2023_new-seis-zones

In this run, the standardized $n = 3$ discretization and smoothing of seismicity were used, and as shown in Figure 2, magnitudes in the range of 7.0–7.5 are assigned to all cells of the seismogenic zones. This is consistent with the NDSHA that accounts for the potential misplacement of historical events and the possibility of future earthquakes to occur near past events.

3.2 Mdesign_new-seis-zones

In this run, we followed the Panza-Rugarli law that suggests to perform the NDSHA with the addition of 0.7 to the magnitudes reported in the seismicity catalogue ($M_{design} = M_{\max} + 0.7$), but with no smoothing applied (Rugarli et al., 2019; Rugarli, 2022; Zhang et al., 2025; Figure. 3).

4. Comparison of Results

The results of run *2023_new-seis-zones* are compared with those of Nemer et al. (2023) by looking at the ratios between the design ground acceleration (DGA) values obtained at each site for the two

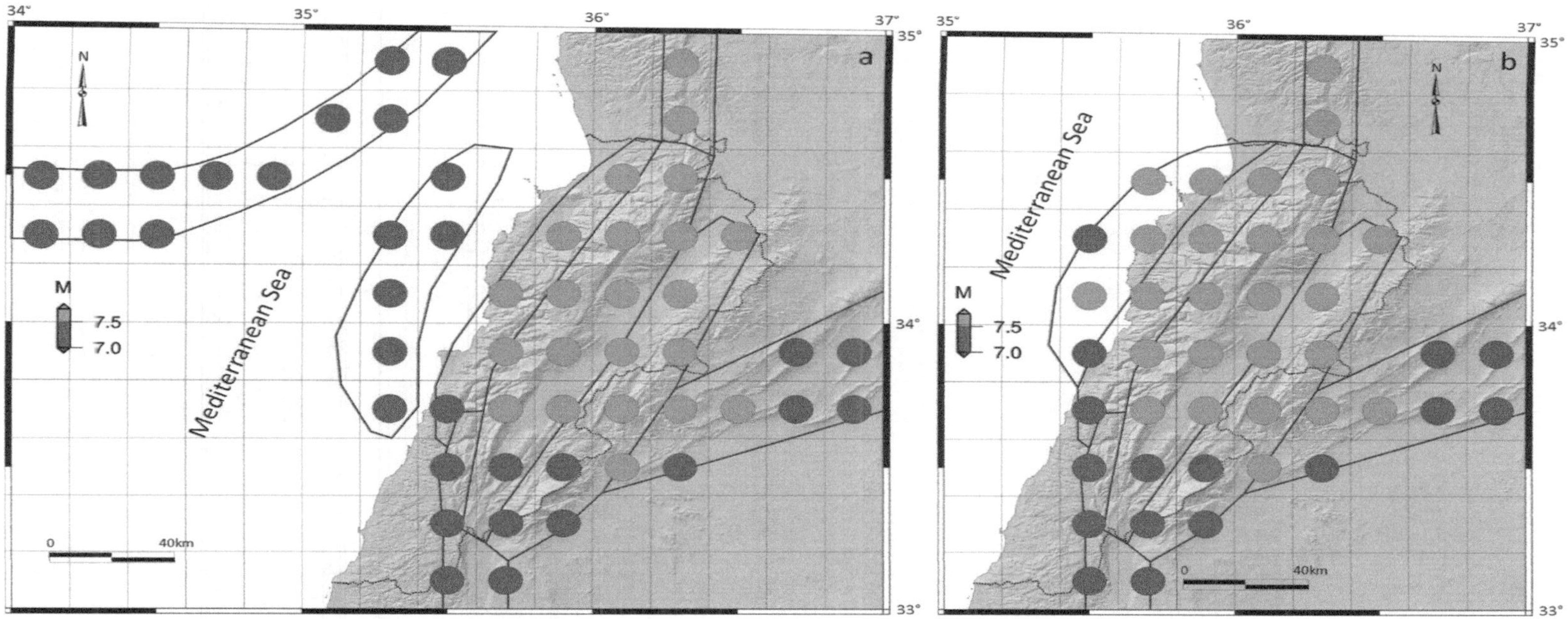

Fig. 2 The study area with the applied discretization and smoothing of the seismicity (of the used catalogue) in 0.2° × 0.2° cells. The smoothing procedure is explained in Nemer et al. (2023). The smoothing window used here is for smoothing radius $n = 3$. M on the scaled legends indicates earthquake magnitude. (a) Seismogenic zones used in the current study. (b) Seismogenic zones used in Nemer et al. (2023); modified from Nemer et al. (2023).

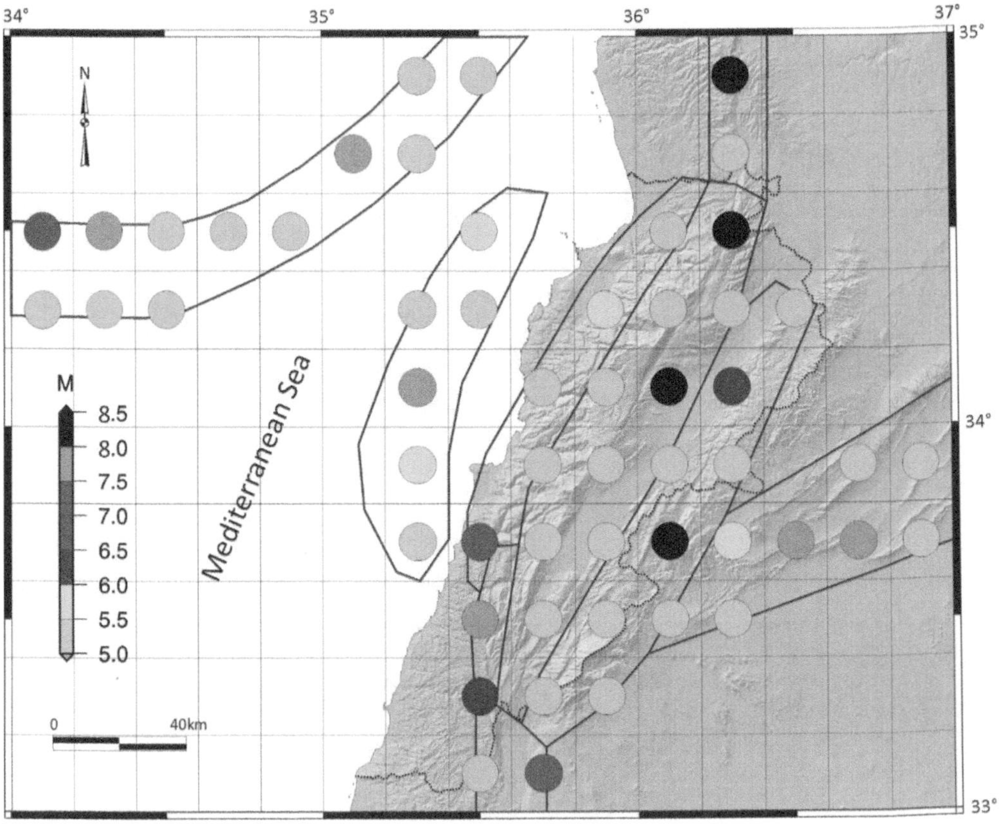

Fig. 3 The study area (seismogenic zones used in the current study) with the applied discretization and smoothing of the seismicity (of the used catalogue) in 0.2° × 0.2° cells. The smoothing procedure is explained in Nemer et al. (2023). The smoothing window used here is for smoothing radius n = 0. M on the scaled legends indicates earthquake magnitude ($M_{design} = M_{max} + 0.7$)

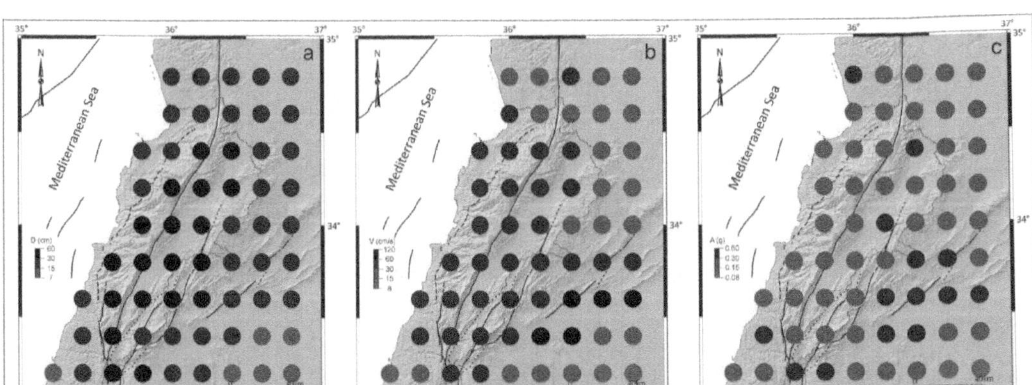

Fig. 4 Seismic hazard maps of the study area from run *2023_new-seis-zones* obtained using the new seismogenic zones with the 2023 model's input. (a) peak ground displacement (PGD) in cm. (b) peak ground velocity (PGV) in cm/s. (c) design ground acceleration (DGA) in g. The main faults, folds, and offshore structures are displayed. On the scaled legends: D is displacement, V is velocity, and A is acceleration. Adapted from Nemer et al. (2023).

runs (Fig. 11 of Nemer et al. (2023) and Figure 4), as described in Panza et al. (2012). The ratios quantify the variations in terms of macroseismic intensity, based on the fact that Cancani (1904) modified the original Mercalli scale into the Mercalli-Cancani-Sieberg scale (MCS) and assigned the peak values of ground movement's acceleration to each grade, so that an increment of one intensity degree roughly corresponds to a doubling of the peak ground acceleration. Under these terms, there is no jump in the macroseismic intensity between the results of the compared models, and therefore we suggest that the change in the offshore seismogenic zones has little influence on the results, very likely because the effect of the large onshore earthquakes dominates the situation in the modeling. Here, we underline that some updates have been introduced to the modeling program since 2023, which could have influenced the minor differences between the results of the two models that are displayed in Fig. 11 of Nemer et al. (2023) and Figure 4.

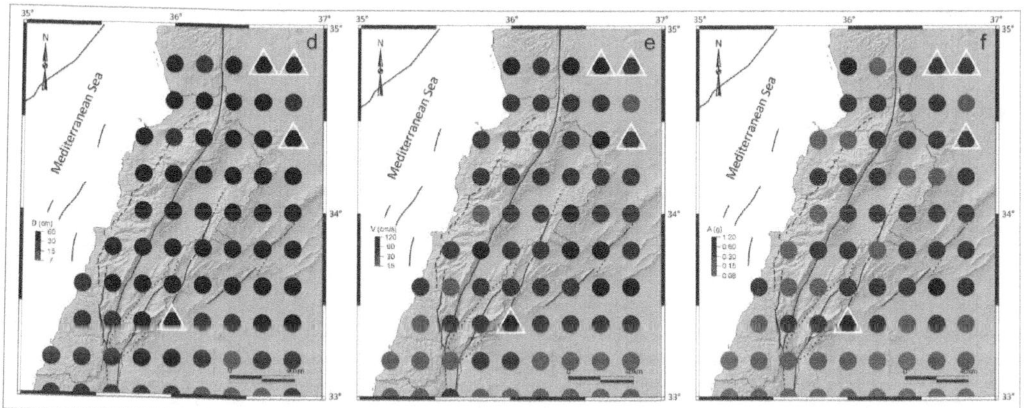

Fig. 5 Seismic hazard maps of the study area from run M_{design}*_new-seis-zones* obtained using the new seismogenic zones with M_{design} applied to the seismicity catalogue; triangular symbols show the sites where M_{design}*_new-seis-zones* leads to a one-degree increase in macroseismic intensity (cf. Figure 4). (d) peak ground displacement (PGD) in cm. (e) peak ground velocity (PGV) in cm/s. (f) design ground acceleration (DGA) in g. The main faults, folds, and offshore structures are displayed. On the scaled legends: D is displacement, V is velocity, and A is acceleration. Adapted from Nemer et al. (2023).

We also compared the results of *2023_new-seis-zones* and M_{design}*_new-seis-zones*, and we found only four sites where macroseismic intensities are increased by one degree for run M_{design}*_new-seis-zones* (Figures 4 and 5). For this run (M_{design}*_new-seis-zones*), by adding 0.7 to the max magnitude in the LRB, we get to $M_{design} = 8.2$ in some places (Figure 3), a value quite hard to justify given the fault dimensions (the largest fault is 170 km long, it has a strike-slip fault mechanism, and it crosses through the restraining bend from south to north; Figure 1) (Wells & Coppersmith, 1994). In effect, if we look back at the earthquakes of 6 February 2023 in Turkiye, the first earthquake (M_w 7.8) took place on a 350-km strike-slip rupture of the East Anatolian Fault, and the second earthquake (M_w 7.6) took place on a 150-km strike-slip rupture (e.g., Nemer, 2024). Nevertheless, the results of run M_{design}*_new-seis-zones* can be used as a worst-case scenario for the study area (Figure 5).

We conclude that the ground motion results of Nemer et al. (2023) remain valid (*their Fig. 11*), although they can be updated following run "*2023_new-seis-zones*" (Figure 4). We add that the neo-deterministic approach is well-suited for conducting parametric tests to evaluate how the uncertainties in the model parameterization may influence the obtained ground shaking scenarios. However, these tests are typically performed at a local scale when generating site-specific ground motion simulations, in which the details of the source rupturing process and the presence of lateral heterogeneities in the geology can be taken into account.

Acknowledgements

This research was supported by a University Research Board grant from the American University of Beirut for project number 27184.

Data availability

Data generated during this study are included in the article or supplementary material. Further inquiries can be directed to the corresponding author.

Declaration of Competing Interest

The authors declare that they have no known financial or personal interests that could influence the results reported in this paper.

References

Cancani, A. (1904). Sur l'emploi d'une double echelle seismique des intesites, empirique et absolue. *Gerl Beitr.*, *2*: 281–283.

Faysal, R., Nemer, T. and Sarieddine, K. (2023). Investigating the geological fault framework offshore Lebanon: Insight into the earthquake geology of the eastern Mediterranean region. *Pure and Applied Geophysics*. https://doi.org/10.1007/s00024-023-03336-5.

Nemer, T.S. (2024). The 6 February 2023 Earthquakes in Southern Türkiye: When Geoscience Meets Public Concern. *Seismological Research Letters*, *95*(4): 2031-2037, https://doi.org/10.1785/0220240002.

Nemer, T.S., Vaccari, F. and Meghraoui, M. (2023). Seismic hazard assessment of the Lebanese Restraining Bend: A neo-deterministic approach. *Pure and Applied Geophysics*. https://doi.org/10.1007/s00024-023-03233-x.

Panza, G.F., La Mura, C., Peresan, A., Romanelli, F. and Vaccari, F. (2012). Seismic hazard scenarios as preventive tools for a disaster resilient society. *Adv. Geophys.*, *53*: 93–165.

Rugarli, P., Vaccari, F. and Panza, G.F. (2019). Seismogenic nodes as a viable alternative to seismogenic zones and observed seismicity for the definition of seismic hazard at regional scale. *Vietnam Journal of Earth Sciences*, *41*: 289–304.

Rugarli, P. (2022). The view of a structural engineer about reliable seismic hazard assessment. In: Panza, G.F., Kossobokov, V.G., Laor, E., and De Vivo, B. (Eds.), *Earthquakes and Sustainable Infrastructure: Neodeterministic (NDSHA) Approach Guarantees Prevention Rather than Cure* (pp. 59–76). Elsevier, Amsterdam, Netherlands.

Salamon, A., Omira, R., Zohar, M. and Baptista, M.A. (2024). Modern outlook on the source of the 551 AD tsunamigenic earthquake that struck the Phoenician (Lebanon) coast. *Natural Hazards*, *120*(9): 8893–8929. https://doi.org/10.1007/s11069-024-06559-4.

Sarieddine, K., Nemer, T.S. and Faysal, R. (2025). Investigating the Messinian salt in the Lebanese Levant Basin: Insights into the geology and tectonics of the eastern Mediterranean region. *Journal of Structural Geology*, *194*: 105368. https://doi.org/10.1016/j.jsg.2025.105368.

Wells, D. and Coppersmith, K. (1994). New empirical relationships among magnitude, rupture length, rupture width, rupture area, and surface displacement. *Bulletin of the Seismological Society of America*, *84*: 974–1002.

Zhang, Y., Rugarli, P., Ji, Z., Jiang, C., Wu, Z. and Panza, G.F. (2025). Falsification of seismic hazard assessment (SHA) by real earthquakes: The design magnitude M_{design} for neo-deterministic seismic hazard assessment (NDSHA). In: Li, Y. et al. (Eds.), *Scientific Investigation of Continental Earthquakes and Relevant Studies*. Higher Education Press Limited Company.

CHAPTER

19

Time-dependent Seismic Hazard Assessment and its Application in Earthquake Early Warning System: A Case Study in the China Seismic Experimental Site (CSES)

Yan Zhang[1*]

Improvements and optimizations of neo-deterministic seismic hazard assessment (NDSHA) in the time domain and its range of application play a critical role in seismic risk reduction. Time-dependent NDSHA (TD-NDSHA) provides an effective framework for characterizing seismic hazard variability in a specific time period, which is essential for enhancing the performance of Earthquake Early Warning Systems (EEWS). By explicitly integrating seismic hazard information into EEWS strategies, TD-NDSHA enables warning decisions to be constrained by the evolving seismic hazard context rather than relying solely on real-time signal characteristics. This combined framework offers a robust approach for improving the protection of critical infrastructures and exposed communities.

1. Introduction

Over the past few years, continuous efforts have been devoted to enhancing neo-deterministic seismic hazard assessment (NDSHA) (e.g., Panza and Bela, 2019; Bela and Panza, 2021; Panza et al., 2022). Among these developments, the progressive incorporation of time-dependent features, together with the integration of NDSHA with complementary earthquake prevention and disaster mitigation tools, has emerged as a key research direction. This line of development aims to move NDSHA beyond a purely scenario-based framework toward an approach that can explicitly account for temporal hazard variability, representing a critical step toward its effective operational and practical application in earthquake disaster mitigation.

With respect to time dependency, the Annual Consultation on the Likelihood of Earthquakes, which has been conducted in China for several decades, represents a practice of particular relevance. This consultation aims to synthesize diverse seismological, geological, and geophysical information to identify earthquake-prone areas over an annual time horizon. Initiated in 1972 by the State Seismological Bureau (SSB, now the China Earthquake Administration, CEA), the Annual

[1] Institute of Geophysics, China Earthquake Administration, Beijing, China.
* Corresponding author: zhangyan@cea-igp.ac.cn

DOI: 10.1201/9781003684121-21

Consultation has been extensively documented in terms of its conceptual framework (e.g., Wu, 1997; Zhu and Wu, 2007), performance evaluation (e.g., Shi et al., 2001; Zhuang and Jiang, 2012), and methodological development (e.g., Zhao et al., 2010; Zhang et al., 2017; Wu, 2021).

Recent efforts have sought to further strengthen this practice by coupling its annual forecasting outcomes with physics-based seismic hazard assessment. In particular, Zhang et al. (2022) integrated the results of the Annual Consultation with TD-NDSHA, enabling annual forecasts to be translated into quantitative hazard estimates and thereby enhancing their relevance for seismic risk reduction. Building on this framework, Zhang et al. (2023) proposed a hazard-driven Earthquake Early Warning System (EEWS) empowered by TD-NDSHA, using the China Seismic Experimental Site (CSES) as a demonstration platform to illustrate the feasibility and potential operational value of this integration (e.g., Li et al., 2022).

With the growing understanding of earthquake physical processes, recent years have witnessed the emergence of time-dependent seismic hazard assessments with finer spatial scales and more clearly defined objectives. Representative examples include the estimation of time-varying seismicity rates based on declustered and non-declustered earthquake catalogs (Montiel-López et al., 2025), as well as the calculation of strong-earthquake occurrence probabilities using the Brownian Passage Time (BPT) model (Xu et al., 2023). In essence, time-dependent seismic hazard assessment aims to incorporate information on the expected evolution of seismic activity within a specified future time window into the hazard evaluation framework. Although the intrinsic uncertainties associated with earthquake forecasting, e.g., the missed events and false alarms, may directly affect seismic hazard assessment results (Papazachos et al., 2018), addressing these uncertainties constitutes one of the pressing challenges faced by modern societies exposed to seismic risk.

In this chapter, we focus on seismic hazard zoning at an annual timescale and explore a time-dependent seismic hazard assessment framework targeting year-scale seismic risk. Furthermore, by integrating the proposed framework with EEWS, we aim to provide a reference for future developments in time-dependent seismic hazard assessment and its practical applications.

2. Time-dependent Neo-deterministic Seismic Hazard Assessment (TD-NDSHA)

As illustrated in Figure 1, within the CSES, the output of the Annual Consultation is expressed as a set of alert regions characterized by their estimated annual seismic hazard, defined in terms of regional boundaries and the expected magnitude range of potentially damaging earthquakes. Although this output still leaves substantial room for improvement, it represents a unique and internally consistent scientific experiment, distinguished by its genuine forward-looking forecasting nature and its intrinsic falsifiability. Conducted continuously for more than five decades, the Annual Consultation constitutes a rare and valuable long-term dataset for earthquake forecast research.

Nevertheless, the information provided by the Annual Consultation, namely the identification of alert regions and associated expected magnitude ranges, remains insufficient for directly supporting timely and actionable risk-reduction measures. Owing in part to this limitation, the consultation results have traditionally not been released to the public in real time. A critical avenue for further development therefore lies in shortening the time lag between forecast generation and societal dissemination, which can be achieved by incorporating reliable and quantitative seismic hazard assessment into the Annual Consultation framework.

The Annual Consultation Meeting is conducted on a regional basis and is typically held around the turn of each year. From a monitoring perspective, the assessment of potential precursory signals is organized into multiple disciplinary categories, including seismology, geomagnetism and geo-electricity, ground deformation and gravity, underground fluids and geochemistry, as well as integrated analyses. To support this process, the China Earthquake Administration (CEA), together with its provincial earthquake agencies, has established and continuously operates a large-scale observational network comprising both permanent and mobile facilities. This network systematically monitors variations in seismicity, crustal deformation, geomagnetic and geo-electric fields, and subsurface hydrological conditions, among others, and is supported by standardized quality-control procedures and unified protocols for routine data processing.

Figure 1 illustrates a representative example of the Annual Consultation output for the CSES, showing the forecast results for the year 2014. In this assessment, for the study area, the alert region was identified as area with an elevated likelihood of strong to major earthquakes during that year (i.e., 2014). Evidence leading to the identification of the alert regions includes[1]:

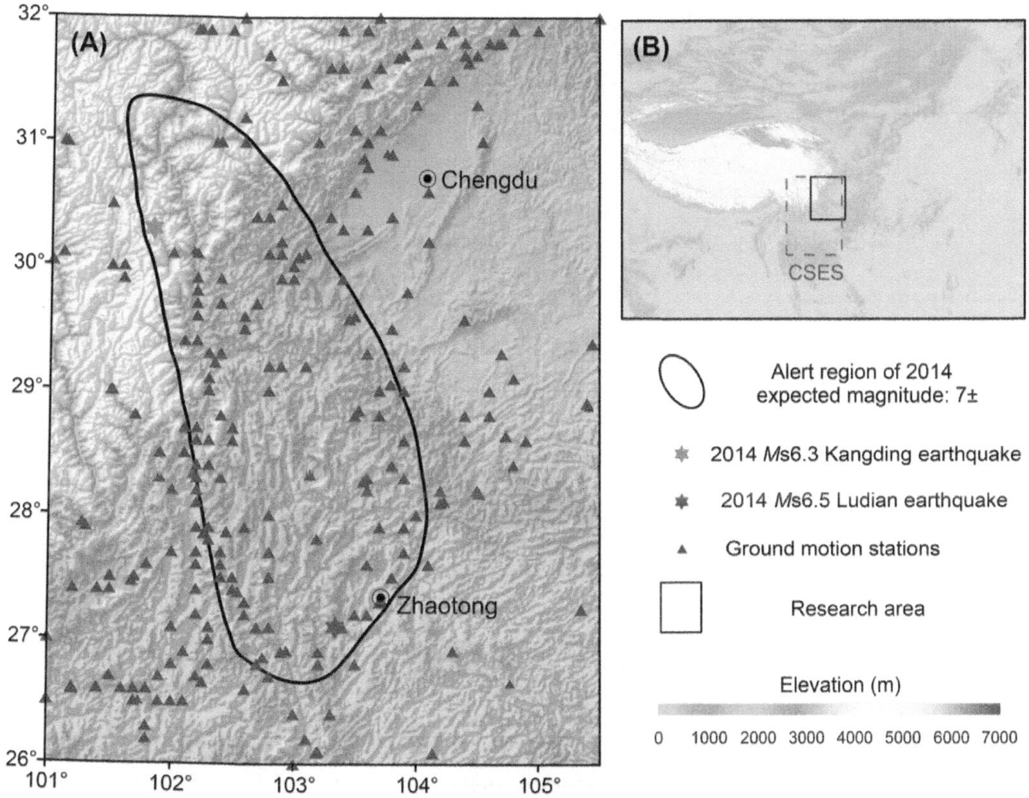

Fig. 1 (A) Alert regions identified by the 2014 Annual Consultation within the China Seismic Experimental Site (CSES; 97.5°E–105.5°E, 21°N–32°N). The alert region characterized by an expected magnitude of approximately 7 is highlighted, within which two strong earthquakes occurred in 2014; (B) Location of the study area relative to the CSES in mainland China.

***Region #1*:** Border of Sichuan and Yunnan, with expected magnitude about 7. Inferred from identified anomalies of seismic gap, increase of micro-seismicity, clustering of small earthquakes, the 'seismic response window'[2], anomalies in apparent stress and focal mechanisms, abnormal signals at several stations after the 2013 Lushan earthquake; abnormal signals revealed by mobile gravity and geomagnetic observation, GPS-revealed shear strain rate variation, 'quasi-synchronization' of anomalies, and clustering of macro-anomalies.

1 China Earthquake Networks Center (CENC) eds., 2013. *Open File of the Annual Consultation on the Likelihood of Earthquakes for the Year 2014*, subject to the Panel Discussion in the Annual Consultation Meeting. Beijing: China Earthquake Administration, in Chinese.

2 Some regions (i.e., the special 'window' here, probably with a peculiar geological structure) in which the increase of seismicity is believed to indicate the increase of the probability of occurrence of bigger earthquakes within a much larger area.

As illustrated in Figure 2, the output of the Annual Consultation can be naturally translated into a set of controlling or scenario earthquakes, which provide physically meaningful inputs for NDSHA (Reiter 1990; Panza et al., 2022). Building upon the long-standing familiarity of engineering practice with deterministic seismic hazard assessment, NDSHA integrates a comprehensive physical description of (a) seismic source processes, (b) seismic wave propagation, and (c) their combined interaction with local site conditions.

By explicitly modeling these components, NDSHA inherently accounts for the tensor nature of earthquake ground motion, rather than reducing ground shaking to scalar intensity measures. As a consequence, it avoids reliance on empirical attenuation relations, commonly expressed as ground-motion prediction equations (GMPEs), whose scalar formulation may inadequately represent the complexity of seismic wavefields. Recent evidence of this limitation is provided by Dhakal (2021), who showed that widely used GMPEs in Japan fail to consistently reproduce the

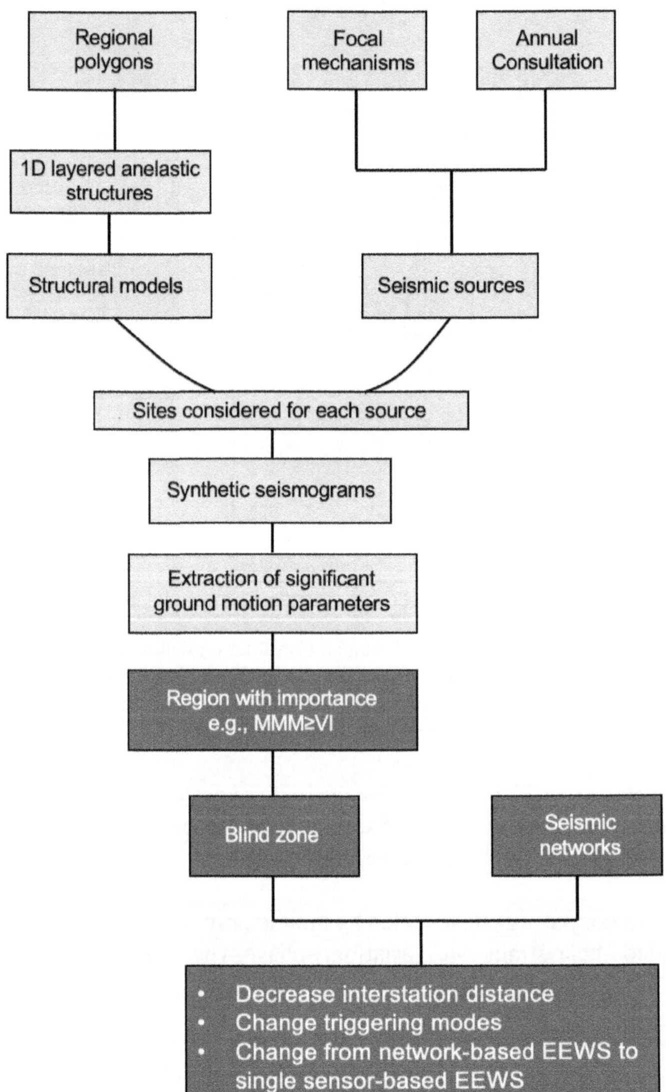

Fig. 2 Procedure of the TD-NDSHA for earthquake emergency

observed maximum intensities of 79 damaging moderate-magnitude earthquakes when compared with reported observations. Such discrepancies are physically unsurprising, as scalar ground-motion descriptors are intrinsically unable to capture the tensor nature of seismic motion (e.g., Aki and Richards, 2009).

3. EEWS empowered by TD-NDSHA

The flowchart shown in Figure 2 further outlines the procedure of TD-NDSHA as applied to earthquake emergency scenarios. Compared with the conventional NDSHA framework, two key distinctions are highlighted. First, the controlling (scenario) earthquakes are derived directly from the Annual Consultation. Second, the resulting ground-shaking predictions are primarily employed to identify regions of particular relevance for earthquake early warning, as highlighted in blue boxes, rather than being used solely for long-term hazard characterization.

Following the formulation of Kuyuk and Allen (2013), we consider a simplified homogeneous medium characterized by constant P- and S-wave velocities, V_p and V_s, respectively, and a point source located at a focal depth h. The time t_p required for the network to detect the earthquake can be expressed as:

$$t_p = T_0 + \frac{R}{V_p} \qquad ...(1)$$

where

$$R = \sqrt{D^2 + h^2} \qquad ...(2)$$

where D denotes the epicenter distance to the farthest station contributing to event location and magnitude estimation. The corresponding S-wave travel time to the target site under consideration is given by:

$$t_s = \frac{R_0}{V_S} \qquad ...(3)$$

where R_0 is the distance from the seismic source to the target site. D_0 is its epicentral distance:

$$D_0 = \sqrt{R_0^2 - h^2} \qquad ...(4)$$

The condition

$$t_p = t_s \qquad ...(5)$$

determines a threshold epicenter distance Z. When

$$D_0 \leq Z \qquad ...(6)$$

If the seismic network detects the event only after the arrival of the S-wave, the issuance of an effective early warning becomes impossible. Under this condition, Eq. (6) defines the early warning blind zone, whose radius Z can be calculated following Kuyuk and Allen (2013) as:

$$Z = \sqrt{V_S^2\, t_p^2 - h^2} \qquad ...(7)$$

The relationship between the blind-zone radius and the interstation distance is shown in Figure 3. To mitigate the extent of the early warning blind zone, Zhang et al. (2023) proposed a hazard-driven mode-switching strategy for earthquake early warning systems (EEWS), in which decision-making is guided by two complementary factors: (1) the increased likelihood of earthquakes as

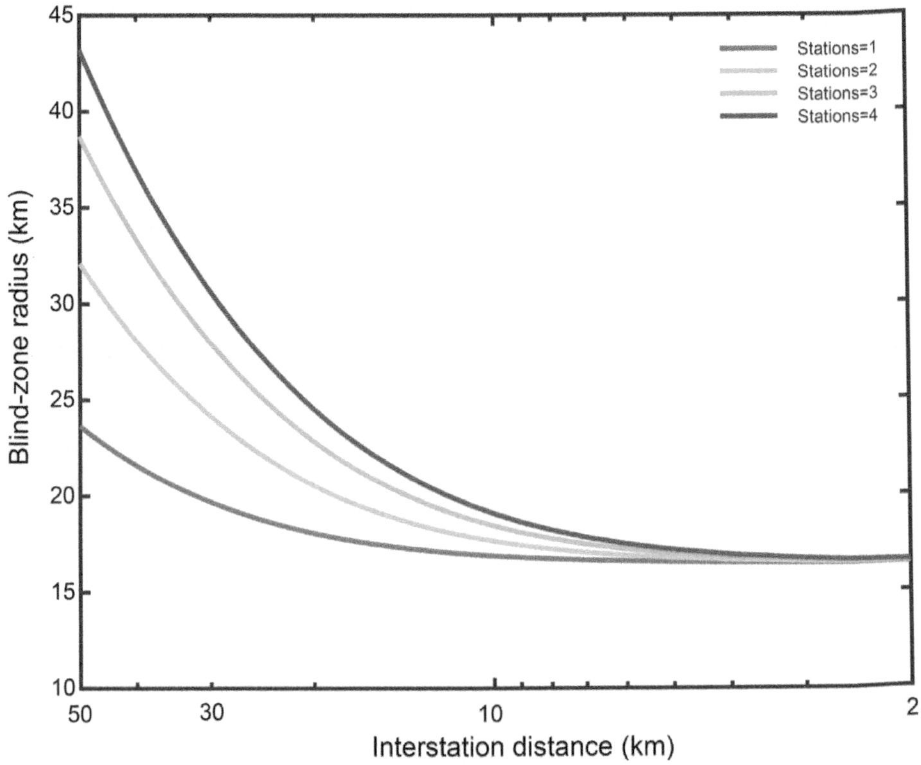

Fig. 3 Relationship between the blind-zone size and interstation distance as a function of the number of stations required for identifying the first P-wave arrival. The figure reproduces the curves proposed by Kuyuk and Allen (2013), using parameters of Vs = 3.4 km/s, Vp = 5.9 km/s and a focal depth of 15 km. It illustrates the dependence of blind-zone size (vertical axis) on interstation distance (horizontal axis).

identified by the Annual Consultation, and (2) the short-term seismic hazard estimated using TD-NDSHA. Within this framework, EEWS is regarded as a preventive tool whose effectiveness can be enhanced when activated in regions and periods characterized by elevated seismic hazard.

Based on the TD-NDSHA results, several switch-on strategies could be used to improve EEWS performance by reducing the size of the blind zone, including (i) increasing the number of available stations, or equivalently reducing the interstation spacing; (ii) modifying the triggering criteria, such as the number of stations required to declare a P-wave detection within the alert algorithm; and (iii) transitioning from a conventional network-based EEWS scheme to a single-sensor-based EEWS approach, thereby bypassing the time-consuming procedures of event location and magnitude determination or prediction.

4. Conclusions and Discussion

TD-NDSHA provides a physically grounded framework for enhancing the performance of EEWS. To illustrate this concept, we present a scenario-based application for the year 2014 in the Sichuan–Yunnan border region of southwest China. Our analysis focuses on a network-based on-site EEWS configuration, in which the existence of an early warning blind zone represents a primary limitation to system effectiveness.

By integrating the results of the Annual Consultation on the Likelihood of Earthquakes with the associated NDSHA, the EEWS can be informed by spatially distributed predictions of ground-motion parameters and expected shaking intensities on an annual time scale. Within this hazard-driven framework, the reduction of the blind-zone extent can be achieved through adaptive mode switching at three levels: (i) increasing the number of stations, or equivalently reducing interstation spacing; (ii) modifying triggering criteria, such as the number of stations required to confirm P-wave detection within the alert algorithm; and (iii) transitioning from a conventional network-based EEWS scheme to a single-sensor-based EEWS approach, thereby shortening processing time by bypassing event location and magnitude determination or prediction.

In this framework, the Annual Consultation serves as a form of intermediate-term, medium-range earthquake prediction, providing information on regions characterized by a temporarily increased likelihood of earthquake occurrence. Other approaches aimed at identifying periods and areas of elevated seismic probability, such as the M8/CN algorithms (Peresan et al. 2011), can likewise be incorporated to offer a more comprehensive representation of the impending seismic hazard.

For the subsequent seismic hazard assessment associated with these temporally elevated-probability regions, NDSHA represents a particularly suitable approach, as it is explicitly grounded in a comprehensive physical description of seismic source processes, seismic wave propagation, and their combined interactions with local site conditions. Overall, the results underscore the potential of TD-NDSHA to bridge seismic hazard assessment and earthquake early warning, providing a robust basis for more context-aware and effective seismic risk reduction.

Funding

This work is supported by the National Natural Science Foundation of China (grant number: 42504084, U25A6028), the Special Fund for Basic Scientific Research in the Institute of Geophysics, China Earthquake Administration (grant number: DQJB25Z03), and China Scholarship Council (No. 202304190023).

Acknowledgements

The author would like to express sincere gratitude to Prof. Giuliano Panza, whose pioneering contributions to Neo-deterministic Seismic Hazard Assessment (NDSHA) constitute a fundamental foundation of this work. The author also gratefully acknowledges Fabio Romanelli (Department of Mathematics, Informatics and Geophysics, University of Trieste), Antonella Peresan (National Institute of Oceanography and Applied Geophysics, CRS-OGS, Udine, Italy), and Prof. Zhongliang Wu (Institute of Earthquake Forecasting, China Earthquake Administration) for their valuable discussions, insights, and long-term support.

References

Aki, K. and Richards, P.G., 2009. *Quantitative Seismology*, 2nd Edition. Mill Valley: University Science Books.

Bela, J. and Panza, G.F., 2021. NDSHA - The new paradigm for RSHA - An updated review. *Vietnam Journal of Earth Sciences*, 43: 111–190.

Dhakal, Y.P., 2021. Strong-motions from damaging moderate magnitude ($5.9 \geq M_W$) earthquakes in Japan recorded by K-NET and KiK-net. *Frontiers in Earth Science*, 9: 618400, DOI: 10.3389/feart.2021.618400.

Kuyuk, H.S. and Allen, R.M., 2013. Optimal seismic network density for earthquake early warning: A case study from California. *Seismological Research Letters*, 84: 946–954.

Li, Y.-G., Zhang, Y. X., Wu, Z.L. eds., 2022. *China Seismic Experimental Site - Theoretical Framework and Ongoing Practice*, Beijing/Basel: Higher Education Press with Springer Nature Publishing.

Montiel-López, D., Molina, S., Galiana Merino, J.J., et al., 2025. Computing the time dependent activity rate using non-declustered and declustered catalogues – a first step towards time-dependent seismic hazard calculations for operational earthquake forecasting. Natural Hazards and Earth System Sciences, 25, 515–539.

Papazachos, C.B., Vamvakairs, D.A., Karakaisis, et al., 2018. Complexity and Time-Dependent Seismic Hazard Assessment: Should We Use Fuzzy, Approximate and Prone-to-Errors Prediction Models to Overcome the Limitations of Time-Independent Models? In: Chelidze et al., *Complexity of Seismic Time Series*. Elsevier.

Panza, G.F. and Bela, J., 2019. NDSHA: a new paradigm for reliable seismic hazard assessment. *Engineering Geology*, 275, 105403.

Panza, G.F., Kossobokov, V.G., Laor, E., De Vivo, B. eds., 2022. *Earthquakes and Sustainable Infrastructure: Neodeterministic [NDSHA] approach guarantees prevention rather than cure*. Amsterdam: Elsevier.

Peresan, A., Zuccolo, E., Vaccari, F., Gorshkov, A. and Panza, G.F., 2011. Neo-deterministic seismic hazard and pattern recognition techniques: Time-dependent scenarios for North-Eastern Italy. *Pure and Applied Geophysics*, 168: 583–607.

Reiter, L., 1990. *Earthquake Hazard Analysis: Issues and Insights*. New York: Columbia University Press.

Shi, Y.L., Liu, J. and Zhang, G.M., 2001. An evaluation of Chinese earthquake prediction, 1990-1998. *Journal of Applied Probability*, 38A: 222–231.

Wu, F.T., 1997. The Annual Earthquake Prediction Conference in China (National Consultative Meeting on seismic tendency). *Pure and Applied Geophysics*, 149: 249–264.

Wu, Z.L., 2021. Improving the annual consultation on the likelihood of earthquakes. *Earthquake Science*, 34: 541–543.

Xu, M.Y., Lu, D.G. and Zhou, W., 2023. Time-dependent probabilistic seismic hazard analysis for seismic sequences based on hybrid renewal process models. Bulletin of the Seismological Society of America, 114, 449–473.

Zhang, S.F., Wu, Z.L. and Jiang, C.S., 2017. Reducing false alarms of annual forecast in the central China north–south seismic belt by reverse tracing of precursors (RTP) using the pattern informatics (PI) 'hotspots'. *Pure and Applied Geophysics*, 174: 2401–2410.

Zhang, Y., Wu, Z.L., Romanelli, F., Vaccari, F., Peresan, A., Zhang, S.F., Jiang, C.S. and Panza, G.F., 2022. Time-dependent seismic hazard assessment based on the Annual Consultation: A case of the China Seismic Experimental Site. *Pure and Applied Geophysics*, 179, 4103–4119.

Zhang, Y., Wu, Z.L., Romanelli, F., Vaccari, F., Peresan, A., Li, J.W. and Panza, G.F., 2023. Earthquake early warning system (EEWS) empowered by Time-dependent Neo-deterministic Seismic Hazard Assessment (T-NDSHA). *Terra Nova*, 25, 230-239.

Zhao, Y.Z., Wu, Z.L., Jiang, C.S. and Zhu, C.Z., 2010. Reverse tracing of precursors applied to the annual earthquake forecast: Retrospective test of the annual consultation in the Sichuan-Yunnan Region of Southwest China. *Pure and Applied Geophysics*, 167: 783–800.

Zhuang, J.C. and Jiang, C.S., 2012. Scoring annual earthquake predictions in China. *Tectonophysics*, 524/525: 155–164.

Zhu, C.Z. and Wu, Z.L., 2007. Annual consultation on likelihood of earthquakes: Chinese approach to the study and practice of earthquake prediction. In: Secretariat of the Organizing Committee, the Third International Conference on Continental Earthquakes - Mechanism, Prediction, Emergency Management & Insurance (eds.), *Collected Papers of the Third International Conference on Continental Earthquakes*. Beijing: Seismological Press, 486–494.

Index

About the Editors

Junbo Jia is an internationally recognized scholar and currently a Senior Principal Engineer at Odfjell Technology, Norway, renowned for his contributions to engineering dynamics, marine engineering, earthquake engineering, and structural and geotechnical engineering across offshore, onshore, and energy-infrastructure applications. He is the sole author of three authoritative Springer monographs – *Essentials of Applied Dynamic Analysis*, *Modern Earthquake Engineering*, and *Soil Dynamics and Foundation Modeling* – each achieving high global readership and becoming essential references in their respective fields. He has also co-edited and co-authored the CRC Press volumes *Engineering Dynamics and Vibrations* and *Structural Mechanics: Analytical and Numerical Approaches for Structural Analysis*. Dr. Jia has delivered numerous invited lectures across continents, including six full-day sessions at the annual OMAE conference, and serves as a guest professor, doctoral thesis examiner, and industry course instructor. Dr. Jia is an elected member of the Norwegian Academy of Technological Sciences and is featured in several international *Who's Who* editions. He has received several international honours, including the Vice Admiral E.L. Cochrane Award from SNAME and the Best Paper Award from *Ships and Offshore Structures*. He currently serves on the editorial boards of several international scientific journals, contributes actively to ISO, Eurocode, and API standards development, and is appointed as the Secretary General of the International Seismic Safety Organization (ISSO).

Zhongliang Wu is a leading authority in earthquake seismology, recognized for his extensive contributions to earthquake physics, seismic networks, and disaster-risk reduction. After earning his B.Sc., M.Sc., and Ph.D. from Peking University, he advanced through key research and leadership roles at the China Earthquake Administration, serving as Deputy Director and later Director of the Institute of Geophysics, and Vice Chairman of its Science and Technology Committee. His research spans earthquake source processes, aftershock risk assessment, and major national programs including the Chinese Seismograph Network, the Wenchuan Fault Scientific Drilling Project, and the ChinArray initiative. Internationally, he has collaborated with Lamont–Doherty Earth Observatory, ICTP, and the CTBT verification program. Within the International Association of Seismology and Physics of the Earth's Interior (IASPEI), he served as First Vice-President and President, shaping global cooperation in seismology. His career is marked by sustained leadership, influential scientific contributions, and broad international engagement.

Mihaela Petrova Kouteva-Guentcheva is a full professor at the University of Architecture, Civil Engineering and Geodesy (UACEG), Bulgaria, and Scientific Committee Chair of the National Earthquake Engineering Centre. She has extensive research experience gained at the Central Laboratory for Seismic Mechanics and Earthquake Engineering, and later at the National Institute of Geophysics, Geodesy and Geography of the Bulgarian Academy of Sciences. She has also worked internationally as an associated member and Advanced School Co-Director at the Abdus Salam International Centre for Theoretical Physics and the University of Trieste. Prof. Kouteva-Guentcheva has been invited as a guest professor and keynote speaker by numerous international organizations and research institutes. She is a member of the working group governed by ISO TC 101 on Sustainable Construction and Digitalization of the Construction Sector in Bulgaria. With broad scientific interests and strong professional expertise, she successfully delivers lecture courses in both, Building Information Modeling (BIM) and natural disaster risk management (NDRM). She has served as editor of the proceedings of the annual international scientific conference of the NATO CMDR COE and is the co-editor of two Springer volumes focused on earthquake hazard and risk assessment.

For Product Safety Concerns and Information please contact our EU representative GPSR@taylorandfrancis.com Taylor & Francis Verlag GmbH, Kaufingerstraße 24, 80331 München, Germany

Batch number: 10731069

Printed by Printforce, the Netherlands